# METHODS IN CELL BIOLOGY

VOLUME 23

*Basic Mechanisms of Cellular Secretion*

# *Advisory Board*

# METHODS IN CELL BIOLOGY

# BIOLOGY

*Prepared under the Auspices of the American Society for Cell Biology*

## VOLUME 23
*Basic Mechanisms of Cellular Secretion*

*Edited by*

## ARTHUR R. HAND AND CONSTANCE OLIVER

LABORATORY OF BIOLOGICAL STRUCTURE
NATIONAL INSTITUTE OF DENTAL RESEARCH
NATIONAL INSTITUTES OF HEALTH
BETHESDA, MARYLAND

1981

## ACADEMIC PRESS
*A Subsidiary of Harcourt Brace Jovanovich, Publishers*

New York    London
Paris   San Diego   San Francisco   São Paulo
Sydney   Tokyo   Toronto

ACADEMIC PRESS, INC.
111 Fifth Avenue, New York, New York 10003

*United Kingdom Edition published by*
ACADEMIC PRESS, INC. (LONDON) LTD.
24/28 Oval Road, London NW1 7DX

LIBRARY OF CONGRESS CATALOG CARD NUMBER: 64–14220

ISBN 0–12–564123–0

PRINTED IN THE UNITED STATES OF AMERICA

81 82 83 84    9 8 7 6 5 4 3 2 1

# CONTENTS

## PART I. PROTEIN SYNTHESIS AND POSTTRANSLATIONAL MODIFICATIONS

4. *Import of Proteins into Mitochondria*
Maria-Luisa Maccecchini

5. *Biosynthesis of Pre-proparathyroid Hormone*
Joel F. Habener, Henry M. Kronenberg, John T. Potts, Jr., and Lelio Orci

6. *Biosynthesis of Insulin and Glucagon*
Howard S. Tager, Donald F. Steiner, and Christoph Patzelt

7. *Synthesis and Processing of Asparagine-Linked Oligosaccharides of Glycoproteins*
Martin D. Snider and Phillips W. Robbins

## 8. Glycosylation Steps Involved in Processing of Pro-Corticotropin-Endorphin in Mouse Pituitary Tumor Cells
*Edward Herbert, Marjorie Phillips, and Marcia Budarf*

## 9. Posttranslational Events in Collagen Biosynthesis
*Jeffrey M. Davidson and Richard A. Berg*

# PART II. TRANSPORT AND PACKAGING IN THE GOLGI REGION

## 10. The Golgi Apparatus: Protein Transport and Packaging in Secretory Cells
*Arthur R. Hand and Constance Oliver*

# PART III. TRANSLOCATION OF SECRETORY GRANULES

# PART IV. EXOCYTOSIS

## PART V. MEMBRANE DYNAMICS

## PART VI. ACTIVATION OF THE SECRETORY RESPONSE

## PART VII. CONCLUSIONS

# LIST OF CONTRIBUTORS

*Numbers in parentheses indicate the pages on which the authors' contributions begin.*

D. Banerjee, The Lindsley F. Kimball Research Institute of the New York Blood Center, 310 East 67th Street, New York, New York 10021 (231)

Philip J. Bassford, Jr.,[1] Department of Microbiology and Molecular Genetics, Harvard Medical School, Boston, Massachusetts 02115 (27)

Jon Beckwith, Department of Microbiology and Molecular Genetics, Harvard Medical School, Boston, Massachusetts 02115 (27)

Hughes Bedouelle, Unité de Programmation Moléculaire et Toxicologie Génétique, Institut Pasteur, 75015 Paris, France (27)

Richard A. Berg,[2] Pulmonary Branch, National Heart, Lung and Blood Institute, National Institutes of Health, Bethesda, Maryland 20205 (119)

N. Borgese, Department of Pharmacology and CNR Center of Cytopharmacology, University of Milan, 20129 Milan, Italy (445)

Brian Bottaro, Department of Oral Diagnosis, University of Connecticut School of Dental Medicine, Farmington, Connecticut 06032 (531)

Marcia Budarf,[3] Department of Chemistry, University of Oregon, Eugene, Oregon 97403 (101)

Anna M. Castle, Section of Cell Biology, Yale University School of Medicine, New Haven, Connecticut 06510 (335)

J. David Castle, Section of Cell Biology, Yale University School of Medicine, New Haven, Connecticut 06510 (335)

Constance P. Christian, Department of Oral Diagnosis, University of Connecticut School of Dental Medicine, Farmington, Connecticut 06032 (531)

Jean-Marie Clément, Unité de Programmation Moléculaire et Toxicologie Génétique, Institut Pasteur, 75015 Paris, France (27)

Y. Clermont, Department of Anatomy, McGill University, Montreal, Quebec, Canada H3A 2B2 (155)

Carl E. Creutz, Section on Cell Biology and Biochemistry of the Clinical Hematology Branch, National Institute of Arthritis, Metabolism and Digestive Diseases, National Institutes of Health, Bethesda, Maryland 20205 (313)

Leslie S. Cutler, Department of Oral Diagnosis, University of Connecticut School of Dental Medicine, Farmington, Connecticut 06032 (531)

Jeffrey M. Davidson,[4] Pulmonary Branch, National Heart, Lung and Blood Institute, Na-

[1] *Present address:* Department of Bacteriology and Immunology, University of North Carolina School of Medicine, Chapel Hill, North Carolina 27514.

[2] *Present address:* Department of Biochemistry, College of Medicine and Dentistry of New Jersey, Rutgers University Medical School, Piscataway, New Jersey 08854.

[3] *Present address:* Department of Molecular Biology, University of California, Berkeley, California 94720.

[4] *Present address:* Department of Pathology, College of Medicine, University of Utah, Salt Lake City, Utah 84132.

tional Institutes of Health, Bethesda, Maryland 20205 (119)

J. S. DAVISON, The Physiological Laboratory, University of Liverpool, Liverpool L69 3BX, United Kingdom (513)

W. W. DOUGLAS, Department of Pharmacology, Yale University School of Medicine, New Haven, Connecticut 06510 (483)

SCOTT D. EMR,[5] Department of Microbiology and Molecular Genetics, Harvard Medical School, Boston, Massachusetts 02115 (27)

MARILYN GIST FARQUHAR, Section of Cell Biology, Yale University School of Medicine, New Haven, Connecticut 06510 (399)

H. DAVID FISCHER, Departments of Pediatrics and Genetics, Washington University School of Medicine, and Division of Medical Genetics, St. Louis Children's Hospital, St. Louis, Missouri 63110 (191)

D. V. GALLACHER, The Physiological Laboratory, University of Liverpool, Liverpool L69 3BX, United Kingdom (513)

ALFONSO GONZALEZ-NORIEGA, Departments of Pediatrics and Genetics, Washington University School of Medicine, and Division of Medical Genetics, St. Louis Children's Hospital, St. Louis, Missouri 63110 (191)

JEFFREY H. GRUBB, Departments of Pediatrics and Genetics, Washington University School of Medicine, and Division of Medical Genetics, St. Louis Children's Hospital, St. Louis, Missouri 63110 (191)

JOEL F. HABENER, Laboratory of Molecular Endocrinology and Endocrine Unit, Massachusetts General Hospital, and Howard Hughes Medical Institute Laboratories, Harvard Medical School, Boston, Massachusetts 02114 (51)

ARTHUR R. HAND, Laboratory of Biological Structure, National Institute of Dental Research, National Institutes of Health, Bethesda, Maryland 20205 (1, 137, 429)

JOE HEDGPETH,[6] Unité de Programmation Moléculaire et Toxicologie Génétique, Institut Pasteur, 75015 Paris, France (27)

EDWARD HERBERT, Department of Chemistry, University of Oregon, Eugene, Oregon 97403 (101)

L. HERMO, Department of Anatomy, McGill University, Montreal, Quebec, Canada H3A 2B2 (155)

SYLVIA T. HOFFSTEIN, Department of Medicine, Division of Rheumatology, New York University School of Medicine, New York, New York 10016 (259)

MAURICE HOFNUNG, Unité de Programmation Moléculaire et Toxicologie Génétique, Institut Pasteur, 75015 Paris, France (27)

ERIC HOLTZMAN, Department of Biological Sciences, Columbia University, New York, New York 10027 (379)

WAYNE L. HUBBELL, Department of Chemistry, University of California, Berkeley, California 94720 (335)

N. IWATSUKI,[7] The Physiology Laboratory, University of Liverpool, Liverpool L69 3BX, United Kingdom (513)

[5] Present address: Cancer Biology Program, NCI Frederick Cancer Research Center, Frederick, Maryland 21701.

[6] Present address: Howard Hughes Medical Institute, Department of Biochemistry, University of California, San Francisco, California 94143.

[7] Present address: Department of Applied Physiology, Tohoku University School of Medicine, Seiryocho 2-1, Sendai, Japan 980.

JAMES D. JAMIESON, Section of Cell Biology, Yale University School of Medicine, New Haven, Connecticut 06510 (547)

HELEN M. KORCHAK, Department of Medicine, Division of Rheumatology, New York University School of Medicine, New York, New York 10016 (461)

HENRY M. KRONENBERG, Laboratory of Molecular Endocrinology and Endocrine Unit, Massachusetts General Hospital, and Howard Hughes Medical Institute Laboratories, Harvard Medical School, Boston, Massachusetts 02114 (51)

R. LAUGIER,[8] The Physiological Laboratory, University of Liverpool, Liverpool L69 3BX, United Kingdom (513)

C. P. LEBLOND, Department of Anatomy, McGill University, Montreal, Quebec, Canada H3A 2B2 (167)

HARVEY F. LODISH, Department of Biology, Massachusetts Institute of Technology, Cambridge, Massachusetts 02139 (5)

MARIA-LUISA MACCECCHINI,[9] Department of Biochemistry, Biocenter, University of Basel, Ch-4056 Basel, Switzerland (39)

J. MELDOLESI, Department of Pharmacology and CNR Center of Cytopharmacology, University of Milan, 20129 Milan, Italy (445)

R. MONTESANO, Institute of Histology and Embryology, University of Geneva Medical School, 1211 Geneva 4, Switzerland (283)

MARVIN NATOWICZ, Departments of Pediatrics and Genetics, Washington University School of Medicine, and Division of Medical Genetics, St. Louis Children's Hospital, St. Louis, Missouri 63110 (191)

CONSTANCE OLIVER, Laboratory of Biological Structure, National Institute of Dental Research, National Institutes of Health, Bethesda, Maryland 20205 (1, 137, 429)

LELIO ORCI, Institute of Histology and Embryology, University of Geneva Medical School, 1211 Geneva 4, Switzerland (51, 283)

RICHARD L. ORNBERG, Section on Functional Neuroanatomy, Laboratory of Neuropathology and Neuroanatomical Sciences, National Institutes of Health, Bethesda, Maryland 20205, and Marine Biological Laboratory, Woods Hole, Massachusetts 02543 (301)

CHRISTOPH PATZELT, Department of Biochemistry, University of Chicago, Chicago, Illinois 60637 (73)

CHRISTOPHER J. PAZOLES,[10] Section on Cell Biology and Biochemistry of the Clinical Hematology Branch, National Institute of Arthritis, Metabolism and Digestive Diseases, National Institutes of Health, Bethesda, Maryland 20205 (313)

G. T. PEARSON, The Physiological Laboratory, University of Liverpool, Liverpool L69 3BX, United Kingdom (513)

A. PERRELET, Institute of Histology and Embryology, University of Geneva Medical School, 1211 Geneva 4, Switzerland (283)

O. H. PETERSEN, The Physiological Laboratory, University of Liverpool, Liverpool L69 3BX, United Kingdom (513)

MARJORIE PHILLIPS,[11] Department of Chemistry, University of Oregon, Eugene, Oregon 97403 (101)

[8]*Present address:* INSERM Unite 31, 46 Chemin de la Gaye, 13009 Marseille, France.

[9]*Present address:* International Minerals & Chemical Corporation, P.O. Box 207, Terre Haute, Indiana 47808.

[10]*Present address:* Pfizer Central Research, Eastern Point Road, Groton, Connecticut 06340.

[11]*Present address:* Laboratory of Toxicology, Harvard School of Public Health, Boston, Massachusetts 02115.

H. G. PHILPOTT, The Physiological Laboratory, University of Liverpool, Liverpool L69 3BX, United Kingdom (513)

HARVEY B. POLLARD, Section on Cell Biology and Biochemistry of the Clinical Hematology Branch, National Institute of Arthritis, Metabolism and Digestive Diseases, National Institutes of Health, Bethesda, Maryland 20205 (313)

MARY PORTER, Department of Biology, Massachusetts Institute of Technology, Cambridge, Massachusetts 02139 (5)

JOHN T. POTTS, JR., Laboratory of Molecular Endocrinology and Endocrine Unit, Massachusetts General Hospital, and Howard Hughes Medical Institute Laboratories, Harvard Medical School, Boston, Massachusetts 02114 (51)

JAMES W. PUTNEY, JR.,[12] Department of Pharmacology, Wayne State University, School of Medicine, Detroit, Michigan 48201 (503)

A. RAMBOURG, Département de Biologie du Commissariat l'Energie Atomique, 91190 Saclay, France (155)

C. M. REDMAN, The Lindsley F. Kimball Research Institute of the New York Blood Center, 310 East 67th Street, New York, New York 10021 (231)

THOMAS S. REESE, Section on Functional Neuroanatomy, Laboratory of Neuropathology and Neuroanatomical Sciences, National Institutes of Health, Bethesda, Maryland 20205, and Marine Biological Laboratory, Woods Hole, Massachusetts 02543 (301)

PHILLIPS W. ROBBINS, Department of Biology and Center for Cancer Research, Mas-

sachusetts Institute of Technology, Cambridge, Massachusetts 02139 (89)

NAPHTALI SAVION,[13] The Department of Biological Chemistry, The Hebrew University of Jerusalem, Jerusalem, Israel (359)

GEORGE A. SCHEELE, Cell Biology Department, The Rockefeller University, New York, New York 10021 (345)

ZVI SELINGER, The Department of Biological Chemistry, The Hebrew University of Jerusalem, Jerusalem, Israel (359)

THOMAS J. SILHAVY,[14] Department of Microbiology and Molecular Genetics, Harvard Medical School, Boston, Massachusetts 02115 (27)

WILLIAM S. SLY, Departments of Pediatrics and Genetics, Washington University School of Medicine, and Division of Medical Genetics, St. Louis Children's Hospital, St. Louis, Missouri 63110 (191)

JAMES E. SMOLEN, Department of Medicine, Division of Rheumatology, New York University School of Medicine, New York, New York 10016 (461)

MARTIN D. SNIDER, Department of Biology and Center for Cancer Research, Massachusetts Institute of Technology, Cambridge, Massachusetts 02139 (89)

DONALD F. STEINER, Department of Biochemistry, University of Chicago, Chicago, Illinois 60637 (73)

THOMAS P. STOSSEL, Hematology–Oncology Unit, Massachusetts General Hospital, and

[12]Present address: Department of Pharmacology, Medical College of Virginia, Virginia Commonwealth University, Richmond, Virginia 23298.

[13]Present address: The Lautenberg Center for General and Tumor Immunology, Hadassah Medical School, The Hebrew University of Jerusalem, Jerusalem, Israel.

[14]Present address: Cancer Biology Program, NCI Frederick Cancer Research Center, Frederick, Maryland 21701.

Department of Medicine, Harvard Medical School, Boston, Massachusetts 02114 (215)

HOWARD S. TAGER, Department of Biochemistry, University of Chicago, Chicago, Illinois 60637 (73)

STUART J. WEISS, Department of Pharmacology, Wayne State University, School of Medicine, Detroit, Michigan 48201 (503)

GERALD WEISSMANN, Department of Medicine, Division of Rheumatology, New York University School of Medicine, New York, New York, 10016 (461)

JOHN A. WILLIAMS, Department of Physiology, University of California, San Francisco, California 94143 (247)

GLENDA M. WRIGHT,[15] Department of Anatomy, McGill University, Montreal, Quebec, Canada H3A 2B2 (167)

S. YU, The Lindsley F. Kimball Research Institute of the New York Blood Center, 310 East 67th Street, New York, New York 10021 (231)

ASHER ZILBERSTEIN, Department of Biology, Massachusetts Institute of Technology, Cambridge, Massachusetts 02139 (5)

[15] *Present address:* Department of Zoology, University of Toronto, Toronto, Ontario, Canada M5S 1A1.

# PREFACE

*Basic Mechanisms of Cellular Secretion* had its beginnings in the Conference sponsored by the National Institute of Dental Research on Basic Mechanisms of Cellular Secretion, held in Annapolis, Maryland, September 17–21, 1979. In planning this conference, it became apparent that there was no publication which addressed itself solely to the various aspects of the secretory process. The study of the secretory process crosses interdisciplinary lines and utilizes all the techniques available to the cell biologist. This book integrates the various methods as they are used to investigate specific aspects of secretion. Each of the six sections of the book deals with a major area of research related to the secretory process. The introductory chapter in each section provides a framework for several shorter chapters dealing with specific experimental systems. The concluding chapter synthesizes the contents of the book into a unified view of secretion, and discusses several areas which warrant further exploration. It is hoped that *Basic Mechanisms of Cellular Secretion* will serve as a source book both for established investigators wishing to broaden the scope of their research efforts and for individuals just beginning to delve into the complexities of the secretory process.

*Basic Mechanisms of Cellular Secretion* is the result of the combined efforts of many individuals. We are especially grateful to Drs. David B. Scott, Director, NIDR, and Marie U. Nylen, Director of Intramural Research, NIDR, for their continued support and encouragement. We would also like to thank the authors for their contributions, and Mrs. Patricia Youmans for her editorial and administrative assistance.

ARTHUR R. HAND
CONSTANCE OLIVER

# Chapter 1

# *Introduction*

## A. R. HAND AND C. OLIVER

*Laboratory of Biological Structures,*
*National Institute of Dental Research,*
*National Institutes of Health,*
*Bethesda, Maryland*

The secretion of proteins is a functional characteristic of virtually all cells. Single-cell organisms respond to and interact with their environment by the secretion of proteins. The cells of higher organisms secrete proteins to create or alter their environment, to influence the activity of other cells in their immediate vicinity or at some distance, or to digest food substances. Thus, it is not surprising that the processes involved in the synthesis and secretion of proteins have been so intensively and extensively studied during the last three decades. The basic steps in protein secretion, as outlined by Palade (1975), are (1) synthesis; (2) segregation; (3) intracellular transport; (4) concentration; (5) intracellular storage; and (6) discharge. The generality of these steps has been confirmed by studies of numerous cells and tissues, and as the title of this volume implies, the basic mechanisms underlying these steps are remarkably similar for organisms as diverse as bacteria and mammals. While the prototype secretory cell, the pancreatic exocrine cell, continues to be widely used, fibroblasts, leukocytes, plasma cells, virus-infected cells, prokaryotes, and isolated organelles have been established as important models for studying various aspects of protein secretion. Thus, much of our knowledge of the secretory process has been gleaned from systems which express their individuality through exaggeration, or deletion, of one or more of the steps.

Although the division of the secretory process into discrete steps is convenient for descriptive and investigative purposes, the interrelationships and interdependence of the various steps are becoming increasingly evident. The development of the signal hypothesis by Blobel and Dobberstein (1975) and its subsequent demonstration in the elaboration of a variety of secretory proteins, has linked the synthesis of secretory proteins to their segregation from the cytoplasm. The cotranslational and posttranslational modifications that alter the molecular architecture of the secretory proteins undoubtedly contribute to the irreversibility of the segregation process. As discussed in various chapters, specific posttranslational modifications may also be important determinants for appropriate intracellular transport, as well as requisite steps for the proper packaging and concentration of certain secretory products. Finally, the last five steps are critically dependent upon the properties and behavior of the different cellular membranes. This is particularly evident for the steps of intracellular transport and discharge. In fact, the recurrent message throughout this volume is that protein secretion cannot be separated from the physical, chemical, and biological properties of cellular membranes. The chapters in Section V directly address many of the problems and issues related to the role of membranes in the secretory process.

The conceptual framework of the secretory process is now well established, and the task that remains is to fill in the specific details. The concluding chapter of the book integrates the other contributions into this framework, and identifies fruitful avenues for future investigations. Having had the opportunity to assemble the chapters for this volume, we are in a unique position to add our own emphasis to the direction of future research on protein secretion. Clearly, the area of membrane structure and function is of major importance if significant advances are to be expected in understanding the secretory process. The specific recognition of one membrane by another is crucial to the orderly transport and discharge of secretory proteins, yet our knowledge of the molecular mechanisms involved is virtually nonexistent. Considerable advances have been made in our understanding of membrane biogenesis, and it now seems clear that both recycling and degradation of membranes occur during the secretory cycle. However, the extent and regulation of these processes, and the mechanisms through which an optimum balance between them is achieved, remain unknown. The involvement of the Golgi apparatus in the transport, posttranslational modification, and packaging of secretory proteins is well recognized, but to date, we are unable to describe the exact route followed by proteins as they move through the Golgi region, the specific localization of the enzymes catalyzing the modification of the proteins, or the mechanisms by which the proteins are sorted, concentrated, and enclosed in their membrane-bound containers. The signal hypothesis has done much to clarify the early events in the segregation of secretory proteins and has provided a general mechanism for the routing of other proteins to specific intracellular compartments. Further work may profitably be focused on the events of signal recognition and channel formation in the endoplasmic reticulum, and the fate and other potential functions of the signal peptide following its cleavage from the protein. Finally, the last decade has witnessed explosive advances in our knowledge of regulatory phenomena occurring at the cell surface, and we are just beginning to appreciate the sequelae of the alterations produced by intracellular regulators, such as cyclic nucleotides, calcium, calmodulin, and other proteins. Elucidation of the linkage between these molecules and the final step in the secretory process, exocytosis, remains as one of the most challenging areas of investigation in the cell biology of secretion.

It is apparent, as one progresses through this volume, that the work described herein has provided a wealth of information in all areas of cell biology, but as many of the authors have made abundantly clear, much new information remains to be revealed by future investigations.

## REFERENCES

Blobel, G., and Dobberstein, B. (1975). *J. Cell Biol.* **67,** 835–851.
Palade, G. E. (1975). *Science* **189,** 347–358.

# Part I.   Protein Synthesis and Posttranslational Modifications

# Chapter 2

# *Synthesis and Assembly of Transmembrane Viral and Cellular Glycoproteins*

HARVEY F. LODISH, ASHER ZILBERSTEIN,
AND MARY PORTER

*Department of Biology,
Massachusetts Institute of Technology,
Cambridge, Massachusetts*

# I.   Introduction

Many cell surface proteins span the phospholipid plasma membrane. Some of these proteins, such as the histocompatibility antigens HLA and H-2, are known to be involved in cell–cell recognition, while others, such as the erythrocyte "Band III" polypeptide, catalyze facilitated transmembrane diffusion of ions. Detailed structural analyses have, at this time, been carried out on only a few of this class of polypeptides, but it is already clear that different proteins interact with the phospholipid bilayer quite differently, particularly with respect to the number of amino acid residues that are embedded in the membrane. Nonetheless, it has become clear recently that several proteins, of quite diverse origin, do have in common a number of key structural characteristics: the surface glycoproteins

5

of vesicular stomatitis, Sindbis, and possibly other viruses; glycophorin, a major erythrocyte surface protein; and the heavy chain of the HLA and H-2 histocompatibility antigens. It is of interest that all of these are structural proteins, and all appear to be involved in cell–cell or virus–cell recognition.

All these polypeptides seem to span the phospholipid bilayer only once. In all of them a region of 20–30 amino acids at the very COOH terminus faces the cytoplasmic surface; this region contains a number of hydrophilic amino acids. Adjacent to this region is a sequence of 20–25 very hydrophobic amino acids, which is believed to be the segment that spans the lipid membrane. The remainder of the polypeptide chain (which varies considerably in size among these polypeptide species) is on the extracytoplasmic surface, as are all the attached carbohydrate chains. Key references for the structure of these proteins are: VSV G protein (Katz *et al.*, 1977; Katz and Lodish, 1979; Lingappa *et al.*, 1978; Morrison and McQuain, 1978; Toneguzzo and Ghosh, 1978); Sindbis (Garoff and Soderland, 1978; Wirth *et al.*, 1977); glycophorin (Tomita and Marchesi, 1975); and HLA and H-2 (Ewenstein *et al.*, 1976; Henning *et al.*, 1976; Springer and Strominger, 1976; Walsh and Crumpton, 1977).

In this article, we shall review our current understanding of several aspects of the biosynthesis of this class of cell surface glycoproteins. First, how is transmembrane asymmetry achieved, and how are these polypeptides inserted into membranes? Second, how and where does glycosylation—in particular, the addition of the common asparagine-linked oligosaccharides—take place? Finally, how and when do these glycoproteins move from their site of synthesis in the rough endoplasmic reticulum, and how do they achieve their final distribution in the cell surface membrane?

We and others have been using the RNA-containing enveloped viruses VSV and Sindbis to probe these problems; these viruses offer a number of advantages in the investigation of these and other problems of membrane biogenesis. Viruses contain few structural proteins, including usually one (VSV: G protein) or two (Sindbis: $E_1$ and $E_2$) integral membrane surface glycoproteins and, in some cases, an internal peripheral membrane protein (VSV: M protein). In general, host-cell protein synthesis is inhibited during viral infection, making the study of specific proteins much easier. In contrast, an average mammalian cell contains several hundred different membrane proteins. Many of these proteins are modified considerably after their synthesis by glycosylation, and by specific proteolytic cleavage; it is not an easy task to follow the synthesis of a specific cellular surface protein. Moreover, these viruses do not contain enough genetic information to encode all the enzymes necessary for their biogenesis and therefore must rely on the host cell for many functions. We thus assume that the mode of biogenesis of these viral glycoproteins is similar to the manner in which cellular glycoproteins of a similar transmembrane orientation are fabricated. Recent work on H-2, HLA, and glycophorin has confirmed this presumption (Dobberstein *et al.*, 1979; Jokinen *et al.*, 1979; Krangel *et al.*, 1979).

## II.   Transmembrane Biosynthesis of the VSV G Protein

VSV encodes five polypeptides, all of which are structural components of the virions, and directs the synthesis of five corresponding mRNA species (Pringle, 1977; Wagner, 1975). Messenger RNAs for four of them (N, NS, M, and L) are translated on free polyribosomes, and the newly made polypeptides are soluble in the cell cytoplasm. By contrast, G mRNA, is exclusively bound to the endoplasmic reticulum (ER), and at all stages of its maturation G itself is bound to membranes (Morrison and Lodish, 1975; Knipe *et al.*, 1977b) (Fig. 1).

Immediately after its synthesis, G spans the ER membrane and is thus transmembrane (Fig. 2). About 30 amino acids at the very COOH terminus remain exposed to the cytoplasm. The balance of the polypeptide, including the $NH_2$ terminus and the two asparagine-linked carbohydrate chains, face the lumen of the ER and are protected from extravesicular protease digestion by the permeability barrier of the ER membrane (Katz *et al.*, 1977; Lingappa *et al.*, 1978; Katz and Lodish, 1979).

The key experiment establishing the transmembrane orientation of G is shown

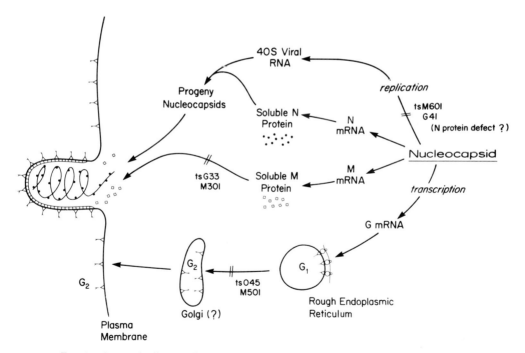

FIG. 1.   Schematic diagram illustrating the pathways of maturation of the major structural proteins of VSV and the proposed site of block in virion assembly for certain temperature-sensitive mutants. From Knipe *et al.* (1977c).

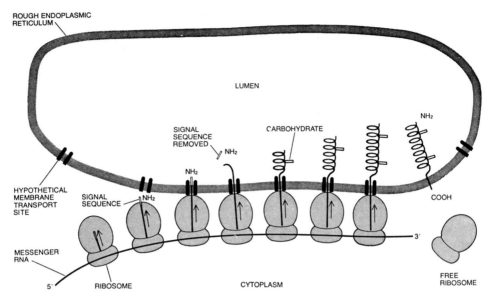

FIG. 2.   Model for synthesis, glycosylation, and transmembrane synthesis of VSV G protein. Among the first 50 amino acid units of G is a "signal sequence" that identifies the protein as one destined to be inserted into the membrane of the rough endoplasmic reticulum. Because of some 40 amino acids remain buried in the ribosome, the signal sequence does not emerge until the polypeptide is about 70 amino acid units long. At that time the signal sequence is recognized by some molecule, presumably a protein, in the membrane of the ER. This hypothetical protein is thought to facilitate the passage of the polypeptide through the lipid bilayer. Once in the lumen of the ER, at least part—16 amino acids—of the signal sequence is removed. The protein continues to elongate, and as it grows it is extruded through the membrane and folds up in the lumen. As it enters, two identical, performed carbohydrate side chains are transferred to it from a lipid carrier (see Fig. 2 in Chapter 7 by M. D. Snider and P. W. Robbins). Proteins secreted by the cell pass all the way through the membrane in this manner, but for reasons that are discussed in the text, G becomes stuck at about the time translation is completed, with some 30 amino acids remaining in the cytoplasm. Thus, the completed glycoprotein has its amino terminus, most of its bulk, and all of its carbohydrates in the lumen of the ER, and a short stub that includes the carboxyl terminus on the cytoplasmic side. Once the protein has folded, it cannot be pulled out of the membrane, nor can it execute a transmembrane "flip-flop"; it is anchored in an asymmetric orientation.

in Fig. 3. In one part of this study, infected cells were labeled for 10 minutes with [$^{35}$S]methionine, and a crude microsomal membrane fraction was prepared. By comparison with marker proteins, it was shown that this preparation (Fig. 3, lane 7) contained only the high-mannose glycosylated (see below) form of the G protein and also various amounts of the other VSV structural proteins—L, N, NS, and M. In the other part of the experiment, crude microsomes from un-labeled cells were allowed to complete synthesis of nascent polypeptides *in vitro* (lanes 1–4). With both preparations of labeled microsomes, essentially all the contaminating cytoplasmic proteins N, NS, L, and M are accessible to digestion by extravesicular trypsin (lanes 4 and 9). [The amount of N that is resistant to trypsin remains the same when the membranes are destroyed by sodium deoxycholate (lanes 2 and 8); it is probably in nucleocapsids.]

In contrast, G is largely resistant to proteolysis, and this protection is afforded by the permeability barrier of membrane vesicles. Trypsin does convert G to a form that migrates slightly faster on polyacrylamide gels; this change in mobility is consistent with a loss of 30–50 amino acids from the polypeptide chain (lanes 4 and 9). All the methionine-containing tryptic peptides found in this fragment are also found in authentic G protein; this establishes that this fragment is indeed

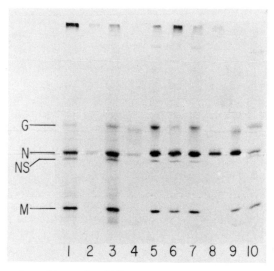

FIG. 3.   Transmembrane biogenesis of VSV glycoprotein (Katz and Lodish, 1979). Lanes 1–4: CHO cells infected with VSV (3 pfu/cell) in the presence of actinomycin D were harvested at 4 hours postinfection. A crude microsomal fraction was prepared and added to a ribosome-free supernatant preparation from uninfected cells and a protein synthesis mixture containing [$^{35}$S]methionine, as previously described (Katz and Lodish, 1979; Wirth et al., 1977). After a 30-minute incubation at 37°C, aliquots were treated for an additional 30 minutes at 37°C with protease and/or detergent as indicated below. All reactions were subsequently treated with soybean trypsin inhibitor at 100 μg/ml for an additional 30 minutes at 37°C before being run on 10% SDS–polyacrylamide gels. Shown is a radioautogram of the dried gel. Additions: (1) 1% DOC (sodium deoxycholate); (2) 1% DOC + trypsin (100 μg/ml); (3) trypsin (100 μg/ml) + soybean trypsin inhibitor (100 μg/ml) added simultaneously; (4) trypsin (100 μg/ml). Lanes 5–9: at 4 hours postinfection, cells were labeled for 10 minutes with [$^{35}$S]methionine. These cells were harvested and broken, and a microsomal fraction was prepared as above. The material was incubated for 30 minutes at 37°C in the presence of protease and/or detergent as indicated below. All reactions were incubated for an additional 30 minutes in the presence of soybean trypsin inhibitor (100 gm/ml) and run on 10% SDS–polyacrylamide gels. Additions: (5) H$_2$O; (6) 1% DOC + soybean trypsin inhibitor (100 μg/ml); (7) trypsin (100 μg/ml) + soybean trypsin inhibitor (100 μg/ml); (8) trypsin (100 μg/ml) + 1% DOC; (9) trypsin (100 μg/ml); (10) marker from VSV-infected cells.

derived from G (Katz *et al.*, 1977; Katz and Lodish, 1979). The resistance of newly made G to protease is dependent on the integrity of a membrane barrier; treatment of the preparation with the detergents sodium deoxycholate (Fig. 3, lanes 2 and 8) or Triton X-100 (Rothman and Lodish, 1977; Rothman *et al.*, 1978) before proteolysis results in essentially complete digestion of G. Direct sequence analysis of this proteolytic fragment, $G_1'$, shows that it contains the amino-terminal sequence of normal, mature VSV G, $G_1$ (see Fig. 4). This and other results (Katz and Lodish, 1979) established that this fragment differs from G only by the loss of the 30–50 COOH-terminal amino acids. Other studies, not reviewed here, prove that this fragment contains both of the Asn-linked oligosaccharide chains found on normal G (Rothman *et al.*, 1978).

A VSV particle, like those of most lipid-containing animal viruses, is formed by budding from the plasma membrane of an infected cell (reviewed in Lenard and Compans, 1974; Lenard, 1978; Wagner, 1975; Pringle, 1977). The transmembrane viral glycoprotein is embedded in the plasma membrane; it becomes, by far, the major protein exposed on the surface of infected cells and on the surface of the virus particle. The overall membrane topology of G is preserved as the protein matures from the ER to the cell surface; the same 30 carboxy-terminal amino acids remain exposed to the cytoplasmic surface, while the carbohydrate chains and the bulk of the polypeptide chain remain extracytoplasmic (Lodish and Porter, 1981).

## III.   Interaction of Glycoprotein mRNA and Microsomal Membranes

The G mRNA appears to be bound to the ER membrane via the nascent G chain, since treatment with puromycin, which causes premature termination of the growing polypeptide, causes release of the G mRNA from the ER. Unlike the case of mRNAs encoding secretory proteins, there apparently is not an ionic linkage between the ribosomes and membranes, since dislodging of the mRNA does not require treatment with solutions of high salt (Lodish and Froshauer, 1977).

Several recent experiments established that the growing G chain is extruded across the ER membrane into the lumen, similar to the way in which a nascent secretory protein is processed.

1. G protein synthesized *in vitro* in the absence of membranes ($G_0$) contains 16 amino acids at the $NH_2$ terminus that are absent from the form of G made either in the presence of ER membranes or by the cell ($G_1$) (Fig. 4) (Lingappa *et al.*, 1978). Most of these residues are highly hydrophobic and resemble in

Ribosome binding site, mRNA

```
met lys cys leu leu tyr leu
 1   2   3   4   5   6   7
```

$G_0$:

```
met lys cys leu leu tyr leu ala phe leu phe ileu his val asn cys lys phe ___ ileu val phe pro
 1   2   3   4   5   6   7   8   9  10  11  12  13  14  15  16  17  18  19  20   21  22  23
```

$G_1$:   
```
         lys phe ___ ileu val phe pro
          1   2   3   4    5   6   7
```

$G_1'$:   
```
         lys phe ___ ___ ___ phe ___
          1   2   3   4   5   6   7
```

$G_2$:   
```
         lys phe trp ileu val phe pro
          1   2   3   4    5   6   7
```

FIG. 4.   Amino-terminal sequences of different forms of the VSV glycoproteins (from Lingappa *et al.*, 1978). VSV $G_0$ is the form synthesized in a wheat germ cell-free system in the absence of membranes; its amino-terminal sequence is identical to that determined from the sequence of the ribosome binding site of G mRNA (Rose, 1977). $G_1$ is the form synthesized in the wheat germ cell-free system in the presence of ER vesicles; $G_1'$ is the product of digestion of $G_1$, contained in ER vesicles, with extravesicular protease (Katz *et al.*, 1977; Rothman and Lodish, 1977). $G_2$ is the virion form of G with the finished carbohydrate chains.

structure and function "signal" peptides found at the $NH_2$ terminus of presecretory proteins (Devillers-Thiery *et al.*, 1975).

2. If a preparation of rough endoplasmic reticulum, from which the endogenous ribosomes have been removed, is added to a wheat germ or reticulocyte cell-free extract that is translating VSV G mRNA, the resultant G protein (1) has lost the $NH_2$-terminal 16 amino acids, (2) contains two "high-mannose" carbohydrate chains attached to Asn residues, and (3) is inserted as a transmembrane protein in the ER with the same orientation as obtains in the infected cell (Katz *et al.*, 1977; Lingappa *et al.*, 1978). During synchronized *in vitro* protein synthesis in wheat germ extracts, the ER vesicles must be added to the protein synthesis reaction before the nascent chain is about 80 amino acids in length in order for the growing molecule to be subsequently inserted into the ER bilayer and to be glycosylated (Rothman and Lodish, 1977). Since about 30–40 of these 80 $NH_2$-terminal residues would, at this key time, still be embedded in the large ribosome subunit, this result establishes that it is the 40–50 most $NH_2$-terminal residues,

containing the 16 cleaved amino acids, that are crucial in directing proper interaction of the nascent chain with the membrane. Presumably, if the nascent chain is of longer length when the membranes are added, the $NH_2$ terminus is folded in such a fashion that it cannot interact with receptors on the ER membrane that recognize this region of the nascent polypeptide.

It is thus clear that insertion of the nascent glycoprotein into the ER membrane is a cotranslational event; it requires the presence of appropriate membrane vesicles during synthesis of the bulk of the polypeptide chain. We assume that, as is the case for nascent secretory proteins, the $NH_2$ terminus is extruded into the lumen of the ER (Palade, 1975), but it must be stated that there is no direct evidence for this point. Why, then, is the nascent G protein not completely extruded into the ER lumen as is a secretory protein; why does it remain embedded in the ER as a transmembrane protein?

The following simple consideration shows that this is not the relevant question; the real problem is to determine why a secreted protein is completely extruded into the lumen of the ER! Consider the situation when synthesis of a molecule of either a secretory protein or VSV G protein has just been completed. The 30 COOH-terminal amino acids will be on the cytoplasmic side of the ER, embedded in the large ribosome subunit. The next 20–30 proximal residues will be spanning the ER membrane (whether they are passing through a proteinaceous channel or through the lipid matrix itself cannot at present be determined). At this point the two ribosome subunits separate from each other and from the mRNA. Presumably the nascent chain is released from the ribosome; in any case there is no motive force of peptide bond formation to push the 50–60 COOH-terminal residues into and across the ER membrane. One can only suppose that the continued extrusion of the COOH terminus of a secretory protein across the ER membrane is driven by the folding of the $NH_2$ terminus of the protein molecule within the lumen; such a polypeptide could not exist stably as a transmembrane protein. In order for a just-completed VSV G protein to achieve its final configuration, on the other hand, it does not have to move at all with respect to the bilayer! It simply remains as a transmembrane polypeptide, anchored by the hydrophobic segment near the COOH terminus!

An interesting feature of this model is that it may explain why a number of diverse polypeptides (VSV G, HLA, glycophorin) all have about 30 amino acid residues at the COOH terminus exposed on the cytoplasmic surface of the membrane (Katz *et al.*, 1977; Katz and Lodish, 1979; Lingappa *et al.*, 1978; Morrison and McQuain, 1978; Toneguzzo and Ghosh, 1978; Garoff and Soderland, 1978; Wirth *et al.*, 1977; Tomita and Marchesi, 1975; Ewenstein *et al.*, 1976; Henning *et al.*, 1976; Springer and Strominger, 1976; Walsh and Crumpton, 1977); it is precisely these residues that were embedded in the 60 S ribosome subunit at the instant the polypeptide chain was terminated. A prediction of this model is that all species of this class of transmembrane proteins that are inserted cotranslationally never have substantially more than 30 amino acids on the cytoplasmic side of the ER membrane.


## IV.   Biogenesis of Two Sindbis Virus Glycoproteins


Sindbis provides a rather different and very important system with which to study biogenesis of transmembrane glycoproteins. Sindbis, like VSV, is a lipid-enveloped virus, but it contains only three structural proteins. Two are envelope glycoproteins, which are integral membrane proteins ($E_1$ and $E_2$). One ($E_2$), like VSV G, spans the lipid membrane with a few amino acids, at the very COOH terminus, exposed to the cytoplasmic surface; $E_1$ is also embedded in the lipid membrane near the COOH terminus, but may not be transmembrane (Garoff and Soderland, 1978; Wirth et al., 1977). The third protein, core (C), is internal to the membrane and, like VSV N, is complexed to the viral RNA genome. In marked contrast to the case of VSV-infected cells, all three of these proteins are synthesized from one polyadenylated mRNA, 26 S, which contains the nucleotide sequences found at the 3' end of the virion 42 S RNA. Both in cell-free systems and in infected cells, a single initiation site is used for the synthesis of all three proteins encoded by the 26 S RNA (Cancedda et al., 1975; Clegg and Kennedy, 1975). The core protein is synthesized first, followed by the two envelope glycoproteins ($E_1$ and $pE_2$, a precursor to $E_2$). C, $E_1$, and $pE_2$ are derived by proteolytic cleavage of the nascent chain. The gene order in Semliki Forest virus, a close relative of Sindbis, and presumably also in Sindbis, is core–$pE_2$–$E_1$ (Lachmi and Kaariainen, 1976). Thus, one RNA encodes two very different types of proteins: a soluble cytoplasmic protein (C), and two integral membrane proteins.

During infection the 26 S RNA is found mainly in membrane-bound polysomes, which synthesize all three virion proteins (Wirth et al., 1977). Attachment of these polysomes to membranes, like those synthesizing VSV G, is mediated by forces similar to those that bind polysomes synthesizing secretory proteins to membranes, since treatment with puromycin will dislodge the polysomes from the membrane vesicles. Vesicles containing Sindbis 26 S RNA in polysomes will direct cell-free synthesis of all three Sindbis structural proteins—C, $pE_2$, and $E_1$. All the newly made C protein is on the outside (cytoplasmic side) of the vesicles, while $E_1$ and $pE_2$ are sequestered in the vesicles. About 30 amino acids of $E_2$ remain exposed to the cytoplasm, as is the case with synthesis of VSV G (Wirth et al., 1977; Garoff et al., 1978).

A model (Wirth et al., 1977) explaining these and other results on the translation of 26 S RNA is shown in Fig. 5. A ribosome begins translating the core

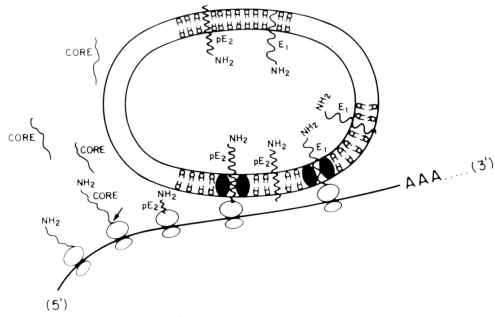

FIG. 5.    Model for synthesis of Sindbis virus proteins, subsequent nascent cleavage, and sequestration of the envelope proteins pE$_2$ and E$_1$ by the ER (Wirth *et al.*, 1977).

protein at the 5′ end of a free 26 S mRNA. As soon as the core protein segment is finished, it is removed by a protease, thus exposing the amino terminus of the nascent envelope protein, pE$_2$. The amino terminus of the pE$_2$ protein initiates an interaction with the membrane and is subsequently transferred into and across the membrane, presumably through a protein channel. As in the case of VSV G protein, this interaction leads to the binding of the polysome to the ER. As the ribosome continues to transverse the mRNA, completed pE$_2$ is cleaved and remains embedded in or sequestered by the membrane. E$_1$ is then translated and inserted into the membrane, again attaching the polysome to the membrane.

Since the same ribosomes synthesize both soluble (C) and membrane (E$_1$ and pE$_2$) proteins, it is clear that the specificity of binding the 26 S RNA to membranes in such a way as to transfer only the glycoproteins into the membrane cannot reside in the ribosomes alone. The 60 S ribosome subunit may interact directly with membrane proteins, but this is not the primary interaction, nor is this sufficient to result in specific insertion of the membrane proteins. Similarly, binding of the mRNA directly to the membrane cannot alone account for the specificity of this interaction. We conclude that the nascent chain of the two glycoproteins—the only other possibility—determines the specific insertion of the membrane proteins and has a major role in the binding of polysomal 26 S RNA to membranes.

# V. Interaction of Nascent Sindbis Glycoprotein and the Endoplasmic Reticulum

Strong support for this model has come recently from two sources. First, Garoff *et al.* (1978) have done *in vitro* translation and ER vesicle addition experiments with the 26 S RNA similar to those we have done with VSV G mRNA (Rothman and Lodish, 1977). In synchronized *in vitro* translation of 26 S mRNA, ER membranes can be added as late as the time when the soluble C protein has just been completed, and still allow subsequent normal membrane insertion of the $E_1$ and $pE_2$ proteins. If, however, membrane addition is delayed beyond this point, after synthesis of an appreciable portion of $pE_2$, the $E_1$ and $pE_2$ proteins subsequently made are not inserted into the ER phospholipid bilayer, nor is $E_1$ cleaved from $pE_2$. This is consistent with the notion that it is the $NH_2$ terminus of nascent $pE_2$ that is crucial in directing interaction of the complex of ribosomes, mRNA, and the growing polypeptide to ER membranes. If the $NH_2$ segment of nascent $pE_2$ is too long, it cannot interact productively with the ER receptors.

Second, we have investigated the properties of a temperature-sensitive mutant of Sindbis virus, ts2 (Wirth *et al.*, 1979). This mutant fails to cleave the structural proteins at the nonpermissive temperature, resulting in the production of a polyprotein of 130,000 molecular weight (referred to as the ts2 protein). The order of the proteins in this polypeptide is presumed to reflect the gene order, which is $NH_2$–core–$pE_2$–$E_1$–COOH (Clegg and Kennedy, 1975; Lachmi and Kaariainen, 1976; Simmons and Strauss, 1974). Therefore, in the ts2 protein, the amino-terminal sequence of each glycoprotein, $E_1$ and $pE_2$, is internal.

According to the predictions of our model, one would not expect the ts2 polyprotein to interact productively with the ER membrane, since the amino terminus of $pE_2$ is buried within the ts2 protein. Thus, it was surprising to us to find all the ts2 protein bound to intracellular membranes (Wirth *et al.*, 1979a). Moreover, contrary to our expectations, all the ts2 26 S mRNA, at either the permissive or the nonpermissive temperature, was bound to membranes, presumably as part of a membrane-bound polyribosome (Wirth *et al.*, 1979). Thus, the crucial question was to ascertain the topologic relationship between the ts2 protein and the isolated membrane vesicles. Is it merely stuck to the cytoplasmic surface of the microsomes, or is it inserted? To determine this, membrane vesicles were isolated from cells infected with ts2 at 41°C and pulse-labeled with [$^{35}$S]methionine for 10 minutes prior to cell lysis. We have previously shown that, in vesicles isolated from the ER, the polysomes are found on the outside of the vesicles, while the lumen is inside (Wirth *et al.*, 1977). These isolated vesicles were digested with low levels of chymotrypsin. If the ts2 protein is entirely on the cytoplasmic side of the ER, it should be totally removed by proteolytic digestion. If the entire protein is inserted through the bilayer into the lumen, then it should be resistant to proteolysis, unless the membrane vesicles are destroyed by pretreatment with detergent. If the protein is partially inserted, then it should be cleaved to a smaller size by proteolysis.

Figure 6 shows that the ts2 protein is absent after proteolytic digestion of the membrane vesicles from pulse-labeled infected cells. No smaller fragments are

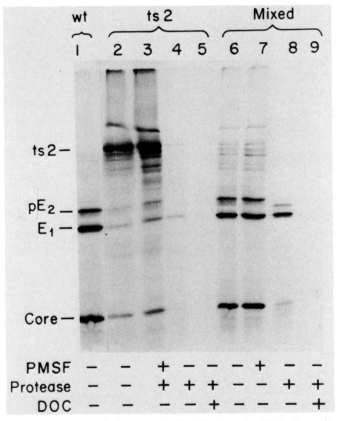

FIG. 6.   Topology of ts2 protein in infected cells (Wirth *et al.*, 1979). One culture of chick embryonic fibroblast cells was infected with Sindbis mutant ts2. A second culture was infected with both ts2 and wild type at a multiplicity of infection of 10 for each virus. After absorption (30 minutes) and a 1-hour incubation at 30°C, the culture was shifted to 41°C. At 4.5 hours postinfection the culture was pulse-labeled for 10 minutes with [$^{35}$S]methionine, and a crude preparation of microsomes was isolated as in Fig. 3. The membrane fractions were resuspended in buffer (Wirth *et al.*, 1979), and aliquots were incubated as detailed below the figure. All incubations were for 30 minutes at 30°C. Chymotrypsin (Worthington, final concentration 10 $\mu$g/ml), phenylmethylsulfonyl fluoride (Sigma, final concentration 20 $\mu$g/ml) and deoxycholate (Sigma, final concentration 1% w/v) were added to the reactions as indicated by "+". At the end of the incubation period, the reactions were heated for 5 minutes at 100°C. The proteins were concentrated by acetone precipitation and resolved by electrophoresis in a 10% SDS–polyacrylamide gel. Shown is an autoradiogram of the dried gel.

found after proteolysis, which could account for the radioactivity associated with the ts2 protein before proteolysis.

As an internal control for the proteolysis experiment, cells were coinfected with wild-type Sindbis and the ts2 mutant at the nonpermissive temperature. As can be seen in Fig. 6, in coinfected cells the ts2 protein, $pE_2$, $E_1$, and core protein are synthesized and localized to microsomes. If membrane vesicles from [$^{35}$S]methionine pulse-labeled cells are digested with chymotrypsin, some 90% of the radioactivity associated with the ts2 and core proteins is lost. By contrast, over 80% of the radioactivity associated with $pE_2$ and $E_1$ is protected from proteolysis. Precursor $E_2$ has an increased electrophoretic mobility, as described previously (Wirth *et al.*, 1977), owing to loss of the COOH-terminal segment (Garoff *et al.*, 1978). If the vesicles are permeabilized with detergent before proteolysis, all the proteins are digested. From these results, we conclude that all of each molecule of ts2 protein is located on the cytoplasmic side of the ER membrane. In this respect, the ts2 protein resembles the core protein and not $pE_2$ or $E_1$ from cells infected with wild-type virus. Although the envelope protein sequences are present, they are not inserted into the membrane and are not protected from proteolysis. This result is consistent with the notion that the proteolytic cleavage between core and $pE_2$ exposes a sequence, presumably at the amino terminus of $pE_2$, that is essential for proper insertion of $pE_2$ into the ER membrane. At least one other proteolytic cleavage does not occur in ts2-infected cells: that between $pE_2$ and $E_1$. Presumably the principal defect of the ts2 mutation is the inhibition of the $C-pE_2$ cleavage; the cleavage between $E_1$ and $pE_2$ probably does not occur, since $pE_2$ is not properly inserted into the membrane.

The majority of the 26 S mRNA is bound to membranes in cells infected with ts2 at the nonpermissive temperature. Presumably this is due to an indirect binding of the mRNA via the nascent ts2 protein, which, itself, binds to membranes (owing to its content of hydrophobic sequences) (Wirth *et al.*, 1979). Thus, the binding of mRNA to membranes may, in fact, be only indirectly related to function. In the case of the ts2 26 S mRNA, there is no detectable insertion of the protein into membranes, yet the mRNA exhibits characteristics similar to the wild-type 26 S RNA, where insertion of the envelope protein does occur. Thus, the presence of mRNA in membrane polysomes may accompany the insertion of a nascent polypetide into the ER, but binding alone is not sufficient to direct this insertion.

## VI.   Movement of Transmembrane Viral Glycoproteins from the Rough Endoplasmic Reticulum to the Plasma Membrane

This process remains one of the most mysterious, least-understood processes in membrane biogenesis. These proteins do not appear on the cell surface until

20–45 minutes after the synthesis of the polypeptide chain, depending on the temperature and the strains of cells and virus employed. The polypeptides move first to an intracellular smooth membrane component, presumably the Golgi, where, as discussed below, glycosylation is completed (Knipe *et al.*, 1977a, b; Kornfeld *et al.*, 1978; Hunt *et al.*, 1978; Robbins *et al.*, 1977). During a late stage in maturation, possibly also in the Golgi, several molecules of fatty acid are covalently linked to the glycoprotein (Schmidt and Schlesinger, 1979). In what manner, and by what force, these integral membrane proteins are channeled, first to the Golgi, and then to the surface, is obscure. A related problem is elucidation of the roles, if any, that glycosylation and lipid addition might play in these processes.

## VII.   Processing of the Asparagine-Linked Oligosaccharide Chains of VSV

The two asparagine-linked "complex" oligosaccharides on each molecule of VSV G (Fig. 7; see also Fig. 2 in Chapter 7 by M. D. Snider and P. W. Robbins) have a structure (Li *et al.*, 1978; Reading *et al.*, 1978) very similar, if not

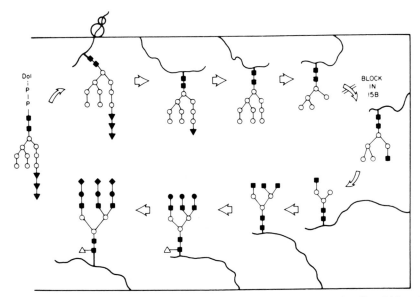

Fig. 7.   Proposed sequence for the synthesis of complex-type oligosaccharides (Kornfeld *et al.*, 1978). Dol = dolichol, the lipid carrier of the oligosaccharide. The symbols are: ■ *N*-acetylglucosamine, ○ mannose, ▼ glucose, ● galactose, ◆ sialic acid, and △ fucose.

identical, to that found on a number of typical secreted glycoproteins, such as immunoglobins (Kornfeld and Kornfeld, 1976).

The biosynthesis of this "complex" type of asparagine-linked oligosaccharide is a multistep process that involves both addition and removal of specific saccharide residues (Fig. 7). A branched oligosaccharide precursor containing three glucose, nine mannose, and two $N$-acetylglucosamine residues is performed on a lipid carrier molecule, localized in the rough endoplasmic reticulum (see Fig. 2 in Chapter 7 by M. D. Snider and P. W. Robbins). This oligosaccharide chain is transferred, en bloc, to the nascent G polypeptide. Synchronized *in vitro* translation studies established that one of the chains is added when the nascent G is about one-third completed, the other when it is about 70% complete (Rothman and Lodish, 1977).

Immediately after transfer to the polypeptide, one or two of the glucose residues are removed. Further processing of the oligosaccharide occurs only 15–20 minutes after synthesis of G, presumably at the time the protein is transferred to the Golgi membrane complex. In a stepwise, concerted set of reactions, the remaining glucose residues and six of the nine mannose residues are removed from the oligosaccharide, and the "peripheral" sugar residues $N$-acetylglucosamine (three residues per chain), galactose (three residues), sialic acid (one to three residues per chain), and fucose (one residue) are added in a stepwise fashion (Fig. 7) Glycosylation is complete about 10 minutes before the G protein reaches the cell surface (Knipe *et al.*, 1977a).

Although Golgi membranes have not been purified, or even studied to any great extent, in the fibroblast cell lines used for virus infections, it is thought that this organelle is the site of these late carbohydrate-processing events. In liver cells the Golgi fraction is greatly enriched for the two $\alpha$-mannosidases and several of the monosaccharide transferases that can function in cell-free systems in these reactions (Bretz *et al.*, 1980; Tabas and Kornfeld, 1979).

A powerful tool in the dissection of these processing steps is a glycosidase, Endo H, which cleaves between the two proximal $N$-acetylglucosamine residues, leaving only one $N$-acetylglucosamine attached to each asparagine. The enzyme will cleave the oligosaccharide chain only if the terminal mannose residues are unsubstituted (Robbins *et al.*, 1977; Tarentino and Maley, 1974); thus, the newly synthesized form of G is a substrate for Endo H, whereas the form in virions is not. The intracellular conversion of G from a form sensitive to Endo H to one resistant thus defines a terminal stage in oligosaccharide maturation. As can be seen in Figs. 8 and 9 (wild-type VSV), at 39°C in Vero cells this conversion occurs synchronously 15 minutes after synthesis of G, and occurs about 5 minutes before the protein reaches the surface.

What, if any, is the function of these oligosaccharides in maturation of G? Tunicamycin is a compound that blocks the first stage in formation of the oligosaccharide–lipid donor; in its presence the G polypeptide is synthesized but

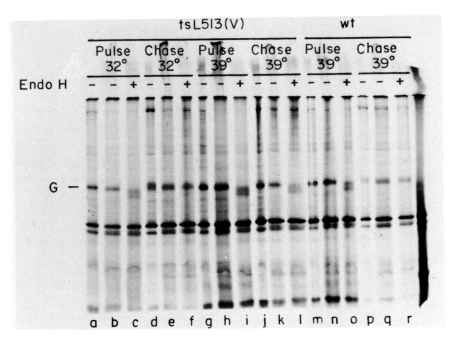

FIG. 8.   Effect of endoglycosidase H on wild-type and ts L513(V) G proteins. Monolayers of Vero cells were infected with 5 pfu/cell of either wild-type VSV (lanes m–r) or ts L513(V) (lanes a–l) and incubated at 32°C (lanes a–f) or 39°C (lanes g–l, m–r) for 4 hours. The cultures were labeled, at the same temperature, for 5 minutes with [35S]methionine. Immediately, the cells from half of the cultures (pulse) were harvested. The remaining monolayers were washed and incubated at the same temperature for an additional 90 minutes in medium containing excess unlabeled methionine. Cell proteins were isolated and portions were digested with endoglycosidase H (indicated by "+") as detailed previously (Rothman *et al.*, 1978). Analysis by gel electrophoresis and autoradiography was as in Fig. 3.

contains no sugar residues. Nonglycosylated G is inserted normally as a transmembrane protein (Rothman *et al.*, 1978) but does not normally mature to the cell surface at any temperature (Leavitt *et al.*, 1977). [There is one interesting exception; nonglycosylated G synthesized by one certain strain of VSV will mature to the cell surface at 32°C, but will not at 37°C (Gibson *et al.*, 1979)]. This indicates that at least the initial, high-mannose oligosaccharides are normally essential for some step in intracellular transport. By contrast, cell lines such as 15B (Fig. 7) have been isolated that are defective in some of the later carbohydrate maturation steps. VSV infection of these cells results in a near-normal yield of virus particles. G protein moves normally to the cell surface and is incorporated normally into virions, even though G lacks most of the peripheral sugar residues galactose and sialic acid (Schlesinger *et al.*, 1976). Apparently these peripheral sugars do not play a key role in intracellular movement of G.

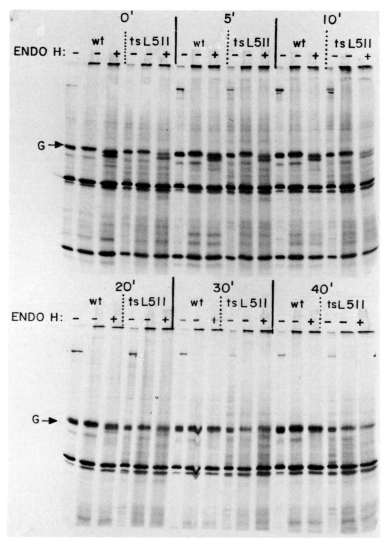

FIG. 9.   Effect of endoglycosidase H on wild-type and ts L511(V) G proteins. Monolayers of Vero cells were infected with wild-type VSV or ts L511(V) and incubated at 39°C. At 4 hours they were pulse-labeled with [$^{35}$S]methionine and, as in Fig. 8, chased with media containing unlabeled methionine for the different times indicated. Digestion with Endo H was as detailed in the legend to Fig. 8.

## VIII. Mutants in G Defective in Maturation

One approach to elucidating the process of intracellular transport is by studying viral ts mutants in the complementation group corresponding to the viral glycoprotein. A number of ts VSV mutants in this structural gene, (V), have been isolated (Pringle, 1977; Knipe *et al.*, 1977c; Lodish and Weiss, 1979). In the ones that have been studied in detail [ts 045(V), ts M501(V), ts L511(V), and ts L513(V)], it is clear that at the nonpermissive temperature the viral G protein is synthesized in normal amounts. The protein receives the normal "high-mannose" sugar side chains (*N*-acetylglucosamine and mannose), is inserted normally as a transmembrane protein into the rough endoplasmic reticulum, and is metabolically stable. However, G does not move to the cell surface. This latter point is demonstrated by the study illustrated in Fig. 10, in which surface G protein is detected by lactoperoxidase-catalyzed iodination of intact infected cells. No surface G is detectable after infection at 39°C by any of the group (V)

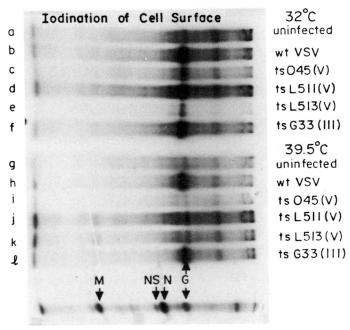

FIG. 10. Iodination of the surface of Vero cells infected with mutants of VSV. Monolayer cultures of Vero cells were infected with a multiplicity of 5 pfu/cell of either wild-type VSV or the indicated ts mutants. After incubation at either 32°C (lanes a–f) or 39.5°C (lanes g–l) for 5 hours, the monolayers were chilled and reacted with [$^{125}$I] and lactoperoxidase as detailed previously (Lodish and Porter, 1980). Total cell membranes were then solubilized and analyzed by SDS–gel electrophoresis.

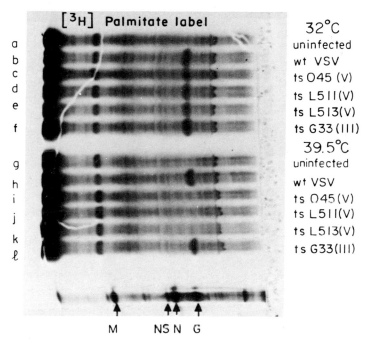

FIG. 11.   Incorporation of [³H]palmitate into mutant VSV G proteins. Monolayers of 5 × 10⁵ Vero cells were grown for 24 hours at 37°C in the presence of 44 μCi of [³H]palmitate, and then infected with wild-type VSV or ts VSC mutants as indicated. Cells were incubated at either 32°C (lanes a–f) or 39°C (lanes g–l) for 4 hours; at 1 hour postinfection an additional 20 μCi of [³H]palmitate was added. At 4 hours the total cellular protein was solubilized and analyzed by SDS–gel electrophoresis; shown is an autoradiogram of the dried gel.

mutants, although surface G is elaborated after infection by wild-type VSV, and after infection by the ts mutants at the permissive temperature, 32°C.

Cell fractionation studies on cells infected by one such mutant, ts 045(V), showed that at the nonpermissive temperature the G polypeptide remained localized in the rough endoplasmic reticulum (Knipe *et al.*, 1977c). Moreover, in the case of three of the four ts mutants studied [ts 045(V), ts M501(V), and ts L513(V)] the oligosaccharides on the G protein do not receive the terminal sugars galactose, sialic acid, and fucose; this is consistent with the primary defect being absence of movement of the mutant protein to the Golgi membranes (A. Zilberstein, M. Snider, and H. F. Lodish, manuscript in preparation). As an example of the fact that maturation of G is blocked at an early stage, Fig. 8 shows that the G protein synthesized at 39°C in cells infected by ts L513(V) remains sensitive to Endo H even 90 minutes after its synthesis. By contrast, wild-type G synthesized at 39°C or 32°C, or ts L513(V) G synthesized at 32°C, becomes resistant to Endo H within 15 minutes of its synthesis (Figs. 8 and 9).

Recently we showed that the G protein synthesized at the nonpermissive temperature in cells infected with ts L511(V) does receive the peripheral sugar residues galactose and sialic acid, with the same kinetics as does wild-type G (Zilberstein *et al.*, 1980). Figure 9, for example, shows that at 39°C the ts L511(V) G protein acquires resistance to Endo H with the same kinetics as does wild-type VSV G; within 15 minutes after synthesis, all the G protein is resistant to this glycosidase. Other studies, not shown here, establish that all terminal oligosaccharides, with the possible exception of fucose, become attached to the ts L511(V) G at 39°C. We conclude that the mutant G protein does move to the Golgi and remains within that organelle, but additional experiments will be required to establish this conclusively.

Why, then, does the ts L511(V) G not move from the Golgi to the cell surface? We do not know, but one possible clue is illustrated in the experiment depicted in Fig. 11. Palmitic acid is not attached to G protein synthesized at 39°C by any ts mutant in gene V, including ts L511(V), but it is attached if the proteins are produced at 32°C. It is possible, as suggested by Schlesinger (Schmidt and Schlesinger, 1979), that covalent addition of fatty acid is essential for some late step in the transport of some glycoproteins to the cell surface and that it is this step that is specifically blocked at 39°C in the case of ts L511(V). It is our hope that sequence analysis of the wild-type and mutant G proteins, and of the G protein from revertants of temperature-sensitive mutants, will shed light on the regions of the G protein that are important in the transport of G from the endoplasmic reticulum to the Golgi, and from the Golgi to the cell surface.

ACKNOWLEDGMENTS

Research in the authors' laboratory was supported by Grants AI 08814 and AM 15322 from the National Institutes of Health.
Asher Zilberstein is a Dr. Chaim Weizmann Postdoctoral Fellow.

REFERENCES

Bretz, R., Bretz, H., and Palade, G. E. (1980). *J. Cell Biol.* **84,** 87–101.
Cancedda, R., Villa-Komaroff, L., Lodish, H. F., and Schlesinger, M. J. (1975), *Cell* **6,** 215–222.
Clegg, J. C. S., and Kennedy, S. I. T. (1975). *J. Mol. Biol.* **97,** 401–411.
Devillers-Thiery, A., Kindt, T., Scheele, G., and Blobel, G. (1975). *Proc. Natl. Acad. Sci. U.S.A.* **72,** 5016–5020.
Dobberstein, B., Garoff, H., and Warren, G. (1979). *Cell* **17,** 759–769.
Ewenstein, B. M., Freed, J. H., Mole, L. E., and Nathenson, S. G. (1976). *Proc. Natl. Acad. Sci. U.S.A.* **73,** 915–918.
Garoff, H., and Soderland, H. (1978). *J. Mol. Biol.* **124,** 535–549.
Garoff, H., Simons, K., and Dobberstein, B. (1978). *J. Mol. Biol.* **124,** 587–600.

Gibson, R., Schlesinger, S., and Kornfeld, S. (1979). *J. Biol. Chem.* **254,** 3600–3607.
Henning, R., Milner, R. J., Reske, K., Cunningham, B., and Edelman, G. M. (1976). *Proc. Natl. Acad. Sci. U.S.A.* **73,** 118–122.
Hunt, L. A., Etchinson, J. R., and Summers, D. F. (1978). *Proc. Natl. Acad. Sci. U.S.A.* **75,** 754–758.
Jokinen, M., Gahmberg, C. G., and Anderson, L. C. (1979). *Nature (London)* **279,** 604–607.
Katz, F. N., and Lodish, H. F. (1979). *J. Cell Biol.* **80,** 416–426.
Katz, F. N., Rothman, J. E., Lingappa, V., Blobel, G., and Lodish, H. F. (1977). *Proc. Natl. Acad. Sci. U.S.A.* **74,** 3278–3282.
Knipe, D., Lodish, H. F., and Baltimore, D. (1977a). *J. Virol.* **21,** 1121–1127.
Knipe, D. M., Baltimore, D., and Lodish, H. F. (1977b). *J. Virol.* **21,** 1128–1139.
Knipe, D. M., Baltimore, D., and Lodish, H. F. (1977c). *J. Virol.* **21,** 1149–1158.
Kornfeld, R., and Kornfeld, S. (1976). *Annu. Rev. Biochem.* **45,** 217–237.
Kornfeld, S., Li, E., and Tabas, I. (1978). *J. Biol. Chem.* **253,** 7771–7778.
Krangel, M. S., Orr, H. T., and Strominger, J. L. (1979). *Cell* **18,** 979–992.
Lachmi, B. E., and Kaarianinen, L. (1976). *Proc. Natl. Acad. Sci. U.S.A.* **73,** 1936–1940.
Leavitt, R., Schlesinger, S., and Kornfeld, S. (1977). *J. Virol.* **21,** 375–385.
Lenard, J. (1978). *Annu. Rev. Biophys. Bioeng.* **7,** 139–165.
Lenard, J., and Compans, R. W. (1974). *Biochim. Biophys. Acta* **344,** 51–94.
Li, E., Tabas, I., and Kornfeld, S. (1978). *J. Biol. Chem.* **253,** 7762–7770.
Lingappa, V., Katz, F. N., Lodish, H. F., and Blobel, G. (1978). *J. Biol. Chem.* **253,** 8667–8670.
Lodish, H. F., and Froshauer, S. (1977). *J. Cell Biol.* **74,** 358–364.
Lodish, H. F., and Porter, M. (1980). *Cell* **19,** 161–170.
Lodish, H. F. and Porter, M. (1981). In preparation.
Lodish, H. F., and Weiss, R. A. (1979). *J. Virol.* **30,** 177–189.
Morrison, T., and Lodish, H. F. (1975). *J. Biol. Chem.* **250,** 6955–6962.
Morrison, T., and McQuain, C. O. (1978). *J. Virol.* **26,** 115–125.
Palade, G. (1975). *Science* **189,** 347–358.
Pringle, C. R. (1977). *Compr. Virol.* **9,** 239–289.
Reading, C. L., Penhoet, E., and Ballou, C. (1978). *J. Biol. Chem.* **253,** 5600–5612.
Robbins, P. W., Hubbard, S. C., Turco, S. J., and Wirth, D. F. (1977). *Cell* **12,** 893–900.
Rose, J. K. (1977). *Proc. Natl. Acad. Sci. U.S.A.* **74,** 3672–3676.
Rothman, J. E., and Lodish, H. F. (1977). *Nature (London)* **269,** 775–780.
Rothman, J. E., Katz, F. N., and Lodish, H. F. (1978). *Cell* **15,** 1447–1454.
Schlesinger, S., Gottlieb, C., Feil, P., Gelb, N., and Kornfeld, S. (1976). *J. Virol.* **17,** 239–246.
Schmidt, M. F. G., and Schlesinger, M. (1979). *Cell* **17,** 813–819.
Simmons, D. T., and Strauss, J. H. (1974). *J. Mol. Biol.* **86,** 397–409.
Springer, T. A., and Strominger, J. L. (1976), *Proc. Natl. Acad. Sci. U.S.A.* **73,** 2581–2485.
Tabas, I., and Kornfeld, S. (1979). *J. Biol. Chem.* **254,** 11655–11663.
Tarentino, A. L., and Maley, F. (1974). *J. Biol. Chem.* **249,** 811–817.
Tomita, M., and Marchesi, V. (1975). *Proc. Natl. Acad. Sci. U.S.A.* **72,** 2964–2968.
Toneguzzo, F., and Ghosh, H. P. (1978). *Proc. Natl. Acad. Sci. U.S.A.* **75,** 715–719.
Wagner, R. R. (1975). *Compr. Virol.* **4,** 1–94.
Walsh, F. S., and Crumpton, M. J. (1977). *Nature (London)* **269,** 307–311.
Wirth, D. F., Katz, F. N., Small, B., and Lodish, H. F. (1977). *Cell* **10,** 253–263.
Wirth, D. F., Lodish, H. F., and Robbins, P. W. (1979). *J. Cell Biol.* **81,** 154–162.
Zilberstein, A., Snider, M., Porter, M., and Lodish, H. (1980). *Cell* **21,** 417–427.

# Chapter 3

# The Genetics of Protein Secretion in Escherichia coli

PHILIP J. BASSFORD, JR.,[1] SCOTT D. EMR,[2]
THOMAS J. SILHAVY,[2] AND JON BECKWITH

*Department of Microbiology and Molecular Genetics,*
*Harvard Medical School,*
*Boston, Massachusetts*

HUGHES BEDOUELLE, JEAN-MARIE CLÉMENT,
JOE HEDGPETH,[3] AND MAURICE HOFNUNG

*Unité de Programmation Moléculaire et*
*Toxicologie Génétique,*
*Institut Pasteur,*
*Paris, France*

[1]Present address: Department of Bacteriology and Immunology, University of North Carolina School of Medicine, Chapel Hill, North Carolina.

[2]Present address: Cancer Biology Program, NCI Frederick Cancer Research Center, Frederick, Maryland.

[3]Present address: Howard Hughes Medical Institute, Department of Biochemistry, University of California, San Francisco, California.

# I.   Introduction

The gram-negative bacterium *Escherichia coli* consists of four major compartments in which its protein is found. (1) The largest fraction of the protein is located in the cytoplasm. (2) The cytoplasmic or inner membrane includes a number of proteins involved in transport of small molecules into the cell, proteins of electron transport and energy transduction, enzymes responsible for lipid biosynthesis, and other processes. (3) The outer membrane contains fewer proteins than the inner membrane, but certain of these represent major cell proteins. These proteins function as receptors for bacteriophage and colicin molecules, and also act as pores for the diffusion of certain hydrophilic substances through this membrane. There is evidence for membrane–membrane attachment sites between the inner and outer membranes. (4) The periplasmic space, or periplasm, which is bounded by the two membranes, includes two classes of proteins. First, there is a series of degradative enzymes such as alkaline phosphatase and ribonuclease. Second, there are the binding proteins, which are involved in the transport of certain small molecules into the cell.

Thus, bacteria have evolved mechanisms, as have eukaryotic cells, for directing proteins to various noncytoplasmic locations. Clearly, prokaryotes differ dramatically from eukaryotes in that they lack the whole set of intracellular organelles involved in secretion, but there are other important differences to note. There is no evidence in the case of *Escherichia coli* and most bacteria for glycoproteins (for an exception, see Mescher and Strominger, 1976). Also, the outer membrane of gram-negative bacteria appears to act as a barrier to secretion of proteins into the growth medium or any extracellular space. No convincing evidence has been presented for any truly secreted proteins in such bacteria. This finding is in contrast with the gram-positive bacteria, which lack an outer membrane and which do secrete proteins into the medium. In this article, we shall refer to the export of bacterial proteins to either the periplasm or the outer membrane as *secretion*.

Despite the enormous differences between prokaryotic and eukaryotic cells, evidence has accumulated over the last several years that the mechanisms by which both classes of cells initiate the secretion of proteins are very similar (see Silhavy *et al.*, 1979, for review of evidence). The following lines of evidence support this conclusion. Nearly all periplasmic or outer membrane proteins that have been studied are made initially as longer polypeptide chain precursors with amino-terminal extensions of 20–26 amino acids. These amino acid sequences are highly hydrophobic. These proteins are synthesized preferentially on membrane-bound ribosomes. The transfer of these proteins through the cytoplasmic membrane is cotranslational. All these studies indicate that the signal hypothesis proposed to explain transfer of proteins into the lumen of the rough endoplasmic reticulum (RER) of eukaryotic cells also applies to the transfer of bacterial proteins across the cytoplasmic membrane in bacterial cells.

In eukaryotic cells, mechanisms exist for sorting out proteins for different locations subsequent to the initiation of transfer through the RER membrane. These locations include the RER itself, lysosomes, the plasma membrane, possibly other cellular compartments, and the extracellular fluid. Analogously, gram-negative bacteria can specifically direct proteins to insert into the cytoplasmic membrane or be exported to the periplasm or outer membrane. Little is known about the signals within such proteins that determine their *ultimate* location.

## II.   Genetics of Protein Secretion

Although the signal hypothesis was developed out of studies in eukaryotic systems, bacteria offer the advantage of ready genetic manipulation for the study of protein secretion. Genetic studies can not only provide direct evidence to support or negate this hypothesis, but can also reveal new features and components of the secretory process. The following are some examples of the kinds of information potentially obtainable by such studies.

### A.   Direct Evidence for the Signal Hypothesis

This hypothesis will be one of the main contributions described in this article. The isolation of strains with mutations in the signal sequence of a secreted protein and an analysis of the effects of these mutations on secretion provide direct support for the proposed function of the signal sequence. If mutations that affect the initiation of secretion are found in portions of the protein other than the signal sequence, then some modification of the hypothesis would be necessary.

### B.   Important Amino Acids and Structural Features in the Signal Sequence

By determining which amino acids are altered in signal sequence mutants, the portions of this sequence that are important to its function would be ascertained (see below).

## C.  Determinants of Ultimate Location of Secreted Proteins

By using both gene fusions (Bassford *et al.*, 1978) and mutant analysis, those components of a gene beyond the signal sequence that determine, for example, whether the protein goes to the periplasm or the outer membrane could be specified.

## D.  Determining the Elements of the Secretory Apparatus

We and others are seeking mutants of *E. coli* that are generally defective in secretion. These mutants could be defective in any of a number of different cellular components (e.g., ribosomal proteins, translation factors, membrane or periplasmic proteins). By identifying the altered component in such mutants, an element of the cell's secretory apparatus would have been identified. In conjunction with *in vitro* studies on protein secretion, the mutants will be important in unraveling the mechanism. In one case, a ribosomal mutation has been shown to affect the secretion process (Emr and Silhavy, 1981).

## E.  The Nature and Role of the Processing Activity(ies)

Several laboratories are also seeking mutants that are lacking the activity(ies) responsible for the processing of precursors of secreted proteins to their mature form. It is hoped that such studies will allow determination of the number of different processing activities in the cell, their specificity, their location, and their exact role in the overall secretion process. For example, Wu and co-workers (Lin *et al.*, 1978) have shown that a mutational alteration in the signal sequence of the *E. coli* lipoprotein prevents the processing of this protein, but not its export to the outer membrane.

## III.  Gene Fusions in the Study of Secretion

We have used the technique of gene fusion to study the export of certain *E. coli* envelope proteins. Much of the information we have obtained comes from work done with two proteins, the *periplasmic* maltose-binding protein (MBP, product of the *malE* gene) and the *outer membrane* bacteriophage λ receptor protein (λrec, product of the *lamB* gene). These two proteins are involved in the transport of the carbon source maltose into the cell (Kellerman and Szmelcman, 1974; Szmelcman *et al.*, 1976). The synthesis of both proteins is inducible by maltose; that is, these proteins are made in high levels only when maltose is present in the growth media (Kellerman and Szmelcman, 1974; Schwartz, 1967;

Hofnung, 1974). Both proteins are known to be synthesized as precursors with typical hydrophobic signal sequences (Randall *et al.*, 1978; Marchal *et al.*, 1980; Bedouelle *et al.*, 1980; see Fig. 4).

We have used the genetic technique developed by Casadaban (1976) to fuse the *malE* and the *lamB* genes to a gene (*lacZ*) that codes for the *cytoplasmic* enzyme, β-galactosidase. β-Galactosidase is a tetrameric enzyme, the monomer of which comprises 1021 amino acids (MW 116,000), making it one of the largest proteins in *E. coli*. Several lines of study have shown that up to at least 27 amino acids can be removed from the amino terminus of this protein and replaced with the amino-terminal sequence of just about any other protein, yielding a hybrid protein that retains nearly normal β-galactosidase enzymatic activity (Bedouelle *et al.*, 1980; Brake *et al.*, 1978; Brickman *et al.*, 1979; Sarthy *et al.*, 1979; Welply *et al.*, 1980). Casadaban's technique allows us to take the *lacZ* gene from its normal position on the bacterial chromosome, insert it within or near a gene of interest (e.g., *malE*) and then select for derivatives of such a strain in which the *lacZ* gene has been fused to *malE* (Fig. 1). The resulting strain produces a hybrid protein in which the amino-terminal sequence of β-galactosidase has been replaced by some portion of the amino-terminal sequence of the MBP.

We can, in this way, obtain a series of bacterial strains that produce hybrid proteins with varying amounts of the amino terminus of the MBP or of the λrec. We can determine the location of the hybrid proteins within the cell by simply assaying various subcellular fractions for β-galactosidase activity. We then see which components of the secreted protein are essential to the export process, by correlating the cellular location at the hybrid protein with the amount of the secreted protein attached to β-galactosidase.

The results of such studies are summarized in Figs. 2 and 3. When only small amino-terminal portions of the λrec are attached to β-galactosidase, the hybrid

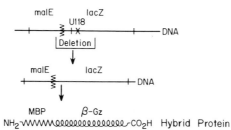

FIG. 1.   The genetic origin of hybrid proteins. The details of this technique are described elsewhere (Bassford *et al.*, 1978, 1979; Casadaban, 1976). The U118 mutation lies at a position in the *lacZ* gene corresponding to amino acid 17 (Zabin *et al.*, 1978). It is a chain-terminating mutation resulting in a *lacZ⁻* phenotype. The deletion replaces the thus defective early portion of the *lacZ* gene with a portion of the *malE* gene corresponding to the amino terminus of the maltose-binding protein.

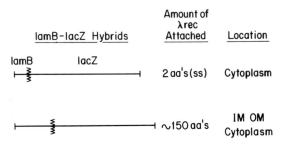

Fig. 2.    Localization of *lamB-lacZ* hybrid proteins. The origin, structure, and characterization of these strains is presented elsewhere (Silhavy *et al.,* 1977; Moreno *et al.,* 1980). IM refers to inner membrane, and OM to outer membrane. The cytoplasmic hybrid contains two amino acids of the λrec signal sequence(s).

protein is found in the cytoplasm. However, when substantial amounts of the amino-terminal sequence of this outer membrane protein replace the normal amino terminus of β-galactosidase, a significant proportion of the hybrid protein is localized to the outer membrane (Silhavy *et al.,* 1977). Thus, given the attachment of sufficient information from the exported protein, a cytoplasmic protein can also be secreted.

Somewhat analogous results were obtained with studies of hybrid proteins consisting of β-galactosidase and the MBP. When the β-galactosidase contains the first 14 amino acids of the 26-amino acid MBP signal sequence, the hybrid protein is exclusively cytoplasmic. As more of the MBP is added, substantial amounts of β-galactosidase activity (and, hence, hybrid protein) appear in the inner membrane. However, rather than being exported to the periplasm (the

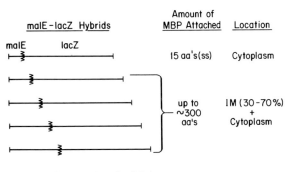

Fig. 3.    Localization of *malE-lacZ* hybrid proteins. The origin, structure, and characterization of these strains is presented elsewhere (Bedouelle *et al.,* 1980; Bassford *et al.,* 1979).

normal location of the MBP), these hybrid proteins are found associated with the cytoplasmic membrane (Bassford *et al.*, 1979). We explain these results as follows. When the entire signal sequence of the MBP is attached to β-galactosidase, the initiation of the secretion process occurs during translation. However, β-galactosidase includes within it certain amino acid sequences that cannot pass through the cytoplasmic membrane, and these cause the hybrid protein to become "stuck" in that membrane. We would predict that these hybrid proteins are transmembrane proteins. There are theoretical reasons for anticipating such a result, based on at least one model for the passage of proteins through membranes (Van Heijne and Blomberg, 1979).

## IV.   Deleterious Effects of Hybrid Protein Synthesis on Bacterial Growth

In the case of strains producing hybrid proteins with substantial amounts of either the MBP or the λrec attached to β-galactosidase, unusual growth properties are observed (Silhavy *et al.*, 1977; Bassford *et al.*, 1979). These strains grow normally in various media where carbon sources other than maltose (e.g., glycerol) are used. However, when these strains are grown in the presence of maltose,[4] and hence hybrid protein synthesis is induced, there are adverse effects on cell growth. Concomitant with the appearance of β-galactosidase activity, cell division is inhibited, the cells elongate to two or three times their normal length, and eventually many of them lyse (Bassford *et al.*, 1979). Several observations have led us to the following explanation for these findings. Those hybrid proteins that have become embedded in the cytoplasmic membrane (most of the MBP–β-galactosidase hybrid proteins and a fraction of the large λrec–β-galactosidase hybrid proteins) have blocked sites in the membrane where a number of proteins are normally secreted. The result is that precursors of these secreted proteins accumulate within the cell. Either the failure to export these proteins or some other resulting membrane defect causes the inhibition of normal cell growth that is observed.

This explanation is supported by several lines of evidence. We have shown that chain-terminating mutations that eliminate the synthesis of the hybrid protein eliminate the maltose-sensitive (Mal[S]) phenotype. Further, we have demonstrated that precursors of several secreted proteins accumulate in the cell when the synthesis of these hybrid proteins is induced. These include the precursors for

---

[4]The *MalE* fusion strains themselves are not inducible by maltose, since the construction of the strain required inactivation of some of the transport genes. In order to induce with maltose, the strains are made diploid with a *mal*[+] copy of these genes.

the λrec, the MBP, and alkaline phosphatase (Bassford *et al.*, 1979; P. Bassford, unpublished results). Finally, the results described below with signal sequence mutants are consistent with this explanation.

## V.   Isolation of MBP and λrec Signal Sequence Mutants

According to our suggestion above, the Mal$^S$ phenotype of certain fusion strains is due to the attempt by the cell to export the hybrid proteins to a location beyond the cytoplasmic membrane. If this explanation is correct, we would predict that mutations that prevent the cell from initiating the secretion process with these specific hybrid proteins should relieve the Mal$^S$ phenotype, yielding maltose-resistant (mal$^R$) bacteria. According to the signal hypothesis, at least some of these mutations should be in the DNA sequence specifying the signal portion of the hybrid proteins.

Following this line of reasoning, we selected Mal$^R$ mutant derivatives of both λrec– and MBP–β-galactosidase fusion strains, which continue to produce the hybrid protein (Emr *et al.*, 1978; Bassford and Beckwith, 1979). The hybrid protein in these Mal$^R$ mutant strains is now found predominantly in the cytoplasm. Further, as anticipated from the signal hypothesis, the mutations map very early in the hybrid gene, in a region that would appear to correspond to the coding regions for the signal sequences of the λrec or the MBP. By genetic recombination, these mutations have been crossed into the wild-type *malE* or *lamB* genes. When the *malE* gene carries such mutations, the strains grow poorly on maltose and export very little MBP to the periplasm (Bassford and Beckwith, 1979). Instead, the MBP accumulates in its *precursor* form in the cytoplasm. Analogously, when presumed signal sequence mutations in the *lamB–lacZ* hybrid gene are crossed into a wild-type *lamB* gene, the location of the λrec is changed from the outer membrane to the cytoplasm (Emr *et al.*, 1978). Again, the cytoplasmic λrec is found in its larger-molecular-weight precursor form.

## VI.   Amino Acid Changes in Signal Sequence Mutants

Thus, it appears that these mutations are affecting the signals that determine the export of the *malE* and *lamB* gene products to the cell envelope. Genetic studies indicated that many of these mutations were due to single base changes and, therefore, that a single amino acid substitution could severely impair the export process. To verify this conclusion and, further, to determine the nature of the amino acid changes, we determined the DNA sequence changes in the *malE* and *lamB* genes of various export-defective mutants. The details of the DNA sequencing procedures are presented elsewhere (Bedouelle *et al.*, 1980; Emr *et al.*, 1980). The wild-type signal sequences of the MBP and the λrec were determined by a combination of DNA and protein sequencing.

Both the messenger RNA sequence (deduced from the DNA sequence) and the amino acid sequence changes for five *malE* mutants and four *lamB* mutants are shown in Fig. 4. First, it should be pointed out that there are analogous features found in most signal sequences, both eukaryotic and prokaryotic, which apply to the pre-MBP and the pre-λrec (Habener *et al.*, 1978; Austen, 1979). In general, signal sequences range in length from 15 to 30 amino acids. Early in the sequence is found at least one and often several positively charged amino acids. These are then followed by a sequence of 10–12 generally nonpolar, highly hydrophobic amino acids. It has been proposed that this hydrophobic region is important to the export process.

Our results show that this hydrophobic region is essential for the initiation of the secretion process. *All* the mutations determined to date cause alterations of amino acids in this region. In eight of nine cases, they result in a change to a charged amino acid. In one instance, a leucine-to-proline change in the MBP signal sequence reduces export. This may be due to the ability of proline to act as a breaker of α-helices.

No mutations so far have been found in other regions of the signal sequence nor in any portion of the mature protein. These results appear to us highly significant by the following reasoning. We have in several cases obtained repeated, independent isolations of mutations that cause the same alteration. In certain cases, this could be due to genetic hot spots. However, hot spots usually occur at a very few sites in a gene (Miller, 1978) and, in this case, do not seem a likely explanation for our results. It is interesting, in particular, that no changes have been observed in those charged amino acids found early in the signal sequence. However, since there are two such amino acids in the λrec signal sequence, and three in that of the MBP, it may require more than one change to seriously alter the secretion pattern.

The results also suggest to us that, within the hydrophobic amino acid sequence, changes from one hydrophobic amino acid to another, or even to a polar noncharged amino acid, do not seriously interfere with secretion. This suggestion comes from an inspection of the codons in which we have observed changes. In the MBP signal sequence, changes to charged amino acids have been observed as a result of mutations in the DNA corresponding to a GCA (alanine) codon, an ACG (threonine) codon, and two AUG (methionine) codons. At the GCA codon, single base changes could give rise to proline, serine, threonine, valine, glycine, or glutamic acid. Of these, only the change to the charged amino acid glutamic acid is obtained. Likewise, at the ACG codon, changes could cause substitution of proline, alanine, serine, lysine, methionine, or arginine. Only the lysine change is found. In the case of the two AUG codons, changes to leucine, valine,

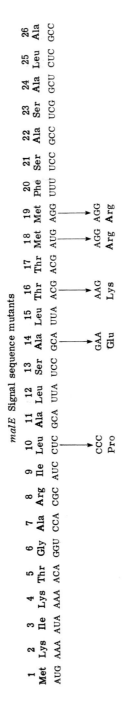

Fig. 4. Mutational alterations in the MBP and λrec signal sequences. The amino acid changes all result in severely reduced ability to export the respective proteins. The techniques and complete data are presented elsewhere (Bedouelle *et al.*, 1980; Emr *et al.*, 1980).

isoleucine, lysine, arginine, or threonine are possible, and yet only the change to arginine is found. It is known that all classes of base changes occur among spontaneous mutations, making it unlikely that this pattern is due to a bias toward specific base changes (Miller, 1978; Coulondre and Miller, 1977). Instead, the pattern strongly suggests that for those residues, at least, only charged amino acids disrupt the signal sequence.

The absence of mutations affecting a particular residue in the hydrophobic region does not necessarily indicate that this residue is not important for secretion. For example, the UUA (Leu) codon at position 15 of the *malE* gene cannot be changed to a codon specifying a charged amino acid by a single base change. No mutations have been found at this site so far. Clearly, a more intensive screening of mutants will be necessary to determine if these preliminary conclusions are correct.

## VII.   Conclusion

The studies presented above provide direct genetic proof for one aspect of the signal hypothesis; that is, since mutations that alter the amino acid composition of a signal sequence prevent the cell from exporting the particular protein, the signal sequence must be essential for the initiation of the secretion process. In addition, the analysis has allowed the demonstration that the hydrophobic region of the signal sequence is a key portion of this sequence. We believe that these results show the utility of the genetic approach. Further genetic studies should help to determine some of the more complex aspects of the secretion mechanism.

### ACKNOWLEDGMENTS

This research has been supported by grants from the Centre National de la Recherche Scientifique (LA 04271 and ACC 4248), the Délégation Générale à la Recherche Scientifique et Technique (ACC 79.7.0664), the North Atlantic Treaty Organization (Grant 1297), and the Institut National de la Santé et de la Recherche Médicale (Groupe Recombinaison et Expression Génétique U.163) to M. H.; the National Institute of General Medical Sciences (GM25524) to T. S.; and a National Science Foundation grant (PCM76-21955) to J. B. P. J. B., Jr., was a postdoctoral fellow of the Helen Hay Whitney Foundation. H. B. is the recipient of a special stipend from the Institut Française du Petrol.

### REFERENCES

Austen, P. M. (1979). *FEBS Lett.* **103,** 308–313.
Bassford, P., and Beckwith, J. (1979). *Nature (London)* **277,** 538–541.

Bassford, P., Beckwith, J., Berman, M., Brickman, E., Casadaban, M., Guarente, L., Saint-Girons, I., Sarthy, A., Schwartz, M., Sherman, H., and Silhavy, T. (1978). *In* "The Operon" (J. H. Miller and W. S. Reznikoff, eds.), pp. 245-261. Cold Spring Harbor Lab., Cold Spring Harbor, New York.

Bassford, P. J., Jr., Silhavy, T. J., and Beckwith, J. R. (1979). *J. Bacteriol.* **139**, 19-31.

Bedouelle, H., Bassford, P. J., Fowler, A. V., Zabin, I., Beckwith, J., and Hofnung, M. (1980). *Nature (London)* **285**, 78-81.

Brake, A. J., Fowler, A. V., Zabin, I., Kania, J., and Müller-Hill, B. (1978). *Proc. Natl. Acad. Sci. U.S.A.* **75**, 4824-4827.

Brickman, E., Silhavy, T. J., Bassford, P. J., Jr., Shuman, H. A., and Beckwith, J. R. (1979). *J. Bacteriol.* **139**, 13-19.

Casadaban, M. (1976). *J. Mol. Biol.* **104**, 541-555.

Coulondre, C., and Miller, J. H. (1977). *J. Mol. Biol.* **117**, 577-606.

Emr, S. D., Hanley-Way, S., and Silhavy, T. J. (1981). *Cell* **23**, 78-88.

Emr, S. D., Schwartz, M., and Silhavy, T. J. (1978). *Proc. Natl. Acad. Sci. U.S.A.* **75**, 5802-5806.

Emr, S. D., Hedgpeth, J., Clément, J.-M., Silhavy, T. J., and Hofnung, M. (1980). *Nature (London)* **285**, 82-85.

Habener, J. F., Rosenblatt, M., Kemper, B., Kronenberg, H. M., Rich, A., and Potts, J. T. (1978). *Proc. Natl. Acad. Sci. U.S.A.* **75**, 2616-2620.

Hofnung, M. (1974). *Genetics* **76**, 169-184.

Kellerman, O., and Szmelcman, S. (1974). *Eur. J. Biochem.* **47**, 139-149.

Lin, J. J. C., Kanazawa, H., Ozols, J., and Wu, H. C. (1978). *Proc. Natl. Acad. Sci. U.S.A.* **75**, 4891-4895.

Marchal, G., Perrin, D., Hedgpeth, J., and Hofnung, M. (1980). *Proc. Natl. Acad. Sci. U.S.A.* **77**, 1491-1495.

Mescher, M. F., and Strominger, J. L. (1976). *J. Biol. Chem.* **251**, 2005-2014.

Miller, J. H. (1978). *In* "The Operon" (J. H. Miller and W. S. Reznikoff, eds.), pp. 31-88. Cold Spring Harbor Lab., Cold Spring Harbor, New York.

Moreno, P., Fowler, A. V., Hall, M., Silhavy, T. J., Zubin, I., and Schwartz, M. (1980). *Nature (London)* **286**, 356-359.

Randall, L. L., Hardy, S. J. S., and Josefsson, L.-G. (1978). *J. Bacteriol.* **139**, 932-939. 1209-1212.

Sarthy, A., Fowler, A., Zabin, I., and Beckwith, J. (1979). *J. Bacteriol.* **139**, 932-939.

Schwartz, M. (1967). *Ann. Inst. Pasteur. Paris* **113**, 685-709.

Silhavy, T., Shuman, H. A., Beckwith, J., and Schwartz, M. (1977). *Proc. Natl. Acad. Sci. U.S.A.* **74**, 5411-5415.

Silhavy, T., Bassford, P. J., and Beckwith, J. (1979). *In* "Bacterial Outer Membrane: Biosynthesis, Assembly and Functions" (M. Inouye, ed.), pp. 203-254. Wiley, New York.

Szmelcman, S., Schwartz, M., Silhavy, T. J., and Boos, W. (1976). *Eur. J. Biochem.* **65**, 13-19.

Van Heijne, G., and Blomberg, C. (1979). *Eur. J. Biochem.* **97**, 175-181.

Welply, J. K., Fowler, A. V., Beckwith, J. R., and Zabin, I. (1980). *J. Bacteriol.* **142**, 732-734.

Zabin, I., Fowler, A. V., and Beckwith, J. R. (1978). *J. Bacteriol.* **133**, 437-438.

# Chapter 4

# *Import of Proteins into Mitochondria*

## MARIA-LUISA MACCECCHINI[1]

*Department of Biochemistry,*
*Biocenter, University of Basel,*
*Basel, Switzerland*

## I.   Introduction

Mitochondria are cell organelles that contain two membranes. The outer mitochondrial membrane is freely permeable to molecules up to 10,000 molecular weight. In contrast, the inner mitochondrial membrane exhibits a highly selective permeability. Although mitochondria have a complete genetic system, they synthesize only on the order of ten very hydrophobic proteins, most of which are subunits of oligomeric complexes, associated with the mitochondrial inner membrane. All other proteins found in the mitochondria are coded by the nucleus and translated in the cytoplasm (Schatz and Mason, 1974). These cytoplasmically made proteins have to be transported into the mitochondria. Depending on the particular cytoplasmically made mitochondrial protein, it must be transferred across one or two mitochondrial membranes. The mechanism of

---

[1]Present address: International Minerals & Chemical Corporation, P.O. Box 207, Terre Haute, Indiana.

this protein import has been under debate for many years (Kadenbach, 1967; Kellems and Buttow, 1972). For several years it was speculated that these proteins might be made on a special subclass of cytoplasmic ribosomes bound to the mitochondrial outer membrane and discharging their nascent polypeptides into the mitochondria by vectorial translation (Kellems and Buttow, 1972). This mechanism would have been analogous to the vectorial translation found for the transfer of proteins into the endoplasmic reticulum in eukaryotes and across the plasma membrane in prokaryotes (Campbell and Blobel, 1976; Inouye *et al.*, 1977).

Alternatively, the import of proteins into mitochondria might be independent of protein synthesis. Transfer of proteins across the mitochondrial membranes would then be a posttranslational event; the protein could perhaps be synthesized on free ribosomes, released into a cytoplasmic pool, and transferred to the organelle in subsequent steps.

In order to see by what mechanism mitochondria solve their import problems, we analyzed three different classes of mitochondrial proteins, whose final locations are the matrix, the inner membrane, and the intermembrane space.

## II.   $F_1$-Subunits Are Made as Larger Precursors

Our initial studies concentrated on the three largest subunits of $F_1$-ATPase ($\alpha$-, $\beta$-, and $\gamma$-subunits) in yeast mitochondria (Maccecchini *et al.*, 1979b). Yeast $F_1$ consists of five nonidentical subunits, which are synthesized on cytoplasmic ribosomes (Schatz and Mason, 1974) and transferred across both mitochondrial membranes to the matrix side of the mitochondrial inner membrane. The biosynthesis of the three subunits was studied in a reticulocyte lysate programmed with yeast RNA and in pulse-labeled spheroplasts. To minimize proteolytic degradation, incorporation of [$^{35}$S]methionine in the reticulocyte lysate and in the pulsed spheroplasts was stopped by adding a mixture of protease inhibitors and boiling the lysate or spheroplast suspension immediately in 3% SDS. The boiled mixtures were diluted to 0.1% SDS with a buffer containing 1% Triton X-100, and the subunits were immunoprecipitated with monospecific antisera against each one of the subunits. The immunocomplexes were adsorbed to fixed *Staphylococcus aureus* cells, extracted from the *S. aureus* cells with SDS, resolved on 10% polyacrylamide gel slabs, and visualized by autoradiography.

All three $F_1$-subunits were initially made as larger precursors both *in vitro* and *in vivo*. Moreover, the larger precursors found in pulse-labeled spheroplasts were converted into their mature forms upon a subsequent chase with unlabeled methionine. This result is shown in Fig. 1 for the $\beta$-subunit of $F_1$-ATPase. Analogous results were obtained for the $\alpha$- and $\gamma$-subunits.

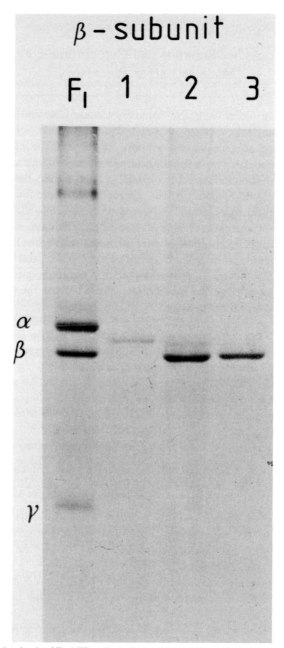

FIG. 1. The β-subunit of F₁-ATPase is made as a larger precursor *in vitro* in a reticulocyte lysate and *in vivo* in pulse-labeled spheroplasts. The F₁β-subunit labeled *in vitro* (track 1), in spheroplasts pulsed for 5 minutes (track 2), and in spheroplasts pulsed for 5 minutes and chased for 45 minutes (track 3). The track labeled F₁ shows the three largest subunits of mature F₁ immunoprecipitated from cells grown for 10 generations in ³⁵SO₄²⁻. Labeling was with [³⁵S]methionine, and the extracts were immunoprecipitated and electrophoresed on a 10% SDS–polyacrylamide gel slab. The labeled polypeptides were visualized by autoradiography.

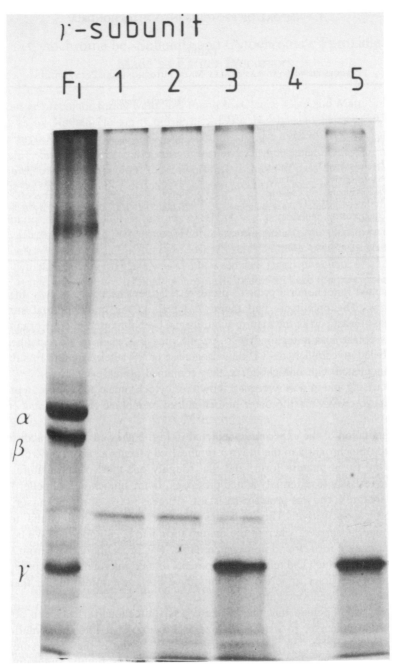

FIG. 2. The precursor to the $F_1\gamma$-subunit is processed and imported *in vitro*. $F_1$, mature $F_1$-ATPase standard; (1) the $\gamma$-subunit was synthesized *in vitro*; (2) after labeling *in vitro*, the reticulocyte lysate was inhibited with cycloheximide and incubated for 60 minutes at 29°C with intact yeast mitochondria (3 mg/ml), freed from mitochondria by centrifugation, and subjected to immunoprecipitation

## V.  Concluding Remarks on Transport

All results obtained thus far indicate that transfer of proteins from the cytosol into mitochondria does not occur by vectorial translation as had been postulated. Most probably, transfer is initiated by the interaction of the polypeptide to be transported with a specific receptor on the outer membrane, followed by the opening of a polar transmembrane channel. Unidirectional transfer could be accomplished by irreversible conversion of the precursor to its mature form in the interior of the organelle. Import into mitochondria, therefore, is independent of protein synthesis, is accompanied by a transient accumulation of discrete precursors, and seems to be coupled to the processing of the precursors in the mitochondria, as we have never been able to find unprocessed precursor forms inside the mitochondria.

A similar mechanism has been demonstrated for the import of the cytoplasmically made small subunit of ribulose-1,5-diphosphate carboxylase into chloroplasts (Dobberstein *et al.,* 1977; Highfield and Ellis, 1978; Chua and Schmidt, 1978). Perhaps also the transport of bacterial toxins across the plasma membrane of prokaryotes and eukaryotes follows a similar mechanism (Neville and Chang, 1978).

The demonstration of a larger precursor to cytochrome c peroxidase contrasts with the observation that cytochrome c, another intermembrane space protein is immediately made as the mature apoprotein (Korb and Neupert, 1978). The apoprotein is then transported into the mitochondria, where it is complemented with the heme group. Cytochrome c may be a counterpart to the secretory protein ovalbumin, whose signal peptide is not cleaved off, while the protein is translated across the endoplasmic reticulum (Palmiter *et al.,* 1978; Lingappa and Blobel, 1979). According to this view a signal sequence of apocytochrome c would recognize a specific receptor on the mitochondrial surface. Unidirectional transport could be driven by the irreversible attachment of heme in the intermembrane space.

## VI.  Energy Dependence of Processing and Transport

In the intact yeast cell, processing of precursors to mitochondrial proteins requires ATP (Nelson and Schatz, 1979). Nathan Nelson exploited the fact that

---

with an anti-γ-subunit serum; (3) same as (2), but the mitochondrial pellet was analyzed; (4) same as (2), except that the lysate–mitochondria mixture was incubated for 20 minutes at room temperature with trypsin and chymotrypsin before removal of the mitochondria; the supernatant (4) and the mitochondrial pellet (5) were then subjected to immunoprecipitation.

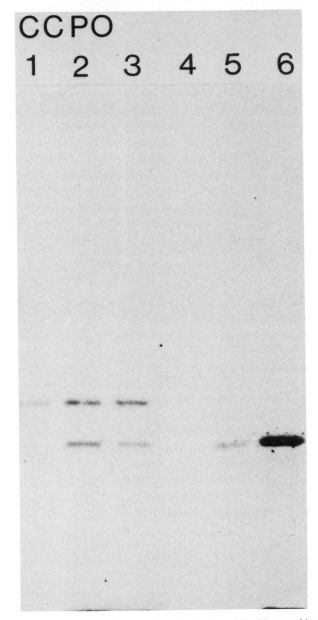

Fig. 3. The precursor to apocytochrome c peroxidase is cleaved and imported into mitochondria *in vitro*. *In vitro* synthesized precursor to apocytochrome c peroxidase was incubated with mitochondria and then protease-treated as described for Fig. 2. An autoradiogram of the dried SDS–polyacrylamide gel slab is shown: (1) apocytochrome c peroxidase synthesized in a reticulocyte lysate; (2) immunoprecipitation from the supernatant after removal of mitochondra; (3) immunoprecipitation from mitochondrial pellet of (2); immunoprecipitation from the supernatant (4) and the pellet (5) after trypsin and chymotrypsin digestion and centrifugation of mitochondria; (6) mature apocytochrome c peroxidase immunoprecipitated from mitochondria of cells that had been grown for 16 hours in the presence of [$^{35}$S]methionine.

the ATP level in the mitochondrial matrix of intact yeast cells can be specifically lowered by combining certain mutations or inhibitors. There are two major pathways by which the mitochondrial matrix can obtain ATP: (1) oxidative phosphorylation by $F_1$-ATPase in the matrix itself; (2) import of glycolytically made ATP from the cytosol via a specific adenine nucleotide transporter located in the mitochondrial inner membrane (Fig. 4).

Oxidative phosphorylation can be blocked either by respiratory inhibitors (cyanide or antimycin A) or by rho⁻ mutations, which cause the loss of several mitochondrial cytochromes as well as the proton channel of the ATPase. The second pathway, the import of ATP from the cytosol, can be inhibited by bongkrekic acid (Klingenberg et al., 1976), which blocks the transporter, or by the nuclear $op_1$ mutation (Kováč et al., 1967; Kolárov et al., 1972), which lowers its affinity for adenine nucleotides. If both oxidative phosphorylation and ATP import are blocked, the matrix can be depleted of ATP without significantly affecting protein synthesis in the cytoplasm. In the experiment depicted in Fig. 5, rho⁻ yeast spheroplasts were pulse-labeled with [³⁵S]methionine in the absence of inhibitors and in the presence of bongkrekic acid. The control cells accumulated exclusively the mature forms, whereas cells treated with the inhibitors

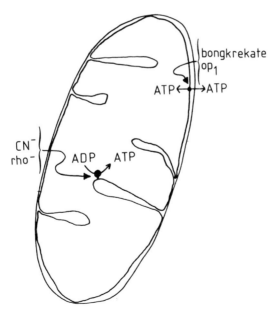

FIG. 4. The two major pathways for supplying ATP to the mitochondrial matrix and experimental possibilities for specifically blocking each of these two pathways in yeast cells. Oxidative phosphorylation can be inhibited by respiratory inhibitors or by extrachromosomal rho⁻ mutation. Import of glycolytically generated ATP from the cytoplasm via the adenine nucleotide transporter is inhibited by bongkrekic acid or the nuclear $op_1$ mutation.

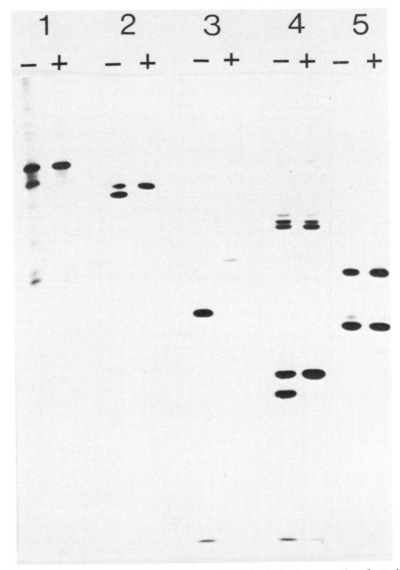

FIG. 5.    ATP depletion of the mitochondrial matrix inhibits the *in vivo* maturation of cytoplasmically made proteins located in the matrix or in the inner mitochondrial membrane, but not in the intermembrane space. In each of two 10-ml Erlenmeyer flasks, 2.5 ml of a suspension of rho⁻ spheroplasts was incubated at 29°C for 60 minutes with gentle shaking. To each flask, 2.5 ml of medium alone (−) or medium containing 73 n$M$ of bongkrekic acid (+) was added. Both aliquots were shaken for another 30 minutes and then pulsed for 20 minutes with 2 mCi of [$^{35}$S]methionine, lysed in boiling SDS, and immunoprecipitated with antisera directed against the following components: (1) $F_1\alpha$-subunit; (2) $F_1\beta$-subunit; (3) $F_1\gamma$-subunit; (4) subunit V of the cytochrome $bc_1$ complex; (5) cytochrome c peroxidase.

accumulated the larger precursors to those polypeptides that are normally transported into the matrix or inserted into the inner membrane. In contrast, no accumulation of precursor to cytochrome c peroxidase was observed. This agrees with the fact that the outer mitochondrial membrane is freely permeable to ATP made in the cytoplasm (Kolárov and Klingenberg, 1974).

The mechanism by which ATP affects processing of mitochondrial precursors is unknown. It could be necessary for transmembrane movement of the precursor, for its processing, or for both. These observations suggest that there might be at least two different processing and/or transport systems. One of them would depend on cytosolic ATP and move proteins into the intermembrane space and possibly into the outer membrane; the other would be dependent on ATP in the matrix and move proteins into the matrix and the inner mitochondrial membrane. It will be interesting to learn whether precursors imported into different compartments interact with distinct receptors on the mitochondrial outer surface, or whether there is only one surface receptor coupled to specific proteases in the different mitochondrial compartments.

## ACKNOWLEDGMENT

This article summarizes some recent studies by our group in Basel. The early phases of this work were done in collaboration with Günter Blobel at The Rockefeller University, New York. I should like to thank Dr. Blobel for the hospitality extended to me during my visit to his laboratory.

## REFERENCES

Campbell, P. N., and Blobel, G. (1976). *FEBS Lett.* **72,** 215–226.
Chua, N.-H., and Schmidt, G. W. (1978). *Proc. Natl. Acad. Sci. U.S.A.* **75,** 6110–6114.
Côté, C., Solioz, M., and Schatz, G. (1979). *J. Biol. Chem.* **254,** 1437–1439.
Dobberstein, B., Blobel, G., and Chua, N.-H. (1977). *Proc. Natl. Acad. Sci. U.S.A.* **74,** 1082–1085.
Highfield, P. E., and Ellis, R. J. (1978). *Nature (London)* **271,** 420–424.
Inouye, S., Wang, S., Sekizawa, J., Halegoua, S., and Inouye, M. (1977). *Proc. Natl. Acad. Sci. U.S.A.* **74,** 1004–1008.
Kadenbach, B. (1967). *Biochim. Biophys. Acta* **134,** 430–442.
Kellems, R. E., and Buttow, R. A. (1972). *J. Biol. Chem.* **247,** 8043–8050.
Klingenberg, M., Riccio, P., Aquila, H., Buchanan, B. B., and Grebe, K. (1976). *In* "The Structural Basis of Membrane Function" (Y. Hatefi and L. Djavadi-Ohaniance, eds.), pp. 293–311. Academic Press, New York.
Kolárov, D., and Klingenberg, M. (1974). *FEBS Lett.* **45,** 320–323.
Kolárov, J., Subik, J., and Kováč, L. (1972). *Biochim. Biophys. Acta* **267,** 465–478.
Korb, H., and Neupert, W. (1978). *Eur. J. Biochem.* **91,** 609–620.
Kováč, L., Lachowicz, T. M., and Slonimski, P. P. (1967). *Science* **158,** 1564–1567.
Lingappa, V., and Blobel, G. (1979). In preparation.
Maccecchini, M.-L., Rudin, Y., and Schatz, G. (1979a). *J. Biol. Chem.* **254,** 7468–7471.

Maccecchini, M.-L., Rudin, Y., Blobel, G., and Schatz, G. (1979b). *Proc. Natl. Acad. Sci. U.S.A.* **76,** 343–347.

Nelson, N., and Schatz, G. (1979). *Proc. Natl. Acad. Sci. U.S.A.* **76,** 4365–4369.

Neville, D. M., and Chang, T.-M. (1978). *Curr. Top. Membr. Transp.* **10,** 66–150.

Palmiter, R. D., Gagnon, J., and Walsh, K. A. (1978). *Proc. Natl. Acad. Sci. U.S.A.* **75,** 94–98.

Schatz, G., and Mason, T. L. (1974). *Annu. Rev. Biochem.* **43,** 51–87.

Scholte, H. R., Weijers, P. J., and Wit-Peeters, E. M. (1973). *Biochim. Biophys. Acta* **291,** 764–773.

# Chapter 5

# Biosynthesis of Pre-proparathyroid Hormone

JOEL F. HABENER, HENRY M. KRONENBERG,
AND JOHN T. POTTS, JR.

*Laboratory of Molecular Endocrinology and Endocrine Unit,*
*Massachusetts General Hospital, and*
*Howard Hughes Medical Institute Laboratories,*
*Harvard Medical School,*
*Boston, Massachusetts*

LELIO ORCI

*Institute of Histology and Embryology,*
*University of Geneva Medical School,*
*Geneva, Switzerland*

## I. Introduction

Parathyroid hormone (PTH), a polypeptide of 84 amino acids whose principal physiological actions are to maintain calcium levels in the extracellular fluid, is synthesized by way of two successive cleavages of NH$_2$-terminal sequences from a larger precursor, pre-proparathyroid hormone (Pre-ProPTH) of 115 amino acids (Habener and Potts, 1978; Habener and Kronenberg, 1978) (Fig. 1). The earliest cleavage occurs predominantly cotranslationally in the rough endoplasmic reticulum (RER) and consists of the removal of an NH$_2$-terminal leader sequence of 25 amino acids, resulting in the formation of an intermediate precursor, proparathyroid hormone (ProPTH) (Habener *et al.*, 1980). In accord with the "signal hypothesis," the leader sequence of Pre-ProPTH appears to serve as a signal that functions in the establishment of a polyribosome–membrane junction via the attachment of the nascent polypeptide to a transport element, as yet unidentified, located in the lipid bilayer of the RER (Blobel and Dobberstein, 1975). The growing polypeptide chain is then transferred in a unidirectional manner into the cisterna of the RER. The second cleavage occurs in the Golgi complex 12–15 minutes later by removal of an NH$_2$-terminal sequence of six amino acids, resulting in the formation of PTH (Habener *et al.*, 1979).

The minimum structure required to exert biological actions of the hormone on target cells in bone and kidney has been shown by structure–activity studies

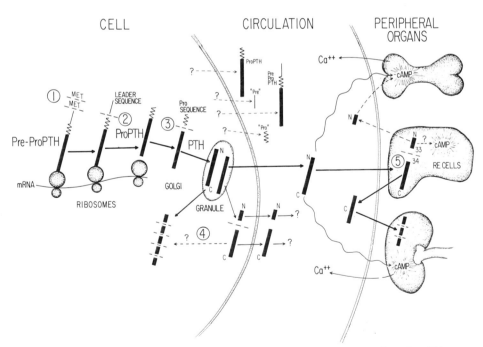

FIG. 1. Schematic depiction of proposed intra- and extracellular cleavages of parathyroid hormone precursors and of parathyroid hormone. Dashed lines denote processes for which evidence is only tentative at the present time. Pre-ProPTH = pre-proparathyroid hormone; ProPTH = proparathyroid hormone; PTH = parathyroid hormone; cAMP = 3',5'-cyclic AMP. (From Habener, 1981.)

involving chemical synthesis of fragments of hormone to reside in a continuous sequence consisting of the first 25 amino acids of the hormone (sequence 1–25) (Tregear *et al.*, 1973) (see Fig. 2).

After release of the hormone from the gland into the circulation, a third highly specific cleavage occurs in liver and perhaps in kidney, resulting in proteolysis of the polypeptide between residues 33 and 34 and several other sites toward the middle of the molecule (Segre *et al.*, 1977). The biological significance of this cleavage is at present unknown. The minimum sequence, however, required for expression of biological action is apparently left intact. Thus, the peripheral cleavage might represent an activation step prior to action of the secreted peptide on target sites (Martin *et al.*, 1978). Fragments of hormone may also enter the circulation from proteolysis within the gland. These processes of late posttranslational or postsecretory proteolysis seem responsible for the heterogeneity of circulating forms of immunoreactive hormone, a phenomenon that complicates interpretation of results achieved with immunoassays of hormone in blood in normal or pathophysiological states (Habener and Segre, 1979). The purpose of this article, however, is, principally, to review the experimental evidence in support of the cellular biosynthetic pathway described above.


## II. Materials and Methods


## A. Incubations of Parathyroid Gland Slices with Labeled Amino Acids

Bovine parathyroid glands obtained at the time of slaughter were immediately placed in ice-cold Earle's Balanced Salt Solution (EBSS) for transport to the laboratory. Slices (0.5–1.0 mm) were made of the glands, and 100 mg of slices was preincubated at 37°C for 45 minutes in 1.0 ml of Eagle's Minimum Essential Medium obtained without leucine or methionine, glutamine, calcium, or magnesium (Grand Island Biological Co., Grand Island, NY) and supplemented with 0.8 m$M$ MgCl$_2$, 0.5 m$M$ CaCl$_2$, and 2 m$M$ glutamine. After the 45-minute preincubation, the slices were "pulse-labeled" by addition of 0.4 ml of a solution consisting of 2 ml of L-[3,4,5-$^3$H]leucine (110 Ci/mmol, 1 mCi/ml) (New England Nuclear, Boston, MA), 0.2 ml of 10× concentrated EBSS, and 0.05 ml of 7.5% sodium bicarbonate or by addition of 0.1 ml of [$^{35}$S]methionine (600–800 Ci/mmol, 8 mCi/ml) (New England Nuclear). After 5 minutes of incubation with the solution containing [$^3$H]leucine, 5 volumes of ice-cold EBSS was added to one of the incubation vessels (5-minute pulse incubation), and 0.05 ml of 0.1 $M$ leucine was added to the remaining vessels (5-minute pulse and chase incubation). Cold EBSS was then added to individual vessels at 10, 15, 25, 40, 55, and 85 minutes of incubation to inhibit further incorporation of radioactive leucine into the slices, and the slices were immediately separated from the media and placed on ice. The cold tissue slices were divided into two approximately equal aliquots; one aliquot was frozen on dry ice for extraction and electrophoretic analyses of radioactive proteins, and the other aliquot was finely minced in a solution of 4% glutaraldehyde, 0.5 m$M$ CaCl$_2$, 0.15 $M$ NaCl, and 0.1 $M$ sodium hydrogen phosphate (pH 7.4) in preparation for electron microscope autoradiography.

In other experiments, slices of parathyroid glands were incubated in Eagle's Minimum Essential Medium as above except without methionine, and pulse-labeling was carried out with [$^{35}$S]methionine (600–900 Ci/mmol) (Amersham Radiochemicals, Arlington Heights, IL) at a level of 1 mCi/ml of medium. Chase incubations were done by addition of 0.05 ml of 0.1 $M$ methionine. Tissues were extracted for electrophoretic analyses as described below.

## B.  Subcellular Fractionation and Limited Proteolysis

Parathyroid gland slices (200 mg) pulse-labeled with [$^{35}$S]methionine for 5 minutes, followed by chase incubations of up to 55 minutes, were homogenized at 2°C in a buffer consisting of 0.25 $M$ sucrose, 50 m$M$ Tris (pH 7.5), 50 m$M$ KCl, and 5 m$M$ MgCl$_2$ (STKM buffer) by ten strokes with a Teflon-glass homogenizer driven at 1200 rpm. The homogenate was centrifuged at 1000 $g$ for 10 minutes. The supernatants (postnuclear) were centrifuged at 105,000 $g$ for 60 minutes (Beckman type-65 rotor, 40,000 rpm), and the pellets were resuspended by gentle homogenization in 0.5 ml of STKM. For each pulse–chase time, 0.1-ml aliquots of these high-speed particulate suspensions were treated as follows: (1) addition of 10 $\mu$l of STKM (control); (2) addition of 10 $\mu$l of a solution of pancreatic chymotrypsin and trypsin (1.0 mg/ml) in STKM; (3) addition of 10 $\mu$l of water and 10 $\mu$l of a 10% solution of Triton X-100; and (4) addition of detergent as in (3), plus enzyme solution as in (2). After a 30-minute incubation at 2°C, 0.5 ml of cold 10% TCA was added, and the acid-insoluble precipitates were collected by centrifugation, dried by washing with acetone–ethyl ether (1:1 vol/vol), and dissolved in 0.1 ml of 0.0625 $M$ Tris–HCl (pH 6.8), 2% sodium dodecyl sulfate (SDS), 10% glycerol, 5% 2-mercaptoethanol, and 0.001% bromphenol blue. Samples (0.025-ml aliquots) were analyzed by electrophoresis on 10–20% polyacrylamide slab gels.

## C.  Extraction and Electrophoretic Analyses of Radioactive Proteins in Parathyroid Gland Slices and in Incubation Media

The frozen parathyroid gland slices were ground to a powder with mortar and pestle and extracted with 2 ml of 8 $M$ urea–0.2 $N$ HCl. The extracts were twice precipitated with cold 10% TCA (after solubilization in 0.2 $N$ NaOH). The final TCA precipitates were suspended in water and lyophilized, and the dried powders were extracted with 1.0 ml of 8 *M* urea–0.15 *N* acetic acid. The radioactive proteins in the urea–acetic acid extracts were analyzed by electrophoresis either on discontinuous 10% polyacrylamide gels containing 8 *M* urea and 0.1 *N* potassium acetate (urea–acetate gels) as described previously or on 10–20% polyacrylamide gradient slab gels containing SDS. The radioactive proteins in the media removed from the tissue slices were precipitated with 10% TCA, as described above, and the acid-insoluble precipitates were extracted with 8 *M* urea–0.15 *N* acetic acid. The radioactive proteins were analyzed by electrophoresis on cylindrical urea–acetate gels.

Amounts of radioactive proteins in cylindrical gels were assessed by scintillation counting of slices prepared from the gels, and amounts in slab gels by densitometric (Zeiss PMQ II) analyses of autoradiographs prepared from the dried gel slabs using Kodak SB-5 film.

## D. Processing of Tissue for Electron Microscope Autoradiography

For each time point of the pulse–chase experiments, three blocks were selected on the basis of the most satisfactory tissue preservation. Pale-gold sections [corresponding to 1000 Å on the thickness scale of Salpeter and Bachmann (1965)] were cut from each selected block with an LKB Ultratome III (LKB Instruments, Inc., Rockville, MD) equipped with a diamond knife. The sections were covered with Ilford L4 emulsion according to the loop method of Caro (1961) and exposed in the dark at 4°C in the presence of Silicagel for 4–6 weeks. Development of the exposed emulsion was done with Phenidon.

From each block, 10 negatives (30 for each time point) were taken at a fixed magnification (×7233) controlled by a calibration grid. The negatives were further enlarged by photographic printing to ×21,700.

To estimate the probability with which the radiation source lies within one given tissue component, we applied to the developed autoradiographs the "95% probability" circle of Whur *et al.* (1969) associated with a morphometric analysis according to Staubli *et al.* (1977) and Weibel (1973). This procedure is needed because a $\beta$-particle will not necessarily hit the silver crystal that lies immediately over the source and because a single developed silver grain often overlaps several structures. The evaluation of developed silver grains and the morphometric study were performed on the same photographic print. In practice, the morphometric data were recorded first, and then the circle method was applied (see below).

## E. Morphometry

The volume density of the different cytoplasmic compartments was determined by the point-counting method (Weibel, 1969), using a multipurpose test screen with 168 points and 84 test lines.

The following cell compartments were considered: the nucleus, the RER (including the nuclear envelope), the Golgi apparatus (including the cytoplasmic matrix immediately surrounding the saccules and vesicles), mitochondria, secretory granules, lysosomes, plasma membrane, and cytoplasm (defined as the cytoplasmic matrix containing none of the membrane compartments delimited above).

## F.  Analysis of Grain Distribution

A circle with a radius of 4.9 mm (corresponding to 2250 Å at the 21,700 magnification of the prints) with five equidistant points on its perimeter was placed around each developed grain (i.e., the grain was considered as the circle's center). The probability that one of the cellular compartments defined above was responsible for the developed silver grain was then estimated as follows (see Fig. 5, for example): because three points of the circle were over RER cisternae, and two points were over mitochondria, the probability that each of these compartments was the source of radiation that generated the particular silver grain was rated 3 in 5, and 2 in 5, respectively.

All the autoradiographic data were based on the analysis of 1500–2000 grains for each time point; the background was 2.5 grains per 1000 $\mu m^2$.

The data were analyzed by determination of the specific radiation label density, which indicates the grain counts per unit volume of a given compartment:

$$R_{vi+} = G_i/P_i \ [5(\sqrt{3/d^2}\ )\ T]$$

where $R_{vi+}$ = the specific radiation label density; $G_i$ = the number of points of the 95% probability circle falling on a compartment $i$; $P_i$ = the number of points of the morphometric test lattice falling on the compartment $i$; $d$ and $T$ are expressed in micrometers; and $R_{vi+}$ represents the number of grains in compartment $i$ per cubic micrometer of compartment $i$.

## G.  Cell-Free Translations of Parathyroid mRNA

Ribonucleic acid was extracted from frozen bovine parathyroid glands that had been stored in liquid nitrogen. The glands were pulverized and extracted several times in a mixture of aqueous (acetate, SDS, EDTA) and organic (phenol, chloroform, isoamyl alcohol) solvents as described previously (Majzoub *et al.*, 1979). Polyadenylated RNA was prepared by affinity binding to oligo(dT)–cellulose (T-3, Collaborative Research, Inc., Waltham, Massachusetts). The final poly(A) RNA was precipitated in 2 volumes of ethanol and redissolved in distilled water at a concentration of approximately 10 $A_{280}$ units/ml.

Cell-free extracts (S-30) were prepared from wheat germ (Roberts and Paterson, 1973). Aliquots (1–5 $\mu$l) of poly(A) RNA were translated in the presence of either [$^{35}$S]methionine or [3,4,5-$^3$H]leucine. In certain of the translations microsomal membranes prepared from canine pancreas (Blobel and Dobberstein, 1975) were added at a concentration of approximately 4 $A_{260}$ units/ml.

## H.  Synthesis of Poly(dC)-Tailed Double-Stranded DNA

Messenger RNA, isolated as described above, was further purified by centrifugation through a 5–20% sucrose gradient (Kronenberg et al., 1977). Fractions with Pre-ProPTH mRNA activity were identified by cell-free translation and precipitated with ethanol. The mRNA (18 $\mu$g) was then incubated in a 2.2-ml reaction mixture containing 169 units of reverse transcriptase, 50 m$M$ Tris–HCl (pH 8.3), oligo(dT) at 27 $\mu$g/ml, 200 $\mu M$ [$\alpha$-$^{32}$P]dATP and dTTP (both 68 mCi/mmol), 900 $\mu M$ dGTP and dCTP, 120 m$M$ KCl, 10 m$M$ dithiothreitol, and 10 m$M$ MgCl$_2$. After 1 hour at 43°C, the reaction mixture was extracted with an equal volume of phenol–choloroform–isoamyl alcohol (25–24:1), and 55 $\mu$g of Escherichia coli tRNA and 0.2 ml of 3 $M$ sodium acetate were added, and the nucleic acids were precipitated with 2 volumes of ethanol. The pellet was treated with 0.3 $M$ NaOH overnight at 37°C, neutralized, and chromatographed over Sephadex G-100.

The complementary DNA (cDNA) was made double-stranded with DNA polymerase I in a 1.5-ml reaction mixture containing 1500 units of DNA polymerase I, 100 m$M$ KPO$_4$ (pH 6.9), 200 $\mu M$ [$\alpha$-$^{32}$P]dATP and dTTP (both 100 mCi/mmol), 1 m$M$ dGTP and dCTP, 10 m$M$ dithiothreitol, and 10 m$M$ MgCl$_2$. After 2 hours at 15°C, the mixture was extracted with phenol–chloroform–isoamyl alcohol, precipitated with ethanol, and passed over a column of Sephadex G-100. The DNA was then digested with S1 nuclease using standard conditions (Efstratiadis et al., 1976), extracted again with phenol–chloroform–isoamyl alcohol, and precipitated with ethanol. The DNA was then centrifuged through a gradient of 5–20% sucrose in 0.8 $M$ NaCl, 8 m$M$ disodium ethylenediaminetetraacetate (EDTA), 10 m$M$ Tris–HCl (pH 7.6) for 8 hours at 49,000 rpm. The largest fraction (30 ng) of DNA was dialyzed against 10 m$M$ Tris–HCl (pH 7.6)/0.1 m$M$ Na$_2$EDTA; 5 $\mu$g of tRNA was added, and the nucleic acids were precipitated with ethanol. Poly(dC) homopolymer extensions, approximately 20 nucleotides long, were added to the DNA by using terminal transferase as described (Roychoudhury et al., 1976). Similarly, poly(dG) homopolymer extensions, 20 nucleotides long, were added to Pst I-cut plasmid pBR 322 DNA after phenol extraction, dialysis, and ethanol precipitation of the DNA.

## I.  Transfection of E. coli χ1776

Poly(dG)-tailed pBR 322 (400 ng) and 30 ng of poly(dC)-tailed parathyroid DNA were annealed in 0.1 $M$ NaCl/10 m$M$ Tris–HCl (pH 7.6)/0.1 m$M$

Na$_2$EDTA for 2 minutes at 65°C and then for 2 hours at 42°C. *Escherichia coli* χ1776 was grown and treated with 70 m*M* MgCl$_2$/30 m*M* CaCl$_2$/40 m*M* Na acetate (pH 5.6) as described (Villa-Komaroff *et al.*, 1978). Transformation was performed in a biological safety cabinet in a P3 physical containment facility according to the then-current National Institutes of Health guidelines.[1]

## J.   DNA Sequence Analysis

After identification of Pre-ProPTH-specific clones by hybrid-arrested translation (Paterson *et al.*, 1977), the largest clone was analyzed by using the DNA sequencing technique of Maxam and Gilbert (1980). DNA fragments generated by restriction endonuclease were 5'-end labeled by using the polynucleotide kinase exchange reaction (Kronenberg *et al.*, 1979). We found this one-step procedure easier to use than the sequential use of alkaline phosphatase and the polynucleotide kinase forward reaction. We did find, however, that the exchange reaction worked well only in the labeling of fragments containing single-stranded 5'-extensions (as opposed to 3'-extensions or flush ends). This requirement limited our choice of restriction endonucleases somewhat.

## III.   Results

Direct translation in heterologous cell-free systems of messenger RNA extracted from parathyroid glands gives Pre-ProPTH as a major product (Habener *et al.*, 1978). The complete amino acid sequence of this product, synthesized in a wheat germ cell-free system, was determined by radiomicrosequencing (Fig. 2). Determination of the nucleic acid sequence of cDNA to the mRNA coding for Pre-ProPTH, prepared by bacterial cloning of a recombinant molecule consisting of the cDNA and plasmid pBR 322, reconfirmed the amino acid sequence of Pre-ProPTH and furthermore, confirmed that Pre-ProPTH represents all the structural information encoded in the gene for PTH (Fig. 3) (Kronenberg *et al.*, 1979). An initiation codon (ATG) was found corresponding to the NH$_2$-terminal methionine of Pre-ProPTH (position −31), and a stop codon (TGA) was found immediately 3' to the codon for glutamine (position 84) at the COOH terminus of PTH. An unexpected, and as yet unexplained, finding in the analyses of the sequence of the cDNA was the presence of a start codon and two stop codons located in the 5' noncoding region of the transcript. All three codons are in phase with the reading frame for the structural sequence of the hormone.

Thus, forms of PTH reported earlier to be larger than Pre-ProPTH must repre-

[1]National Institutes of Health, *Federal Register*, July 7, 1976.

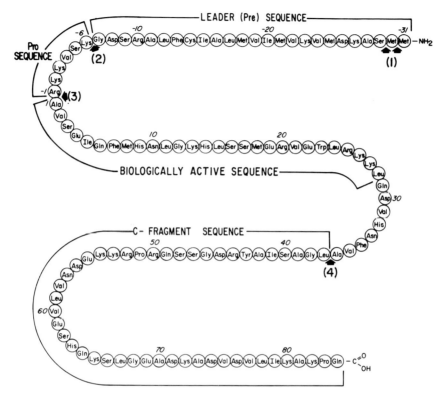

FIG. 2. Structure of bovine pre-proparathyroid hormone. The amino acid sequence was determined by automated Edman degradation on the product of parathyroid mRNA translation in a wheat germ cell-free system. Arrows indicate cleavage sites cleaved by processing enzymes in the parathyroid gland (1–3) and in the liver after secretion of the hormone (4). The biologically active region of the molecule is flanked by sequences not required for activity on target organ receptors.

sent aggregates of the hormone or other artifacts of binding of the hormone to larger macromolecules in blood or tissue extracts or, alternatively, they are translated from an mRNA different from the one that predominates in the bovine parathyroid gland.

To investigate the subcellular sites in the parathyroid cell where the biosynthetic precursors undergo specific proteolytic cleavages, we examined, by electrophoresis, the kinetics of the disappearance of labeled Pre-ProPTH and the conversion of labeled ProPTH to PTH, and, by electron-microscope autoradiography, the spatiotemporal migration of autoradiographic grains in bovine parathyroid gland slices incubated with [³H]leucine for 5 minutes (pulse incubation) followed by incubations with unlabeled leucine for periods up to 85 minutes (chase incubations). Electrophoretic analyses showed that Pre-ProPTH disappeared rapidly (by 5 minutes) and that conversion of ProPTH to PTH was first

TAG CAG CTG ATG CTT TCT CAA AGT TGA GTA AAC CTG AGA AGG CTG ATA AAT TGA GCT GCT AAT ACA TTT
1　　　　　10　　　　　20　　　　　30　　　　　40　　　　　50　　　　　60

　　　　　　　　　　　　　　　　　　　　　　　-31
　　　　　　　　　　　　　　　　　　　　met met ser ala lys asp met val lys val met ile
GAA AGA AGA TTG TAT CCT AAG ACG TGT GTT AAT ATG ATG TCT GCA AAA GAC ATG GTT AAG GTA ATG ATT
70　　　　　80　　　　　90　　　　　100　　　　　110　　　　　120　　　　　130

　　　　　　　　　　　　　　　　　　　-6　　　　　　　　　　-1 +1
val met leu ala ile cys phe leu ala arg ser asp gly lys ser val lys lys arg ala val ser glu
GTC ATG CTT GCC ATC TGT TTT CTT GCA AGA TCA GAT GGG AAG TCT GTT AAG AAG AGA GCT GTG AGT GAA
140　　　　　150　　　　　160　　　　　170　　　　　180　　　　　190　　　　　200

　　　　　　　　　　　　　　　　　　　　　　　　　　　+22
ile gln phe met his asn leu gly lys his leu ser ser met gly arg val glu trp leu arg lys lys leu
ATA CAG TTT ATG CAT AAC CTG GGC AAA CAT CTG AGC TCC ATG GAA AGA GTG GAA TGG CTG CGG AAA AAG CTA
210　　　　　220　　　　　230　　　　　240　　　　　250　　　　　260　　　　　270

gln asp val his asn phe val ala leu gly ala ser ile ala tyr arg asp gly ser ser gln arg pro
CAG GAT GTG CAC AAC TTT GTT GCC CTT GGA GCT TCT ATA GCT TAC AGA GAT GGT AGT TCC CAG AGA CCT
280　　　　　290　　　　　300　　　　　310　　　　　320　　　　　330　　　　　340

arg lys lys glu asp asn val leu val glu ser his gln lys ser leu gly glu ala asp lys ala asp
CGA AAA AAG GAA GAC AAT GTC CTG GTT GAG AGC CAT CAG AAA AGT CTT GGA GAA GCA GAC AAA GCT GAT
350　　　　　360　　　　　370　　　　　380　　　　　390　　　　　400　　　　　410

　　　　　　　　　　　　　　　+84
val asp val leu ile lys ala lys pro gln stop
GTG GAT GTA TTA ATT AAA GCT AAA CCC CAG TGA AAA CAG ATA TGA TCA GAT CA
420　　　　　430　　　　　440　　　　　450　　　　　460　　　　　470

FIG. 3. Nucleic acid sequence of a 470 base pair cDNA prepared by molecular cloning of a recombinant plasmid (pBR 322) in *E. coli* (strain χ1776). Note start codon ATG (10–12) and two stop codons TGA (25–27 and 52–54) within the 5′ noncoding region of the transcript. (From Kronenberg *et al.*, 1979.)

detectable at 15 minutes and was completed by 30 minutes (Fig. 4). By 5 minutes, 85% of the autoradiographic grains was confined to the rough endoplasmic reticulum (RER) (Fig. 5). Autoradiographic grains increased rapidly in number in the Golgi region after 15 minutes of incubation (Fig. 6); from 15 to 30 minutes, there was a migration of label to secretory vesicles still in the Golgi region, and then to mature secretory granules outside the Golgi area (Fig. 7). At later times of incubation (30–90 minutes), labeled secretion granules migrated to the periphery of the cell and to the plasma membrane, in correlation with the release of PTH first detected by 30 minutes. We conclude that proteolytic conversion of Pre-ProPTH to ProPTH takes place in the RER and that subsequent conversion of ProPTH to PTH occurs in the Golgi complex (Fig. 8).

Evidence that the enzymatic activity responsible for the conversion of Pre-ProPTH to ProPTH resides in the membranes of the endoplasmic reticulum has come from comparisons of the products resulting from the translations of parathyroid messenger RNA in cell-free systems devoid of microsomal membranes compared with those containing such membranes (Blobel and Dobberstein, 1975; Dorner and Kemper, 1978). Pre-Pro-PTH is the only hormonal product synthesized in membrane-free translation systems derived from extracts of wheat germ or reticulocytes, whereas both Pre-ProPTH and ProPTH appear as products of synthesis under conditions in which the translation system is supplemented with microsomal membranes (Fig. 9) (Dorner and Kemper, 1978; Kemper *et al.*, 1974). Additional evidence indicates that the enzyme responsible for the cleavage of Pre-ProPTH to form ProPTH is localized within the membrane. Only polypeptide chains undergoing synthesis and vectorial discharge through the membrane are cleaved to form ProPTH. Pre-ProPTH added to membrane-containing cell-free extracts of either ascites tumor cells (Habener *et al.*, 1975) or parathyroid glands (Habener *et al.*, 1977) remains intact and is not converted either to ProPTH or to PTH under conditions in which ProPTH is readily converted to PTH.

Early events in the cellular synthesis and transfer into membrane-limited compartments of Pre-ProPTH and ProPTH were investigated further and in greater

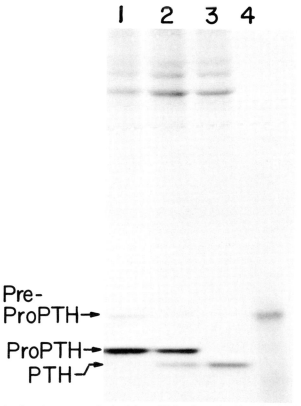

FIG. 4. Electrophoretic patterns obtained from SDS-polyacrylamide gels of labeled proteins in extracts of parathyroid gland slices incubated for the times indicated. The pulse incubations were done with [³H]leucine; channel 1, 5-minute pulse; channel 2, 5-minute pulse followed by 10-minute chase; channel 3, 5-minute pulse followed by 25-minute chase; channel 4, products of translation of parathyroid gland mRNA in a wheat germ cell-free system. Radioactivity was measured by autoradiography of a dried gel slab.

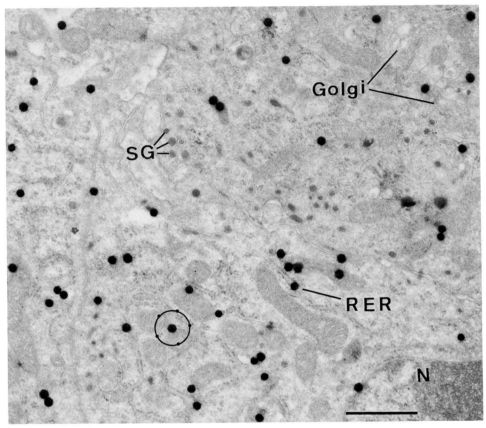

FIG. 5.   Autoradiograph of parathyroid cells at the end of the 5-minute [³H]leucine pulse. The three main intracellular membrane compartments of the parathyroid cell are identified: rough endoplasmic reticulum (RER), Golgi apparatus, and secretory granules (SG). At this time point, many autoradiographic grains (black dots) relate to RER cisternae. Around one developed grain, the probability circle, used for quantitative evaluation of the labeling, has been drawn. On this circle, the five equidistant points are represented: Three of five points fall on RER profiles, whereas two of five are over mitochondrial profiles. N, nucleus. Bar, 1 μm. (From Habener *et al.*, 1979.)

detail by electrophoretic analyses of newly synthesized proteins in subcellular fractions of parathyroid gland slices pulse-labeled for 0.5–5 minutes with [³⁵S]methionine. During these short times of incubation, both Pre-ProPTH and ProPTH were confined to the microsomal fraction, which was defined as the particulate fraction obtained after sedimentation of a postnuclear supernatant at 105,000 *g* (Fig. 10). Labeled Pre-ProPTH and ProPTH were detected in a 30-second interval between 0.5 and 1.0 minute of incubation. The radioactivity in Pre-ProPTH became relatively constant between 3 and 5 minutes, whereas the radioactivity in ProPTH increased markedly over this period. When corrected for the known content of methionine in the prehormone (five residues) and the prohormone (two residues), we found four times as much radiolabeled prohormone as prehormone between 0.5 and 1.0 minute of synthesis. Moreover, a small amount of labeled prohormone but not of pre-prohormone was observed by autoradiofluorographic analysis on an acrylamide gel slab of a detergent extract of tissues that had been pulse-labeled for 0.5 minute (Habener *et al.*, 1980). Evidence for the sequestration of labeled protein into endoplasmic reticulum compartments was obtained by showing that treatment of the microsomal fraction with a mixture of chymotrypsin and trypsin resulted in the degradation of the

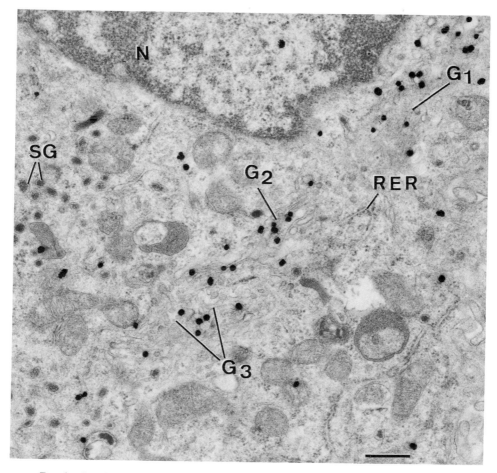

FIG. 6. Parathyroid cell 10 minutes after the end of the pulse. At this stage, three Golgi regions (G₁, G₂, G₃) are preferentially labeled with developed autoradiographic grains. Rough endoplasmic reticulum cisternae (RER) and secretory granules (SG) are indicated. N, nucleus. Bar, 0.5 μm. (From Habener *et al.*, 1979.)

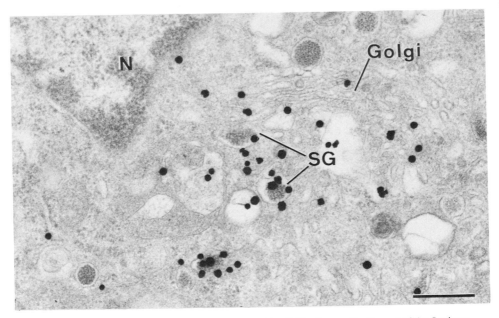

FIG. 7.   Detail of the Golgi region of a parathyroid cell 10 minutes after the end of the 5-minute pulse. The Golgi cisternae are labeled with several developed grains, as well as secretory granules (SG) associated with Golgi cisternae and distinguished by their relatively large size and wide halo surrounding the granule core. Such granules are interpreted as "maturing" secretory vesicles. N, nucleus. Bar, 0.5 μm. (From Habener *et al.*, 1979.)

prehormone but not of the prohormone (Fig. 11); complete degradation of both pre-prohormone and prohormone occurred only after lysis of microsomal membranes with Triton X-100. Approximately 50% of pre-prohormone and 25% of prohormone were released from the microsomes by their extraction with 1.0 *M* KCl, whereas 80–90% of each hormonal precursor was released by treatment with Triton X-100 (Habener *et al.*, 1980).

These results are consistent with the signal hypothesis (Blobel and Dobberstein, 1975; Habener *et al.*, 1980), inasmuch as demonstration of membrane-bound, proteolysis-resistant labeled prohormone before the appearance of pre-prohormone (0.5 minute) by 0.5–1 minute of synthesis indicates that cleavage of the leader sequence of Pre-ProPTH and transfer of prohormone into the cisternal space of the RER occurs concomitantly with the growth of the nascent polypeptide chain (cotranslational processing). Appearance of membrane-sequestered ProPTH appears to take place without prior entry of Pre-ProPTH into the cisternal space, further suggesting that proteolytic removal of the leader peptide occurs during transfer of the polypeptide through the lipid bilayer. A model of cotranslational processing does not account for the labeled Pre-ProPTH observed in the gland slices. Therefore, an alternative explanation must be proposed. We propose that the Pre-ProPTH represents a small number of chains that escape processing during their growth and do not transfer across the membrane (Habener *et al.*, 1980). Evidence for mechanism of such a transfer arrest is supported by the finding that Pre-ProPTH is only partly extracted from the microsomes by treatment with 1.0 *M* KCl, a finding consistent with a model whereby a substantial

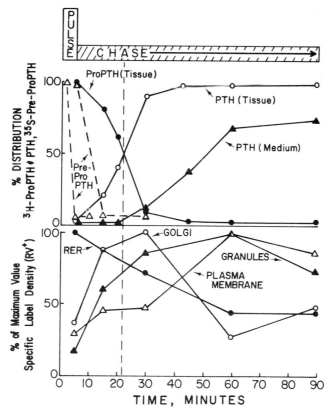

FIG. 8.   Summary of pulse–chase incubations with bovine parathyroid gland slices. Parathyroid gland slices were pulse-labeled with [³H]leucine for 5 minutes, after which the incubations were continued in the presence of unlabeled leucine (chase incubations) for the times indicated. Upper graph: Distributions of [³H]leucine-labeled ProPTH and PTH in parathyroid tissues and in incubation medium as determined by polyacrylamide gel electrophoresis. Open and closed circles indicate percentage distributions of PTH and ProPTH within tissues, where the sum of PTH and ProPTH at each time is taken as 100%. Open triangles indicate Pre-ProPTH in tissue where amount at end of pulse incubation is taken as 100%. Closed triangles indicate percent of total PTH secreted into the incubation medium. Lower graph: Distribution of specific label density in the subcellular compartments of parathyroid cells analyzed by quantitative electron microscope autoradiography. (From Habener *et al.*, 1979.)

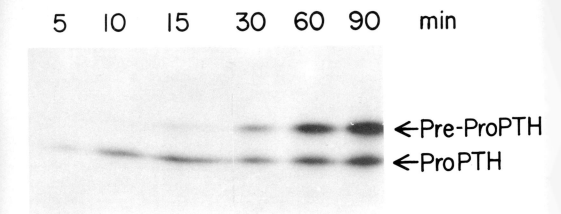

FIG. 9. Autoradiofluorogram of products of translation of parathyroid mRNA in a wheat germ cell-free system containing microsomal membranes prepared from canine pancreas. Note processing to ProPTH at early times of the reaction, followed at later times by accumulation of unprocessed Pre-ProPTH.

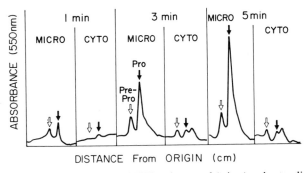

FIG. 10. Labeled Pre-ProPTH and ProPTH in microsomal (micro) and cytosolic (cyto) fractions prepared from parathyroid gland slices pulse-labeled for 1–5 minutes with [$^{35}$S]methionine. Patterns are densitometric tracings of autoradiograms prepared from a 10–20% gradient polyacrylamide gel containing SDS. Only regions of the gel in the region of migration of the hormonal precursors are shown.

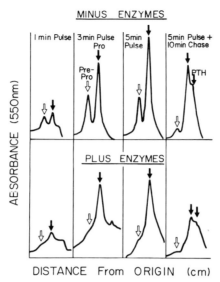

FIG. 11.   Evidence for rapid sequestration of newly synthesized ProPTH into cisterna of RER. Microsomes were prepared from parathyroid gland slices that were pulse-labeled for 1–5 minutes followed by a chase-incubation of 10 minutes. Aliquots of microsomes were subjected to limited proteolysis with a mixture of trypsin and chymotrypsin. Labeled proteins were analyzed by electrophoresis on 10–20% gradient acrylamide gels containing SDS. Autoradiograms of the gel slabs are shown. Minus enzymes, no enzymes added; plus enzymes, enzymes added.

fraction of the Pre-ProPTH is only partially inserted into the membranes before it is cleaved to form ProPTH.

## IV.   Discussion

As a result of these studies, we envision a model whereby the initiation of the synthesis of Pre-ProPTH occurs on polyribosomes located within the cell matrix (Figs. 12 and 13). When the growing polypeptide chain is approximately 20–30 amino acids in length, the $NH_2$-terminus of the chain first emerges from the large subunit of the ribosome, and, at this time, the two $NH_2$-terminal methionines of Pre-ProPTH are removed by a putative methionyl amino peptidase. As the nascent chain continues to grow, the hydrophobic amino-terminal sequence ("leader," "pre," or "signal" sequence) of Pre-ProPTH emerges and associates with the membrane of the endoplasmic reticulum in accord with the "signal" hypothesis originally proposed by Blobel and his co-workers.

Newly synthesized polypeptide chains are vectorially discharged into the cisternal space of the endoplasmic reticulum. By this means, the proteins that are to be secreted are segregated from other proteins that are to be retained within the cell. The secretory proteins are then transported within membrane-limited compartments to the Golgi region of the cell, where they are incorporated into either secretory vesicles or granules. The protein within granules is then either stored within the cell or transported to the cell membrane and released by exocytosis into the extracellular fluid in response to the appropriate stimulus.

In the experiments involving very short pulse-labeling times, newly synthesized Pre-ProPTH and ProPTH were demonstrated during the earliest 30-second period in which labeled proteins could be detected (between 30 and 60 seconds). The labeled ProPTH thus formed was found almost exclusively in association with the microsomal fraction of the parathyroid cell. Furthermore, rapid sequestration of the prohormone within the interior of microsomal vesicles at this early time of synthesis was evidenced when the prohormone was shown to be largely resistant to limited proteolysis or extraction by 1.0 $M$ KCl, yet was released readily after disruption of the microsomes by treatment with Triton X-100.

An intriguing finding in these studies was that, during the 30-second interval (between 30 and 60 seconds) of protein labeling, the amounts of labeled ProPTH were at least four times as great as the amounts of labeled Pre-ProPTH when appropriate corrections for actual relative amounts of these two polypeptides were made, based on their known contents of methionine (Habener *et al.*, 1977). This observation strongly indicates that the formation of ProPTH via the cleavages of the $NH_2$-terminal leader sequence from the initial translation product,

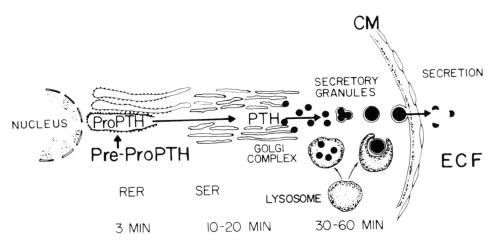

FIG. 12.   Schematic diagram of proposed intracellular translocation and processing of parathyroid hormone precursors. CM = cell membrane. ECF = extracellular fluid.

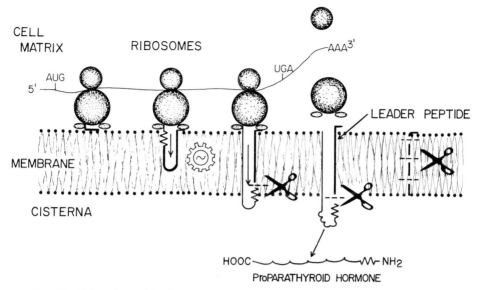

FIG. 13.   Schematic model of events proposed in the unidirectional transport of the nascent polypeptide chain across the lipid bilayer. The schema indicates cotranslational processing; i.e., cleavage of the leader peptide occurs before growth of the nascent chain is completed.

Pre-ProPTH, must occur, to a large extent, during growth of the nascent polypeptide chain (cotranslational processing) rather than after completion of the polypeptide chain (posttranslational processing).

These observations and conclusions derived from our studies of PTH biosynthesis are in agreement with those proposed by a number of other workers who have reported evidence for cotranslational processing of presecretory proteins in systems other than the parathyroid gland. Of particular note in this regard are the studies of Blobel and Dobberstein (1975), who, in a series of "readout" experiments involving translations of immunoglobulin mRNA under cell-free conditions utilizing microsomal membranes, demonstrated removal of the leader sequence before completion of the growth of the polypeptide chain. Similarly, Spielman and Bancroft (1977) and Boime et al. (1977) observed cotranslational processing in cell-free systems supplemented with membranes and primed with mRNAs coding for growth hormone and placental lactogen. In addition to our studies using intact cells, Patzelt et al. (1978) studied proinsulin synthesis in intact pancreatic islets, with results qualitatively similar to ours. They observed a rapid accumulation of $^3$H-labeled bovine proinsulin, greater than the amounts of pre-proinsulin accumulated during pulse-labeling periods of 1–20 minutes.

The evidence pointing to the formation of ProPTH during nascent chain growth, via cleavage of the NH$_2$-terminal leader sequence from the chains, leaves undetermined the subcellular site and pathway of the Pre-ProPTH, which were observed in these and earlier (Habener *et al.*, 1976; Habener and Potts, 1979) pulse-labeling studies. An apparently reasonable explanation for this phenomenon is that the Pre-ProPTH detected represents a small fraction of the polypeptide chains that escape processing during their growth. Such an occurrence could result from a small but finite degree of inefficiency of the responsible transport and processing enzymatic systems believed to exist within the membrane, resulting in an arrest of the transfer of the newly synthesized chains across the membrane. It is difficult to envision a model whereby actual transfer of the completed Pre-ProPTH chain takes place across the membrane into the cisternal space where enzymatic processing then takes place posttranslationally or even fails to occur, inasmuch as (unlike our findings for ProPTH) we found no evidence that the pre-prohormone was resistant to limited proteolysis. Therefore, by this criterion, Pre-ProPTH could not be sequestered within microsomes.

Another possible explanation for the apparent survival of unprocessed Pre-ProPTH chains is that these particular chains are synthesized on polyribosomes that fail to attach to the membrane and, as a consequence, are released into the cell matrix (heterotopic synthesis). This explanation, however, must take into account the observation that the Pre-ProPTH was found in the subcellular fractions in association with the microsomes and not in the cell sol. It must be considered, however, that, if released into the cell matrix, the Pre-ProPTH might adsorb nonspecifically to the microsomes.

The fact that all the Pre-ProPTH was found to be susceptible to proteolytic digestion under limiting conditions indicates that, in the "transfer-arrested" chains as well as in any chains adsorbed peripherally on the microsomes, there must be some portions exposed to the aqueous environment and available for proteolysis. Clearly, much additional study will be required to provide a more accurate and detailed model of the processess involved and to understand the nature of the recognition and transport elements in the membrane, the forces involved in the discharge of polypeptide chains across the membrane, and the nature of the enzymatic recognition and cleavage of the leader peptide from Pre-ProPTH.

REFERENCES

Blobel, G., and Dobberstein, B. (1975). *J. Cell Biol.* **67**, 835–851.
Boime, I, Szczesna, E., and Smith, D. (1977). *Eur. J. Biochem.* **73**, 515–520.
Caro, L. G. (1961). *J. Biophys. Biochem. Cytol.* **10**, 37–45.
Dorner, A. J., and Kemper, B. (1978). *Biochemistry* **17**, 5550–5555.
Efstratiadis, A., Kafatos, F. C., Maxam, A., and Maniatis, T. (1976). *Cell* **7**, 279–288.
Habener, J. F. (1981). *Annu. Rev. Physiol.* **43**, 211–223.
Habener, J. F., and Kronenberg, H. M. (1978). *Fed. Proc., Fed. Am. Soc. Exp. Biol.* **37**, 2561–2566.

Habener, J. F., and Potts, J. T., Jr. (1978). *N. Engl. J. Med.* **299**, 580–585; 635–644.
Habener, J. F., and Potts, J. T., Jr. (1979). *Endocrinology (Baltimore)* **104**, 265–275.
Habener, J. F., and Segre, G. V. (1979). *Ann. Intern. Med.* **91**, 782–785.
Habener, J. F., Kemper, B., Potts, J. T., Jr., and Rich, A. (1975). *Biochem. Biophys. Res. Commun.* **67**, 1114–1121.
Habener, J. F., Potts, J. T., Jr., and Rich, A. (1976). *J. Biol. Chem.* **251**, 3893–3899.
Habener, J. F., Chang, H. T., and Potts, J. T., Jr. (1977). *Biochemistry* **16**, 3910–3917.
Habener, J. F., Rosenblatt, M., Kemper, B., Kronenberg, H. M., Rich, A., and Potts, J. T., Jr. (1978). *Proc. Natl. Acad. Sci. U.S.A.* **75**, 2616–2620.
Habener, J. F., Amherdt, M., Ravazzola, M., and Orci, L. (1979). *J. Cell Biol.* **80**, 715–731.
Habener, J. F., Maunus, R., Dee, P. C., and Potts, J. T., Jr. (1980). *J. Cell Biol.* (in press).
Kemper, B., Habener, J. F., Mulligan, R. C., Potts, J. T., Jr., and Rich, A. (1974). *Proc. Natl. Acad. Sci. U.S.A.* **71**, 3731–3735.
Kronenberg, H. M., Roberts, B. E., Habener, J. F., Potts, J. T., Jr., and Rich, A. (1977). *Nature (London)* **267**, 804–807.
Kronenberg, H. M., McDevitt, B. E., Majzoub, J. A., Nathans, J., Sharp, P. A., Potts, J. T., Jr., and Rich, A. (1979). *Proc. Natl. Acad. Sci. U.S.A.* **76**, 4981–4985.
Majzoub, J. A., Kronenberg, H. M., Potts, J. T., Jr., Rich, A., and Habener, J. F. (1979). *J. Biol. Chem.* **254**, 7449–7455.
Martin, K. J., Reitag, J. J., Conrades, M. B., Hruska, K. A., Klahr, S., and Slatopolsky, E. (1978). *J. Clin. Invest.* **62**, 256–261.
Maxam, A. M., and Gilbert, W. (1980). *Methods Enzymol.* (in press).
Paterson, B. M., Roberts, B. E., and Kuff, E. L. (1977). *Proc. Natl. Acad. Sci. U.S.A.* **74**, 4370–4374.
Patzelt, C., Labrecque, A. D., Duguid, J. R., Carroll, R. J., Keim, P. S., Heinrikson, R. L., and Steiner, D. F. (1978). *Proc. Natl. Acad. Sci. U.S.A.* **75**, 1260–1264.
Roberts, B. E., and Paterson, B. M. (1973). *Proc. Natl. Acad. Sci. U.S.A.* **70**, 2330–2334.
Roychoudhury, R., Jay, E., and Wu, R. (1976). *Nucleic Acids Res.* **3**, 101–116.
Salpeter, M. M., and Bachmann, L. (1965). *Symp. Int. Soc. Cell Biol.* **4**, 23–39.
Segre, G. V., Niall, H. D., Sauer, R. T., and Potts, J. T., Jr. (1977). *Biochemistry* **16**, 2417–2427.
Spielman, L. L., and Bancroft, F. C. (1977). *Endocrinology (Baltimore)* **101**, 651–658.
Staubli, W., Schweizer, W., Suter, J., and Weibel, E. R. (1977). *J. Cell Biol.* **74**, 665–689.
Tregear, G. W., van Rietschoten, J., Greene, E., Keutmann, H. T., Niall, H. D., Reit, B., Parsons, J. A., and Potts, J. T., Jr. (1973). *Endocrinology (Baltimore)* **93**, 1349–1353.
Villa-Komaroff, L., Efstratiadis, A., Broome, S., Lomedico, P., Tizard, R., Naber, S., Chick, W. L., and Gilbert, W. (1978). *Proc. Natl. Acad. Sci. U.S.A.* **75**, 3727–3731.
Weibel, E. R. (1969). *Int. Rev. Cytol.* **26**, 235–302.
Weibel, E. R. (1973). *Princ. Tech. Electron Microsc.* **3**, 238–313.
Whur, P., Herscovics, A., and Leblond, C. P. (1969). *J. Cell Biol.* **43**, 289–311.

# Chapter 6

# *Biosynthesis of Insulin and Glucagon*

## HOWARD S. TAGER, DONALD F. STEINER, AND CHRISTOPH PATZELT

*Department of Biochemistry, University of Chicago, Chicago, Illinois*

## I. Introduction

The discovery of proinsulin some 13 years ago (Steiner *et al.*, 1967; Steiner, 1977) initiated an important area of investigation concerning the biosynthesis of peptide hormones via the selective proteolytic conversion of higher-molecular-weight precursors. These posttranslational modifications, which result in the cleavage of the hormone from a longer amino acid sequence, occur during the biosynthesis of a great number of bioactive peptides and often yield complex mixtures of hormone-related forms (for reviews, see Tager and Steiner, 1974; Tager *et al.*, 1975; and Chapter 5 by J. F. Habener *et al.* and Chapter 8 by E. Herbert *et al.* in this volume for detailed examples). Although the sites for proteolytic cleavage in many hormone precursors are remarkably similar, the pathway for conversion of each precursor appears to follow a unique course that leads to the production of the correct hormonal product. In many cases, the nonhormonal sequence is secreted along with the hormone product, and measurements of its formation and release have provided important clues regarding endocrine cell function as well as precursor processing and hormone biosynthesis.

More recent investigations have demonstrated that proteolytic modification plays an additional role in the biosynthesis of most peptides and proteins destined for secretion (Milstein *et al.*, 1972; Devillers-Thiery *et al.*, 1975; Blobel and

Dobberstein, 1975). An NH$_2$-terminal signal peptide or leader sequence on the prohormone structure participates in the segregation of the precursor from the cytosolic compartment to the lumen of the rough endoplasmic reticulum prior to both its passage to the Golgi and its conversion to final products (Kemper *et al.,* 1976; Chan *et al.,* 1976). Further studies of hormone biosynthesis at the level of nucleic acid structure and function have suggested that the initial mRNA transcripts of many genes—including those for insulin (Lomedico *et al.,* 1979; Duguid and Steiner, 1978; Bell *et al.,* 1980)—are larger molecules, which undergo posttranscriptional processing to their mature forms. Thus, the biosynthesis of many peptide hormones appears to require several stages of macromolecular processing, which involve the participation of both proteases and nucleases. Investigations of peptide hormone biosynthesis in the broadest sense thus rely heavily on the techniques of cell biology, protein chemistry, and nucleic acid chemistry. This contribution will summarize aspects of our understanding of the biosynthesis of the islet cell hormones insulin and glucagon.

## II.   Insulin Biosynthesis

Until 1967, it was widely believed that insulin was formed by the combination of separately synthesized A and B chains. The discovery of proinsulin (Steiner *et al.,* 1967) disproved this hypothesis, however, and it is now known that a connecting peptide attaches the COOH terminus of the B chain to the NH$_2$ terminus of the A chain in the single-chain precursor (Steiner *et al.,* 1969; Steiner, 1977). As shown diagrammatically at the top of Fig. 1, the folded proinsulin molecule has correctly formed disulfide bonds both within the A-chain region and between the A-chain and B-chain regions. The C peptide connects the two insulin chains by pairs of dibasic amino acid residues (Lys-Arg and Arg-Arg at the A and B chains, respectively) and forms a bridge between the two portions of the insulin molecule (Steiner *et al.,* 1969; Chance, 1971). Although insulin itself displays a high degree of structural conservation during animal evolution, the C-peptide region shows a great deal of interspecies variability in terms of both its length and its primary structure. The region contains between 21 (dog) and 31 (human, rat, and horse) amino acid residues (Oyer *et al.,* 1971; Tager and Steiner, 1972). Studies on these and other C peptides have shown a high degree of mutation acceptance throughout their sequences (Peterson *et al.,* 1972; Steiner, 1977). Notwithstanding the usual appearance of acidic residues at their NH$_2$ termini (forming the partial proinsulin sequence X-Arg-Arg-Glu or X-Arg-Arg-Asp) and the appearance of glutamine residues at their COOH termini (giving the partial sequence Gln-Lys-Arg-Gly), conserved structures are difficult to identify.

Since the crystal structure of insulin suggests that the NH$_2$ terminus of the A

**proinsulin**

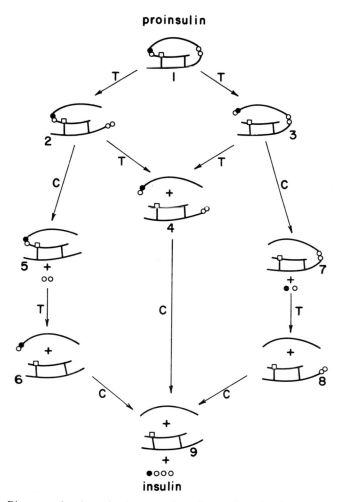

FIG. 1. Diagrammatic scheme for the conversion of proinsulin to insulin. The figure shows the branched pathway by which proinsulin is converted to insulin plus C peptide by trypsin-like (T) and carboxypeptidase B-like (C) enzymes. The dibasic amino acid residues at the conversion sites are illustrated by circles: ○, Arg; ●, Lys.

chain and the COOH terminus of the B chain could be bridged by a connector only 8 Å long (Blundell *et al.*, 1972), in one respect the C peptide seems longer than necessary. Studies showing the efficient oxidation of cysteine residues of both reduced proinsulin (Steiner and Clark, 1968) and model insulin analogs in which $Lys^{B29}$ and $Gly^{A1}$ are bridged by small chemical linkers (Brandenberg and Wollmer, 1973; Busse *et al.*, 1974) suggest that the C peptide maintains an important role in reducing the folding event to an unimolecular process. As previously suggested, however, the "extra" length of the C peptide may permit the nascent peptide to span the large ribosomal subunit and microsomal membrane during the biosynthesis and segregation of the insulin precursor (Patzelt *et al.*, 1978a).

The conversion of proinsulin to insulin by limited proteolysis, as illustrated in Fig. 1, has been examined both by pulse–chase, biosynthetic labeling experiments (Steiner *et al.*, 1969; Tager and Steiner, 1979) and by isolation and characterization of presumed biosynthetic intermediates (Steiner *et al.*, 1969; Chance, 1971). Once proinsulin has entered the cisterna of the rough endoplasmic reticulum and correct disulfide bond formation has occurred, the precursor is transferred in an energy-requiring step to the Golgi apparatus and from there to the secretion granule fraction of the $\beta$ cell (Steiner, 1977). Although conversion begins during the formation of these granules, it proceeds, usually to about 95% completion, during secretion granule maturation. Since the order of the proinsulin molecule is given by $NH_2$-B chain-Arg-Arg-C peptide-Lys-Arg-A chain-COOH, and since both pairs of dibasic amino acids must be removed during the formation of insulin and C peptide, two enzyme activities are required for the overall process. The model predicts a trypsin-like enzyme, which would cleave proinsulin either at the connecting region between the B chain and C peptide (Fig. 1, compound 2) or between the C peptide and the A chain (compound 3), and a carboxypeptidase B-like enzyme to remove COOH-terminal basic residues, producing compounds 5 and 7, respectively. Since intermediates related in structure to those illustrated in Fig. 1 have been isolated, it appears that both courses for conversion occur in normal tissue. The further conversion of intermediates 5 and 7 then proceeds by the actions of trypsin-like and carboxypeptidase B-like enzymes on the contralateral sides of the respective molecules, producing insulin, C peptide, and four basic amino acid residues. The C peptide is retained within the secretion granule and is cosecreted with insulin upon appropriate stimulation of the $\beta$ cell (Rubenstein *et al.*, 1977).

Although a mixture of trypsin and carboxypeptidase B is effective in converting proinsulin to insulin plus C peptide in the test tube (Kemmler *et al.*, 1971), the nature of the enzymes actually involved in cellular conversion is not known. Lysed secretion granules disclose the presence of a carboxypeptidase B-like activity (Kemmler *et al.*, 1973; Zühlke *et al.*, 1976), but fail to demonstrate clearly a trypsin-like activity. Although cathepsin B has been proposed as the natural endopeptidase, recent studies on proparathyroid hormone have shown that the purified enzyme does not have the appropriate cleavage specificity (MacGregor *et al.*, 1979). Similarly, experiments suggesting glandular kallikrein as the converting enzyme (Ole-Moi *et al.*, 1979) have not yet explored in detail the chemistry of the proposed conversion. The complexity of the conversion process illustrated in Fig. 1 is further heightened, at least for the rat, pig, and human (Tager *et al.*, 1973; Chance, 1971; DeHaen *et al.*, 1978), by the occurrence of a chymotrypsin-like cleavage in the C-peptide region of proinsulin prior to conversion by trypsin- and carboxypeptidase B-like enzymes. In many cases, more than 15 intermediates are either known or can be inferred to participate in the overall process by which proinsulin is converted to insulin.

Studies on the translation of insulin mRNA extracted from rat islets in cell-free systems have shown that the product is not proinsulin, but is a larger peptide, which bears a 24-residue, $NH_2$-terminal extension on the prohormone sequence (Chan et al., 1976). This molecule, called preproinsulin, has also been identified during the cell-free translation of insulin mRNA from cattle, anglerfish, sea raven, hagfish, and humans (Lomedico et al., 1977; Shields and Blobel, 1977; Tager et al., 1980a). The $NH_2$-terminal extensions (called both leader sequences and signal peptides) contain a preponderance of hydrophobic amino acids and undoubtedly play a crucial role in the vectorial translocation of peptides destined for secretion to the cisterna of the rough endoplasmic reticulum (see Milstein et al., 1972; Blobel and Dobberstein, 1975; Steiner et al., 1980). After the removal of the leader sequence by as yet unidentified proteolytic enzymes or "signal peptidases," proinsulin is transferred to subsequent cellular compartments and eventually to secretion granules for further processing.

Two models for the translocation event have been developed: one early model proposes the passage of the leader sequence into the microsomal lumen (Blobel and Dobberstein, 1975), whereas the other strongly suggests a looping of the sequence through the microsomal membrane so that its $NH_2$ terminus remains in the cytosolic compartment (Steiner et al., 1980). In both cases, cleavage of the leader sequence from the secretory protein is generally believed to be a cotranslational rather than a posttranslational event. Thus, proinsulin and proinsulin intermediates represent the vast majority of the early biosynthetic products of the pancreatic $\beta$ cell. Recent studies using rat islets incubated for very short periods with radioactive amino acids, however, have demonstrated the cellular synthesis of preproinsulin (Patzelt et al., 1978b). Figure 2 illustrates the separation of radioactively labeled islet proteins by SDS slab gel electrophoresis. Although preproinsulin appears as a minor component, this 11,500-dalton peptide (identified by both peptide mapping and sequence determination) is clearly separable from 9000-dalton proinsulin. As expected, the biosynthesis of preproinsulin in isolated islets is highly sensitive to the concentration of extracellular glucose (Fig. 2) and is converted to proinsulin with a half-time of only about 1 minute (Patzelt et al., 1978b). Preproinsulin has also been detected during biosynthetic experiments using catfish islets, where the processing of the presecretory form appears to occur over a more extended period (Albert and Permutt, 1979).

Although little is known about the insulin gene at the level of transcription, much structural information has recently accumulated on both the gene and its corresponding mRNA. Insulin mRNA isolated from X-ray-induced $\beta$-cell tumors of the rat is known to contain a 5' 7-methylguanosine pyrophosphate or "cap"

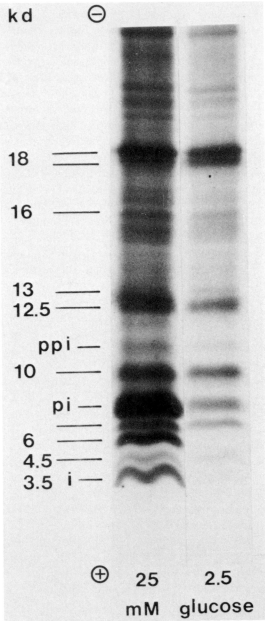

FIG. 2.  Synthesis of proinsulin and preproinsulin by rat islets. Islets were incubated with
[³⁵S]methionine at 25 m*M* or 2.5 m*M* glucose for 60 minutes, and the newly synthesized proteins
were separated by SDS–polyacrylamide gel electrophoresis. The figure shows a fluorogram of the
dried slab gel. Molecular weights of the major proteins are shown in kilodaltons. Preproinsulin,
identified by sequence analysis and tryptic mapping, is labeled ppi, and proinsulin is labeled pi.

structure and a 3' tail of polyadenylic acid (Duguid *et al.*, 1976). The sequences of most of the coding regions of the mRNAs for the two nonallelic insulin genes of the rat (as determined by analysis of the corresponding complementary DNAs) are also known (Ulbrich *et al.*, 1977; Chan *et al.*, 1979). Determination of the structures of the rat insulin genes (Lomedico *et al.*, 1979) and of the human insulin gene (Bell *et al.*, 1980) have provided important new information: all three of these genes contain an intervening sequence in the 5' untranslated region, whereas the rat insulin II gene and the human insulin gene contain an additional intervening sequence within the C-peptide region. Interestingly, these second intervening sequences occur at the same position in the two species, but the sequence for the human (786 base pairs) is longer than that for the rat (499 base pairs). These sequences predict the existence of larger mRNA precursors, which would mature by excision and ligation to their smaller, translationally effective cytoplasmic forms; such larger forms of insulin mRNA have been detected in X-ray-induced $\beta$-cell tumors of the rat (Duguid and Steiner, 1978).

The process of information transfer during the biosynthesis of insulin is presented schematically in Fig. 3. As outlined, transcription of the gene followed by

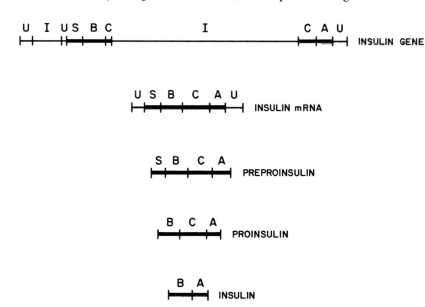

Fig. 3. Information transfer from the insulin gene to insulin. The uppermost entry illustrates the structure of the human insulin gene as determined by Bell and co-workers (1980). Noncoding regions are shown as lines, and coding regions as heavier bars. Transcription of the gene and maturation of the mRNA precursor leads to excision of the intervening sequences (I) and formation of insulin mRNA. Translation of the mRNA leads to formation of preproinsulin and cotranslational processing to proinsulin. Finally, posttranslational processing leads to the formation of insulin. Letter codes identify the different portions of the gene and its products: U, untranslated region; I, intervening sequence; S, signal peptide region; B, B-chain region; C, C-peptide region; A, A-chain region.

maturation of the mRNA precursor results in the loss of the intervening sequences. Translation of the mature mRNA on the rough endoplasmic reticulum, accompanied by cotranslational segregation, results in loss of the $NH_2$-terminal leader sequence. Finally, posttranslational conversion of proinsulin to insulin results in the loss of the C peptide. Thus, several stages of macromolecular processing—involving nucleases as well as proteases—are required for the eventual expression of the hormone product. Studies of abnormal, insulin-related proteins have disclosed two probable mutations in the human insulin gene. The first apparently occurs at a DNA sequence coding for one of the dibasic amino acid conversion sites of proinsulin and results in the secretion of a split proinsulin intermediate (Gabbay *et al.*, 1979; Robbins *et al.*, 1981); the second occurs within the B-chain region of the gene and results in the secretion of an abnormal insulin with a leucine for phenylalanine substitution at position B24 (Tager *et al.*, 1979, 1980b). Bell and co-workers (1980) have also found sequence variability within the 3' untranslated region of the human insulin gene. Although our understanding of gene function is still at an early stage, it is probable that additional mutations in the insulin gene (or in the genes for the processing enzymes) will be found, and determination of their consequences will provide important tools for the further study of insulin biosynthesis.

# III.   Glucagon Biosynthesis

Although studies on pancreatic glucagon biosynthesis were initiated shortly after the discovery of insulin, a clear understanding of the pathway has emerged only recently. A number of investigators have examined the heterogeneity of glucagon-related peptides in islet tissues of both mammals (Rigopoulou *et al.*, 1970; Hellerström *et al.*, 1972; O'Connor and Lazarus, 1976) and fish (Noe and Bauer, 1971; Traketellis *et al.*, 1975), but the sizes and immunological properties of these forms vary considerably (see Tager and Steiner, 1974, for a review). An early indication of proglucagon structure arose from an examination of crystalline glucagon for higher-molecular-weight, hormone-related peptides (Tager and Steiner, 1973). As illustrated in Fig. 4, one such peptide contains the primary structure of glucagon with an eight-residue, COOH-terminal extension. Since the COOH-terminal sequence is connected to the hormone structure by a pair of dibasic amino acid residues (Lys-Arg), both trypsin- and carboxypeptidase B-like enzymes would be necessary to convert this proglucagon fragment to glucagon. The use of trypsin and carboxypeptidase B digestions to probe the structures of glucagon-related peptides is shown schematically in the lower part of Fig. 4. Development of an antiserum specifically directed against the COOH-terminal tryptic fragment of glucagon (Tager and Markese, 1979) permits the

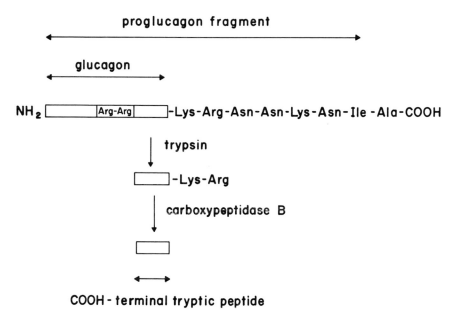

**FIG. 4.** Structure of a proglucagon fragment. The upper portion of the figure illustrates the structure of a fragment of proglucagon containing the entire sequence of glucagon at its NH₂ terminus (open bar; arginine residues shown explicitly) and an eight-residue extension at its COOH terminus. The lower portion of the figure shows how sequential digestion of the fragment with trypsin and carboxypeptidase B releases the COOH-terminal tryptic peptide of glucagon. (See text for details.)

radioimmunometric detection of the unlabeled fragment and the immunoprecipitation of the biosynthetically labeled fragment after enzyme digestion of larger forms. These procedures thus allow an immunochemical mapping of glucagon-related peptides at femtomole to picomole concentrations.

Studies of larger glucagon-related peptides using both centrally directed and COOH-terminally directed antisera have revealed a high degree of heterogeneity in tissue forms (Tager and Markese, 1979). Application of both direct immunoassays and immunoassays after trypsin or trypsin plus carboxypeptidase B digestion (see Fig. 4) to gel-filtered pancreatic extracts showed that (1) the tissue contains 12,000- and 8000-dalton peptides that react with centrally directed antiglucagon sera, but not with the antiserum directed against the COOH-terminal tryptic fragment of the hormone; (2) a mixture of trypsin plus carboxypeptidase B (but neither enzyme alone) releases the immunoreactive COOH-terminal tryptic fragment of glucagon from both higher-molecular-weight peptides; and (3) the tissue contains, in addition, a 9000-dalton peptide that reacts with both antisera equally well. These studies thus suggested that the 8000- and 12,000-molecular-weight peptides contain the COOH-terminal sequence of glucagon, but that this sequence is extended from its COOH terminus by a trypsin-sensitive site. On the other hand, the 9000-molecular-weight peptide appears to lack the COOH-terminal extension.

Our conclusions on the structures of the higher-molecular-weight forms of pancreatic glucagon were confirmed in two ways. First polyacrylamide gel electrophoresis showed that digestion of the 9000-dalton peptide with trypsin resulted in a fragment having the electrophoretic mobility of the native COOH-terminal fragment of glucagon, whereas digestion with trypsin plus carboxypeptidase B was necessary to obtain the fragment from the two COOH-terminally extended peptides. Second, we found that the immunoreactive determinant of the 9000-dalton peptide was degraded by carboxypeptidase A alone, but that, as predicted from the sequence shown in Fig. 4, a mixture of carboxypeptidases A and B was necessary to degrade the determinant in the 8000- and 12,000-dalton forms. A careful determination of the rates of degradation of the 8000- and 12,000-dalton peptides by the carboxypeptidases further suggested that they bear COOH-terminal extensions of both similar length and similar amino acid composition (Tager and Markese, 1979). Thus, the different molecular weights of these two peptides likely result from alterations in their $NH_2$-terminal regions. In all, we have identified five glucagon-containing peptides of pancreas ranging in molecular weight from 12,000 to 3500. Many of these forms have $NH_2$-terminal and/or COOH-terminal extensions on the hormone sequence. Their structures suggest both their probable roles as intermediates in the biosynthesis of glucagon and a likely course by which they might be converted to the hormone in the biosynthetic scheme (Tager et al., 1980b).

Studies of radiolabeled amino acid incorporation into islet proteins have often yielded confusing results with regard to glucagon biosynthesis. Investigations using either avian or mammalian islets have resulted in the assignment of peptides ranging in molecular weight from 9000 to >50,000 as the glucagon precursor, and in many cases conversion of the putative precursor to glucagon has been difficult to prove (for reviews, see Tager and Steiner, 1974; Tager, 1980). Nevertheless, in a number of well-designed studies, Noe and his co-workers have identified a 12,000-dalton peptide as anglerfish proglucagon (Noe and Bauer, 1971, 1975) and have proposed a biosynthetic conversion that might progress through both a 9000-dalton intermediate and a short, COOH-terminally extended peptide such as the one illustrated in Fig. 4 (Noe, 1977). Although little information is available on the structures of these peptides, the kinetics of their processing is consistent with their proposed roles, and their molecular weights are similar to those of the major glucagon-related peptides identified in mammalian pancreas.

We recently examined the biosynthesis of glucagon in rat pancreatic islets using SDS-polyacrylamide gel electrophoresis to analyze the biosynthetically labeled products. Our methods, which included (1) very short periods of pulse-labeling followed by longer periods of chase, (2) disintegration of islet tissue directly in SDS-containing electrophoresis buffer, (3) fluorography of SDS slab gels for the localization of labeled products, and (4) analysis of peptides by both immunological and chemical means, were designed to achieve the most sensitive results. Figure 5 illustrates the results from an experiment in which rat islets were incubated with [$^{35}$S]methionine for 2 minutes (the pulse) and then with unlabeled methionine for the periods indicated (the chase). Although proinsulin remains a major product during these incubations, our attention was directed toward higher-molecular-weight peptides, which had the kinetics of appearance and

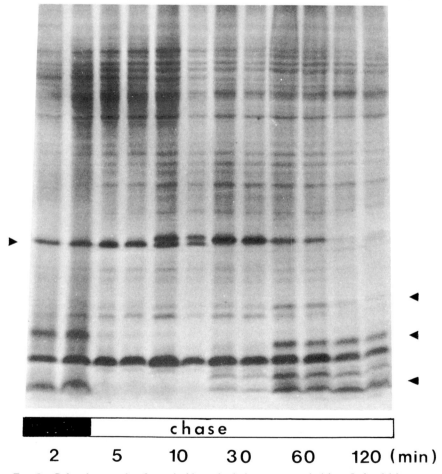

FIG. 5. Pulse-chase study of protein biosynthesis in rat pancreatic islets. Isolated islets were pulsed with [$^{35}$S]methionine for 2 minutes and were chased with the unlabeled amino acids for the indicated periods. The figure shows a fluorogram of an SDS slab gel used to separate the radioactively labeled products. (See text for further details.)

disappearance consistent with what would be expected for a peptide hormone precursor; that is, these peptides should be synthesized rapidly and should be processed to smaller forms with a half-life of about 1 hour. Two such peptides were identified. The first has a molecular weight of about 12,500 and was later shown to be prosomatostatin (Patzelt *et al.*, 1980); the second (indicated by the left-hand arrow in Fig. 5) has a molecular weight of about 18,000 and has been identified as the glucagon precursor (Patzelt *et al.*, 1979; Tager *et al.*, 1980b).

Experiments that identified the 18,000-dalton peptide as proglucagon can be summarized as follows. (1) Of the approximately ten rapidly synthesized islet cell peptides and proteins examined (irrespective of the kinetics of their subsequent processing during the period of chase) only the 18,000-dalton peptide was specifically immunoprecipitated by a variety of antiglucagon sera. (2) Digestion of this peptide with trypsin plus carboxypeptidase B, but not by either enzyme alone, resulted in an immunoprecipitable, methionine-labeled fragment, which had the electrophoretic mobility of the COOH-terminal tryptic fragment of glucagon. (3) Two-dimensional mapping of the phenylalanine-labeled, 18,000-dalton peptide after trypsin digestion showed that it contained the $NH_2$-terminal and central tryptic peptides of pancreatic glucagon. These results, as well as others, suggested that the hormone sequence within the precursor contained both $NH_2$-terminal and COOH-terminal extensions and that these extensions were abutted by trypsin-sensitive sites. The proposed structure is thus consistent with the structures of the smaller, glucagon-containing peptides identified in pancreas extracts. Interestingly, the 18,000-molecular-weight precursor appears to undergo a posttranslational modification to a slightly larger form during short periods of chase: both peptides of the resulting 18,000-dalton doublet (see Fig. 5) contain the structure of the hormone (Patzelt *et al.*, 1979; Tager *et al.*, 1980b). Although the nature of this modification remains unknown, failure of either of these precursor forms to be labeled by radioactive sugars or to bind to any of a variety of lectins suggests that the modification is not related to glycosylation.

Further studies on the processing of the 18,000-dalton glucagon precursor have suggested the existence of an approximately 13,000-dalton intermediate of conversion (identified by the uppermost, right-hand arrow in Fig. 5) and a 10,000-dalton product (middle right-hand arrow) as well as glucagon itself (lowermost, right-hand arrow). The 10,000-dalton peptide, like glucagon, appears only later in the pulse–chase experiment. Although the 10,000-dalton peptide does not contain the structure of glucagon, peptide mapping studies have shown that it is indeed present in the 18,000-dalton precursor (Patzelt *et al.*, 1979). It thus appears that this peptide represents the $NH_2$-terminal portion of proglucagon and that it remains as a stable product after the conversion of the precursor to the hormone. A tentative scheme for the conversion of proglucagon to glucagon is presented in Fig. 6. The figure illustrates at the top the initial posttranslational modification of proglucagon resulting in an apparent increase in its molecular

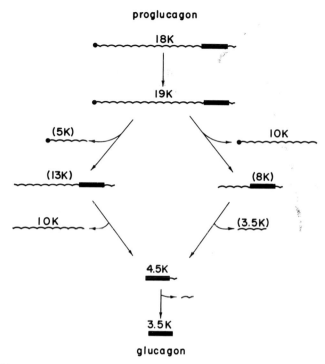

FIG. 6. Diagrammatic scheme for the conversion of proglucagon to glucagon. The solid bars indicate the sequence of glucagon within the larger forms, and the wavy lines indicate the nonhormonal extensions on that sequence. Approximate molecular weights for the intermediates (in kilodaltons) are shown above the pictorial representations.

weight. The lower portion of the figure illustrates a branched pathway for the conversion based on our biosynthetic studies and on our identification of related peptides in extracts of pancreatic tissue. Whether the 10,000-dalton peptide is cleaved from the precursor or from a subsequent intermediate of conversion is not yet known. Although this question and the firm identification of biosynthetic intermediates in the conversion of proglucagon to glucagon are important matters for further study, the nonterminal position of glucagon within the precursor, the complexity of the conversion process, and the requirements for both trypsin-like and carboxypeptidase B-like enzymes in the mechanism for conversion are now recognized.

Tissue heterogeneity of glucagon-related peptides is by no means unique to the pancreas. Glucagon-like immunoreactive peptides are also known to occur throughout the gastrointestinal tract, and it is only recently that their structural relationships to pancreatic glucagon have been clarified. The immunoreactivity of these forms with only selected antiglucagon sera appears to result from the fact that they, like some peptides of the pancreas, bear COOH-terminal extensions on the glucagon sequence (Tager and Markese, 1979). In terms of their immunoreactivity with centrally directed antisera, their immunoreactivity with COOH-terminally directed antisera only after digestion with trypsin and carboxypeptidase B, their molecular weights, and their electrophoretic mobilities, the major glucagon-containing peptides of the intestine are indistinguishable from their 8000- and 12,000-molecular-weight pancreatic counterparts (Tager and Markese, 1979). In addition, a 12,000-dalton glucagon-like peptide of porcine intestine bears a COOH-terminal amino acid sequence identical (except for an apparent inversion) to that of the proglucagon fragment shown in Fig. 4 (Sundby *et al.*, 1976; Jacobsen *et al.*, 1977). Thus, although the biosynthetic intermediates illustrated in Fig. 6 represent only minor components of the glucagon-like peptides of pancreas, they represent the vast majority of the glucagon-like peptides of intestine. These structural studies suggest that the biosynthetic precursors for pancreatic glucagon and for intestinal glucagon-containing peptides are identical, but that proteolytic modifications resulting in the loss of the COOH-terminal extension are lacking in intestinal tissue (Tager and Markese, 1979; Tager *et al.*, 1980b). Whether or not such tissue-specific processing plays a regulatory role remains to be seen. Nevertheless, the specificity and complexity of macromolecular processing during the biogenesis of glucagon, as during the biogenesis of insulin, suggest the participation of both highly evolved and generally applied biosynthetic mechanisms.

ACKNOWLEDGMENTS

Research performed in the authors' laboratories was supported by Grants AM 18347, AM 13914, and AM 20295 from the National Institutes of Health. H.S.T. is the recipient of Research Career Development Award AM 00145.

REFERENCES

Albert, S. G., and Permutt, A. (1979). *J. Biol. Chem.* **254**, 3483–3491.
Bell, G. I., Pictet, R. L., Rutter, W. J., Cordell, B., Tischer, E., and Goodman, H. M. (1980). *Nature (London)* **284**, 26–32.
Blobel, G., and Dobberstein, B. (1975). *J. Cell Biol.* **67**, 852–862.
Blundell, T., Dodson, G., Hodgkin, D., and Mercola, D. (1972). *Adv. Protein Chem.* **26**, 279–402.
Brandenberg, D., and Wollmer, A. (1973). *Hopper-Seyler's Z. Physiol. Chem.* **354**, 613–623.
Busse, W. D., Hansen, S. R., and Carpenter, F. H. (1974). *J. Am. Chem. Soc.* **96**, 5947–5950.
Chan, S. J., Keim, P., and Steiner, D. F. (1976). *Proc. Natl. Acad. Sci. U.S.A.* **73**, 1964–1968.
Chan, S. J., Noyes, B. E., Agarwal, K. L., and Steiner, D. F. (1979). *Proc. Natl. Acad. Sci. U.S.A.* **76**, 5036–5040.
Chance, R. (1971). *In* "Diabetes" (R. R. Rodriguez and J. Vallance-Owen, eds.), pp. 292–305. Excerpta Medica, Amsterdam.

DeHaen, C., Little, S. A., May, J. M., and Williams, R. H. (1978). *J. Clin. Invest.* **62**, 727–732.

Devillers-Thierry, A., Kindt, T., Scheele, G., and Blobel, G. (1975). *Proc. Natl. Acad. Sci. U.S.A.* **72**, 5016–5020.

Duguid, J. R., and Steiner, D. F. (1978). *Proc. Natl. Acad. Sci. U.S.A.* **75**, 3249–3253.

Duguid, J. R., Steiner, D. F., and Chick, W. L. (1976). *Proc. Natl. Acad. Sci. U.S.A.* **73**, 3539–3543.

Gabbay, K. H., Bergenstal, R. H., Wolf, J., Mako, M. E., and Rubenstein, A. H. (1979). *Proc. Natl. Acad. Sci. U.S.A.* **76**, 2881–2885.

Hellerström, C., Howell, S. L., Edwards, J. C., Andersson, A., and Ostenson, C.-G. (1974). *Biochem. J.* **140**, 13–23.

Jacobsen, H., Demandt, A., Moody, A. J., and Sundby, F. (1977). *Biochim. Biophys. Acta* **493**, 452–459.

Kemmler, W., Peterson, J. D., and Steiner, D. F. (1971). *J. Biol. Chem.* **246**, 6786–6791.

Kemmler, W., Steiner, D. F., and Borg, J. (1973). *J. Biol. Chem.* **248**. 4544–4551.

Kemper, B., Habener, J., Ernst, M. D., Potts, J., and Rich, A. (1976). *Biochemistry* **15**, 15–19.

Lomedico, P., Rosenthal, N., Efstratiadis, A., Gilbert, W., Kolodner, R., and Tizard, R. (1979). *Cell* **18**, 545–558.

Lomedico, P. T., Chan, S. J., Steiner, D. F., and Saunders, G. F. (1977). *J. Biol. Chem.* **252**, 7971–7978.

MacGregor, R. R., Hamilton, J. W., Kent, G. N., Shofstall, R. E., and Cohn, D. V. (1979). *J. Biol. Chem.* **254**, 4428–4433.

Milstein, C., Brownlee, G. G., Harrison, T. M., and Mathews, M. B. (1972). *Nature (London), New Biol.* **239**, 117–120.

Noe, B. D. (1977). *In* "Glucagon: Its Role in Physiology and Clinical Medicine" (P. P. Foa, J. S. Bajaj, and W. L. Foa, eds.), pp. 31–41. Springer-Verlag, Berlin and New York.

Noe, B. D., and Bauer, G. E. (1971). *Endocrinology (Baltimore)* **89**, 642–651.

Noe, B. D., and Bauer, G. E. (1975). *Endocrinology (Baltimore)* **97**, 868–877.

O'Connor, K. J., and Lazarus, N. R. (1976). *Biochem. J.* **156**, 265–277.

Ole-Moi, Y. O., Pinkus, G. S., Spragg, J., and Austen, K. F. (1979). *N. Engl. J. Med.* **300**, 1289–1294.

Oyer, P. E., Cho, S., Peterson, J. D., and Steiner, D. F. (1971). *J. Biol. Chem.* **246**, 1375–1386.

Patzelt, C., Chan, S. J., Duguid, J., Horton, G., Keim, P., Heinrikson, R. L., and Steiner, D. F. (1978a). *In* "Regulatory Proteolytic Enzymes and Their Inhibitors" (S. Magnusson, ed.), pp. 69–78. Pergamon, Oxford.

Patzelt, C., Labrecque, A. D., Duguid, J. R., Carroll, R. J., Keim, P. S., Heinrikson, R. L., and Steiner, D. F. (1978b). *Proc. Natl. Acad. Sci. U.S.A.* **75**, 1260–1264.

Patzelt, C., Tager, H. S., Carroll, R. J., and Steiner, D. F. (1979). *Nature (London)* **282**, 260–266.

Patzelt, C., Tager, H. S., Carroll, R. J., and Steiner, D. F. (1980). *Proc. Natl. Acad. Sci. U.S.A.* **77**, 2410–2414.

Peterson, J. D., Nerlich, S., Oyer, P. E., and Steiner, D. F. (1972). *J. Biol. Chem.* **247**, 4866–4871.

Rigopoulou, D., Valverde, I. Marco, J., Faloona, G., and Unger, R. H. (1970). *J. Biol. Chem.* **245**, 496–501.

Robbins, D. C., Blix, P. M., Rubenstein, A. H., Kanazawa, Y., Kosaka, K., and Tager, H. S. (1981). *Nature (London)* **291**, 679–681.

Rubenstein, A. H., Steiner, D. F., Horwitz, D. L., Mako, M. E., Block, M. B., Starr, J. I., Kuzuya, H., and Melani, F. (1977). *Recent Prog. Horm. Res.* **33**, 435–468.

Shields, D., and Blobel, G. (1977). *Proc. Natl. Acad. Sci. U.S.A.* **74**, 2059–2063

Steiner, D. F. (1977). *Diabetes* **26**, 322–340.

Steiner, D. F., and Clark, J. L. (1968). *Proc. Natl. Acad. Sci. U.S.A.* **60**, 622–629.

Steiner, D. F., Cunningham, D. D., Spigelman, S., and Aten, B. (1967). *Science* **157**, 697–700.

Steiner, D. F., Clark, J. L., Nolan, C., Rubenstein, A. H., Margoliash, E., Aten, B., and Oyer, P. E. (1969). *Recent Prog. Horm. Res.* **25,** 207–282.

Steiner, D. F., Quinn, P. S., Patzelt, C., Chan, S. J., Marsh, J., and Tager, H. S. (1980). *In* "Cell Biology: A Comprehensive Treatise" (L. Goldstein and D. M. Prescott, eds.), Vol. 4, pp. 175–201. Academic Press, New York.

Sundby, F., Jacobson, H., and Moody, A. J. (1976). *Horm. Metab. Res.* **8,** 366–371.

Tager, H. S. (1980). *In* "Glucagon: Physiology, Pathophysiology and Morphology of the Pancreatic A-Cells" (R. H. Unger and L. Orci, eds.), pp. 39–54. American Elsevier, New York.

Tager, H. S., and Markese, J. (1979). *J. Biol. Chem.* **254,** 2229–2233.

Tager, H. S., and Steiner, D. F. (1972). *J. Biol. Chem.* **247,** 7936–7940.

Tager, H. S., and Steiner, D. F. (1973). *Proc. Natl. Acad. Sci. U.S.A.* **70,** 2321–2325.

Tager, H. S., and Steiner, D. F. (1974). *Annu. Rev. Biochem.* **43,** 509–538.

Tager, H. S., Emdin, S. O., Clark, J. L., and Steiner, D. F. (1973). *J. Biol. Chem.* **248,** 3476–3482.

Tager, H. S., Rubenstein, A. H., and Steiner, D. F. (1975). *In* "Methods in Enzymology" (B. W. O'Malley and J. G. Hardman, eds.), Vol. 37, Part B, pp. 326–345. Academic Press, New York.

Tager, H. S., Given, B., Baldwin, D., Mako, M., Rubenstein, A., Olefsky, J., Kolterman, O., Kobayashi, M., and Poucher, R. (1979). *Nature (London)* **281,** 122–125.

Tager, H. S., Patzelt, C., Assoian, R. K., Chan, S. J., Duguid, J. R., and Steiner, D. F. (1980a). *Ann. N.Y. Acad. Sci.* **343,** 133–147.

Tager, H. S., Thomas, N., Assoian, R., Rubenstein, A., Olefsky, J., and Kaiser, E. T. (1980b). *Proc. Natl. Acad. Sci. U.S.A.* **77,** 3181–3185.

Trakatellis, A. C., Tada, K., Yamaji, K., and Gondiki-kouidov, P. (1975). *Biochemistry* **14,** 1508–1512.

Ulbrich, A., Shine, J., Chirguin, J., Pictet, R., Tischer, E., Rutter, W. J., and Goodman, H. M. (1977). *Science* **196,** 1313–1319.

Zühlke, H., Steiner, D. F., Lernmark, Å., and Lipsey, C. (1976). *Ciba Found. Symp.* [N.S.] **41,** 183–195.

# Chapter 7

# Synthesis and Processing of Asparagine-Linked Oligosaccharides of Glycoproteins

MARTIN D. SNIDER AND PHILLIPS W. ROBBINS

*Department of Biology and Center for Cancer Research,*
*Massachusetts Institute of Technology,*
*Cambridge, Massachusetts*

## I. Introduction

The asparagine-linked oligosaccharides of cell surface and extracellular glycoproteins are a heterogeneous class of molecules that can be divided into two basic types (reviewed by Kornfeld and Kornfeld, 1976). High-mannose structures (Fig. 1) have a $Man_3GlcNAc_2$ inner core at the reducing end and two to six additional mannose residues in a peripheral three-branched structure. Complex oligosaccharides (Fig. 1) also have a $Man_3GlcNAc_2$ inner core, and in addition contain a fucose residue and two to four terminal trisaccharide branches contain-

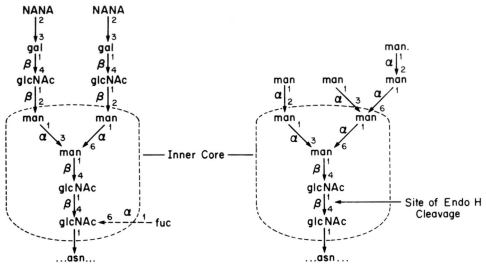

Complex Glycopeptide
Sindbis S1

High-Mannose Glycopeptide
Sindbis S4

FIG. 1.   Representative complex and high-mannose glycopeptides. The structures are those proposed for Sindbis virus glycopeptides (Burke, 1976). The common Man$_3$GlcNAc$_2$ inner-core regions
are shown by the dashed line. The arrow marks the site of Endo H cleavage of high-mannose
oligosaccharides.

ing N-acetylglucosamine, galactose, and sialic acid. Both types of oligosaccharide show microheterogeneity; a number of related structures are usually
found in a single purified glycoprotein preparation. The number of mannose
residues in high-mannose oligosaccharides is variable, whereas complex
oligosaccharides show variability in sialic acid and fucose content (Kornfeld and
Kornfeld, 1976).

The presence of the Man$_3$GlcNAc$_2$ inner core in the two oligosaccharides
suggested that their synthetic pathways might share certain steps. Recent work in
a number of laboratories has established that in fact the two types of oligosac-

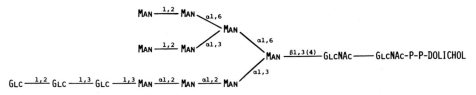

FIG. 2.   Proposed structure of the largest oligosaccharide–lipid (Li et al., 1978; Liu et al., 1979).

charides do arise from a common lipid-linked precursor. This brief review will discuss the three stages in the synthesis of asparagine-linked carbohydrates: (1) synthesis of the large lipid-linked oligosaccharide, (2) glycosylation of asparagine residues by the transfer of oligosaccharide from lipid to protein, and (3) processing of protein-linked oligosaccharides to yield mature high-mannose and complex glycopeptides.

## II.   Oligosaccharide–Lipid

### A.   Synthesis

The structure of the largest oligosaccharide–lipid is shown in Fig. 2 (Li *et al.,* 1978; Liu *et al.,* 1979). The oligosaccharide is linked through pyrophosphate to dolichol, a polyprenol of 16–20 isoprene units (Waechter and Lennarz, 1976) and has the typical three-branched $Man_9GlcNAc_2$ structure. Surprisingly, the oligosaccharide also contains three residues of glucose, a sugar not found in mature asparagine-linked glycopeptides, at the nonreducing end of one of the oligomannose branches. These glucose residues appear to be important in the function of this molecule in protein glycosylation (see Section III, A).

Oligosaccharide–lipid is synthesized from dolichol phosphate and the nucleotide sugars UDP-GlcNAc, GDP-Man, and UDP-Glc (reviewed by Waechter and Lennarz, 1976). The first step, the addition of GlcNAc-phosphate to dolichol phosphate, yields GlcNAc-pyrophosphoryl-dolichol (Fig. 3). The antibiotic

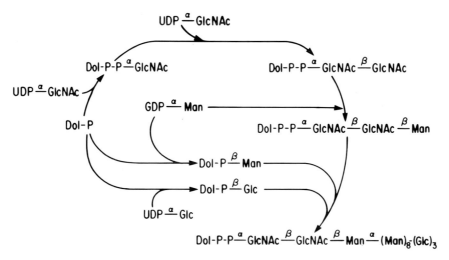

FIG. 3.   Pathway of synthesis of oligosaccharide–lipid.

tunicamycin inhibits this first step of oligosaccharide–lipid synthesis (Takatsuki *et al.,* 1975); in so doing it blocks the synthesis of both types of asparagine-linked oligosaccharide. A second *N*-acetylglucosamine and a mannose residue are added from the nucleotide sugars to yield the trisaccharide–lipid ManGlcNAc$_2$-pyrophosphoryl-dolichol. The rest of the oligomannose structure is then built up, either by direct transfer from GDP-Man, or from Man-phosphate-dolichol, a lipid-linked intermediate that is synthesized from the nucleotide sugar. Finally, the addition of the three glucose residues, from Glc-phosphate-dolichol, which is made from UDP-Glc, yields full-size oligosaccharide–lipid.

The oligosaccharide Glc$_3$Man$_9$GlcNAc$_2$, like all high-mannose oligosaccharides, is sensitive to the enzyme endo-$\beta$-*N*-acetylglucosaminidase H (Endo H) from *Streptomyces plicatus,* which cleaves between the two *N*-acetylglucosamine residues of lipid- and peptide-linked oligosaccharides (Fig. 1). Endo H has been extremely useful in studying the synthesis of asparagine-linked oligosaccharides for two reasons. First, it is specific for high-mannose oligosaccharides (Tarentino *et al.,* 1974). Second, the use of Endo H has allowed direct comparison of lipid- and protein-derived oligosaccharides because cleavage by the enzyme results in the release of free oligosaccharides.

## B.   Transmembrane Asymmetry of Oligosaccharide–Lipid Synthesis

The sugar nucleotides, which are the starting materials for oligosaccharide synthesis, are water-soluble molecules of the cytoplasm. Since newly made glycoproteins, which are the products of the pathway, are sequestered in the lumen of the endoplasmic reticulum, it is clear that, at some point during synthesis, sugar residues, either nucleotide-, lipid-, or peptide-linked, must cross the membrane. Isolated microsomes, sealed vesicles derived from the endoplasmic reticulum, which maintain the proper orientation (DePierre and Ernster, 1977), have been used to investigate this problem *in vitro.* Hanover and Lennarz (1979) have recently reported the localization of an early intermediate, GlcNAc$_2$-pyrophosphoryl-dolichol, to the luminal face of hen oviduct microsomal vesicles. However, our recent results with rat liver microsomal vesicles suggest that oligosaccharide–lipid is assembled on the cytoplasmic face of the endoplasmic reticulum (Snider *et al.,* 1980). The enzymes catalyzing three early steps—the synthesis of Glc-phosphate-dolichol, Man-phosphate-dolichol, and GlcNAc$_2$-pyrophosphoryl-dolichol—as well as the enzymes catalyzing the final steps, the transfer of glucose from Glc-phosphate-dolichol to oligosaccharide–lipid, were examined. All these activities were labile to protease digestion in sealed microsomal vesicles. Since only the cytoplasmic face of the membrane was accessible to protease under these conditions, presumably the active sites of these enzymes reside on the cytoplasmic face.

Since oligosaccharide–lipid appears to be synthesized on the cytoplasmic face of the endoplasmic reticulum, it is unclear whether sugar residues cross the membrane as oligosaccharide–lipid or as glycopeptide. To distinguish between these possibilities, the site, in the membrane, of protein glycosylation is currently being investigated.

# III. Protein Glycosylation

## A. Oligosaccharide–Lipid Donor

A growing amount of evidence suggests that protein asparagine residues are glycosylated by the *en bloc* transfer of oligosaccharide from lipid to protein. This process has been well documented *in vitro* (Waechter and Lennarz, 1976). That glycosylation occurs by this route *in vivo* is suggested by mannose labeling studies of cultured cells (Hubbard and Robbins, 1979). After short labeling times (see Figs. 5A and 6A), only large oligosaccharides, of the same size as those that are lipid-linked, are found in the glycoprotein fraction. In addition, the antibiotic tunicamycin, which blocks oligosaccharide–lipid synthesis, inhibits protein glycosylation both *in vitro* and *in vivo* (Struck and Lennarz, 1977). These results are all consistent with the addition of oligosaccharide as a unit, but not with the stepwise addition of sugar residues to protein.

Experiments *in vitro* have shown that the largest oligosaccharide, $Glc_3Man_9GlcNAc_2$, is the preferred substrate for transfer from lipid to protein. When an oligosaccharide–lipid mixture containing roughly equimolar amounts of the di- and triglucosyl species was incubated briefly *in vitro* with a membrane preparation from cultured cells, nearly all the oligosaccharide transferred to protein was the triglucosyl species (Fig. 4) (Turco and Robbins, 1979). Similar *in vitro* experiments have shown that glucose-containing oligosaccharide was transferred from lipid to protein nine times as fast as was the oligosaccharide containing only mannose and $N$-acetylglucosamine residues (Turco *et al.*, 1977).

Recent experiments have suggested that the glycosyltransferase that catalyzes protein glycosylation may recognize only the three glucose residues rather than the entire oligosaccharide structure. M. J. Spiro *et al.* (1979) have reported that the removal of peripheral mannose residues from oligosaccharide–lipid by a $\alpha$-mannosidase treatment does not affect the rate of protein glycosylation. In addition, in class E thy-1 mutants of lymphoma cells, $Glc_3Man_5GlcNAc_2$ is the largest oligosaccharide–lipid found (Chapman *et al.*, 1979). However, glycopro-

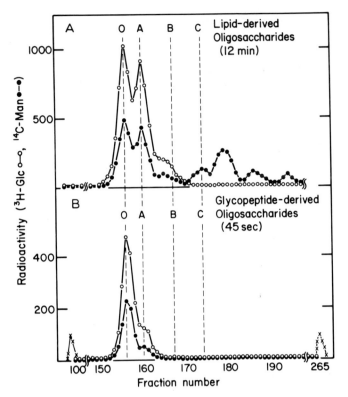

FIG. 4. Preferential transfer of $Glc_3Man_9GlcNAc_2$ from lipid to protein. (A) A sample of [$^3$H]Glc-, [$^{14}$C]Man-oligosaccharide–lipid mixture was subjected to mild acid hydrolysis and treated with Endo H. The resultant oligosaccharides were subjected to gel filtration chromatography on Biogel P4 (Turco and Robbins, 1979). (B) The oligosaccharide–lipid mixture was incubated for 45 seconds with a membrane preparation from Nil 8 cultured cells. The glycoprotein fraction was digested with Pronase and treated with Endo H. The resulting oligosaccharides were analyzed by gel filtration on Biogel P-4 (Turco and Robbins, 1979). ✕, void (left) and inclusion (right) volumes, ○, [$^3$H]Glc-oligosaccharides; ● [$^{14}$C]Man-oligosaccharides. Oligosaccharides with 3, 2, 1, and 0 glucose residue are denoted by O, A, B, and C, respectively.

tein levels in these cells are comparable to the parental type (Trowbridge et al., 1978).

## B.    Peptide Acceptor

Surveys of amino acid sequences around glycosylated asparagines of glycoproteins has revealed that glycopeptides always contain the tripeptide Asn-X-$^{Ser}_{Thr}$ (Marshall, 1974). Recent studies on the specificity of protein glycosylation by microsomal membranes *in vitro* have shown that no other protein sequence information is necessary; the tripeptides Asn-Leu-Thr and Asn-Gly-Ser could be glycosylated, as long as both amino and carboxy terminal were blocked (Hart *et al.*, 1979). In addition, there is evidence that protein asparagine residues are glycosylated during the synthesis of the polypeptide. In the cases that have been examined, asparagine residues were glycosylated while the polypeptide was attached to the ribosome (Keily *et al.*, 1976; Rothman and Lodish, 1977; Sefton, 1977). Results from *in vitro* protein translation studies also suggest that protein glycosylation is tightly coupled to the movement of the nascent chain across the microsomal membrane, since the only glycosylated peptides are those that have been transported across the membrane (Katz *et al.*, 1977; Lingappa *et al.*, 1978).

# IV. Glycoprotein Processing

## A. Common Precursor of High-Mannose and Complex Glycopeptides

Because the oligosaccharide that is transferred to the nascent polypeptide is different from any of the mature glycopeptides, protein-linked oligosaccharides must undergo modification or processing. The extent of this processing recently became clear with the demonstration that complex as well as high-mannose structures arise from the common precursor $Glc_3Man_9GlcNAc_2$ (Robbins *et al.*, 1977; Tabas *et al.*, 1978; Hunt *et al.*, 1978). The G protein of vesicular stomatitis virus, a membrane glycoprotein that has only complex oligosaccharides (Reading *et al.*, 1978), was studied in these experiments. Figure 5 shows a pulse–chase experiment in which glycopeptides from mannose-labeled G protein were treated with Endo H (Hubbard and Robbins, 1979). In a 10-minute pulse (Fig. 5A), the glycopeptides were cleaved by Endo H; on gel filtration, the resulting oligosaccharides migrated in the position of glycosylated high-mannose oligosaccharides. However, after 60 minutes of chase (Fig. 5B), the G protein glycopeptides were resistant to Endo H and migrated in the position of the complex glycopeptides from virions. Examination at intermediate times (not shown) revealed the gradual conversion of high-mannose to complex glycopeptides, establishing the precursor–product relationship.

## B. Kinetics of Glycoprotein Processing

The kinetics of glycoprotein processing have been studied in several types of cultured cells (Hubbard and Robbins, 1979; Kornfeld *et al.*, 1978). Examination

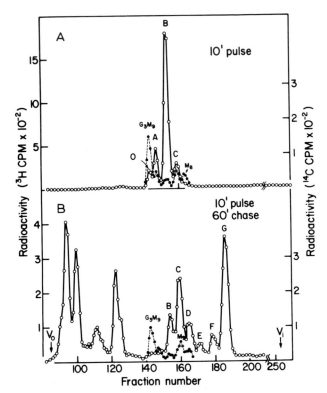

FIG. 5. High-mannose glycopeptides are precursors to complex glycopeptides. Vesicular stomatitis virus-infected secondary chick embryo fibroblasts were labeled with [2-³H]mannose under conditions where the viral G protein is the only glycoprotein labeled. Cells were harvested after 10 minutes of labeling (A), or after a 60-minute chase (B). The glycoprotein fraction was digested with Pronase, treated with Endo H, and analyzed by gel filtration on Biogel P4 (Hubbard and Robbins, 1979). ○, Oligosaccharides from [³H]Man-labeled infected cells; ●, [¹⁴C]Man-oligosaccharide marker. Abbreviations are (O) $Glc_3Man_9GlcNAc$; (A) $Glc_2Man_9GlcNAc$; (B) $GlcMan_9GlcNAc$; (C) $Man_9GlcNAc$; (D) $Man_8GlcNAc$; (E) $Man_7GlcNAc$; (F) $Man_6GlcNAc$; (G) $Man_5GlcNAc$.

of glycopeptide-derived oligosaccharides labeled with mannose for various lengths of time (Fig. 6) shows that the first processing event, the stepwise removal of the three glucose residues, begins soon after protein glycosylation. After 2.5 minutes, the first glucose residue has largely been removed, and by 10 minutes, glycopeptides that have lost all three residues begin to accumulate. In contrast, the removal of mannose residues, probably in stepwise fashion as well, begins after 10–30 minutes of labeling. This late removal is in good agreement with the kinetics of appearance of Endo H-resistant glycopeptides in pulse–chase experiments.

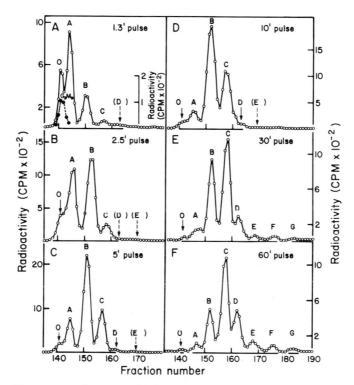

FIG. 6. Time course of mannose labeling of glycopeptide-derived high-mannose oligosaccharides. Secondary chick embryo fibroblasts were labeled with [2-³H]mannose for the indicated times. Then the cells were harvested, and the glycoprotein fraction was digested with Pronase and treated with Endo H. The resulting oligosaccharides were analyzed by gel filtration on BioGel P4 (Hubbard and Robbins, 1979). ○, [³H]Man-oligosaccharides from labeled cells; ●, [¹⁴C]Man-oligosaccharide marker present in all samples but shown only in (A). The abbreviations are the same as in Fig. 5.

## C. Enzymes Involved in Glycoprotein Processing

### 1. GLUCOSIDASES

The rapid removal of glucose residues *in vivo* makes it likely that this process occurs in the same subcellular location as the initial glycosylation—namely, the endoplasmic reticulum. Grinna and Robbins (1979) have recently described glucosidases in rat liver microsomal membranes, which catalyze the sequential cleavage of the three glucose residues from $Glc_3Man_9GlcNAc_2$ *in vitro*. Glucosidase I removes the outermost residue, and the inner two are cleaved by glucosidase II. These membrane-bound enzymes, found in both rough and smooth microsomes, are located on the lumenal face of the microsomal membrane, in the same compartment as their glycoprotein substrates. This lumenal location was indicated by the findings that glucosidase activities were both latent and resistant to proteolysis in sealed microsomal vesicles. Both enzymes appear to be specific for glucose-containing high-mannose oligosaccharides; when these oligosaccharide substrates were treated with jack bean $\alpha$-mannosidase, which removes mannose residues from the nonglucosylated oligomannosyl branches (Fig. 2), activity of the glucosidases was greatly reduced (L. Grinna, personal communication). Similar glucosidase activities have been described in other tissues (Ugalde et al., 1978; Chen and Lennarz, 1978; R. G. Spiro et al., 1979).

## 2. MANNOSIDASES

The late removal of mannose residues from glycoproteins suggests that this occurs in the Golgi apparatus. A number of $\alpha$-mannosidases, which might be responsible for this processing of mannose residues, have recently been identified. Two mannosidases, specific for $\alpha 1-2$ linkages, appear to function in the initial processing of zero to four mannose residues from high-mannose oligosaccharides (Opheim and Touster, 1978; Tabas and Kornfeld, 1979). The resulting products are mature high-mannose glycopeptides containing five to nine mannose residues. The smallest of these, $Man_5GlcNAc_2$, is also the direct precursor of complex oligosaccharides. A third mannosidase acts to remove the two noncore mannose residues during the high-mannose to complex conversion (see below).

## 3. TERMINAL GLYCOSYLTRANSFERASES

The enzymes responsible for the addition of the peripheral sugars—$N$-acetylglucosamine, galactose, sialic acid, and fucose—from the respective sugar nucleotides, have all been localized to the Golgi apparatus (Schachter et al., 1970; Munro et al., 1975). Tabas and Kornfeld (1978) have recently shown that the conversion of high-mannose to complex oligosaccharides begins with the addition of an $N$-acetylglucosamine residue to the $Man_5GlcNAc_2$ structure. The two noncore mannose residues are then removed by the mannosidase mentioned above, and a second $N$-acetylglucosamine is added. Maturation of complex oligosaccharides concludes with the completion of the terminal trisaccharide branches and the addition of fucose. High-mannose and complex glycopeptides in the Golgi apparatus are then ready for transport to the cell surface, or packaging in secretory granules.

A great deal of progress has been made toward understanding the synthesis of asparagine-linked oligosaccharides. In addition to structural details of various lipid- and peptide-linked intermediates of the pathway, several interesting problems remain. Little is known of the enzymes involved, and their organization in the membrane is unclear. Moreover, the information, which presumably resides in the structure of the peptide, which determines whether the initial glycosylation product becomes high-mannose or complex, remains to be understood.

## ACKNOWLEDGMENTS

Portions of this work were supported by Grant CA14142 from the National Cancer Institute. M.D.S. is a postdoctoral fellow of the National Institute of Allergy and Infectious Diseases. We thank L. Grinna and M. Rosner for helpful comments and D. Young for typing the manuscript.

## REFERENCES

Burke, D. J. (1976). Ph.D. Thesis, State University of New York at Stony Brook.
Chapman, A., Trowbridge, I. S., Hyman, R., and Kornfeld S. (1979). *Cell* **17,** 509-516.
Chen, W. W., and Lennarz, W. J. (1978). *J. Biol. Chem.* **253,** 5780-5785.
DePierre, J. W., and Ernster, L. (1977). *Annu. Rev. Biochem.* **46,** 201-262.
Grinna, L. S., and Robbins, P. W. (1979). *J. Biol. Chem.* **254,** 8814-8818.
Hanover, J. A., and Lennarz, W. J. (1979). *J. Biol. Chem.* **254,** 9237-9246.
Hart, G. W., Brew, K., Grant, G. A., Bradshaw, R. A., and Lennarz, W. J. (1979) *J. Biol. Chem.* **254,** 9747-9753.
Hubbard, S. C., and Robbins, P. W. (1979). *J. Biol. Chem.* **254,** 4568-4576.
Hunt, L., Etchison, J., and Summers, D. (1978). *Proc. Natl. Acad. Sci. U.S.A.* **75,** 754-758.
Katz, F. N., Rothman, J. E., Lingappa, V. R., Blobel, G., and Lodish, H. F. (1977). *Proc. Natl. Acad. Sci. U.S.A.* **74,** 3278-3282.
Keily, M. L., McKnight, G. S., and Schimke, R. T. (1976). *J. Biol. Chem.* **251,** 5490-5495.
Kornfeld, R., and Kornfeld, S. (1976). *Annu. Rev. Biochem.* **45,** 217-237.
Kornfeld, S., Li, E., and Tabas, I. (1978). *J. Biol. Chem.* **253,** 7771-7778.
Li, E., Tabas, I., and Kornfeld, S. (1978). *J. Biol. Chem.* **253,** 7762-7770.
Lingappa, V. R., Shields, D., Woo, S. L. C., and Blobel, G. (1978). *J. Cell Biol.* **79,** 567-572.
Liu, T., Stetson, B., Turco, S. J., Hubbard, S. C., and Robbins, P. W. (1979). *J. Biol. Chem.* **254,** 4554-4559.
Marshall, R. D. (1974). *Biochem. Soc. Symp.* **40,** 17-27.
Munro, J. R., Narasimhan, S., Wetmore, S., Riordan, J. R., and Schachter, H. (1975). *Arch. Biochem. Biophys.* **169,** 269-277.
Opheim, D. J., and Touster, O. (1978). *J. Biol. Chem.* **253,** 1017-1023.
Reading, C. L., Penhoet, E. E., and Ballou, C. E. (1978). *J. Biol. Chem.* **253,** 5600-5612.
Robbins, P. W., Hubbard, S. C., Turco, S. J., and Wirth, D. F. (1977). *Cell* **12,** 893-900.
Rothman, J. E., and Lodish, H. F. (1977). *Nature (London)* **269,** 775-780.
Schachter, H., Jabbal, I., Hudgin, R. L., Pinteric, L., McGuire, E. J., and Roseman, S. (1970). *J. Biol. Chem.* **245,** 1090-1100.
Sefton, B. M. (1977). *Cell* **10,** 659-668.
Snider, M. D., Sultzman, L. A., and Robbins, P. W. (1980). *Cell* **21,** 385-392.
Spiro, M. J., Spiro, R. G., and Bhoyroo, V. D. (1979). *J. Biol. Chem.* **254,** 7668-7674.
Spiro, R. G., Spiro, M. J., and Bhoyroo, V. D. (1979). *J. Biol. Chem.* **254,** 7659-7667.

Struck, D. K., and Lennarz, W. J. (1977). *J. Biol. Chem.* **252,** 1007–1013.

Tabas, I., and Kornfeld, S. (1978). *J. Biol. Chem.* **253,** 7779–7786.

Tabas, I., and Kornfeld, S. (1979). *Fed. Proc., Fed. Am. Soc. Exp. Biol.* **38,** 291.

Tabas, I., Schlesinger, S., and Kornfeld, S. (1978). *J. Biol. Chem.* **253,** 716–722.

Takatsuki, A., Kohno, K., and Tamura, G. (1975). *Agric. Biol. Chem.* **39,** 2080–2091.

Tarentino, A. L., Plummer, T. J., Jr., and Maley, F. (1974). *J. Biol. Chem.* **249,** 818–824.

Trowbridge, I. S., Hyman, R., and Masauskas, C. (1978). *Cell* **14,** 21–32.

Turco, S. J., and Robbins, P. W. (1979). *J. Biol. Chem.* **254,** 4560–4567.

Turco, S. J., Stetson, B., and Robbins, P. W. (1977). *Proc. Natl. Acad. Sci. U.S.A.* **74,** 4411–4414.

Ugalde, R. A., Staneloni, R. J., and Leloir, L. F. (1978). *FEBS Lett.* **91,** 209–212.

Waechter, C. J., and Lennarz, W. J. (1976). *Annu. Rev. Biochem.* **45,** 95–112.

METHODS IN CELL BIOLOGY, VOLUME 23

# Chapter 8

# Glycosylation Steps Involved in Processing of Pro-Corticotropin-Endorphin in Mouse Pituitary Tumor Cells

EDWARD HERBERT, MARJORIE PHILLIPS,[1]
AND MARCIA BUDARF[2]

*Department of Chemistry,*
*University of Oregon,*
*Eugene, Oregon*

## I. Introduction

It has recently been demonstrated that $\beta$-endorphin, ACTH, and $\alpha$- and $\beta$-melanocyte-stimulating hormones (MSH) are all derived from the same precursor protein (pro-ACTH-endorphin) in the pituitary (Mains *et al.*, 1977; Roberts and Herbert, 1977a,b; Nakanishi *et al.*, 1977). The complete amino acid

---

[1]Present address: Laboratory of Toxicology, Harvard School of Public Health, Boston, Massachusetts.
[2]Present address: Department of Molecular Biology, University of California, Berkeley, California.

sequence of the precursor in bovine pituitary has been determined by Nakanishi *et al.* (1979), using recombinant DNA technology. The same technology has revealed a portion of the amino acid sequence of the precursor in mouse pituitary cells (Roberts *et al.*, 1979b). A model of the structure of the precursor based on the results of these and other studies is shown in Fig. 1. The form of the precursor that is synthesized in the cell-free system under the direction of messenger RNA is approximately 260 amino acids long. It has a signal sequence at the N terminus (pre sequence). β-Lipotropin (β-LPH) is located at the carboxy terminus of the molecule, and ACTH is adjacent to β-LPH near the middle of the molecule, leaving a stretch of approximately 100 amino acids with no known hormone function. The precursor has been detected in the pituitaries of a variety of vertebrates including mouse (Mains *et al.*, 1977; Roberts and Herbert, 1977b); bovine (Nakanishi *et al.*, 1977); rat (Seidah *et al.*, 1978; Crine *et al.*, 1978; Mains and Eipper, 1979); camel (Kimura *et al.*, 1979), and toad (Loh, 1979; Loh and Gainer, 1979). It has also been found in human ectopic tumors (Bertagna *et al.*, 1978); bovine hypothalamus (Liotta *et al.*, 1979; Krieger *et al.*, 1980), and human placenta (Krieger *et al.*, 1980; Odagiri *et al.*, 1979).

The precursor protein has been found in both the anterior and intermediate lobes of the pituitary (Roberts *et al.*, 1978; Eipper and Mains, 1978a). Although the structure of the precursor protein appears to be quite similar in anterior and intermediate lobes of the pituitary, it is processed proteolytically to different hormones in the two lobes, as illustrated in the diagram in Fig. 2 (Scott *et al.*, 1974, 1976; Mains and Eipper, 1975; Kraicer, 1977; Roberts *et al.*, 1978; Briaud *et al.*, 1978; Eipper and Mains, 1978b; Seidah *et al.*, 1978; Crine *et al.*, 1979). The major hormones produced in the anterior lobe are α-(1–39)ACTH (4.5K ACTH) and a glycosylated form of α-(1–39)ACTH (13K ACTH), β-LPH, and the N-terminal fragment of the precursor. The major hormones produced in the intermediate lobe are α-MSH (identical to the first 13 amino acids of ACTH with an acetylated N terminus and an amidated C terminus), β-endorphin [also N-acetylated β-endorphin and degradation products of β-endorphin (Smyth *et al.*, 1979)], α-(18–39)ACTH, known as CLIP (Scott *et al.*, 1973), and the N-terminal fragment of the precursor (Eipper and Mains, 1978b). Very little Leu- or Met-enkephalin has been found in either lobe of the pituitary (Mains and Eipper, 1979; unpublished results of authors).

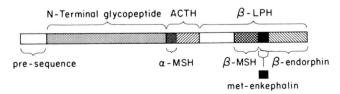

FIG. 1.   Model of the structure of pro-ACTH-endorphin synthesized in a cell-free system under the direction of messenger RNA.

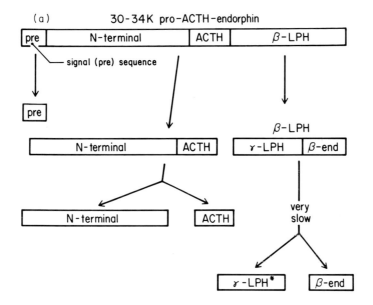

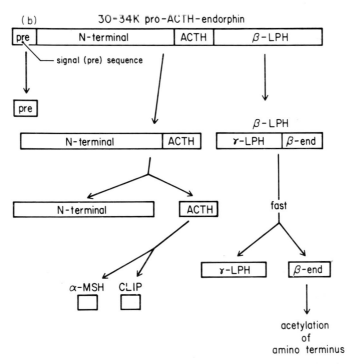

FIG. 2. Model of processing of pro-ACTH-endorphin in anterior lobe (upper chart) and intermediate lobe (lower chart) of mouse and rat pituitary.

The release of hormones is also regulated differently in the anterior and intermediate lobes of the pituitary. Secretion of ACTH and $\beta$-LPH peptides in the anterior pituitary is regulated by a hypothalamic peptide (or peptides) known as corticotropin-releasing hormone(s) or CRH (Vale *et al.*, 1978; Allen *et al.*, 1978). CRH has proved to be an elusive peptide to identify. CRH release in the hypothalamus is stimulated by pain stress (Guillemin *et al.*, 1955). CRH is carried to the anterior pituitary by a system of portal blood vessels (Howe, 1973), where it stimulates the release of ACTH. ACTH stimulates the production of glucocorticoids, which, in turn, inhibit the release of ACTH and $\beta$-LPH peptides in the pituitary (Vale *et al.*, 1978; Guillemin *et al.*, 1977; Allen *et al.*, 1978; Krieger *et al.*, 1979). Corticotrophic cells in the anterior pituitary, which comprise 2–4% of the cells in this lobe, do not appear to have any direct synaptic input (Bergland and Page, 1979). On the other hand, intermediate lobe cells of rat pituitary, which comprise almost all the cells in this lobe, are innervated by dopaminergic neurons from the hypothalamus (Scott and Baker, 1975). Intermediate lobe cells appear to be activated to secrete their hormone products by "neurogenic" (psychological) stress (Moriarty *et al.*, 1975). However, the origin and nature of the signals that trigger this response are not well understood.

## II.   Approaches Used to Study Proteolytic Processing and Glycosylation of the Precursor in Mouse Pituitary Tumor Cells

The AtT-20/$D_{16v}$ mouse pituitary tumor cell line has been used as a model system to study the details of synthesis and processing of the precursor because of the large quantities of ACTH and $\beta$-LPH it produces and because it responds to CRH, vasopressin, and glucocorticoids in the same manner as do mouse anterior pituitary cells (Herbert *et al.*, 1978; Allen *et al.*, 1978; Paquette *et al.*, 1979). The $D_{16v}$ cell line grows as a monolayer in culture with a generation time of 24 hours in Dulbecco Vogt Minimum Essential Medium (DVMEM) supplemented with 10% horse serum (Herbert *et al.*, 1978).

Glycosylation and proteolytic processing of the precursor have been studied by continuous labeling and pulse-labeling of cells with radioactive amino acids and sugars. This is done as follows: the cells are labeled with amino acids and (or) sugars, extracted, and immunoprecipitated with antisera specific for each domain of the precursor molecule (ACTH, $\beta$-endorphin, or the N-terminal region) (Roberts *et al.*, 1978). Proteins in the immunoprecipitate are then separated by sodium dodecyl sulfate (SDS) tube or slab gel electrophoresis, eluted from the gels, and digested with trypsin, chymotrypsin, or cyanogen bromide. The digests are analyzed by paper electrophoresis, chromatography, or gel electrophoresis. These procedures have been described in detail by Roberts *et al.* (1978) and Phillips *et al.* (1981). Using this approach we can identify fragments derived from each domain of the precursor molecule and determine their content of glucosamine, galactose, fucose, and mannose. This allows us to relate the order of addition of each of the sugars to the time course of each proteolytic cleavage event.

Three distinct glycosylated forms of the precursor are detected in AtT-20 cells by SDS tube gel and slab gel electrophoresis as reported by Roberts *et al.* (1978). These forms have molecular weights of 29,000 (29K), 32,000 (32K), and 34,000 (34K). Tryptic peptide mapping studies show that the three forms have a very similar peptide backbone but differ in their content of oligosaccharides. Mapping of glucosamine-labeled tryptic peptides by paper electrophoresis at pH 6.5 and chromatography shows that there are three different glycopeptides present in the precursor: a neutral, a basic, and an acidic glycopeptide (Roberts *et al.*, 1978). The 29K, 32K, and 34K forms have variable amounts of the neutral and basic tryptic glycopeptides. The 32K form has, in addition to the neutral glycopeptide, an acidic glycopeptide derived from the ACTH region of the molecule (Eipper and Mains, 1978b). The basic and neutral glycopeptides are derived from the N-terminal region of the precursor (Phillips *et al.*, 1981). Other studies have shown that all the oligosaccharides are attached to the peptide backbone via *N*-acetylglucosamine and are of the complex type (see Fig. 7 in Chapter 2 by H. F. Lodish *et al.*).

The signal for attachment of an N-glycosidic-linked carbohydrate to Asn in a protein is the sequence Asn-x-Ser or Asn-x-Thr, where x is any amino acid. The amino acid sequence of the precursor from bovine pituitary has been determined by Nakanishi *et al.* (1979), using recombinant DNA technology as mentioned above. This analysis shows that there is only one potential attachment site for N-glycosidically linked oligosaccharides. This site is in the N-terminal region of the molecule (Asn-Gly-Ser-residues 65–67). Thus, either the mouse precursor has two additional carbohydrate attachment sites not found in the bovine precursor, or else the three tryptic glycopeptides observed by paper electrophoresis and chromatography are variants of the same glycopeptide, due possibly to incomplete digestion with trypsin or to the presence of variable amounts of sialic acid. Experiments have been done to determine if tryptic digestion is complete and if neuraminidase treatment of the peptides alters their electrophoretic mobility by removing sialic acid residues (M. Phillips, M. Budarf, and E. Herbert, unpublished results). Taken together, these studies indicate that there are three different oligosaccharide side chains present in the 29–34K complex of precursor molecules.

A model of the structure of the three precursor forms is presented in Fig. 3. According to this model the 29K and 32K forms of pro-ACTH-endorphin have an oligosaccharide in the N-terminal region of the molecule. In addition, the 32K form has a second oligosaccharide attached to the ACTH region of the molecule. The structure of the 34K form is not as well understood. It is clearly missing the oligosaccharide in the ACTH region of the molecule, but we are not sure how

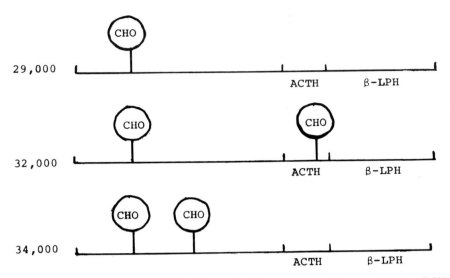

FIG. 3.   Postulated structure of 29K, 32K, and 34K forms of pro-ACTH-endorphin in AtT-20/D$_{16v}$ cells. CHO, carbohydrate side chains.

many N-terminal oligosaccharides it has. The structure of the 34K form depicted in Fig. 3 with two N-terminal oligosaccharides is hypothetical. We predict the existence of this form to explain the origin of a fragment in AtT-20 cells that has the peptide structure of the N-terminal region of the precursor and contains two oligosaccharides (18K N-terminal fragment).

Additional evidence for the presence of three different oligosaccharide side chains in the 29–34K pool of pro-ACTH-endorphin molecules has been obtained by studying the processing of the molecule in detail. A summary of this work with an emphasis in the glycosylation steps involved is presented below.

## III.   Intermediate and End Products of Processing of Pro-ACTH-Endorphin

Pulse–chase studies with radioactive amino acids suggest that the 29K form of pro-ACTH-endorphin is the precursor of the 32K and 34K forms of this protein in AtT-20 cells (Roberts *et al.*, 1978). If each of these forms is processed proteolytically as shown in the diagram in Fig. 2, one would predict that the first cleavage between β-LPH and ACTH would give rise to three different ACTH intermediates and β-LPH.

All the expected intermediates have been detected in extracts of AtT-20 cells by immunoprecipitation with ACTH-specific antisera followed by SDS gel electrophoresis and tryptic peptide mapping (Roberts *et al.*, 1978; Phillips *et al.*, 1981). These intermediates have apparent molecular weights of 26K, 23K, and 21K. Their tryptic glycopeptide content is given in Table I.

Cleavage of the ACTH intermediates (Fig. 2) should yield two forms of $\alpha$-(1–39)ACTH (a glycosylated and an unglycosylated form) and one or more glycosylated forms of the N-terminal fragment. These forms have also been detected in AtT-20 cells (Roberts *et al.*, 1978; Phillips *et al.*, 1981).

Two forms of $\alpha$-(1–39)ACTH were first identified by Eipper and Mains (1976) and later by Roberts *et al.* (1978). These forms have been referred to as 4.5K ACTH (unglycosylated) and 13K ACTH (glycosylated) according to their electrophoretic mobilities on SDS gels (Eipper and Mains, 1976). However, it has been shown by Eipper and Mains that 13K ACTH is actually $\alpha$-(1–39)ACTH with a single oligosaccharide attached to Asn residue 29. The 13K ACTH has a molecular weight of about 6500 as determined by gel filtration in guanidinium–HCl (Eipper and Mains, 1977).

N-terminal fragments have been detected in AtT-20 cells and in pituitary tissue by use of antisera to the N-terminal region of the precursor (Mains and Eipper, 1978; Roberts *et al.*, 1978; Phillips *et al.*, 1981). Phillips *et al.* (1981) have identified two proteins in immunoprecipitates of AtT-20 cells (immunoprecipitated with an N-terminal antiserum) as N-terminal fragments. These proteins have apparent molecular weights of 18K and 16K as determined by SDS tube gel and slab gel electrophoresis and 11K and 9K as determined by gel filtration on guanidinium-HCl columns (Phillips *et al.*, 1981). Both proteins contain the tryptic peptides present in the N-terminal region of pro-ACTH-endorphin but not

TABLE I

TRYPTIC GLYCOPEPTIDE SUMMARY

| | Basic tryptic glycopeptide $(R_{Lys} = 0.1)^a$ | Neutral tryptic glycopeptide $(R_{Lys} = 0)$ | Acidic $\alpha$-(22–39) tryptic glycopeptide |
|---|---|---|---|
| 34K ACTH-endorphin | + | + | − |
| 32K ACTH-endorphin | + | ± | + |
| 29K ACTH-endorphin | ± | + | − |
| 23K ACTH | ± | + | + |
| 21K ACTH | − | + | − |
| 13K ACTH | − | − | + |
| 18K N-terminal glycoprotein | + | + | − |
| 16K N-terminal glycoprotein | − | + | − |
| "β-LPH" | − | − | − |
| 3.5K β-endorphin | − | − | − |

[a] $R_{Lys}$ = mobility relative to lysine after paper electrophoresis at pH 6.5 (Roberts *et al.*, 1978).

in the ACTH or β-LPH regions of this molecule. The 18K form contains both the neutral and basic glycopeptides present in pro-ACTH-endorphin and in ACTH intermediates (Phillips *et al.*, 1981), whereas the 16K form has only the neutral tryptic glycopeptide. These results suggest that the 18K protein is the N-terminal fragment with two oligosaccharides and the 16K protein is the N-terminal fragment with one oligosaccharide. Further support for this conclusion is that the difference in molecular weight (as determined by gel filtration in guanidinium–HCl) between the two proteins is approximately equal to the size of an oligosaccharide chain. These results are compatible with the existence of pro-ACTH-endorphin molecules that contain two N-terminal oligosaccharides and pro-ACTH-endorphin molecules that contain one N-terminal oligosaccharide (Fig. 3). Further structural analysis of the tryptic glycopeptides is necessary to verify this conclusion.

The above results are compatible with the model of processing shown in Fig. 4. Since the 32K form of pro-ACTH-endorphin and the 23K ACTH intermediate both have an N-terminal oligosaccharide and an oligosaccharide attached to the ACTH region of the molecule, it appears likely that the 32K form of pro-ACTH-endorphin gives rise to 23K ACTH and β-LPH and that the 23K form of ACTH subsequently gives rise to the 16K N-terminal fragment and the glycosylated form of α-(1–39)ACTH (13K ACTH). The 29K form of pro-ACTH-endorphin is processed in a manner similar to that of the 32K form, giving rise to unglycosylated α-(1–39)ACTH and β-LPH. The processing of 34K pro-ACTH-endorphin is not very well understood because we do not know how many oligosaccharides it contains. If it has two N-terminal oligosaccharides as depicted in the model in Fig. 3, then it might be the precursor to 26K ACTH. The 26K form of ACTH would then give rise to unglycosylated α-(1–39)ACTH and the 18K N-terminal fragment. An alternative to the model of processing shown in Fig. 4 is that the 29K, 32K, and 34K proteins are different gene products with different amino acid sequences. To distinguish this alternative from the one depicted in Fig. 4 would require complete amino acid sequence determination of these forms of pro-ACTH-endorphin.

As noted in Fig. 4, the precursor is synthesized on membrane-bound ribosomes (Roberts and Herbert, 1977a) with an N-terminal signal sequence. The first form of the precursor synthesized in the cell (29K pro-ACTH-endorphin) is missing the signal sequence (Policastro *et al.*, 1981). Therefore, the signal sequence is very likely removed prior to completion of synthesis of the polypeptide chain as in the case of other secretory proteins (Blobel and Dobberstein, 1975; Jackson and Blobel, 1977). Another interesting point is that glycosylation also appears to take place prior to completion of the peptide chain, since 29K pro-ACTH-endorphin has sugars attached to it (Roberts *et al.*, 1978).

The major end products of processing of pro-ACTH-endorphin in AtT-20 cells are glycosylated and unglycosylated forms of α-(1–39)ACTH, two or three

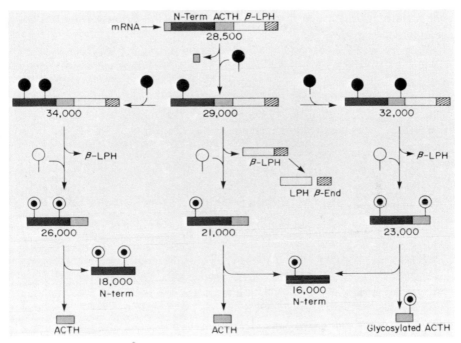

FIG. 4. Detailed model of processing of pro-ACTH-endorphin in AtT-20/D$_{16v}$ Cells. □,Removal of signal sequence; ●, mannose-rich core oligosaccharide; ○, trimming of mannose core and addition of peripheral branch sugars—e.g., galactose, fucose, and sialic acid; ◎, processed oligosaccharide side chains.

glycosylated forms of the N-terminal fragments, and β-LPH. (We find that β-LPH is converted very slowly to β-endorphin.) These products are found in stoichiometric amounts (Mains and Eipper, 1978; Roberts et al., 1978). All the end products are secreted by AtT-20 cells (Allen *et al.*, 1978; R. G. Allen *et al.*, unpublished results).

The reasons for the synthesis of glycosylated and unglycosylated forms of ACTH is not clear. Gasson (1979) has shown that the two forms of α-(1–39) ACTH stimulate steroidogenesis to about the same extent (on a molar basis) in cultures of mouse adrenal cortical cells. ACTH intermediates, N-terminal fragments, and pro-ACTH-endorphin are much less active than α-(1–39)ACTH as steroidogenic agents in this system (Gasson, 1979). An important difference between the two forms of ACTH is that the glycosylated form is much more stable in serum than the unglycosylated form (R. G. Allen *et al.*, unpublished results). This is interesting because particular sugar residues at the ends of oligosaccharides are known to influence the stability of glycoproteins and the rate of clearance of glycoproteins in the liver (Ashwell and Morell, 1977). Perhaps the state of glycosylation determines whether ACTH is long-acting or short-acting in the circulation.

Thus far, we have observed that the two forms of $\alpha$-(1–39)ACTH are produced in roughly equal amounts in AtT-20 cells and in monolayer cultures of anterior pituitary cells, whether the cells are in the basal state (no regulators added), stimulated state (stimulated by CRH), or inhibited state (inhibited by dexamethasone) (Allen *et al.*, 1978; Roberts *et al.*, 1979a). Perhaps there are other conditions under which the production of the two forms of ACTH can be uncoupled.

# IV.　Details of Glycosylation Steps Involved in Processing of Pro-ACTH-Endorphin

Recent investigations of the biosynthesis of the complex type of oligosaccharides of vesicular stomatitus virus G protein have shown that the first step in glycosylation is synthesis of a lipid-linked high-mannose-type core containing peripheral glucose residues (Tabas *et al.*, 1978; Hunt *et al.*, 1978; Robbins *et al.*, 1977; see also Chapter 2 by H. F. Lodish *et al.* and Chapter 7 by M. D. Snider and P. W. Robbins).

The lipid-linked core is then attached to the protein, the glucose and mannose residues are trimmed back, and stepwise attachment of the branch sugars—*N*-acetylglucosamine, galactose, fucose, and sialic acid—occurs (see Fig. 7 in Chapter 2 by H. F. Lodish *et al.*). This highly ordered process generates many oligosaccharide intermediates. If this type of oligosaccharide processing applies to pro-ACTH-endorphin, then one should be able to detect the intermediates generated and to relate the time course of proteolytic processing events to the time course of glycosylation and secretion events.

To study processing of oligosaccharides, AtT-20 cells were incubated with different radioactive sugars. Pro-ACTH-endorphin, ACTH intermediates, and N-terminal fragments were purified from cell extracts and culture medium by immunoprecipitation and SDS gel electrophoresis. The purified components were digested with pronase (Phillips *et al.*, 1981) and the pronase glycopeptides were analyzed by (1) gel filtration to determine their size, (2) susceptibility to various glycosidases ($\alpha$-mannosidase, glucosidases, and endoglucosidases) to determine if terminal mannose or glucose residues are present, and (3) sugar composition. Phillips *et al.* (1981) also analyzed the time course of labeling of the oligosaccharides with different radioactive sugars.

Gel filtration analysis of pronase glycopeptides derived from intracellular glycoproteins indicates that the oligosaccharides attached to pro-ACTH-endorphin are smaller and more heterogeneous in size than those attached to

ACTH intermediates and N-terminal fragments (Table II). [$^3$H]Mannose-labeled pronase glycopeptides derived from intracellular and secreted proteins were digested with either $\alpha$-mannosidase or endoglycosidase H, and the radioactive products were analyzed by gel filtration. As shown in Table II, glycopeptides derived from intracellular forms of pro-ACTH-endorphin are susceptible to these enzymes, suggesting that pro-ACTH-endorphin molecules contain high-mannose-type oligosaccharides and terminal mannose residues. Therefore, oligosaccharides attached to these proteins are in a fairly early stage of processing. The failure to detect fucose in these oligosaccharides (Table II) supports the view that addition of outer branch sugars is not complete in the 29K, 32K, and 34K forms of pro-ACTH-endorphin.

Oligosaccharides derived from intracellular and secreted intermediates and N-terminal fragments and secreted forms of pro-ACTH-endorphin contain fucose and galactose and are not susceptible to digestion with $\alpha$-mannosidase and endoglycosidase H. The latter proteins very likely contain more completely processed oligosaccharides with smaller mannose cores and more outer branch sugars. These results suggest that, by the time the first proteolytic cleavage of pro-ACTH-endorphin occurs, processing of the oligosaccharides is nearly complete. This is depicted in the processing scheme in Fig. 4.

We have found that the first proteolytic cleavage occurs within 15–25 minutes of labeling of pro-ACTH-endorphin with radioactive amino acids. From the results of studies with other systems, one would estimate that these events occur during passage of these proteins through smooth endoplasmic reticulum and Golgi apparatus (Schacter *et al.*, 1970; Sharon, 1975; Tulsiani *et al.*, 1977; Wagner and Cynkin, 1971). Both of the latter structures have been observed in AtT-20 cells by electron microscopy (unpublished results of M. Budarf, and personal communication from S. Sabol, National Institutes of Health).

# V.  Use of the Glycosylation Inhibitor, Tunicamycin, to Determine If Cleavage Sites Are Specified by Oligosaccharides

Glycosylation of proteins that contain N-glycosidically linked oligosaccharides can be prevented by the antibiotic, tunicamycin, which inhibits the formation of $N$-acetylglucosamine-lipid linked intermediates (Takatsuki *et al.*, 1971; Tkacz and Lampén, 1975; Kuo and Lampén, 1976). When AtT-20 cells are treated with tunicamycin (5 $\mu$g/ml) for 10 hours, [$^3$H]glucosamine incorporation into protein is inhibited by 96–98%, whereas [$^{35}$S]Met incorporation into protein is inhibited by 20–30%. The intracellular levels of ACTH and endorphin in tunicamycin-treated cells are 80% of control values. In contrast, the levels of ACTH immunoactivity in the tissue culture medium as determined by radioim-

## TABLE II

RESULTS OF STRUCTURAL STUDIES ON OLIGOSACCHARIDES FROM INTRACELLULAR FORMS OF ACTH-ENDORPHIN PROTEINS[a,b]

| | Molecular weight oligosaccharide | Susceptibility to digestion by | | Branch chain sugars present | | Core sugars present | |
|---|---|---|---|---|---|---|---|
| | | α-Mannosidase | Endoglycosidase H | Fucose | Galactose | Glucosamine | Mannose |
| 29 + 32K pro-ACTH-endorphin | 1900–2600 | + | + | – | + | + | + |
| N-terminal fragments | 2700–2900 | – | – | + | + | + | + |
| ACTH intermediates (21 + 23K) | 2900–3100 | – | – | + | + | + | + |
| 13K ACTH | 2900–3100 | – | – | + | + | + | + |

[a] AtT-20 cells were labeled with an [3]H-sugar and/or [[35]S]Met as described by Roberts et al. (1978). The ACTH-endorphin proteins and N-terminal fragments were purified by immunoprecipitation with specific antisera and fractionated by SDS slab gel or tube electrophoresis (Phillips et al., 1981). The labeled components were eluted from the gels and identified by mapping [35]S[Met]-containing tryptic peptides (Roberts et al., 1978). Each glycosylated protein was digested with pronase, and pronase glycopeptides were analyzed by gel filtration on Bio-Gel P-4 (200–400 mesh) columns equilibrated with 50 m$M$ NH$_4$OCOCH$_3$, pH 6.0, 0–02% NaN$_3$. Elution was carried out with the equilibration buffer. The column was calibrated with bacitracin, pronase glycopeptides from ovalbumin, fetuin and bovine thyroglobulin, and two oligosaccharides, raffinose and stachyose. The standards were located by the phenol-H$_2$SO$_4$ assay method of Ashwell (1966). After purification, pronase glycopeptides labeled with various sugars were digested with either α-mannosidase or endoglycosidase H, and the products were separated and identified by the gel filtration method described above. Each fraction eluted from the column was counted by scintillation spectrometry to determine its content of [3]H and/or [35]S.

[b] All the forms of pro-ACTH-endorphin, ACTH intermediates, and N-terminal fragments present in the culture medium were found to contain fucose in addition to galactose, mannose, and glucosamine. None of these forms were susceptible to digestion with α-mannosidase or endoglycosidase.

munoassay show a marked reduction with tunicamycin, roughly paralleling inhibition of glycosylation by tunicamycin. This suggests that secretion and/or stability of the ACTH proteins in the culture medium may be dependent upon glycosylation.

AtT-20 cells were treated with tunicamycin (5 $\mu$g/ml) for 10 hours and then incubated with [$^{35}$S]Met for 2 hours. ACTH and endorphin proteins were purified from cell extracts and culture medium by immunoprecipitation and SDS slab gel electrophoresis as described by M. Budarf and E. Herbert (unpublished results). Densitometer scans of autoradiographs of the gels (Fig. 5) show that, in both the ACTH and the endorphin immunoprecipitates from cells (charts A and C, respectively), the 29–32K pro-ACTH-endorphin proteins, seen in the control cultures, are absent in the tunicamycin-treated cells, and instead a prominent new protein band is present with an increased mobility and an apparent molecular weight of 26,000 (26K). The ACTH intermediates (21–23K proteins) and glycosylated $\alpha$-(1–39)ACTH (13K) are either absent or greatly reduced in tunicamycin-treated cells (chart A, Fig. 5). In place of 21–23K ACTH intermediates, a set of proteins can be seen with increased mobility and apparent molecular weight of 17–18K.

When the culture medium from tunicamycin-treated [$^{35}$S]Met-labeled cells was analyzed in the same way as cell extracts, proteins with increased mobility were also found (charts B and D, Fig. 5), indicating that the novel forms of ACTH seen in tunicamycin-treated cells are secreted.

Other experiments have shown that the ACTH-endorphin proteins made in tunicamycin-treated cells do not incorporate [$^{3}$H]glucosamine under conditions that give good incorporation of [$^{3}$H]glucosamine into these proteins in untreated cells (M. Budarf and E. Herbert, unpublished results). These results demonstrate that none of the ACTH or endorphin proteins synthesized in tunicamycin-treated cells are glycosylated.

SDS slab gel electrophoresis (Fig. 5A) also shows that pro-ACTH-endorphin made in tunicamycin-treated cells (26K protein) is smaller than the pro-ACTH-endorphin made in a reticulocyte cell-free system under the direction of AtT-20 RNA by a molecular weight of 2500. This is the molecular weight difference one would expect if pro-ACTH-endorphin were missing a signal sequence known to be present in the cell-free product. Sequencing work done in collaboration with Drs. Seidah and Chrétien of the Clinical Research Institute of Montreal shows that the cell-free product contains a signal sequence of 26 amino acids at the amino terminus not found in pro-ACTH-endorphin isolated from cell extracts (Policastro et al., 1981). It seems probable that the difference in molecular weight between the cell-free product and the 26K protein is due to cleavage of the signal sequence. Demonstration of this cleavage is normally very difficult, because the addition of carbohydrate offsets the decrease in molecular weight that accompanies removal of the signal sequence.

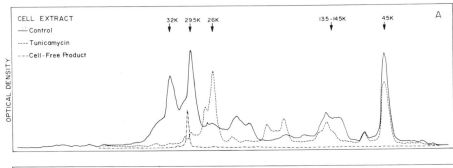

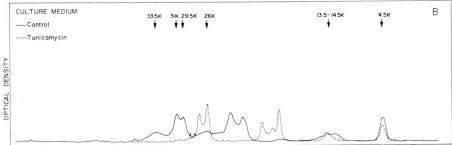

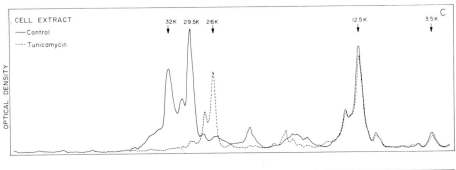

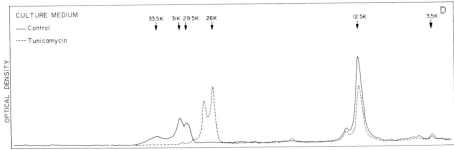

## VI. Structure of the ACTH-Endorphin Peptides from Tunicamycin-Treated Cells

The [$^{35}$S]Met-labeled 26K protein from tunicamycin-treated cells and 29K pro-ACTH-endorphin from control cells were digested with trypsin, and the digests were fractionated by paper electrophoresis at pH 6.5. The results in Fig. 6A show that the Met-labeled tryptic peptides derived from the two proteins are very similar, indicating that they have similar amino acid sequences. Furthermore, since the Met peptides of pro-ACTH-endorphin are derived from the three major domains of the molecule, [the $\beta$-(61-69)LPH peptide, the $\alpha$-(1-8)ACTH peptide, and the N-terminal peptide], this result also shows that the 26K protein is not missing any major protein sequences.

Figure 6B shows that [$^{35}$S]tryptic peptide maps of the ACTH intermediates from control cells (21K and 23K ACTH) and from tunicamycin-treated cells (17-18K protein) are very similar, providing further evidence that $\beta$-LPH can be correctly cleaved out of the 26K protein and that the 17-18K protein from tunicamycin-treated cells has the same peptide backbone as do the glycosylated ACTH intermediates.

Finally, Edman degradation was performed on [$^3$H]phenylalanine-labeled 4.5K protein from tunicamycin-treated cells and control cells. The results show that in both cases phenylalanine is in position 7, as expected for $\alpha$-(1-39)ACTH (Herbert *et al.*, 1980).

These results clearly demonstrate that unglycosylated pro-ACTH-endorphin can be processed normally to unglycosylated products (4.5K ACTH and a $\beta$-LPH-like component) and that the unglycosylated forms of ACTH and endorphin can also be secreted.

An important question to ask at this point is how much of the unglycosylated form of pro-ACTH-endorphin is processed to 4.5K ACTH and "$\beta$-LPH." Results of pulse-chase experiments with radioactive amino acids suggest that more than half of the radioactivity incorporated into 26K pro-ACTH-endorphin in tunicamycin-treated cells is processed normally.

FIG. 5. Effect of tunicamycin on cellular and secreted forms of ACTH and endorphin. Microwells were pretreated for 10.5 hours in the presence (5 $\mu$/ml) or absence of tunicamycin. At the end of this period, culture medium was removed and 50 $\mu$l of minus Met, minus horse serum (Dulbecco-Vogt Minimal Essential Medium), containing 125 $\mu$Ci of [$^{35}$S]Met was added to each microwell. Tunicamycin treatment was continued at the same concentration as pretreatment. After 1.5 hours of incubation, the culture medium was removed and frozen, and the cells were extracted as described by Roberts *et al.* (1978). ACTH- and endorphin-containing proteins were immunoprecipitated and analyzed on 12.5% polyacrylamide slab gels. Dried gels were autoradiographed, and the films were scanned with a microdensitometer (Grant Instruments). (A) ACTH-containing proteins in cell extracts; (B) ACTH-containing proteins in culture medium; (C) endorphin-containing proteins in cell extracts; (D) endorphin-containing proteins in culture medium.

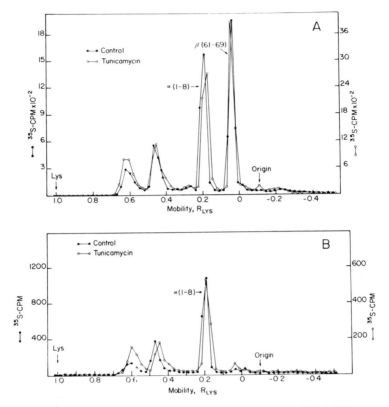

FIG. 6.   Tryptic peptides of 30K and 26K pro-ACTH-endorphin, 21–23K ACTH intermediates, and 17–18K ACTH. Microwells were pretreated and labeled with [³⁵S]Met as described in Fig. 1. Culture medium was immunoprecipitated with ACTH antiserum, and ACTH-containing proteins were separated by tube gel electrophoresis as described by Roberts and Herbert (1977a). Eluates containing the pro-ACTH-endorphin proteins and ACTH intermediates were pooled and digested with trypsin. The tryptic peptides were analyzed by paper electrophoresis, pH 6.5. (A) 30K pro-ACTH-endorphin is indicated by solid circles, and 26K pro-ACTH-endorphin from tunicamycin-treated cultures by open circles. (B) 21–23K ACTH intermediates from control cultures are solid circles; 17–18K ACTH from tunicamycin-treated culture are open circles.

These results differ from those of Loh and Gainer (1979), who found that tunicamycin inhibits the glycosylation of a 32,000-molecular-weight form of pro-ACTH-endorphin in toad intermediate lobe. However, the unglycosylated precursor was apparently unstable, and intracellular processing did not occur normally. One explanation for the difference in results is the difference in the cell systems used. However, preliminary studies with tunicamycin-treated rat intermediate lobe cultures show that unglycosylated pro-ACTH-endorphin is processed to normal unglycosylated products. Another possible explanation for the difference in results might be differences in methods of analysis. Obviously, further work is required to settle this issue.

## ACKNOWLEDGMENT

This work was supported by National Institutes of Health Grant AM 16879.

## REFERENCES

Allen, R. G., Herbert, E., Hinman, M., Shibuya, H., and Pert, C. B. (1978). *Proc. Natl. Acad. Sci. U.S.A.* **75,** 4972-4976.

Ashwell, G. (1966). *In* "Methods of Enzymology" (E. F. Neufeld and V. Ginsburg, eds.), Vol. 8, 93-94. Academic Press, New York.

Ashwell, G., and Morell, A. G. (1977). *Top. Biochem. Sci.* **76,** 281.

Bergland, R., and Page, R. (1979). *Science,* **204,** 18-26.

Bertagna, X. Y., Nicholson, W. E., Sorenson, G. D., Pettengill, O. S., Mount, C. D., and Orth, D. (1978). *Proc. Natl. Acad. Sci. U.S.A.* **75,** 5160-5164.

Blobel, G., and Dobberstein, B. (1975). *J. Cell Biol.* **67,** 852-862.

Briaud, B., Koch, B., Lutz-Bucher, B., and Mialhe, C. (1978). *Neuroendocrinology* **25,** 47-63.

Budarf, M. I., and Herbert, E. (submitted).

Crine, P., Gianoulakis, C., Seidah, N. G., Gossaud, F., Pezella, P. D., Lis, M., and Chrétien, M. (1978). *Proc. Natl. Acad. Sci. U.S.A.* **75,** 4719-4723.

Eipper, B. A., and Mains, R. E. (1975). *Biochemistry* **14,** 3836-3844.

Eipper, B. A., and Mains, R. E. (1978a). *J. Supramol. Struct.* **8,** 247.

Eipper, B. A., and Mains, R. E. (1978b). *J. Biol. Chem.* **253,** 5732-5744.

Eipper, B. A., Mains, R. E., and Guenzi, D. (1976). *J. Biol. Chem.* **251,** 4121-4126.

Gasson, J. (1979). *Biochemistry* **18,** 4215.

Guillemin, R., Vargo, T., Rossier, J., Minick, S., Ling, N., Rivier, C., and Vale, W. (1977). *Science* **197,** 1367-1370.

Herbert, E., Allen, R. G., and Paquette, T. P. (1978). *Endocrinology (Baltimore)* **102,** 218-226.

Herbert, E., Budarf, M., Phillips, Rosa, P., Policastro, P., Oates, E., Roberts, J. L., Seidah, N. G., and Chrétien, M. (1980). *Ann. N.Y. Acad. Sci.* (in press).

Hunt, L. A., Etchinson, J. R., and Summers, D. F. (1978). *Proc. Natl. Acad. Sci. U.S.A.* **75,** 754-758.

Jackson, R. C., and Blobel, G. (1977). *Proc. Natl. Acad. Sci. U.S.A.* **74,** 5598-5602.

Kimura, S., Lewis, R. V., Gerber, L. D., Brink, L., Rubinstein, M., Stein, S., and Udenfriend, S. (1979). *Proc. Natl. Acad. Sci. U.S.A.* **76,** 1756-1760.

Kraicer, J., Gosbee, J. L., and Bencosme, S. A. (1973). *Neuroendocrinology* **11,** 156-176.

Krieger, D. T., and Liotta, A. S. (1979). *Science* **205,** 366.

Krieger, D. T., Liotta, A. S., Hauser, H., and Brownstein, M. J. (1979). *Endocrinology (Baltimore)* **105,** 737-742.

Krieger, D. T., Liotta, A. S., Brownstein, M. J., and Zimmerman, E. A. (1980). *Recent Prog. Horm. Res.* **36,** 272-344.

Kuo, S. C., and Lampén, J. O. (1976). *Arch. Biochem. Biophys.* **172,** 574-581.

Liotta, A. S., Gildersleeve, D., Brownstein, M. J., and Krieger, D. T. (1979). *Proc. Natl. Acad. Sci. U.S.A.* **76,** 1448-1452.

Loh, Y. P. (1979). *Proc. Natl. Acad. Sci. U.S.A.* **76,** 796–800.

Loh, Y. P., and Gainer, H. (1979). *Endocrinology (Baltimore)* **105,** 474–487.

Mains, R. E., and Eipper, B. A. (1979). *J. Biol. Chem.* **254,** 7885–7894.

Mains, R. E., Eipper, B. A., and Ling, N. (1977). *Proc. Natl. Acad. Sci. U.S.A.* **74,** 3014–3018.

Moriarty, G. C., Halmi, N. S., and Moriarty, C. M. (1975). *Endocrinology (Baltimore)* **96,** 1426–1436.

Nakanishi, S., Inoue, A., Taii, S., and Numa, S. (1977). *FEBS Lett.* **84,** 105–109.

Nakanishi, S., Inoue, A., Kita, T., Nakamura, M., Chang, A. C. Y., Cohen, S. N., and Numa, S. (1979). *Nature (London)* **278,** 423–427.

Odagiri, E., Sherrell, B. J., Mount, C. D., Nicholson, W. E., and Orth, D. N. (1979). *Proc. Natl. Acad. Sci. U.S.A.* **76,** 2027–2032.

Paquette, T. L., Hinman, M., and Herbert, E. (1979). *Endocrinology (Baltimore)* **104,** 1211–1216.

Phillips, M., Budarf, M., and Herbert, E. (1981). *Biochemistry* **20,** 1666–1675.

Policastro, P., Phillips, M., Oates, E., Herbert, E., Roberts, J. L., Seidah, N., and Chrétien, M. (1981). *Eur. J. Biochem.* **116,** 255–259.

Robbins, P. W., Hubbard, S. C., Turco, S. J., and Wirth, D. F. (1977). *Cell* **12,** 893–900.

Roberts, J. L., and Herbert, E. (1977a). *Proc. Natl. Acad. Sci. U.S.A.* **74,** 4826–4830.

Roberts, J. L., and Herbert, E. (1977b). *Proc. Natl. Acad. Sci. U.S.A.* **74,** 5300–5304.

Roberts, J. L., Phillips, M., Rosa, P. A., and Herbert, E. (1979a). *Biochemistry* **17,** 3609–3618.

Roberts, J. L., Seeburg, P., Shine, J., Herbert, E., Baxter, J. D., and Goodman, H. M. (1979b). *Proc. Natl. Acad. Sci. U.S.A.* **76,** 2153–2157.

Roberts, J. L., Budarf, M. L., Baxter, J. D., and Herbert, E. (1979c). *Biochemistry* **18,** 4907–4915.

Schachter, H., Jabbal, I., Hudgin, R. L., Pinteria, L., McGuire, E. J., and Roseman, S. (1970). *J. Biol. Chem.* **245,** 1090–1100.

Scott, A. P., Ratcliffe, J. G., Rees, L. H., Bennett, H. P. J., Lowry, P. J., and McMartin, C. (1973). *Nature (London), New Biol.* **244,** 65–67.

Scott, A. P., Lowry, P. J., Ratcliffe, J. G., Rees, L. H., and Landon, J. (1974). *J. Endocrinol.* **61,** 355–367.

Scott, A. P., Lowry, P. J., and van Wimersma Greidanus, T. B. (1976). *J. Endocrinol.* **70,** 197–205.

Seidah, N. G., Gianoulakis, C., Crine, P., Lis, M., Benjannet, S., Routhier, R., and Chrétien, M. (1978). *Proc. Natl. Acad. Sci. U.S.A.* **75,** 3153–3157.

Sharon, N. (1975). ''Complex Carbohydrates.'' Addison-Wesley, Reading, Massachusetts.

Smyth, D. G., Massey, D. E., Zakarian, S., and Finnie, M. D. A. (1979). *Nature (London)* **279,** 252–254.

Tabas, I., Schlesinger, S., and Kornfeld, S. (1978). *J. Biol. Chem.* **253,** 716–722.

Takatsuki, A., Arima, K., and Tamura, G. (1971). *J. Antibiot.* **24,** 215–223.

Tkacz, J. S., and Lampén, J. O. (1975). *Biochem. Biophys. Res. Commun.* **65,** 248–257.

Tulsiani, D. R. P., Opheim, D. V., and Touster, O. (1977). *J. Biol. Chem.* **252,** 3227–3233.

Vale, W., and Rivier, C. (1977). *Fed. Proc., Fed. Am. Soc. Exp. Biol.* **36,** 2094–2099.

Vale, W., Rivier, C., Yang, L., Minick, S., and Guillemin, R. (1978). *Endocrinology (Baltimore)* **103,** 1910–1917.

Waechter, C. J., and Lennarz, W. J. (1976). *Annu. Rev. Biochem.* **45,** 95–112.

Wagner, R. R., and Cynkin, M. A. (1971). *J. Biol. Chem.* **246,** 143–151.

# Chapter 9

# Posttranslational Events in Collagen Biosynthesis

JEFFREY M. DAVIDSON[1] AND RICHARD A. BERG[2]

*Pulmonary Branch,*
*National Heart, Lung and Blood Institute,*
*National Institutes of Health,*
*Bethesda, Maryland*

## I. Introduction

Collagen is the major protein of the extracellular matrix; it is synthesized and secreted by fibroblasts and other cell types as a large-molecular-weight precursor protein, procollagen. Procollagen differs from collagen in that it has additional nonhelical polypeptide domains extending from both its COOH and $NH_2$ termini. Before and after secretion, procollagen undergoes extensive posttranslational modifications that modulate the physicochemical properties of this molecule, the most important of which is the association of three polypeptide chains into a

---

[1]Present address: Department of Pathology, College of Medicine, University of Utah, Salt Lake City, Utah 84132.

[2]Present address: Department of Biochemistry, College of Medicine and Dentistry of New Jersey, Rutgers University Medical School, Piscataway, New Jersey 08854.

triple-helical conformation. Several recent comprehensive reviews have appeared on procollagen structure and biosynthesis (Fessler and Fessler, 1978; Prockop *et al.*, 1979; Harwood, 1979; Bornstein and Traub, 1979; Bornstein and Sage, 1980); as a result, this review will emphasize more current contributions to these topics. The intent of this review is to describe the known posttranslational modifications of the major types of procollagen and to relate these modifications of the protein to secretory processes in the cells that synthesize them. Indeed, the extent and nature of posttranslational modifications may represent a major difference among the known genetic types of procollagen.

## II.   Genetic Types of Collagen

Although there are at least nine genes coding for collagen polypeptide chains (Bornstein and Sage, 1980), only five procollagen trimers have been adequately described in terms of their chain composition: types I, II, III, IV, and type I trimer. By far the largest amount of information has been obtained for type I procollagen, the major procollagen produced in skin, bone, tendon, and lung (Fessler and Fessler, 1978; Prockop *et al.*, 1979; Harwood, 1979; Bornstein and Traub, 1979; Bornstein and Sage, 1980). It is composed of three polypeptide chains, two pro $\alpha1(I)$ chains and one pro $\alpha2$ chain, linked together by five interchain disulfide bonds in the COOH propeptide region of the molecule (Olsen *et al.*, 1977) (Fig. 1). Type II procollagen, the major procollagen molecule of cartilaginous tissue, is composed of three identical pro $\alpha1(II)$ chains, having interchain disulfide bonds in the COOH propeptide region of the procollagen molecule. Type III procollagen consists of three pro $\alpha1(III)$ chains and codistributes in most tissues with type I procollagen. It is present to a larger extent, relative to type I procollagen, in blood vessels, internal organs, and fetal skin, and to a lesser extent in adult skin (Epstein, 1974; Mayne *et al.*, 1977, 1978; Gay and Miller, 1978). Type III procollagen contains disulfide bonds between the $NH_2$-terminal propeptides, the COOH-terminal propeptides, and the collagen chains near their COOH termini (Fessler and Fessler, 1978). Type IV procollagen is present in most basement membrane structures underlying epithelial and endothelial cells; it consists of three pro $\alpha1(IV)$ chains. However, this procollagen may be heterogeneous and comprise two subtypes of different chain composition (Bornstein and Sage, 1980; Timpl *et al.*, 1978, 1979; Kresina and Miller, 1979; Sage *et al.*, 1979; Glanville *et al.*, 1979; Tryggvason *et al.*, 1980; Raj *et al.*, 1979). Type IV procollagen contains interchain disulfide bonds, but their exact location is as yet unknown (Fessler and Fessler, 1978; Prockop *et al.*, 1979; Harwood, 1979; Bornstein and Traub, 1979; Bornstein and Sage, 1980). In addition to the four types of procollagens described, there is evidence for a

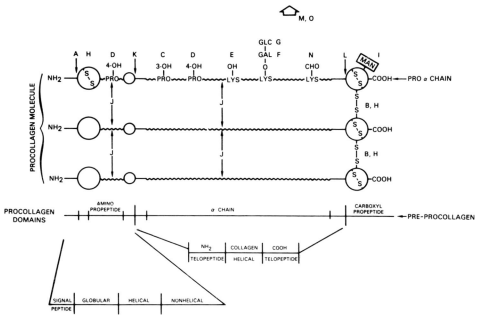

FIG. 1. Upper: Structural features and posttranslational modifications of an idealized pre-procollagen molecule. Compact, globular regions are drawn as circles; helical regions as wavy lines. Side-chain modifications are indicated on only one chain of the model for simplicity. The positions of each modification within a domain are purely schematic, and certain modifications may not occur in all collagen types (see Table I). Letters refer to posttranslational steps as outlined in Table II: A, signal peptide cleavage; B, chain association (one or both propeptides); C, prolyl-3-hydroxylation; D, prolyl-4-hydroxylation; E, lysyl hydroxylation; F, galactosylhydroxylysyl transfer; G, glucosylgalactosylhydroxylysyl transfer; H, disulfide bond formation; I, lipid-linked oligosaccharide transfer; J, triple-helix formation; K, $NH_2$-propeptide cleavage; L, COOH-propeptide cleavage; M, fibril formation (involving more than one molecule); N, lysyl oxidation; O, crosslinking (involving two chains). Lower: Structural domains of an idealized pre-procollagen chain. The three major regions of the chain can be further divided into subregions having distinct physicochemical properties. Linear distances are not proportional to actual numbers of amino acid residues in these models.

type I trimer, consisting of [pro $\alpha 1(I)]_3$, which occurs with type I procollagen, but to a lesser extent. This molecule has been proposed as a separate gene product (Little and Church, 1978), but it is more likely an alternative molecular association for pro $\alpha 1(I)$ chains (Mayne *et al.*, 1975; Naranayan and Page, 1976; Moro and Smith, 1977; Jimenez *et al.*, 1977; Munksgaard *et al.*, 1978; Wohllebe and Carmichael, 1978; Crouch and Bornstein, 1978; Uitto, 1979). Type V collagen has been described (Burgeson *et al.*, 1976; Benya *et al.*, 1977; Rhodes and Miller, 1978; Bentz *et al.*, 1978; Sage and Bornstein, 1979) and has been shown to consist of $\alpha A$ or $\alpha 2(V)$ and $\alpha B$ or $\alpha 1(V)$ and, most recently, of $\alpha C$ or $\alpha 3(V)$ polypeptide chains (Bornstein and Sage, 1980; Sage and Bornstein,

1979). The molecular arrangement of these chains has not been established (Rhodes and Miller, 1978; Bentz *et al.*, 1978; Sage and Bornstein, 1979; von der Mark and von der Mark, 1979), so it may be any trimeric combination of the three chains. Type V procollagen has not yet been identified.

Although there are many distinct procollagen trimeric molecules in connective tissue, most of the evidence so far obtained indicates that the chains in their "pro" form are all of approximately the same molecular weight, from 150,000 to 200,000 (Clark *et al.*, 1975; Fessler and Fessler, 1978; Fessler and Fessler, 1979; Heathcote *et al.*, 1978; Timpl *et al.*, 1978, 1979; Bornstein and Traub, 1979; Crouch and Bornstein, 1979; Harwood, 1979; Prockop *et al.*, 1979; Raj *et al.*, 1979; Bornstein and Sage, 1980; Tryggvason *et al.*, 1980). Each of the pro $\alpha$ chains of all five procollagen types contains a large core region, the collagen domain (Fig. 1). This domain consists of a repeating Gly-X-Y triplet, where X is frequently proline and Y is frequently 4-hydroxyproline (Bornstein and Traub, 1979). This region of the molecule as well as a small domain within the $NH_2$-terminal propeptide (Rohde *et al.*, 1979; Hörlein *et al.*, 1979; Pesciotta *et al.*, 1980) can fold into the triple-helical conformation characteristic of collagen. The stringent primary structural requirements for stability of the collagen domain among the various types of collagen most likely account for their similarities in amino acid composition and sequence (Bornstein and Traub, 1979).

Although compositional variation makes a contribution to the biological properties of procollagen types (Prockop *et al.*, 1979; Bornstein and Traub, 1979; Bornstein and Sage, 1980), one of the major differences in the genetic types of procollagens isolated from various connective tissues is the extent of their posttranslational modifications. These include hydroxylation of prolyl and lysyl residues, glycosylation, disulfide bonding, and proteolytic processing. Some of the differences in the extent to which the collagens are modified are indicated in Table I. For example, type II collagen has considerably more carbohydrate than have types I and III collagens, and it has an increased amount of hydroxylysine. Type III collagen has a larger proportion of 4-hydroxyproline than type I collagen. Type IV collagen from some tissues has a higher proportion of 3-hydroxyproline than have any of the first three types of collagen, and it has the highest proportion of carbohydrate and hydroxylysine. Type V collagen appears to be similar to type IV collagen in terms of the extent of posttranslational modifications. This chemical resemblance may reflect the localization and function of both types IV and V in certain basement membrane structures (Gay and Miller, 1978; Timpl *et al.*, 1979; Sage *et al.*, 1979; Rhodes and Miller, 1978).

A second set of major posttranslational differences between the genetic types of procollagen is the relative rates or extents of removal of propeptides from the procollagen molecule. Types I and II procollagen are cleaved to their respective collagens most rapidly (Fessler and Fessler, 1978). Type I trimer is processed similarly in embryonic chick bone (J. M. Davidson and L. S. G. McEneany,

TABLE I

EXTENT OF POSTTRANSLATIONAL MODIFICATION OF THE KNOWN COLLAGENS

| Type | Chain[a] | 3-Hyp[b] | 4-Hyp[c] | Hyl[d] | Glycosylation[e] (%) | Carbohydrate[f] | NH₂ cleavage | COOH cleavage | References[g] |
|---|---|---|---|---|---|---|---|---|---|
| I | α1(I) | 0.9 | 48.9 | 27.0 | 15.3 | 0.9 | + | + | 1 |
|  | α2 | 1.1 | 47.7 | 37.5 | 18.4 | 1.8 | + | + | 1 |
| I trimer | α1(I) | 11.4 | 53.2 | 60.0 | — | — | + | + | 2, 3 |
| II | α1(II) | 1.98 | 44.5 | 38.9 | 39.1 | 9 | + | + | 4 |
| III | α1(III) | — | 53.8 | 14.3 | 18.0 | 0.9 | − | + | 5 |
| IV | α1(IV) or αC | 0.2–4.9 | 56.6–64.9 | 85.1–89.3 | 61.2–92.0 | 30–46 | ?/− | ?/− | 6–8 |
| IV | α1(IV) or αD | 0–0.9 | 59.1–59.7 | 88.4–89.4 | 79.5–85.5 | 31–38 | ?/− | ?/− | 6, 7 |
| V | α2(V) or αA | 0.9–4.3 | 44.0–47.4 | 63.6–75.4 | 65 2 | 32 | ?/+ | ?/+ | 1, 9, 10, 11 |
| V | α1(V) or αB | 0.9–4.3 | 44.0–47.4 | 63.6–75.4 | 65 2 | 32 | ?/+ | ?/+ | 1, 9, 10, 11 |
| V | α3(V) or αC | 0.98 | 47.7 | 74.1 | 54 7 | 23.5 | ? | ? | 1, 9, 10 |

[a] The nomenclature for types IV and V collagens is in flux. The reader is referred to Bornstein and Sage (1980). Data represent average values of ranges from several sources.

[b] Ratio of 3-hydroxyproline (3-Hyp) to total imino acids × 100.

[c] Ratio of 4-hydroxyproline (4-Hyp) to total imino acids × 100.

[d] Ratio of hydroxylysine (Hyl) to lysine plus Hyl × 100.

[e] Ratio of glucosylgalactosyl hydroxylysine plus glactosyl hydroxylysine to total hydroxylysine × 100.

[f] Total residues of glucose plus galactose per 1000 amino acid residues.

[g] References are as follows: (1) Burgeson et al. (1976); (2) Crouch and Bornstein (1978); (3) Davidson, J. M., and McEneany, L. S. G., unpublished results; (4) Miller and Lunde (1973); (5) Chung and Miller (1974); (6) Kresina and Miller (1979); (7) Sage et al. (1979); (8) Dehm and Kefalides (1978); (9) Mayne et al. (1978); (10) Sage and Bornstein (1979); (11) Kumamoto and Fessler (1980).

unpublished results). Type III procollagen is cleaved much more slowly (Fessler and Fessler, 1977; Goldberg, 1977; Sodek and Limeback, 1979) and probably mainly at the COOH terminus (J. H. Fessler, personal communication). The processing of type IV procollagen is either of limited extent (Timpl *et al.*, 1978; Tryggvason *et al.*, 1980) or absent under certain *in vitro* culture conditions (Bornstein and Sage, 1980; Sage *et al.*, 1979; Clark *et al.*, 1975; Heathcote *et al.*, 1978; Crouch and Bornstein, 1979). Type V procollagen is reported to be processed in at least two steps to chains somewhat larger than α(V) chains (Kumamoto and Fessler, 1980).

## III. Posttranslational Modifications

The genetic types of collagens appear to share similar posttranslational steps in their biosynthesis (see Table II). Procollagen is synthesized on membrane-bound ribosomes and is segregated across the membrane of the endoplasmic reticulum and into the intracisternal space (Fig. 2). Pro α(I) chains that have been synthesized by cell-free translation have been shown to contain a signal sequence at their amino termini (Palmiter *et al.*, 1979; Graves *et al.*, 1979; Olsen and Berg, 1979), as has been observed for many other secretory proteins (Lingappa *et al.*, 1978). Entry into the cisternal space is apparently accompanied by cleavage of the signal peptide (Palmiter *et al.*, 1979). The cleavage of the signal sequence may occur as more than one step, since the reported size of the procollagen "signal" is unusually large (Palmiter *et al.*, 1979; Graves *et al.*, 1979; Olsen and Berg, 1979) compared with the signal peptides described for other secretory proteins (Lingappa *et al.*, 1978).

Once the procollagen chains are within the intracisternal space, they are subjected to hydroxylation of selected peptidyl prolyl residues in the collagenous domains by the membrane-associated enzymes prolyl-4-hydroxylase and prolyl-3-hydroxylase (for a review, see Kivirikko and Myllylä, 1981). Certain lysyl residues in the collagenous domain are also subjected to hydroxylation by lysyl hydroxylase (Kivirikko and Myllylä, 1981). Both prolyl hydroxylases and lysyl hydroxylase have identical cofactors, $Fe^{2+}$ and ascorbate, and identical cosubstrates, oxygen and α-ketoglutarate. All three enzymes have specific requirements for amino acid sequences in the region of their target prolyl or lysyl residues. The extent of hydroxylation of a given residue is variable, however, and may depend on the rate of triple-helix formation (see below). O-Glycosylation of hydroxylysine occurs by the sequential action of hydroxylysylgalactosyl transferase and galactosylhydroxylysylglucosyl transferase (Kivirikko and Myllylä, 1979; Butler, 1978). The substrate requirements for

TABLE II

ENZYMATIC STEPS AND SEQUENCE OF EVENTS IN PROCOLLAGEN BIOSYNTHESIS

| Steps and events | Enzyme | Function |
|---|---|---|
| Segregation | ?[a,b] | Entry to endoplasmic reticulum |
| Cleavage of presequence (A) | Signal peptidase | Removal of hydrophobic leader sequence |
| Chain association (B) | Self-assembly? | Alignment of chains |
| 3-Hydroxyprolyl residues (C) | Prolyl-3-hydroxylase | ? |
| 4-Hydroxyprolyl residues (D) | Prolyl-4-hydroxylase | Stabilization of triple helix |
| Hydroxylysyl residues (E) | Lysyl hydroxylase | Site for O-glycosylation Stabilization of cross-links[c] |
| Galactosylhydroxylysyl residues (F) | Hydroxylysyl galactosyl transferase | ? |
| Glucosylgalactosylhydroxylysyl residues (G) | Galactosylhydroxylysyl glucosyl transferase | ? |
| Intra- and interchain disulfide bonds (H) | Disulfide isomerase? | Stabilization of propeptides |
| $(Man)_n$ $(GlcNAc)_2$ Asn residues (I) | Lipid-linked oligosaccharide transferase | Secretion? |
| Triple helix (J) | Self-assembly? | Formation of collagen structure |
| Translocation | ?[d] | Transport to Golgi |
| Packaging | ?[d] | Formation of secretory vesicles |
| Exocytosis | ?[d] | Secretion |
| Cleavage of $NH_2$ prosequence (K) | $NH_2$-procollagen peptidase | Fibril morphology |
| Cleavage of COOH prosequence (L) | COOH-procollagen peptidase | Fibril morphology |
| Fibril formation (M) | ?[e] | Matrix morphogenesis |
| Allysine, hydroxyallysine (N) | Lysyl oxidase | Site of crosslink formation |
| Crosslink formation (O) | ?[e] | Stabilization of fibers |

[a] Probably involves recognition of signal sequence by membrane proteins.
[b] ? represents uncertain or unknown enzyme or functions.
[c] Probable.
[d] Energy-dependent requires microtubules and probably microfilaments.
[e] Probably self-assembly.

these enzymes are less well characterized than those for prolyl and lysyl hydroxylase. It is known, however, that all five enzymes require procollagen to be in a nonhelical conformation in order to serve as a substrate. Besides the O-glycosylation of hydroxylysyl residues, at least one other oligosaccharide containing N-acetylglucosamine and mannose is transferred to an asparagine residue contained with the COOH-terminal propeptide extensions of both pro α1 and pro α2 chains of type I procollagen (Olsen et al., 1977; Clark, 1979; Murphy

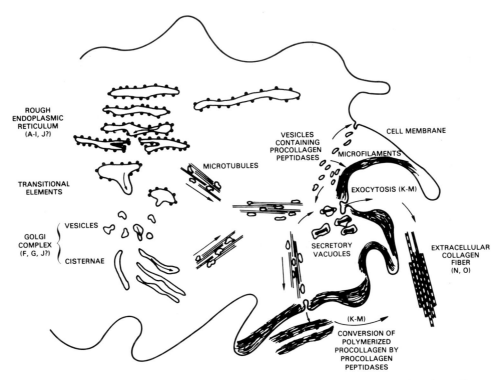

FIG. 2. A schematic diagram of a fibroblast showing the intracellular translocation of procollagen molecules. Letters indicate the probable location of posttranslational modification described in Table II and shown in Fig. 1. As indicated, it is still unknown whether individual molecules of procollagen are secreted, polymerized, and then converted to collagen or whether the conversion to collagen occurs before polymerization. Adapted with permission from Bornstein and Ehrlich (1973).

*et al.*, 1975). Type II procollagen also contains mannose, although the distribution and stoichiometry of the carbohydrate are less well understood. There is one report (Guzman *et al.*, 1978) of mannose in both $NH_2$ and COOH propeptides.

The relationship between chain association and interchain disulfide bond formation is uncertain (Fessler and Fessler, 1978; Prockop *et al.*, 1976). Although the $NH_2$ propeptide was initially envisioned as a registration peptide (Speakman, 1971), the more recently identified carboxy-terminal extension is currently favored as the site of initial interchain recognition (Rosenbloom *et al.*, 1976).

A microsomal enzyme, disulfide isomerase (Anfinsen, 1973; Freedman *et al.*, 1978), has been suggested to catalyze the exchange of disulfide bonds and may promote the covalent linkage between and perhaps within the propeptide chains (Harwood, 1979; Harwood and Freedman, 1978). The hydroxylation of 80–120 prolyl residues is required for the collagenous domains of the procollagen chains to align correctly and undergo the conformational change from random coil to triple helix. Underhydroxylated procollagen does not form a stable triple helix at body temperature (Prockop *et al.*, 1976). This folding is probably facilitated by interchain disulfide bonds (Fessler and Fessler, 1978; Prockop *et al.*, 1979; Harwood, 1979; Bornstein and Traub, 1979; Bornstein and Sage, 1980). However, chain association and disulfide bond formation are processes that can proceed in the absence of hydroxylation (Prockop *et al.*, 1976).

Procollagen molecules are translocated from the cisternae of the rough endoplasmic reticulum to the Golgi apparatus. Subcellular fractionation (Harwood *et al.*, 1975) suggests that O-glycosylation reactions accompany the procollagen molecules into the Golgi apparatus (Kivirikko and Myllylä, 1979; Oohira *et al.*, 1979), particularly in the more glycosylated chains of type II procollagen (Table I, Fig. 2). From the Golgi apparatus, procollagen molecules are translocated to the extracellular space, probably by way of condensing vacuoles and secretory vesicles (Weinstock and Leblond, 1974; Karim *et al.*, 1979). An alternative secretory pathway for a variable portion of newly synthesized procollagen molecules consists in intracellular degradation followed by the secretion of degraded peptides (Bienkowski *et al.*, 1978a,b; Baum *et al.*, 1980).

Either concomitantly with or shortly after exocytosis, some types of procollagen molecules undergo proteolytic cleavage to remove amino-terminal and/or carboxy-terminal propeptide extensions. These reactions are catalyzed by separate enzymes (Duksin *et al.*, 1978; Tuderman *et al.*, 1978; Leung *et al.*, 1979; Davidson *et al.*, 1979; Kessler and Goldberg, 1978) and are independent events (Davidson *et al.*, 1979; Morris *et al.*, 1979). Although processing is probably accomplished by neutral metalloenzymes (Duksin *et al.*, 1978; Tuderman *et al.*, 1978; Leung *et al.*, 1979), a similar conversion can occur in the presence of acid proteases (Davidson *et al.*, 1979).

Once in the extracellular space, procollagen molecules undergo polymerization and fibril formation to an extent probably dictated by both the genetic type of collagen and its association with other matrix components (Fessler and Fessler, 1978; Prockop *et al.*, 1979; Harwood, 1979; Bornstein and Traub, 1979; Bornstein and Sage, 1980). A role for the telopeptides has been implicated in fibril formation of type I collagen *in vitro* (Gelman *et al.*, 1979b), and it is unknown whether the procollagen extensions serve a similar role *in vivo*.

Accompanying the process of fibril formation is the oxidation of certain lysyl and hydroxylysyl residues by lysyl oxidase (Siegel, 1979) to produce aldehydes that serve as reactive groups for covalent cross-linking between collagen molecules (Fessler and Fessler, 1978; Prockop *et al.*, 1979; Harwood, 1979; Bornsteinf and Traub, 1979; Bornstein and Sage, 1980; Tanzer, 1976). The process of fibrillogenesis (Gelman *et al.*, 1979a,b; Veis *et al.*, 1979) is thought to occur in the pericellular space; morphological evidence has demonstrated association of procollagen molecules into aggregates, protofibrils, or fibrillar subassemblies within the cells prior to secretion (Weinstock and Leblond, 1974; Karim *et al.*, 1979; Trelstad and Hayashi, 1979; Bruns *et al.*, 1979).

## IV.   Role of Posttranslational Modifications in the Secretory Process

### A.   Segregation

The presence of a signal peptide on a nascent procollagen chain is probably essential for its transmembrane transport into the endoplasmic reticulum; thus, it initiates the segregation of procollagen molecules into an extracytoplasmic compartment of the cell. Although the unique example of ovalbumin biosynthesis suggests that removal of the signal peptide is not essential for continued post-translational modification or secretion of some proteins (Palmiter *et al.*, 1978), cleavage of the signal sequence of procollagen is catalyzed by signal peptidase preparations (Palmiter *et al.*, 1979; Olsen and Berg, 1979). The "signal" peptide for procollagen I is two to five times as large as that of similar regions in other secretory proteins (Palmiter *et al.*, 1979; Graves *et al.*, 1979; Olsen and Berg, 1979). The significance of this additional material at the amino terminus is not well understood.

### B.   Translocation

Entry of nascent procollagen polypeptide chains into the rough endoplasmic reticulum (Olsen and Prockop, 1974; Nist *et al.*, 1975) is followed by a succession of events whose order is not clearly distinguishable: peptidyl prolyl and lysyl hydroxylation, peptidyl hydroxylysyl and asparaginyl glycosylation, and disulfide bond formation. Hydroxylation and disulfide bonding will both apparently make contributions to the rate of formation and stability of the triple-helical structure of procollagen (Prockop *et al.*, 1976, 1979; Fessler and Fessler, 1978; Bruckner *et al.*, 1978; Bornstein and Traub, 1979). Both events can apparently occur on nascent polypeptides (Rosenbloom *et al.*, 1976; Veis and Brownell, 1977), but hydroxylation ceases once the molecule assumes a triple-helical conformation, as does the O-glycosylation of selected hydroxylysyl residues (Prockop *et al.*, 1976, 1979; Fessler and Fessler, 1978; Harwood, 1979; Bornstein and Traub, 1979; Bornstein and Sage, 1980; Kivirikko and Myllylä, 1979, 1981).

Disulfide bonds appear to be important in the assembly of the procollagen molecule and are possibly required for the stabilization of chain association subsequent to the formation of the triple-helical conformation in the collagenous domain (Rosenbloom *et al.*, 1976; Christner *et al.*, 1975; Uitto and Prockop,

1973). For example, the renaturation of thermally denatured procollagen is substantially accelerated by the presence of interchain disulfide bonds (Bruckner *et al.*, 1978; Fessler *et al.*, 1974; Bornstein, 1974). Types I and II procollagens contain interchain disulfide bonds between their COOH-terminal propeptides, while type III procollagen contains disulfide bonds among the COOH-terminal propeptides, the $NH_2$-terminal propeptides, and also within its collagenous domains (J. H. Fessler and Fessler, 1978; Bentz *et al.*, 1978; L. I. Fessler and Fessler, 1979). These additional covalent linkages at the $NH_2$ terminus may contribute to the apparent lack of $NH_2$-propeptide cleavage in procollagen III (J. H. Fessler, personal communication).

There is a direct correlation between the rate of secretion of procollagen from cells and the level of hydroxylation of prolyl residues (Blanck and Peterkofsky, 1975). The hydroxylation of approximately 100 residues of proline in the collagenous domains of type I procollagen is required for it to be triple-helical at body temperature (Prockop *et al.*, 1976). The absence of hydroxyprolyl residues in procollagen, which can be caused by inhibiting prolyl hydroxylase activity in cells and tissues, results in the inability of the procollagen molecule to form a triple helix under physiological conditions, and may explain the relationship between hydroxylation and secretion (Jimenez and Yankowski, 1978). Procollagen that is nonhelical is secreted with a rate constant that is only one tenth of the rate constant for the secretion of helical procollagen (Kao *et al.*, 1977, 1979).

Concomitant with the hydroxylation of prolyl residues is the hydroxylation of lysyl residues and the enzymatic transfer of carbohydrates to selected hydroxylysyl acceptors in the molecule. The extent of this modification is variable among procollagen types (Table I). The transfer of galactose and glucose to hydroxylysyl residues in procollagen, as well as the synthesis of hydroxylysine per se in procollagen, may not be required for procollagen to be secreted from cells, since in inherited human lysyl hydroxylase deficiency (Ehlers–Danlos type VI) hydroxylysine-deficient procollagen I is secreted from fibroblasts (Quinn and Crane, 1979).

Although each of the preceding reactions can occur only on unfolded procollagen chains, subcellular fractionation studies suggest that the enzymes associated with these events are distributed in several subcellular compartments, including the smooth endoplasmic reticulum and the Golgi apparatus (Fessler and Fessler, 1978; Prockop *et al.*, 1979; Harwood, 1979; Bornstein and Traub, 1979; Bornstein and Sage, 1980; Kivirikko and Myllylä, 1979; Harwood *et al.*, 1975; Oohira *et al.*, 1979). These data thus imply that triple-helix formation, particularly in non-type I procollagens, is perhaps retarded, while these more extensively modified procollagens are in the process of intracellular transit from the site of synthesis to the site of secretion (Fessler and Fessler, 1978; Prockop *et al.*, 1979; Bornstein and Sage, 1980). The final extent of posttranslational reactions is probably dictated by (1) the levels of posttranslational enzymes relative to the activity of prolyl hydroxylase, (2) the subcellular distribution of posttranslational enzymes, (3) the rate of synthesis of the polypeptide chains, and (4) the rate of chain association (Prockop *et al.*, 1976). If helix formation is prevented, type I procollagen is synthesized with increased amounts of hydroxylysine and glycosylated hydroxylysine (Oikarinen *et al.*, 1976; Uitto *et al.*, 1978). Such a mechanism may explain why the type I trimer, which is secreted by amniotic fluid cells, appears to be processed to a greater extent as evidenced by a higher 3- and 4-hydroxyproline and hydroxylysine content relative to type I collagen (Crouch and Bornstein, 1978).

Glycosylation of asparagine residues in the COOH extension propeptides of type I procollagen is accomplished via a dolichol phosphate intermediate, which introduces a high-mannose type of oligosaccharide onto each chain. The oligosaccharide can be cleaved with endoglucosaminidase H (Clark, 1979), and its synthesis is sensitive to tunicamycin (Duksin and Bornstein, 1977; Tanzer *et al.*, 1977; Housley *et al.*, 1980). The significance of the presence of a high-mannose oligosaccharide is presently unclear, since conflicting data exist on the secretion and stability of procollagen synthesized in the presence of the antibiotic tunicamycin, which prevents the transfer of such a saccharide to procollagen and other glycoproteins. Studies have shown that this drug (or combination of drugs) can indeed prevent glycosylation of procollagen but may (Housley *et al.*, 1980) or may not (Duksin and Bornstein, 1977) inhibit the synthesis and secretion of procollagen. Further studies have shown that cleavage of the propeptide extensions is unimpaired in underglycosylated procollagen and suggest that glycosylation of procollagen proteases may be more affected (Duksin *et al.*, 1978). Other studies have suggested that the major effect of the oligosaccharides is to protect secreted glycoproteins from haphazard degradation (Olden *et al.*, 1978). Undoubtedly, more highly purified preparations of this inhibitor may help to resolve the discrepancies of these findings.

The subsequent processes in procollagen secretion involve transport into the Golgi, packaging, and translocation of secretory vesicles (Fig. 2). The initial process has been shown to be perturbed by the $Na^+$ ionophore, monensin (Uchida *et al.*, 1979). The latter processes, which probably require the integrity of microtubules and microfilaments, can be arrested by local anesthetics (Eichhorn and Peterkofsky, 1979), colchicine, vinblastine (Dehm and Prockop, 1972; Ehrlich and Bornstein, 1972), cytochalasin B (Ehrlich and Bornstein, 1972), and uncouplers of oxidative phosphorylation (Kruse and Bornstein, 1975). Thus, procollagen is translocated by an ion- and energy-dependent mechanism common to the export of other secretory proteins (Jamison and Palade, 1977).

## C.   Secretion

A simplistic view assumes that procollagen molecules are secreted as individual entities into the extracellular matrix prior to further processing and polymerization. However, there is morphological evidence from several laboratories that the partial assembly or polymerization of procollagen molecules may precede secretion (Weinstock and Leblond, 1974; Karim et al., 1979; Trelstad and Hayashi, 1979; Bruns et al., 1979). Intracellular assemblies of procollagen have been identified in several tissues by electron microscopy and by immunocytochemistry.

The role of intracellular degradation in the secretion of procollagen is currently under exploration. It is known that a fraction of newly synthesized procollagen is degraded in cells and secreted as small peptides (Bienkowski et al., 1978a,b; Baum et al., 1980). Under conditions that prevent procollagen from becoming triple-helical, intracellular degradation of newly synthesized collagen increases severalfold (Berg et al., 1980). Intracellular degradation may be part of a quality control mechanism in which cells monitor the conformation of procollagen such that triple-helical procollagen is secreted intact by cells and nonhelical ''defective'' collagen is degraded by lysosomal proteases (Berg et al., 1980). Further evidence for such a concept has recently been obtained by using antibodies to procollagen to stain lysosomes in odontoblasts (Karim et al., 1979).

The relationship between the secretion of types I and II procollagen and the cleavage of the amino-terminal propeptides indicates that the procollagen molecules are undergoing proteolytic cleavage either concomitant with or immediately after secretion from cells in an intact matrix (Morris et al., 1975; Davidson et al., 1975; Uitto and Lichtenstein, 1976). However, recent findings have shown that acid proteases are also capable of processing mature procollagen (Davidson et al., 1979) and may suggest an alternative site of procollagen processing. In embryonic chick calvaria synthesizing type I procollagen in vitro, the removal of the $NH_2$-terminal propeptide probably occurs at the time of secretion of procollagen into the matrix (Morris et al., 1975), thereby releasing amino-terminal propeptides into the culture medium (Fessler and Fessler, 1978; Prockop et al., 1979; Bornstein and Traub, 1979). An $NH_2$-terminal protease has been recovered from extracellular fluids as well as tissues (Duksin et al., 1978; Tuderman et al., 1978; Leung et al., 1979), and thus it may continue to process procollagen I both during and after secretion from tendon fibroblasts.

In the chick frontal bone the COOH-terminal propeptides of type I procollagen are probably removed stepwise in a pericellular location subsequent to cleavage of the $NH_2$-terminal propeptide (Davidson et al., 1979; Morris et al., 1979; Goldberg and Sherr, 1973; Goldberg et al., 1975; Davidson et al., 1977). In contrast, recent evidence indicates that in the culture medium of smooth muscle cells the COOH-terminal propeptide is removed prior to the $NH_2$-terminal propeptide (Burke et al., 1977). The cleavage of either extension in type I procollagen does not depend on the presence of the other (Tuderman et al., 1978; Davidson et al., 1979; Morris et al., 1979). Type II procollagen may be cleaved at random (Uitto, 1977). Thus, the sequence of cleavage may be dictated by the proportion and location of the proteases. Moreover, the rate of cleavage in cell culture is a function of both cell type (Taubman and Goldberg, 1976) and cell density (J. M. Davidson, unpublished observations). Although both peptides are rapidly removed from type I procollagen in organ culture, the removal of the COOH-terminal propeptides from type III procollagen can occur up to 24 hours after secretion of the procollagen from the cells of the chick aorta (Fessler and Fessler, 1978), and processing of type III procollagen in cell culture is likewise retarded (Goldberg, 1977; Burke *et al.*, 1977). Processing of type V procollagen is also a relatively slow reaction (Kumamoto and Fessler, 1980).

Current evidence would suggest that amino-terminal cleavage may not occur in type III procollagen (Fessler and Fessler, 1978; Crouch and Bornstein, 1979), and type IV procollagen is either unprocessed (Heathcote *et al.*, 1978; Minor *et al.*, 1976) or cleaved to a limited extent during secretion (Tryggvason *et al.*, 1980). Even for type I procollagen the removal of the propeptide extensions is not required for procollagen to polymerize (Leung *et al.*, 1979). The morphology of the resulting fibers is markedly altered, however, if either the $NH_2$- or COOH-terminal propeptides of type I procollagen are not removed (Goldberg *et al.*, 1975; Lenaers *et al.*, 1972).

Postsecretion extracellular events, which may not be directly related to the secretion of the procollagen molecule, involve the oxidation of lysyl residues to aldehydes, polymerization of collagen fibers, and cross-linking of the collagen chains and molecules. Although the cross-linking of collagen molecules is thought to occur after secretion, the intracellular procollagen polymers or aggregates may present an opportunity for the oxidation of lysyl residues and the cross-linking of collagen to occur concomitant with secretion (Fessler and Fessler, 1978; Leung *et al.*, 1979; Siegel, 1979; Fessler *et al.*, 1977).

# V.   Disorders of Collagen Secretion

Limited information exists on the naturally occurring disorders of procollagen processing that affect secretion (Prockop *et al.*, 1979; Bornstein and Byers, 1980). Scurvy, or ascorbic acid deficiency, results in the synthesis of underhydroxylated procollagen, which is secreted at a lower rate (Kao *et al.*, 1979) and is more readily degraded intracellularly (Berg *et al.*, 1981). Heine and Schaeg (1977) have described a patient in whom paracrystalline structures identified as procollagen accumulate in the endoplasmic reticulum of skin fibroblasts. Byers *et al.* (1979a,b) have shown an apparent intracellular, membrane-bound accumulation of type III procollagen and reduced amounts of extracellular type III collagen in patients with a form of Ehlers–Danlos type IV syndrome. Since type I procollagen is relatively unaffected, it is possible that a primary structural defect in type III procollagen renders it incapable of normal intracellular translocation.

Two defects in $NH_2$-terminal cleavage have been described: in dermatosparaxis, a genetic disease in sheep and cattle, $NH_2$-terminal protease activity is reduced and the animals develop fragile skin and tendons (Fessler and Fessler, 1978; Prockop *et al.*, 1979; Lenaers *et al.*, 1972). In a similar human disorder, Ehlers–Danlos type VII syndrome (Lichtenstein *et al.*, 1973), the $NH_2$-terminal processing of pro $\alpha2$ chains is impaired, probably as the result of a structural defect in the substrate or the enzyme (Lichtenstein *et al.*, 1973; Steinmann *et al.*, 1980). It has been suggested that such defects may reduce the effectiveness of a feedback mechanism that involved the $NH_2$-terminal propeptide as a regulator (Krieg *et al.*, 1978; Wiestner *et al.*, 1979; Paglia *et al.*, 1979).

Our current approach to describing the roles of posttranslational modification *vis-à-vis* procollagen secretion relies largely on cellular poisons and relatively rare genetic defects. The more precise analysis of new, genetically defined disorders of procollagen secretion that can be maintained or chemically induced in cultured fibroblasts may provide the best approach for further understanding the importance of posttranslational modifications in the secretory process.

### ACKNOWLEDGMENTS

We thank Drs. Steven Rennard, Paul Bornstein, and Ronald G. Crystal for their helpful comments.

### REFERENCES

Anfinsen, C. B. (1973). *Science* **181,** 223–230.
Baum, B. J., Moss, J., Breul, S., Berg, R. A., and Crystal, R. G. (1980). *J. Biol. Chem.* **255,** 2843–2847.
Bentz, H., Bächinger, H.-P., Glanville, R., and Kühn, K. (1978). *Eur. J. Biochem.* **92,** 563–567.
Benya, P. D., Padilla, S. R., and Nimni, M. E. (1977). *Biochemistry* **16,** 865–872.
Berg, R. A., Schwartz, M. L., and Crystal, R. G. (1980). *Proc. Natl. Acad. Sci. U.S.A.* **77,** 4746–4750.
Bienkowski, R. S., Cowan, M. J., McDonald, J. A., and Crystal, R. G. (1978a). *J. Biol. Chem.* **253,** 4356–4363.
Bienkowski, R. S., Baum, B. J., and Crystal, R. G. (1978b). *Nature (London)* **276,** 413–416.
Blanck, T. J. J., and Peterkofsky, B. (1975). *Arch. Biochem. Biophys.* **171,** 259–267.
Bornstein, P. (1974). *J. Supramol. Struct.* **2,** 108–120.
Bornstein, P., and Byers, P. (1980). *In* "Duncan's Diseases of Metabolism" (P. K. Bondy and L. E. Rosenberg, eds.), 8th ed., pp. 1089–1153. Saunders, Philadelphia, Pennsylvania.
Bornstein, P., and Ehrlich, H. P. (1973). *In* "Biology of Fibroblast" (E. Kulonen and J. Pikkarainen, eds.), pp. 321–338. Academic Press, New York.
Bornstein, P., and Sage, H. (1980). *Annu. Rev. Biochem.* **49,** 957–1003.
Bornstein, P., and Traub, W. (1979). *In* "The Proteins" (H. Neurath and R. L. Hill, eds.), 3rd ed., Vol. 4, pp. 411–632. Academic Press, New York.
Bruckner, P., Bächinger, H.-P., Timpl, R., and Engel, J. (1978). *Eur. J. Biochem.* **90,** 595–603.
Bruns, R. R., Hulmes, D. J. S., Therrien, S. F., and Gross, J. (1979). *Proc. Natl. Acad. Sci. U.S.A.* **76,** 313–317.
Burgeson, R. E., El Adli, F. A., Kaitila, I. I., and Hollister, D. W. (1976). *Proc. Natl. Acad. Sci. U.S.A.* **73,** 2579–2583.

Burke, J. M., Balian, G., Ross, R., and Bornstein, P. (1977). *Biochemistry* **16**, 3243–3249.
Butler, W. T. (1978). *In* "Glycoconjugates" (M. I. Horowitz and W. Pigman, eds.), Vol. 2, pp. 79–85. Academic Press, New York.
Byers, P. H., Holbrook, K. A., Smith, L. T., Bornstein, P., McGillivray, B., and MacLeod, P. M. (1979a). *Clin. Res.* **29**, 512A.
Byers, P. H., Holbrook, K. A., McGillivray, B., MacLeod, P. M., and Lowry, R. B. (1979b). *Hum. Genet.* **47**, 141–150.
Christner, P., Carpousis, A., Harsch, M., and Rosenbloom, J. (1975). *J. Biol. Chem.* **19**, 7623–7630.
Chung, E. H. and Miller, E. J. (1974). *Science* **183**, 1200–1201.
Clark, C. C. (1979). *J. Biol. Chem.* **254**, 10798–10802.
Clark, C. C., Tomichek, E. A., Koszalka, T. R., Minor, R. R., and Kefalides, N. A. (1975). *J. Biol. Chem.* **250**, 5259–5267.
Crouch, E., and Bornstein, P. (1978). *Biochemistry* **17**, 5499–5509.
Crouch, E., and Bornstein, P. (1979). *J. Biol. Chem.* **254**, 4197–4204.
Davidson, J. M., McEneany, L. S. G., and Bornstein, P. (1975). *Biochemistry* **14**, 5188–5194.
Davidson, J. M., McEneany, L. S. G., and Bornstein, P. (1977). *Eur. J. Biochem.* **81**, 349–355.
Davidson, J. M., McEneany, L. S. G., and Bornstein, P. (1979). *Eur. J. Biochem.* **100**, 551–558.
Dehm, P., and Kefalides, N. A. (1978). *J. Biol. Chem.* **253**, 6680–6686.
Dehm, P., and Prockop, D. J. (1972). *Biochim. Biophys. Acta* **264**, 375–382.
Duksin, D., and Bornstein, P. (1977). *J. Biol. Chem.* **252**, 955–962.
Duksin, D., Davidson, J. M., and Bornstein, P. (1978). *Arch. Biochem. Biophys.* **185**, 326–332.
Ehrlich, H. P., and Bornstein, P. (1972). *Nature (London), New Biol.* **238**, 257–260.
Eichhorn, J. H., and Peterkofsky, B. (1979). *J. Cell Biol.* **81**, 26–42.
Epstein, E. H., Jr. (1974). *J. Biol. Chem.* **249**, 3225–3231.
Fessler, J. H., and Fessler, L. I. (1978). *Annu. Rev. Biochem.* **47**, 129–162.
Fessler, J. H., Doege, K. J., Siegel, R. C., and Fessler, L. I. (1977). *Fed. Proc., Fed. Am. Soc. Exp. Biol.* **36**, 680.
Fessler, L. I., and Fessler, J. H. (1979). *J. Biol. Chem.* **254**, 233–239.
Fessler, L. I., Rudd, C., and Fessler, J. H. (1974). *J. Supramol. Struct.* **2**, 103–107.
Freedman, R. B., Newell, A., and Walklin, C. M. (1978). *FEBS Lett.* **88**, 49–52.
Gay, S., and Miller, E. J. (1978). "Collagen in the Physiology and Pathology of Connective Tissue." Fischer, Stuttgart.
Gelman, R. A., Williams, B. R., and Piez, K. A. (1979a). *J. Biol. Chem.* **254**, 180–186.
Gelman, R. A., Poppke, D. C., and Piez, K. A. (1979b). *J. Biol. Chem.* **254**, 11741–11745.
Glanville, R. W., Rauter, A., and Fietzek, P. (1979). *Eur. J. Biochem.* **95**, 383–389.
Goldberg, B. (1977). *Proc. Natl. Acad. Sci. U.S.A.* **74**, 3322–3325.
Goldberg, B., and Sherr, C. J. (1973). *Proc. Natl. Acad. Sci. U.S.A.* **70**, 361–365.
Goldberg, B., Taubman, M. B., and Radin, A. (1975). *Cell* **4**, 45–50.
Graves, P. N., Olsen, B. R., Fietzek, P. O., Monson, J. M., and Prockop, D. J. (1979). *Fed. Proc., Fed. Am. Soc. Exp. Biol.* **38**, 620.
Guzman, N. A., Graves, P. N., and Prockop, D. J. (1978). *Biochem. Biophys. Res. Commun.* **84**, 691–698.
Harwood, R. (1979). *Int. Rev. Connect. Tissue Res.* **8**, 159–226.
Harwood, R., and Freedman, R. B. (1978). *FEBS Lett.* **88**, 46–48.
Harwood, R., Grant, M. E., and Jackson, D. S. (1975). *Biochem. J.* **152**. 291–302.
Heathcote, J. G., Sear, C. H. J., and Grout, M. E. (1978). *Biochem. J.* **176**, 283–294.
Heine, H., and Schaeg, G. (1977). *Virchows Arch. A: Pathol. Anat. Histol.* **376**, 89–94.
Hörlein, D., Fietzek, P. P., Wachter, E., Lapière, C. M., and Kühn, K. (1979). *Eur. J. Biochem.* **99**, 31–38.
Housley, T. J., Rowland, F. N., Ledger, P. W., Kaplan, J., and Tanzer, M. L. (1980). *J. Biol. Chem.* **255**, 121–128.

Jamison, J. D., and Palade, G. E. (1977). In "International Cell Biology 1976-1977" (B. R. Brink ley and K. R. Porter, eds.), pp. 308-317. Rockefeller Univ. Press, New York.

Jimenez, S. A., and Yankowski, R. (1978). J. Biol. Chem. 253, 1420-1426.

Jimenez, S. A., Bashey, R. I., Beneditt, M., and Yankowski, R. (1977). Biochem. Biophys. Res. Commun. 78, 1354-1361.

Kao, W. W.-Y, Berg, R. A., and Prockop, D. J. (1977). J. Biol. Chem. 252, 8391-8397.

Kao, W. W.-Y, Prockop, D. J., and Berg, R. A. (1979). J. Biol. Chem. 254, 2234-2243.

Karim, A., Cournil, I., and Leblond, C. P. (1979). J. Histochem. Cytochem. 27, 1070-1083.

Kessler, E., and Goldberg, B. (1978). Anal. Biochem. 86, 463-469.

Kivirikko, K. I., and Myllylä, R. (1979). Int. Rev. Connect. Tissue Res. 8, 23-72.

Kivirikko, K. I., and Myllylä, R. (1981). In "The Enzymology of Post-translational Modification of Proteins" (R. B. Freedman and H. C. Hawkins, eds.). Academic Press, New York (in press).

Kresina, T., and Miller, E. J. (1979). Biochemistry 18, 3089-3097.

Krieg, T., Hörlein, D., Wiestner, M., and Müller, P. (1978). Arch. Dermatol. Res. 263, 171-180.

Kruse, N. J., and Bornstein, P. (1975). J. Biol. Chem. 250, 4841-4847.

Kumamoto, C. A., and Fessler, J. H. (1980). Proc. Natl. Acad. Sci. U.S.A. 77, 6434-6438.

Lenaers, A., Ansay, M., Nusgens, B. V., and Lapière, C. M. (1972). Eur. J. Biochem. 23, 533-543.

Leung, M. K. K., Fessler, L. I., Greenberg, D. B., and Fessler, J. H. (1979). J. Biol. Chem. 254, 224-232.

Lichtenstein, J. R., Martin, G. R., Kohn, L. D., Byers, P. H., and McKusick, V. A. (1973). Science 182, 298-299.

Lingappa, V. R., Katz, F. N., Lodish, H. F., and Blobel, G. (1978). J. Biol. Chem. 253, 8667-8670.

Little, C. D., and Church, R. L. (1978). Arch. Biochem. Biophys. 190, 632-639.

Mayne, R., Vail, M., Mayne, P., and Miller, E. J. (1975). Proc. Natl. Acad. Sci. U.S.A. 72, 4511-4515.

Mayne, R., Vail, M. S., Miller, E. J., Blose, S. H., and Chacko, S. (1977). Arch. Biochem. Biophys. 181, 462-469.

Mayne, R., Vail, M. S., and Miller, E. J. (1978). Biochemistry 17, 446-452.

Miller, E. J., and Lunde, L. G. (1973). Biochemistry 12, 3153-3159.

Minor, R. R., Clark, C. C., Strause, E. L., Koszalka, T. R., Brent, R. L., and Kefalides, N. A. (1976). J. Biol. Chem. 251, 1789-1794.

Moro, L., and Smith, B. D. (1977). Arch. Biochem. Biophys. 182, 33-41.

Morris, N. P., Fessler, L. I., Weinstock, A., and Fessler, J. H. (1975). J. Biol. Chem. 251, 7137-7143.

Morris, N. P., Fessler, L. I., and Fessler, J. H. (1979). J. Biol. Chem. 254, 11024-11032.

Munksgaard, E. C., Rhodes, M., Mayne, R., and Butler, W. T. (1978). Eur. J. Biochem. 82, 609-617.

Murphy, W. H., von der Mark, K., McEneany, L. S. G., and Bornstein, P. (1975). Biochemistry 14, 3243-3250.

Naranayan, A. S., and Page, R. C. (1976). J. Biol. Chem. 251, 5464-5471.

Nist, C., von der Mark, K., Hay, E. D., Olsen, B. R., Bornstein, P., Ross, R., and Dehm, P. (1975). J. Cell Biol. 65, 75-87.

Oikarinen, A., Anttinen, H., and Kivirikko, K. I. (1976). Biochem. J. 160, 639-645.

Olden, K., Pratt, R. M., and Yamada, K. M. (1978). Cell 13, 461-473.

Olsen, B. R., and Berg, R. A. (1979). Symp. Soc. Exp. Biol. 33, 57-58.

Olsen, B. R., and Prockop, D. J. (1974). Proc. Natl. Acad. Sci. U.S.A. 71, 2033-2037.

Olsen, B. R., Guzman, N. A., Engel, J., Condit, C., and Aase, S. (1977). Biochemistry 16, 3030-3037.

Oohira, A., Nogami, H., Ktsukiko, K., Kimata, K., and Suzuki, S. (1979). J. Biol. Chem. 254, 3576-3583.

Paglia, L., Wilczek, J., Diaz de Leon, L., Hörlein, D., Martin, G. R., and Müller, P. (1979). *Biochemistry* **18**, 5030-5034.

Palmiter, R. D., Gagnon, J., and Walsh, K. A. (1978). *Proc. Natl. Acad. Sci. U.S.A.* **75**, 94-98.

Palmiter, R. D., Davidson, J. M., Gagnon, J., Rowe, D. W., and Bornstein, P. (1979). *J. Biol. Chem.* **254**, 1433-1436.

Pesciotta, D. M., Silkowitz, M. H., Fietzek, P. P., Graves, P. N., Berg, R. A., and Olsen, B. R. (1980). *Biochemistry* **19**, 2447-2454.

Prockop, D. J., Berg, R. A., Kivirikko, K. I., and Uitto, J. (1976). *In* "Biochemistry of Collegen" (G. N. Ramachandran and A. H. Reddi, eds.), pp. 163-273. Plenum, New York.

Prockop, D. J., Kivirikko, K. I., Tuderman, L., and Guzman, N. A. (1979). *N. Engl. J. Med.* **301**, 13-23, 75-85.

Quinn, R. S., and Krane, S. M. (1979). *Biochim. Biophys. Acta* **585**, 589-598.

Raj, C. V. S., Freeman, I. L., Church, R. L., and Brown, S. I. (1979). *Invest. Ophthalmol. Visual Sci.* **18**, 75-84.

Rhodes, R. K., and Miller, E. J. (1978). *Biochemistry* **17**, 3442-3448.

Rohde, H., Wachter, E., Richter, W. J., Bruckner, P., Helle, O., and Timpl, R. (1979). *Biochem. J.* **179**, 631-642.

Rosenbloom, J., Endo, R., and Harsch, M. (1976). *J. Biol. Chem.* **251**, 2070-2076.

Sage, H., and Bornstein, P. (1979). *Biochemistry* **18**, 3815-3822.

Sage, H., Woodbury, R. G., and Bornstein, P. (1979). *J. Biol. Chem.* **254**, 9893-9900.

Siegel, R. C. (1979). *Int. Rev. Connect. Tissue Res.* **8**, 73-118.

Sodek, J., and Limeback, H. F. (1979). *J. Biol. Chem.* **254**, 10496-10502.

Speakman, P. T. (1971). *Nature (London)* **229**, 241-243.

Steinmann, B., Tuderman, L., Peltonen, L., Martin, G. R., McKusick, V. A., and Prockop, D. J. (1980). *J. Biol. Chem.* **255**, 8887-8893.

Tanzer, M. L. (1976). *In* "Biochemistry of Collagen" (G. N. Ramachandran and A. H. Reddi, eds.), pp. 137-157. Plenum, New York.

Tanzer, M. L., Rowland, F. N., Murray, L. W., and Kaplan, J. (1977). *Biochim. Biophys. Acta* **500**, 187-196.

Taubman, M. B., and Goldberg, B. (1976). *Arch. Biochem. Biophys.* **173**, 490-494.

Timpl, R., Martin, G. R., Bruckner, P., Wick, G., and Weidemann, H. (1978). *Eur. J. Biochem.* **84**, 43-52.

Timple, R., Bruckner, P., and Fietzek, P. P. (1979). *Eur. J. Biochem.* **95**, 255-263.

Trelstad, R. L., and Hayashi, K. (1979). *Dev. Biol.* **71**, 228-242.

Tryggvason, K., Gehron-Robey, P., and Martin, G. R. (1980). *Biochemistry* **19**, 1284-1289.

Tuderman, L., Kivirikko, K. I., and Prockop, D. J. (1978). *Biochemistry* **17**, 2948-2954.

Uchida, N., Smilowitz, H., and Tanzer, M. L. (1979). *Proc. Natl. Acad. Sci. U.S.A.* **76**, 1868-1872.

Uitto, J. (1977). *Biochemistry* **16**, 3421-3429.

Uitto, J. (1979). *Arch. Biochem. Biophys.* **192**, 371-379.

Uitto, J., and Lichtenstein, J. R. (1976). *Biochem. Biophys. Res. Commun.* **71**, 60-67.

Uitto, J., and Prockop, D. J. (1973). *Biochem. Biophys. Res. Commun.* **55**, 904-911.

Uitto, V.-J., Uitto, J., Kao, W. W.-Y., and Prockop, D. J. (1978). *Arch. Biochem. Biophys.* **185**, 214-221.

Veis, A., and Brownell, A. G. (1977). *Proc. Natl. Acad. Sci. U.S.A.* **74**, 902-905.

Veis, A., Miller, A., Liebovich, S. J., and Traub, W. (1979). *Biochim. Biophys. Acta* **576**, 88-98.

von der Mark, H., and von der Mark, K. (1979). *FEBS Lett.* **99**, 101-105.

Weinstock, M., and Leblond, C. P. (1974). *J. Cell Biol.* **60**, 92-127.

Wiestner, M., Krieg, T., Hörlein, D., Glanville, R. W., Fietzek, P. P., and Müller, P. K. (1979). *J. Biol. Chem.* **254**, 7016-7023.

Wohllebe, M., and Carmichael, D. J. (1978). *Eur. J. Biochem.* **92**, 183-188.

# Part II.   Transport and Packaging in the Golgi Region

## Chapter 10

# The Golgi Apparatus: Protein Transport and Packaging in Secretory Cells

## ARTHUR R. HAND AND CONSTANCE OLIVER

*Laboratory of Biological Structure,*
*National Institute of Dental Research,*
*National Institutes of Health,*
*Bethesda, Maryland*

## I.   Introduction

For many years the Golgi apparatus has been recognized as playing an important role in the secretory process. Light microscopic and early electron microscopic studies (reviewed by Beams and Kessel, 1968) supplied evidence that secretory granules originate in the Golgi region. Radioautographic studies showed that the Golgi apparatus receives newly synthesized protein from the rough endoplasmic reticulum (Van Heyningen, 1964; Caro and Palade, 1964; Jamieson and Palade, 1967b) and that it is an important site of carbohydrate (Neutra and Leblond, 1966; Droz, 1966; Whur et al., 1969; Zagury et al., 1970; Haddad et al., 1971) and sulfate (Lane et al., 1964; Berg and Young, 1971) addition to the secretory product. Analyses of Golgi-enriched subcellular fractions of various secretory cells (Schachter *et al.*, 1970; Fleischer *et al.*, 1969; Morré *et al.*, 1969; Ronzio, 1973; Brandan and Fleischer, 1980) have substantiated the presence of the enzymatic machinery required to carry out a number of the processes attributed through morphological observations to the Golgi apparatus.

The purpose of this chapter is to examine in detail the structural basis of secretory protein transport and packaging by the Golgi apparatus. Recent evidence summarized here points to distinct structural, chemical, and functional specialization within the Golgi region, which may be altered or modulated by changes in cellular activity. Insofar as they bear on the processes of protein transport and packaging, aspects of membrane dynamics and the segregation and sorting of various synthetic products are also discussed. More extensive treatments of these latter subjects can be found in other sections of this book (see chapters by E. Holtzman, M. G. Farquhar, and W. S. Sly *et al.*).

## II.  Transport from the Endoplasmic Reticulum to the Golgi Apparatus

Newly synthesized secretory proteins, following their segregation within the cisternal space of the endoplasmic reticulum (ER), migrate toward the Golgi region of the cell. Numerous morphological studies have suggested that the transfer of secretory proteins from the ER to the Golgi apparatus occurs via small vesicles that bud from ribosome-free portions of the ER, termed transitional ER, adjacent to the cis face of the Golgi apparatus (Zeigel and Dalton, 1962; Caro and Palade, 1964; Friend, 1965; Jamieson and Palade, 1967a) (Fig. 1). In many cells these transfer vesicles may be coated, presumably with clathrin (Jamieson and

---

FIG. 1.  Golgi region of a serous demilune cell of the rat sublingual salivary gland. The Golgi apparatus consists of several stacks of five to seven saccules (S), numerous vesicles, and immature secretory granules (IG) of variable size and density. The rough endoplasmic reticulum (ER) approaches the cis face (c) of the stack of saccules, giving rise to the peripheral transfer vesicles (arrowheads). Cisternae of rough ER also are present near the trans side of the saccules. The cis saccules usually have a wider lumen and a less dense content than the trans saccules. The immature granules are located at the trans face (t) and frequently have membranous blebs (arrows) suggestive of vesicle fusion or formation. G, mature secretory granule. Magnification 30,000×.

This and the subsequent electron micrographs, except for Figs. 2 and 5, are of tissue fixed by vascular perfusion of a glutaraldehyde–formaldehyde fixative (Karnovsky, 1965), postfixed in osmium tetroxide, stained in block with uranyl acetate, and embedded in Spurr (1969) epoxy resin. Thin sections were stained with lead citrate or doubly stained with uranyl acetate and lead citrate. Figures 3, 4, and 6–11 are from tissues sectioned at 75 $\mu$m on a Smith–Farquhar TC-2 tissue sectioner and incubated for cytochemical demonstration of enzyme activity prior to OsO$_4$ postfixation. Scale bar on each micrograph equals 0.25 $\mu$m.

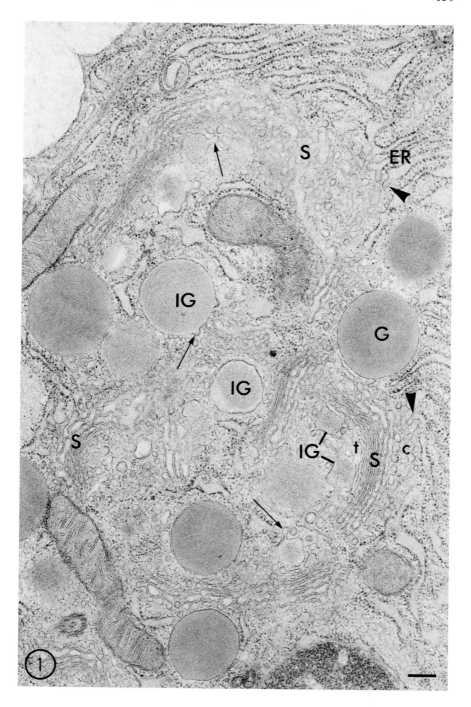

Palade, 1967a; Weinstock and Leblond, 1974; Winkler, 1977; Palade and Fletcher, 1977; Rothman *et al.*, 1980). Secretory proteins have been localized in transfer vesicles by cytochemical techniques (Herzog and Miller, 1970, 1972; Hand and Oliver, 1977a,b). "Smooth microsomes," believed to consist primarily of transfer vesicles, have been isolated from pancreatic tissue and contain newly synthesized protein in pulse–chase labeling experiments (Jamieson and Palade, 1967a). Additional evidence for vesicular transport has been obtained from experiments employing metabolic and other cellular poisons. Inhibition of cellular respiration and oxidative phosphorylation block the transfer of newly synthesized proteins from the rough ER to the Golgi apparatus (Jamieson and Palade, 1968; Chu *et al.*, 1977), and in pulse–chase radioautographic studies labeled proteins remain in the rough ER but do not accumulate in condensing vacuoles or secretory granules (Jamieson and Palade, 1968). Disruption of microtubular function by colchicine or vinblastine results in the formation of intracisternal granules in the rough ER (Kelly *et al.*, 1978), accumulation of vesicles at the cis face of the Golgi apparatus (Patzelt *et al.*, 1977; Malaisse-Lagae *et al.*, 1979; see also Chapter 16 by J. A. Williams), and a decrease in the conversion of proparathyroid hormone (Chu *et al.*, 1977) and proinsulin (Malaisse-Lagae *et al.*, 1979) to parathyroid hormone and insulin, respectively, a process believed to be initiated in the Golgi apparatus. These findings suggest that energy-requiring processes, such as membrane fission and fusion related to vesicle formation, and directed movement, possibly mediated by microtubules, are necessary for transport of secretory proteins from the rough ER to the Golgi apparatus.

In contrast, studies of hepatocytes (Claude, 1970; Ovtracht *et al.*, 1973) and of isolated hepatocyte Golgi apparatus (Morré *et al.*, 1971a; Ovtracht *et al.*, 1973) suggest that smooth tubular continuities may exist between the rough ER and the Golgi apparatus. The apparent lack of effect of microtubular drugs on rough ER–Golgi apparatus transport in hepatocytes (see Chapter 15 by Redman *et al.*) also suggests that alternative transport mechanisms may exist in these cells. Additionally, in certain conditions in ameba (Flickinger, 1969; Wise, 1972) and possibly other cells, secretory products and membrane may reach the Golgi apparatus through conversion of ER cisternae into Golgi saccules at the cis face.

At present, the weight of the evidence favors vesicular transport of secretory proteins between the rough ER and the Golgi apparatus. However, various other methods may be involved in certain cell types, under certain physiological conditions, and for specific secretory, lysosomal, or membrane proteins.

## III.   Transport through the Golgi Apparatus

The saccules of the Golgi apparatus appear to be the primary recipients of the secretory proteins synthesized by the rough ER. The proximity of the transitional

ER and transfer vesicles to the cis face of the Golgi apparatus and the usual dilated, irregular appearance of the cis Golgi saccule (Fig. 1) suggest that the transfer vesicles either fuse with this saccule, depositing their content in it, or coalesce to form the cis saccule. These interpretations are supported by the demonstration of cytochemical similarities between the transfer vesicles and the cis Golgi saccule—i.e., staining of both structures after osmium impregnation (Friend, 1969; Hand and Oliver, 1977a) (Fig. 2). Secretory proteins, as judged by cytochemical and immunocytochemical staining, are present in saccules at all levels of the Golgi apparatus (Essner, 1971; Herzog and Fahimi, 1976; Hand and Oliver 1977a,b; Kraehenbuhl *et al.*, 1977; Geuze *et al.*, 1979; Broadwell *et al.*,

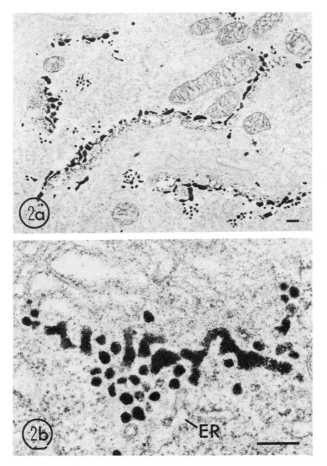

FIG. 2. Golgi apparatus of acinar cells of the rat exorbital lacrimal (a) and parotid (b) glands following OsO₄ impregnation for 48 hours at 40°C (Friend and Murray, 1965). Osmium deposits fill the cis Golgi saccule and peripheral vesicles between the ER and the Golgi apparatus. Magnification: (a) 14,000×; (b) 44,000×.

1979) (Fig. 3). Radioautographic studies of cells in which transport is slow suggest that the cis side of the Golgi apparatus becomes labeled prior to accumulation of labeled secretory proteins at the trans face (Castle *et al.,* 1972; Flickinger, 1974; Slot *et al.,* 1976). Nevertheless, some of the transfer vesicles may fuse with saccules at various levels of the Golgi apparatus, or even directly with the forming secretory granules (Jamieson and Palade, 1967b; Palade, 1975) (Fig. 1), bypassing the saccules altogether. Studies by Jamieson and Palade (1971b) suggest that the destination of the transfer vesicles may also be influenced by the physiological state of the cell.

Since secretory granule formation in most cells is restricted to the trans face of the Golgi apparatus, the secretory proteins must migrate from the cis to the trans saccules. One proposed mechanism for this transport is the migration of whole saccules and their content from their site of formation at the cis face to the site of granule formation at the trans face (Mollenhauer and Whaley, 1963; Neutra and Leblond, 1966; Franke *et al.,* 1971; Susi *et al.,* 1971; Weinstock and Leblond, 1974; Slot *et al.,* 1974). The migration of the membrane would be accompanied by its modification from ER-like at the cis face, to plasma membrane-like at the trans face (Grove *et al.,* 1968; Morré *et al.,* 1971b, 1974, 1979; Mollenhauer *et al.,* 1976; Morré and Ovtracht, 1977). However, the concept of ''membrane flow'' has been challenged on the basis of compositional differences among the membranes of the various intracellular compartments (Meldolesi *et al.,* 1971a,b; Bergeron *et al.,* 1973) and on the rates of synthesis and turnover of membrane proteins as opposed to content proteins (Meldolesi, 1974a,b; Wallach *et al.,* 1975; Winkler, 1977). The shuttling of vesicles between subcellular compartments and specific mechanisms to prevent intermixing of membrane components during transient interactions have been proposed (Meldolesi, 1974b; Palade,

---

FIG. 3.    Golgi region of an acinar cell of the rat exorbital lacrimal gland, incubated for endogenous peroxidase activity at pH 7.0 (Hand and Oliver, 1977a). Reaction product is located in the rough ER, Golgi saccules (S), and secretory granules (G), indicating the presence of secretory protein. A gradient of reaction product occurs in the Golgi saccules; the deposits are heavier in the trans saccules than in the cis saccules. At the trans face, note the absence of reaction product in many of the cisternae (arrowheads) and vesicles that have been identified as GERL in acid phosphatase and thiamine pyrophosphatase preparations. A small segment of GERL (arrow) appears slightly reactive. Magnification 44,000×.

FIG. 4.    Golgi region of a lacrimal acinar cell, prepared as in Fig. 3. Peroxidase reaction product is present in the Golgi saccules (S) and an immature granule (IG), but GERL (arrowhead) is poorly reactive, even though it is in direct continuity with the immature granule. A cisterna of ER adjacent to GERL and the immature granule is strongly reactive. Magnification 57,000×.

FIG. 5.    Golgi region of a rat parotid acinar cell, fixed in glutaraldehyde–formaldehyde and embedded in glycol methacrylate (Leduc and Bernhard, 1967) without $OsO_4$ postfixation. The thin section was stained with 1% phosphotungstic acid in 10% HCl (Rambourg, 1971) to reveal the presence of glycoproteins. The trans Golgi saccules (tS), GERL (arrowheads), and secretory granules (G) are strongly stained. Magnification 59,000×.

1975). Thus, secretory proteins would enter and leave the various compartments unaccompanied by their transporting membranes. Nonetheless, certain membrane components such as the G- protein of vesicular stomatitis virus (Katz *et al.*, 1977; see also Chapter 2 by H. F. Lodish *et al.*) and glycophorin A (Jokinen *et al.*, 1979) are known to be inserted into the membrane of the rough ER and to

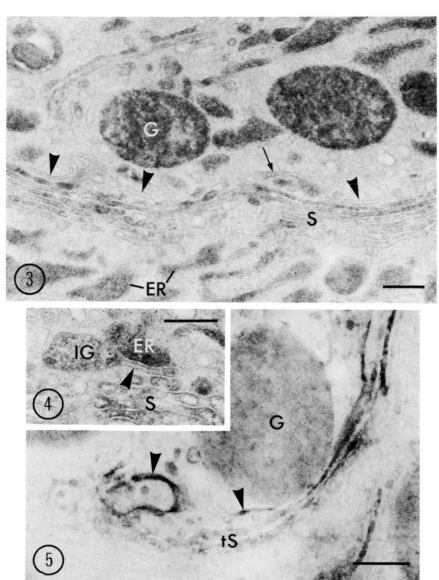

move to the Golgi apparatus and finally to the plasma membrane. Although the flow of intact segments of membrane between or within intracellular compartments is currently being questioned, the unidirectional movement of certain membrane components clearly does occur. Whether or not similar mechanisms contribute to the transport of secretory proteins remains to be established.

Another possibility for movement of content from one saccule to the next is through direct continuities between adjacent saccules (Castle *et al.*, 1972). The existence of such continuities has proved difficult to verify. Rambourg *et al.* (see Chapter 11) have suggested that an intersaccular network of anastomosing tubules may serve to interconnect saccules at various levels within the same or different Golgi stacks. While inherently attractive, this arrangement conceivably might further complicate certain aspects of the transport process, such as maintenance of a concentration gradient of secretory product from the cis to the trans saccules (Fig. 3), regulation of sequential posttranslational modifications, and the routing of proteins to specific intracellular destinations. Further three-dimensional studies of the Golgi apparatus of secretory cells may help to resolve these issues.

## IV.   Secretory Granule Formation

Once the secretory proteins reach the trans face of the Golgi apparatus, they are concentrated and packaged in membrane-delimited vesicles, vacuoles, or granules. This process serves to protect the cytoplasmic constituents from potent hydrolytic enzymes, and it permits economical storage and discharge in cells that secrete intermittently. The mechanisms involved in granule formation are poorly understood. Major gaps in our knowledge include the route of transport from the

---

FIG. 6.   Golgi region of a rat parotid acinar cell, 3 hours after *in vivo* secretory stimulation with isoproterenol; incubated for the demonstration of acid phosphatase activity, pH 5.0, using cytidine monophosphate (CMP) as substrate (Novikoff, 1963). Reaction product is present in immature secretory granules (IG) and the narrow cisternal profiles (arrowheads) at the trans face of the Golgi apparatus. The morphology, cytochemistry, and location of this cisternal structure characterize it as GERL (Novikoff, 1976; Hand and Oliver, 1977a,b). The Golgi saccules (S) are free of reaction product. Magnification 32,000×.

FIGS. 7 and 8.   Golgi region of lacrimal acinar cells, 1 hour (Fig. 7) and 3 hours (Fig. 8) after *in vivo* secretory stimulation with pilocarpine; incubated for acid phosphatase localization as in Fig. 6. Reaction product is present in GERL (arrowheads) and immature secretory granules (IG) forming from it. In Fig. 7 the reactive immature granule appears to be forming from a dilated saccule (arrows) adjacent to the trans face of the Golgi apparatus (cf. Figs. 9 and 10, transitional saccule). No reaction product is present in the Golgi saccules (S). Note the cisternae of ER adjacent to GERL and the immature granules in Fig. 8. Magnification: Fig. 7, 55,000×; Fig. 8, 58,000×.

Golgi saccules to the granules, the source of the membranes utilized in granule formation, the extent of mixing of products destined for different intracellular compartments and the mechanisms by which they are ultimately sorted out, and the potential for further chemical or structural modification of the packaged secretory products.

Recent morphological and cytochemical studies of several exocrine and endocrine cells (Novikoff, 1976; Novikoff and Novikoff, 1977; Novikoff *et al.*, 1977;

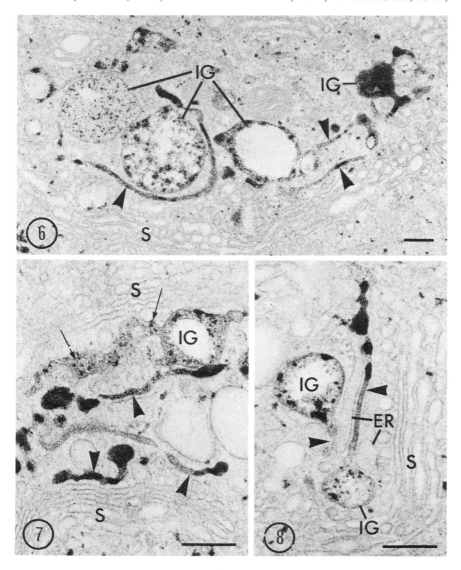

Hand and Oliver, 1977a,b, 1980; Oliver *et al.,* 1980; Broadwell and Oliver, 1981) have provided new information regarding the structural relationships at the trans face of the Golgi apparatus. In many cells the secretory granules form in continuity with an acid phosphatase-positive membrane system at the trans face (Figs. 6–8). The forming granules (condensing vacuoles) also have acid phosphatase activity, the smaller ones usually being more reactive than the larger ones. The structure and cytochemical reactivity of this membrane system are consistent with that described for GERL (*G*olgi–*E*ndoplasmic *R*eticulum–*L*ysosome) in other cells by Novikoff and his co-workers (Novikoff, 1964, 1976; Novikoff *et al.,* 1971). Although initially regarded as a specialized region of smooth ER (Novikoff, 1964; Novikoff *et al.,* 1971), membranous continuities between GERL and the ER have not been established with certainty in secretory cells.

In spite of the evidence for the involvement of GERL in granule formation, the route of transport of secretory proteins between the trans Golgi saccule and the forming granule has not been clearly defined. Cytochemical and immunocytochemical studies indicate that secretory proteins are absent from GERL or present only in limited amounts (Hand and Oliver, 1977a,b; Broadwell *et al.,* 1979; Geuze *et al.,* 1979) (Fig. 3), regardless of the fact that forming granules are in continuity with GERL and are strongly reactive (Fig. 4). In contrast, GERL stains strongly for the presence of glycoprotein (Fig. 5). The unexpected pattern of secretory protein distribution revealed by cytochemical techniques could occur if granule formation is initiated by the trans saccule, with the subsequent segregation of secretory protein in the granule and conversion of the saccule to GERL (Hand and Oliver, 1977a,b, 1980; Oliver *et al.,* 1980). The presence of a "transitional saccule" (Figs. 7, 9, and 10) at the trans face, which is slightly dilated and contains both thiamine pyrophosphatase and acid phosphatase reaction product, is suggestive of such a conversion. Additional support comes from studies of stimulated cells (Broadwell and Oliver, 1981; Oliver *et al.,* 1980; Hand and Oliver, 1980), which demonstrate modulations in the distribution of thiamine pyrophosphatase activity—in particular, the presence of reaction product in forming secretory granules (Fig. 9, inset, and Fig. 11) and GERL-like cisternae. A vesicular transport mechanism (Winkler, 1977) or an undetected tubular network between the trans saccule and the forming granules are other alternatives. Finally, the difficulty in detecting secretory proteins in GERL conceivably may be due to a cytochemical artifact—for example, inaccessibility of the secretory proteins to certain reagents or substrates (Broadwell *et al.,* 1979). Unequivocal evidence for any of these possibilities is lacking; thus, at present, the route taken by secretory proteins from the trans saccule to the forming granules during their growth and maturation remains obscure.

The origin of the membranes enclosing the secretory granules has been a topic of considerable interest (see discussions in the chapters by E. Holtzman, M. G.

Farquhar, and C. Oliver and A. R. Hand). Although most morphological studies indicate that granule membranes are derived from the Golgi apparatus (or GERL), the biochemical evidence for extensive membrane reutilization suggests that the granule membranes may be part of a pool of membrane shuttling between the Golgi apparatus and the cell surface (Meldolesi, 1974b; Palade, 1975). The recent work of Herzog and Farquhar (1977) and Farquhar (1978) (see also Chapter 25 by M. G. Farquhar) provides morphological evidence that, after exocytosis, secretory granule membranes may be returned to the Golgi apparatus for reuse during the formation of new granules. Considering the present uncertainties regarding the route of protein transport between the saccules and the forming granules, and the apparent role of GERL in granule formation, it will be important to delineate the specific structures in the Golgi region (e.g., saccules, GERL, and/or forming granules) with which the returning membranes fuse, to obtain quantitative data on the amount of recycled membrane as opposed to *de novo* synthesized membrane incorporated into new granules, and to determine the site(s) where *de novo* synthesized membrane enters the cycle.

The presence of acid phosphatase, a lysosomal marker enzyme, in GERL of secretory cells suggests that GERL may represent a site of intracellular sorting between lysosomal and secretory proteins. As described by Sly (see Chapter 13 by W. S. Sly *et al.*), the sorting mechanism could involve receptors associated with the membrane of GERL that bind specific recognition markers on the lysosomal proteins. Acid phosphatase activity in immature secretory granules (Figs. 6–8) may indicate imperfect sorting, which allows escape of some enzyme into the granules. In other cells, such as neurons (Novikoff, 1964; Novikoff *et al.*, 1971) and hepatocytes (Novikoff *et al.*, 1966; Essner and Oliver, 1974), GERL is involved in the formation of lysosomes. However, in exocrine cells such as parotid, lacrimal, and pancreatic acinar cells, the relationship between GERL and lysosomes has not been established. Further, it is not clear whether acid phosphatase observed cytochemically in GERL and immature granules represents lysosomal or secretory enzyme, or if it has a specific, but as yet unidentified, role in granule formation and maturation. Other hydrolases present in lysosomes of these cells, including nonspecific esterase, trimetaphosphatase, and aryl sulfatase (Hand, 1972; Doty *et al.*, 1977; Oliver, 1980, and unpublished observations), have not been demonstrated in GERL or immature secretory granules.

In light of the recent studies demonstrating intracellular degradation of newly synthesized secretory products, notably collagen (Bienkowski *et al.*, 1978) and prolactin (Shenai and Wallis, 1979), the incorporation of hydrolytic enzymes into secretory granules may be of considerable importance. The presence of these enzymes could provide a mechanism for regulating, at the stage of intracellular storage, the quantity of secretory protein available for discharge. Such a mechanism could function independently from, or in conjunction with,

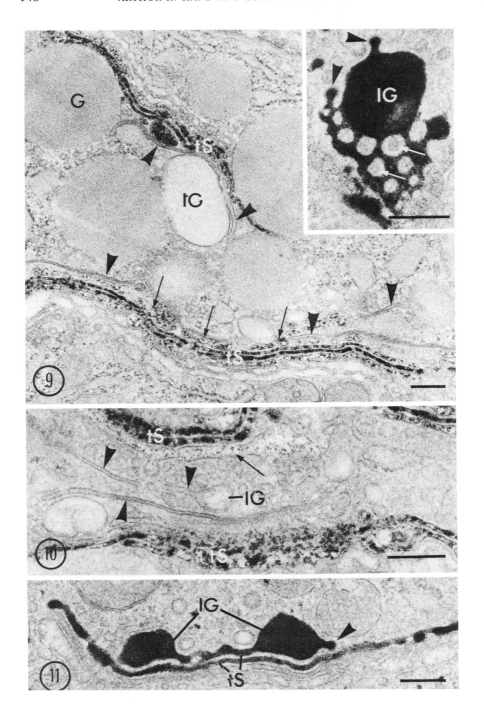

crinophagic processes for secretory granule degradation (Smith and Farquhar, 1966; Hand, 1972). Additionally, certain secretory proteins undergo specific intracellular proteolytic modifications, such as the conversion of precursor molecules to active hormones (Steiner *et al.*, 1974; Gainer *et al.*, 1977; see also chapters by H. S. Tager *et al.* and E. Herbert *et al.*). These modifications could be effected by proteases present in GERL or included in the secretory granules during their formation.

# V. Maturation of Secretory Granules

In most secretory cells the granules forming in the Golgi region are clearly distinguishable from the mature granules stored in the cytoplasm. Morphological differences include their size, shape, and the density and organization of their content (Fig. 1). Specific cytochemical features (Figs. 6-8), the potential for further processing of secretory product, and changes in the composition and/or organization of their limiting membrane (Orci *et al.*, 1980), possibly related to their ability to undergo exocytosis (Kramer *et al.*, 1978; Hand and Oliver, 1980), suggest important functional differences between forming and mature granules. Thus, a number of critical processes apparently occur simultaneously during granule maturation.

The increase in density of the granule content appears to be due to the conden-

---

Fig. 9. Golgi region of an acinar cell of the rat exorbital lacrimal gland, incubated for thiamine pyrophosphatase (TPPase) activity at pH 7.2 (Novikoff and Goldfischer, 1961). Reaction product is present in the trans Golgi saccules (tS). Note the presence at the trans face of a slightly dilated transitional saccule with fewer deposits of reaction product (arrows), and the narrow cisternal profiles of GERL (arrowheads), which are free of reaction product. IG, immature granule; G, mature secretory granule. Magnification: 36,000×. Inset: Golgi region of an acinar cell of the rat parotid gland, 10 hours after *in vivo* secretory stimulation by isoproterenol; incubated for TPPase localization. An immature granule (IG) containing reaction product is forming from the trans Golgi saccule, seen in face view. Note the coated vesicles (arrowheads) attached to the immature granule and the fenestrated portion of the saccule. Small vesicles or tubules (arrows) of unknown origin are present within the fenestrations of the trans saccule. Magnification 64,000×.

Fig. 10. Golgi region of a lacrimal acinar cell, incubated for TTPase localization. Reaction product is present in the trans Golgi saccules (tS); weak activity is also present in the dilated transitional saccule at the trans face (arrow). No reaction product is present in the numerous vesicles or GERL (arrowheads). Note the continuity of GERL with the small immature granule (IG). Magnification 61,000×.

Fig. 11. Golgi region of a rat parotid acinar cell, 10 hours after *in vivo* secretory stimulation with isoproterenol; incubated for TPPase localization. Reaction product is present in the trans Golgi saccules (tS) and in two small immature granules (IG) forming from the trans saccule. A coated vesicle (arrowhead) is attached to one of the immature granules. An adjacent immature granule is unreactive. Magnification 48,000×.

sation and concentration of the initially dilute product of the rough ER. Work by Jamieson and Palade (1971a) established that condensation is a passive process, occurring in the absence of energy input, and most likely resulting from the formation of large, osmotically inactive molecular aggregates. Such aggregates could result from the interaction of secretory proteins with high-molecular-weight sulfated anions (Berg and Young, 1971; Palade, 1975; Reggio and Palade, 1978; Zanini *et al.*, 1980; Giannattasio *et al.*, 1980), soluble acidic lipoproteins (Koenig, 1974), or metal ions (Wallach and Schramm, 1971; Clemente and Meldolesi, 1975). The mechanism of aggregate formation probably varies between cell types and presumably depends on the chemical and ionic properties of the specific secretory proteins.

The factors that regulate the development of the final mature form of the secretory granules are unknown. The size, shape, and substructure of mature granules are relatively uniform and frequently characteristic for specific cell types. In contrast, the forming granules may show wide variations in these features. The forming granules may achieve a larger size than the mature granules, thus necessitating a reduction in surface membrane area with maturation. Coated vesicles have been suggested as one mechanism for achieving the withdrawal of membrane from the immature granules (Weinstock and Leblond, 1974) (Figs. 9 and 11). Changes in granule substructure may reflect underlying changes in molecular organization of the secretory material during maturation. For example, the development of secretory granules in collagen-producing cells begins with a tangle of fine threads in dilated Golgi saccules and progresses to bundles of parallel filaments as the pro-$\alpha$ chains assemble into procollagen molecules (Weinstock and Leblond, 1974; see also Chapter 12 by C. P. Leblond and G. M. Wright). The development of other highly ordered substructures, such as those found in the salivary glands of various mammals (Tandler and MacCallum, 1972; Tandler and Erlandson, 1972), may be related to compositional variations of the secretory material or possibly to the presence of intragranular reticular structures (Ermak and Rothman, 1978).

# VI.    Conclusions

The Golgi apparatus clearly plays a central role in secretory cells. Its functions include the transport, modification, and packaging of secretory material, the segregation and sorting of products destined for different intracellular compartments, and the receipt and rerouting of various membranous components. The heterogeneity of the Golgi apparatus, demonstrated in structural, cytochemical, and biochemical studies, suggests that specialized regions of the Golgi apparatus may be responsible for specific functions.

The identification of GERL at the trans face of the Golgi apparatus and its apparent role in secretory granule formation adds additional complexity to our understanding of the functioning of the Golgi apparatus. Regardless of whether or not GERL should be differentiated from the trans Golgi saccule of secretory cells, it is clear that a number of significant events in the transport and packaging of secretory proteins occur at the interface between the trans saccule and GERL. The available evidence also suggests that the segregation of secretory products from lysosomal constituents, in addition to some aspects of membrane circulation, may be functions of GERL. Future studies should be directed toward defining the structural relationship of GERL to the Golgi apparatus, and its origin from and/or continuity with other subcellular compartments. Additionally, studies of the functional role of the Golgi apparatus in secretory cells, especially in regard to secretory protein modification and transport, as well as membrane reutilization, must take into account the potential contributions of GERL.

## REFERENCES

Beams, H. W., and Kessel, R. G. (1968). *Int. Rev. Cytol.* **23,** 209–276.

Berg, N. B., and Young, R. W. (1971). *J. Cell Biol.* **50,** 469–483.

Bergeron, J. J. M., Ehremeich, J. H., Siekevitz, P., and Palade, G. E. (1973). *J. Cell Biol.* **59,** 73–88.

Bienkowski, R. S., Cowan, M. J., McDonald, J. A., and Crystal, R. G. (1978). *J. Biol. Chem.* **253,** 4356–4363.

Brandan, E., and Fleischer, B. (1980). *J. Cell Biol.* **87,** 200a.

Broadwell, R. D., and Oliver, C. (1981). *J. Cell Biol.* (in press).

Broadwell, R. D., Oliver, C., and Brightman, M. W. (1979). *Proc. Natl. Acad. Sci. U.S.A.* **76,** 5999–6003.

Caro, L. G., and Palade, G. E. (1964). *J. Cell Biol.* **20,** 473–495.

Castle, J. D., Jamieson, J. D., and Palade, G. E. (1972). *J. Cell Biol.* **53,** 290–311.

Chu, L. L. H., MacGregor, R. R., and Cohn, D. V. (1977). *J. Cell Biol.* **72,** 1–10.

Claude, A. (1970). *J. Cell Biol.* **47,** 745–766.

Clemente, F., and Meldolesi, J. (1975). *J. Cell Biol.* **65,** 88–102.

Doty, S. B., Smith, C. E., Hand, A. R., and Oliver, C. (1977). *J. Histochem. Cytochem.* **25,** 1381–1384.

Droz, B. (1966). *C. R. Hebd. Seances Acad. Sci.* **262,** 1766–1768.

Ermak, T. H., and Rothman, S. S. (1978). *J. Ultrastruct. Res.* **64,** 98–113.

Essner, E. (1971). *J. Histochem. Cytochem.* **19,** 216–225.

Essner, E., and Oliver, C. (1974). *Lab. Invest.* **30,** 596–607.

Farquhar, M. G. (1978). *J. Cell Biol.* **77,** R35–R42.

Fleischer, B., Fleischer, S., and Ozawa, H. (1969). *J. Cell Biol.* **43,** 59–79.

Flickinger, C. J. (1969). *J. Cell Biol.* **43,** 250–262.

Flickinger, C. J. (1974). *Anat. Rec.* **180,** 427–448.

Franke, W. W., Morré, D. J., Deumling, B., Cheetham, R. D., Kartenbeck, J., Jarasch, E. D., and Zentgraf, H. W. (1971). *Z. Naturforsch., B: Anorg. Chem., Org. Chem., Biochem., Biophys., Biol.* **26B,** 1031–1039.

Friend, D. S. (1965). *J. Cell Biol.* **25,** 563–576.

Friend, D. S. (1969). *J. Cell Biol.* **41,** 269–279.

Friend, D. S., and Murray, M. J. (1965). *Am. J. Anat.* **117,** 135–150.

Gainer, H., Sarne, Y., and Brownstein, M. J. (1977). *J. Cell Biol.* **73,** 366–381.

Geuze, J. J., Slot, J. W., and Tokuyasu, K. T. (1979). *J. Cell Biol.* **82,** 697–707.

Giannattasio, G., Zanini, A., Rosa, P., Meldolesi, J., Margolis, R. K., and Margolis, R. U. (1980). *J. Cell Biol.* **86,** 273–279.

Grove, S. N., Bracker, C. E., and Morré, D. J. (1968). *Science* **161,** 171–173.

Haddad, A., Smith, M. D., Herscovics, A., Nadler, N. J., and Leblond, C. P. (1971). *J. Cell Biol.* **49,** 856–882.

Hand, A. R. (1972). *Am. J. Anat.* **135,** 71–92.

Hand, A. R., and Oliver, C. (1977a). *J. Cell Biol.* **74,** 399–413.

Hand, A. R., and Oliver, C. (1977b). *Histochem. J.* **9,** 375–392.

Hand, A. R., and Oliver, C. (1980). *J. Cell Biol.* **87,** 304a.

Herzog, V., and Fahimi, H. D. (1976). *Histochemistry* **46,** 273–286.

Herzog, V., and Farquhar, M. G. (1977). *Proc. Natl. Acad. Sci. U.S.A.* **74,** 5073–5077.

Herzog, V., and Miller, F. (1970). *Z. Zellforsch. Mikrosk. Anat.* **107,** 403–420.

Herzog, V., and Miller, F. (1972). *J. Cell Biol.* **53,** 662–680.

Jamieson, J. D., and Palade, G. E. (1967a). *J. Cell Biol.* **34,** 577–596.

Jamieson, J. D., and Palade, G. E. (1967b). *J. Cell Biol.* **34,** 597–615.

Jamieson, J. D., and Palade, G. E. (1968). *J. Cell Biol.* **39,** 589–603.

Jamieson, J. D., and Palade, G. E. (1971a). *J. Cell Biol.* **48,** 503–522.

Jamieson, J. D., and Palade, G. E. (1971b). *J. Cell Biol.* **50,** 135–158.

Jokinen, M., Ghamberg, C. G., and Andersson, L. C. (1979). *Nature (London)* **279,** 604–607.

Karnovsky, M. J. (1965). *J. Cell Biol.* **27,** 137a.

Katz, F. N., Rothman, J. E., Knipe, D. M., and Lodish, H. F. (1977). *J. Supramol. Struct.* **7,** 353–370.

Kelly, R. B., Oliver, C., and Hand, A. R. (1978). *Cell Tissue Res.* **195,** 227–237.

Koenig, H. (1974). *Adv. Cytopharmacol.* **2,** 273–301.

Kraehenbuhl, J. P., Racine, L., and Jamieson, J. D. (1977). *J. Cell Biol.* **72,** 406–423.

Kramer, M. F., Geuze, J. J., and Strous, G. J. A. M. (1978). *Ciba Found. Symp.* [N.S.] **54,** 25–51.

Lane, N., Caro, L., Otero-Vilardebó, L. R., and Godman, G. C. (1964). *J. Cell Biol.* **21,** 339–351.

Leduc, L., and Bernhard, W. (1967). *J. Ultrastruct. Res.* **19,** 196–199.

Malaisse-Lagae, F., Armherdt, M., Ravazzola, M., Sener, A., Hutton, J. C., Orci, L., and Malaisse, W. J. (1979). *J. Clin. Invest.* **63,** 1284–1296.

Meldolesi, J. (1974a). *J. Cell Biol.* **61,** 1–13.

Meldolesi, J. (1974b). *Philos. Trans. R. Soc. London, Ser. B* **268,** 39–53.

Meldolesi, J., Jamieson, J. D., and Palade, G. E. (1971a). *J. Cell Biol.* **49,** 130–149.

Meldolesi, J., Jamieson, J. D., and Palade, G. E. (1971b). *J. Cell Biol.* **49,** 150–158.

Mollenhauer, H. H., and Whaley, W. G. (1963). *J. Cell Biol.* **17,** 222–225.

Mollenhauer, H. H., Hass, B. S., and Morré, D. J. (1976). *J. Microsc. Biol. Cell.* **27,** 33–36.

Morré, D. J., and Ovtracht, L. (1977). *Int. Rev. Cytol., Suppl.* **5,** 61–188.

Morré, D. J., Merlin, L. M., and Keenan, T. W. (1969). *Biochem. Biophys. Res. Commun.* **37,** 813–819.

Morré, D. J., Keenan, T. W., and Mollenhauer, H. H. (1971a). *Adv. Cytopharmacol.* **1,** 159–182.

Morré, D. J., Mollenhauer, H. H., and Bracker, C. E. (1971b). *In* "Origin and Continuity of Cell Organelles" (J. Reinert and H. Ursprung, eds.), pp. 82–126. Springer-Verlag, Berlin and New York.

Morré, D. J., Keenan, T. W., and Huang, C. M. (1974). *Adv. Cytopharmacol.* **2,** 107–125.

Morré, D. J., Kartenbeck, J., and Franke, W. W. (1979). *Biochim. Biophys. Acta* **559,** 71–152.

Neutra, M., and Leblond, C. P. (1966). *J. Cell Biol.* **30,** 119–136.

Novikoff, A. B. (1963). *In* "Lysosomes" (A. V. S. de Reuck and M. P. Cameron, eds.), pp. 36–73. Little, Brown, Boston, Massachusetts.

Novikoff, A. B. (1964). *Biol. Bull. (Woods Hole, Mass.)* **127,** 358.

Novikoff, A. B. (1976). *Proc. Natl. Acad. Sci. U.S.A.* **73,** 2781–2787.

Novikoff, A. B., and Goldfischer, S. (1961). *Proc. Natl. Acad. Sci. U.S.A.* **47,** 802–810.

Novikoff, A. B., and Novikoff, P. M. (1977). *Histochem. J.* **9,** 525–551.

Novikoff, A. B., Roheim, P. S., and Quintana, N. (1966). *Lab. Invest.* **15,** 27–49.

Novikoff, A. B., Mori, M., Quintana, N., and Yam, A. (1977). *J. Cell Biol.* **75,** 148–165.

Novikoff, P. M., Novikoff, A. B., Quintana, N., and Hauw, J.-J. (1971). *J. Cell Biol.* **50,** 859–886.

Oliver, C. (1980). *J. Histochem. Cytochem.* **28,** 78–81.

Oliver, C., Auth, R. E., and Hand, A. R. (1980). *Am. J. Anat.* **158,** 275–284.

Orci, L., Miller, R. G., Montesano, R., Perrelet, A., Amherdt, M., and Vassalli, P. (1980). *Science* **210,** 1019–1021.

Ovtracht, L., Morré, D. J., Cheetham, R. D., and Mollenhauer, H. H. (1973). *J. Microsc. (Paris)* **18,** 87–102.

Palade, G. E. (1975). *Science* **189,** 347–358.

Palade, G. E., and Fletcher, M. (1977). *J. Cell Biol.* **75,** 371a.

Patzelt, C., Brown, D., and Jeanrenaud, B. (1977). *J. Cell Biol.* **73,** 578–593.

Rambourg, A. (1971). *Int. Rev. Cytol.* **31,** 57–114.

Reggio, H. A., and Palade, G. E. (1978). *J. Cell Biol.* **77,** 288–314.

Ronzio, R. A. (1973). *Biochim. Biophys. Acta* **313,** 286–295.

Rothman, J. E., Bursztyn-Pettegrew, H., and Fine, R. E. (1980). *J. Cell Biol.* **86,** 162–171.

Schachter, H., Jabbal, I., Hudgin, R. L., Pinteric, L., McGuire, E. J., and Roseman, S. (1970). *J. Biol. Chem.* **245,** 1090–1100.

Shenai, R., and Wallis, M. (1979). *Biochem. J.* **182,** 735–743.

Slot, J. W., Geuze, J. J., and Poort, C. (1974). *Cell Tissue Res.* **155,** 135–154.

Slot, J. W., Geuze, J. J., and Poort, C. (1976). *Cell Tissue Res.* **167,** 147–165.

Smith, R. E., and Farquhar, M. G. (1966). *J. Cell Biol.* **31,** 319–347.

Spurr, A. R. (1969). *J. Ultrastruct. Res.* **26,** 31–43.

Steiner, D. F., Kemmler, W., Tager, H. S., and Rubenstein, A. H. (1974). *Adv. Cytopharmacol.* **2,** 195–205.

Susi, F. R., Leblond, C. P., and Clermont, Y. (1971). *Am. J. Anat.* **130,** 251–268.

Tandler, B., and Erlandson, R. A. (1972). *Am. J. Anat.* **135,** 419–434.

Tandler, B., and MacCallum, D. K. (1972). *J. Ultrastruct. Res.* **39,** 186–204.

Van Heyningen, H. E. (1964). *Anat. Rec.* **148,** 485–497.

Wallach, D., and Schramm, M. (1971). *Eur. J. Biochem.* **21,** 433–437.

Wallach, D., Kirshner, N., and Schramm, M. (1975). *Biochim. Biophys. Acta* **375,** 87–105.

Weinstock, M., and Leblond, C. P. (1974). *J. Cell Biol.* **60,** 92–127.

Whur, P., Hersovics, A., and Leblond, C. P. (1969). *J. Cell Biol.* **43,** 289–311.

Winkler, H. (1977). *Neuroscience* **2,** 657–683.

Wise, G. E. (1972). *Z. Zellforsch. Mikrosk. Anat.* **126,** 431–436.

Zagury, D., Uhr, J. W., Jamieson, J. D., and Palade, G. E. (1970). *J. Cell Biol.* **46,** 52–63.

Zanini, A., Giannattasio, G., Nussdorfer, G., Margolis, R. K., Margolis, R. U., and Meldolesi, J. (1980). *J. Cell Biol.* **86,** 260–272.

Zeigel, R. F., and Dalton, A. J. (1962). *J. Cell Biol.* **15,** 45–54.

METHODS IN CELL BIOLOGY, VOLUME 23

# Chapter 11

# Three-Dimensional Structure of the Golgi Apparatus

## A. RAMBOURG

*Département de Biologie du*
*Commissariat l'Energie Atomique,*
*Saclay, France*

## Y. CLERMONT AND L. HERMO

*Department of Anatomy, McGill University,*
*Montreal, Quebec, Canada*

## I. Introduction

In 1898, Golgi discovered in the Purkinje cell of the owl a cytoplasmic structure that could be impregnated selectively with silver salts. This structure appeared in the light microscope as a network surrounding the nucleus of the cell and was given the name "appareil réticulaire interne" (internal reticular apparatus) to distinguish it from an external reticulum, which covered the outer surface of neurons (Golgi, 1898a,b).

In thin sections prepared for electron microscopy, the reticular network of the nerve cell originally described by Golgi and nowadays referred to as the Golgi

155

apparatus appeared to be composed of separate units, each consisting of a stack of closely apposed saccules surrounded by vesicles (for reviews, see Beams and Kessel, 1968; Favard, 1969; Whaley, 1976). In an early electron microscopic study of the Golgi apparatus of zooflagellates, Grassé (1957) suggested, in addition, that the stack of saccules displayed a morphological polarity; one side of the stack, because of its proximity to the parabasal filament, was called the "proximal face," whereas the other side was given the name "distal face." Later, the proximal face, thought to arise by fusion of transitional sheets or vesicles originating from the rough endoplasmic reticulum, was referred to as the "forming face," and the opposite side, which in glandular cells gave rise to secretion granules, was called the "mature face" (Mollenhauer and Whaley, 1963). More recently, the terminology has been modified, with the proximal, forming face being now designated as the "cis face," and the distal, mature face as the "trans face" (Ehrenreich *et al.*, 1973).

Coinciding with this morphological polarity, there also exists a histochemical polarity; the enzyme thiamine pyrophosphatase has been located preferentially in one or two saccules located on the trans face of the stack (for a review, see Novikoff and Novikoff, 1977), and a staining gradient that reaches a maximum on the trans face can be shown by using techniques detecting carbohydrates (see review, Rambourg, 1971). In contrast, a prolonged osmication of various tissues produces a selctive labeling of one or more saccules located on the "cis face" (Friend and Murray, 1965).

## II.   Architecture of the Golgi Apparatus in Neurons

New insight into the overall architecture of the Golgi apparatus was gained when thick sections of various tissues were examined in the electron microscope after a selective impregnation of the Golgi elements (Rambourg, 1969). Thus, when the osmic acid-impregnated cis element was examined in 1-$\mu$m-thick sections, it took on the appearance of a continuous polygonal network of tubules (Fig. 1). This network, which could be observed in several types of cells such as

---

FIG. 1.   A 1-$\mu$m-thick section of small nerve cell of a trigeminal ganglion impregnated with osmium and examined at 1000 kV. The impregnated cis element of the Golgi apparatus forms a tight network of anastomotic tubules referred to as the "primary structure" of the Golgi apparatus. The arrow indicates a less intensely impregnated part of the network. Magnification 30,000×. From Rambourg *et al.* (1974).

FIG. 2.   A 5-$\mu$m-thick section of small nerve cell of a spinal ganglion, impregnated with osmium and examined at 1000 kV. The osmiophilic portion of the Golgi apparatus extends over a large area of the cytoplasm in the form of an extensive continuous network reminiscent of the internal reticular apparatus described by Golgi. In this type of cell, this perinuclear network was referred to as the "secondary structure" of the Golgi apparatus. Slender projections of the perinuclear network toward the unstained nucleus (N) are indicated by arrowheads. Magnification 10,000×.

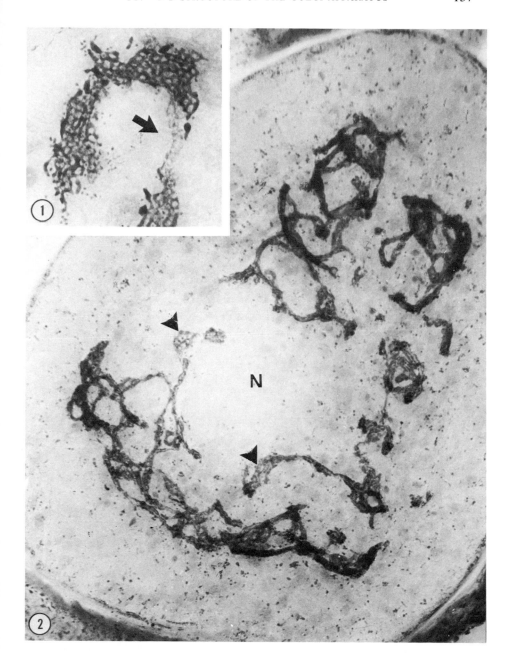

neurons, Leydig, and Sertoli cells (Rambourg *et al.*, 1974), was referred to as the primary network or structure of the Golgi apparatus. When sections of increasing thickness (2–7 μm) were examined in the high-voltage electron microscope, the osmiophilic portion of the Golgi apparatus was seen to extend over large areas of the cytoplasm in the form of an extensive continuous structure reminiscent of the internal reticular aparatus described by Golgi (1898a,b). This structure was referred to as the secondary network or structure of the Golgi apparatus (Rambourg *et al.*, 1974). In contrast to the primary structure, which was similar in these three cell types, the secondary structure of the osmiophilic portion of the Golgi apparatus varied between cell types. In ganglion nerve cells, it consisted of a perinuclear network sending slender projections toward the nucleus and wider expansions toward the cell surface (Fig. 2), whereas in the Leydig cell, it formed an ovoid mass located at one pole of the nucleus. P. M. Novikoff *et al.* (1971) and A. B. Novikoff *et al.* (1977) reached a similar conclusion in an investigation of the thiamine pyrophosphatase-containing element located on the trans face of the Golgi apparatus of the neuron. Whereas a model for the overall tridimensional architecture of the Golgi apparatus has been proposed in the case of small neurons of the dorsal root ganglia (P. M. Novikoff *et al.*, 1971), as was recently pointed out by A. B. Novikoff *et al.* (1977), parts of this reconstruction were speculative, since the middle elements of the Golgi stacks were neither impregnated with osmium nor reacted for thiamine pyrophosphatase, and thus several aspects of the three-dimensional organization of the various components of the Golgi apparatus remained to be elucidated.

## III.   The Golgi Apparatus of Sertoli Cells

In 1976, Thiéry and Rambourg developed a technique in which the endoplasmic reticulum and the Golgi apparatus were impregnated preferentially and were sharply outlined against an unstained background, thus permitting their selective visualization in thick sections. This technique was subsequently employed in an investigation of the three-dimensional organization of the various components of the Golgi apparatus in a cell type that does not produce secretory granules—i.e., the Sertoli cell (Rambourg *et al.*, 1979). Moreover, the results obtained on thick sections by this technique were compared with results obtained on thin sections of material postfixed in potassium ferrocyanide-reduced osmium (Karnovsky, 1971), a technique that preserves and stains the intracytoplasmic membranes of various organelles.

When thick sections (1.0–3.0 μm) stained by the technique of Thiéry and Rambourg (1976) were examined at low magnification, the whole Golgi apparatus of the Sertoli cell was seen to form a continuous network of interconnected wavy ribbon or platelike structures extending from the juxtanuclear region

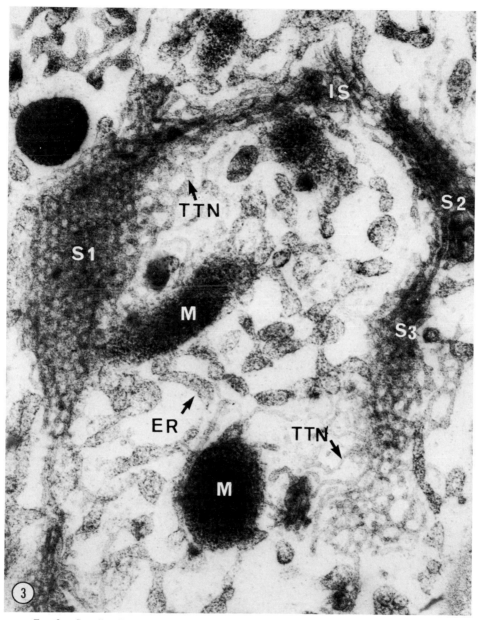

Fig. 3.   Sertoli cell impregnated with the Ur-Pb-Cu technique (Thiéry and Rambourg, 1976). Section 0.5 μm thick and examined at 100 kV. In this portion of the fully impregnated Golgi apparatus, saccular regions (S₁, S₂, S₃) are interconnected by tubular intersaccular regions (IS) to form a continuous structure. The faintly stained tubes of the trans tubular network (TTN) are intermingled with the loosely interconnected elements of the endoplasmic reticulum (ER). M, mitochondria. Magnification 32,000×. From Rambourg *et al.* (1979).

in a secreting cell—i.e., the spermatid, which is involved in the formation of the acrosome (Hermo *et al.*, 1979a,b). When examined at low magnification, the whole Golgi apparatus in young spermatids appeared as a roughly spherical mass located at one pole of the nucleus. At later stages, when the acrosomic granule approached the nucleus to form the acrosome, the Golgi apparatus became hemispherical, with its concave aspect oriented toward the nucleus (Figs. 6 and 7). At higher magnification, when the plane of section was perpendicular to the nuclear surface, two regions could be distinguished in the Golgi apparatus: a peripheral or cortical zone in which saccular regions were interconnected by intersaccular regions, and a central or medullary zone containing tubular and vesicular elements of various sizes and staining properties (Figs. 6 and 7). In the saccular regions of the cortical zone, as in the case of the Sertoli cell, there was on the cis face of the stack of saccules a tight polygonal meshwork of tubules corresponding to the cis (osmiophilic) element of the Golgi apparatus best seen in surface views (inset, Fig. 7). Underlying it there were several closely apposed saccules perforated with pores (S, Figs. 6 and 7). On the trans face of the stack, the trans tubular network as seen in Sertoli cells was absent, but one or two flattened membranous elements were present, the lumen of which was wider than that of the other saccules of the stack (TE, Fig. 7). In the intersaccular regions of the cortex, anastomotic systems of tubules were present serving to interconnect adjacent stacks of saccules (Fig. 7). As in the Sertoli cell, these tubules intertwined, interdigitated, and established connections between saccules located in adjacent stacks or at different levels of the same stack (Fig. 7). Although numerous cisternae of the endoplasmic reticulum were closely applied to the cis face of the Golgi apparatus (Figs. 6 and 7), they were also seen to branch and cross the cortical zone through the intersaccular regions to reach the medulla. In

FIG. 6.    Spermatid impregnated with the Ur-Pb-Cu technique. Thin section, examined at 80 kV. The Golgi apparatus forms a hemispherical mass located over the developing acrosomic system (A) closely applied to the surface of the nucleus. The periphery of the hemisphere or cortical region (CO) is made up of several stacks of saccules (S), whereas the central core or medullary region (M) contains various vesicular and tubular profiles. ER elements (ER) are present on the cis face of the Golgi apparatus where they cap the stacks of saccules. In the medullary region, they are interspersed among other elements or are in close relationship with the trans face of the stack of saccules (arrowhead). Oblique arrows point to faintly stained trans saccular elements. Magnification 36,000×.

FIG. 7.    Thin section of a spermatid fixed with potassium ferrocyanide-reduced osmium showing saccular regions (S) interconnected by intersaccular connecting regions (IS). In the latter, wide membranous tubules arising from the edges of the saccules branch and intertwine with one another. On the cis face, the osmiophilic element is indicated by horizontal arrows. On the trans face, two or three trans saccular elements (TE) with a lumen wider than that of the other saccules of the stack may be identified. Cisternae of the endoplasmic reticulum (ER) are seen to cap the cis face of the Golgi apparatus. On the trans face, an ER cisterna (arrowhead) is closely applied to a saccule of the Golgi stack. In the insert at lower right, the cis osmiophilic element (C) may be observed in front view, whereas the wide membranous tubules of the intersaccular region can be seen to intertwine with each other (arrows). Magnification 45,000×.

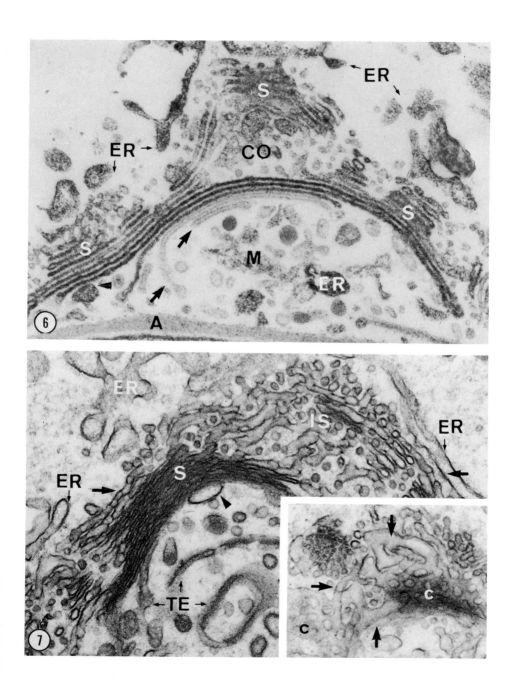

this location, they were often closely applied to the trans elements of the Golgi stacks but never in continuity with them (Fig. 7). Finally, ER cisternae were frequently encountered within the stacks themselves interposed between the saccules.

# V.   Summary and Conclusions

In two diagrammatic representations, the Golgi apparatus in the Sertoli cell (Fig. 8a) and spermatid (Fig. 8b) may be compared. In both cases, the backbone

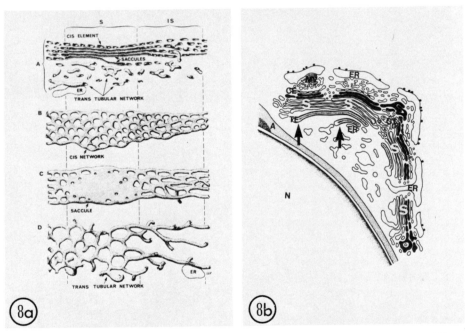

FIG. 8.   Diagrams illustrating the morphological appearance of the various components of the Golgi apparatus in the Sertoli cell (8a), and a young spermatid (8b). In both cases, the main component or backbone of the Golgi apparatus is made up of alternating saccular (S) and intersaccular connecting regions (IS). Drawings B and C in diagram 8a are three-dimensional face views of the cis (B) and saccular (C) elements present in both cell types. The trans tubular network observed in the Sertoli cell (8a, A and D) but not in spermatids is represented in three dimensions as a tubular network with wide meshes. In spermatids it is replaced by tubular and vesicular elements of the medulla. Relations between ER and Golgi elements, which in the Sertoli cell (8a, ER, A and D) are restricted to the trans face of the Golgi apparatus, are more numerous in the spermatid, where they are observed not only on both faces of the continuous backbone but also in the intersaccular regions (8b, ER). In 8b, trans saccular elements with a wider lumen than that of the other saccules are indicated by vertical arrows; A and N refer, respectively, to the acrosome and nucleus of the spermatid. Two saccules in 8b were labeled in black to show their continuity throughout the cortex of the Golgi apparatus.

of the Golgi apparatus is made up of saccular regions interconnected by intersaccular regions composed of anastomotic tubules. On the cis face, the cis-osmiophilic element displays a similar appearance in both types of cells. In contrast, the trans tubular network observed in the Sertoli cell on the trans face of the Golgi apparatus is replaced in the spermatid by elements of the medullary zone, the three-dimensional organization of which remains to be elucidated. Finally, relations between ER and Golgi elements, which in the Sertoli cell are restricted to the trans face of the Golgi backbone, are more numerous in the spermatid, where they are observed not only on both faces of this organelle but also in the intersaccular and saccular regions themselves.

In conclusion, the Golgi apparatus forms a continuous structure made up of alternating saccular and intersaccular connecting regions. In these regions the saccules and bridging tubules form the main component or backbone of the Golgi apparatus. On the cis face, the osmiophilic part of the Golgi apparatus, which forms a tubular network, is generally present in most cell types examined. In contrast, the structure of the trans face, which is relatively simple in a non-secreting cell, such as the Sertoli cell, may reach a high degree of complexity as in the case of spermatids. Similarly, the relationship between ER and Golgi elements may vary according to the cell type. Thus, although the general organization of the Golgi apparatus may be common to many cell types, the relative development of its various parts as well as its relationship with other cell organelles may vary between cell types.

## ACKNOWLEDGMENTS

The work done in the Department of Anatomy, McGill University, was supported by the Medical Research Council of Canada. This investigation was sponsored by a France–Quebec exchange program.

## REFERENCES

Beams, H. W., and Kessel, R. G., (1968). *Int. Rev. Cytol.* **23,** 209–276.
Ehrenreich, J. H., Bergeron, J. J. M., Siekevitz, P., and Palade, G. E. (1973). *J. Cell Biol.* **59,** 45–72.
Favard, P. (1969). *In* "Handbook of Molecular Cytology" (A. Lima de Faria, ed.), pp. 1130–1155. North-Holland Publ., Amsterdam.
Friend, D. S., and Murray, M. (1965). *Am. J. Anat.* **117,** 135–150.
Golgi, C. (1898a). *Arch. Ital. Biol.* **30,** 60–71.
Golgi, C. (1898b). *Arch. Ital. Biol.* **30,** 278–286.
Grassé, P. D. (1957). *C. R. Hebd. Seances Acad. Sci.* **245,** 1278–1281.
Hermo, L., Clermont, Y., and Rambourg, A. (1979a). *Anat. Rec.* **193,** 243–256.
Hermo, L., Rambourg, A., and Clermont, Y. (1979b). *Anat. Rec.* **193,** 564.
Karnovsky, M. J. (1971). *Proc. Am. Soc. Cell Biol., 11th, 1971* Abstract 284, p. 146.

Mollenhauer, H. H., and Whaley, W. G. (1963). *J. Cell Biol.* **17,** 222–226.

Novikoff, A. B., and Novikoff, P. M. (1977). *Histochem. J.* **9,** 525–551.

Novikoff, A. B., Mori, M., Quintana, N., and Yam, A. (1977). *J. Cell Biol.* **75,** 148–165.

Novikoff, P. M., Novikoff, A. B., Quintana, N., and Hauw, J. J. (1971). *J. Cell Biol.* **50,** 859–886.

Rambourg, A. (1969). *C. R. Hebd. Seances Acad. Sci., Ser. D* **269,** 2125–2127.

Rambourg, A. (1971). *Int. Rev. Cytol.* **31,** 57–114.

Rambourg, A., Clermont, Y., and Marraud, A. (1974). *Am. J. Anat.* **140,** 27–46.

Rambourg, A., Clermont, Y., and Hermo, L. (1979). *Am. J. Anat.* **154,** 455–476.

Thiéry, G., and Rambourg, A. (1976). *J. Microsc. Biol. Cell.* **26,** 103–106.

Whaley, W. G. (1976). ''The Golgi Apparatus,'' Cell Biol. Monogr., Vol. 2. Springer Verlag, Berlin and New York.

# Chapter 12

# Steps in the Elaboration of Collagen by Odontoblasts and Osteoblasts

## C. P. LEBLOND AND GLENDA M. WRIGHT[1]

*Department of Anatomy, McGill University,*
*Montreal, Quebec, Canada*

## I.  Introduction

Many cells, in addition to fibroblasts, are capable of producing fibrils of collagen. Both the odontoblasts associated with dentin and the osteoblasts associated with bone are active collagen producers. This was demonstrated with tritium-labeled glycine (Carneiro and Leblond, 1959) and proline (Leblond, 1963)—the two most abundant amino acids in collagen. After injection of either amino acid, the odontoblasts of rat incisor teeth incorporate the label and, after about a half hour, release it into the proximal portion of predentin, which, a day or so later, becomes dentin. Similarly, the osteoblasts of growing alveolar bone pick up labeled glycine (Carneiro and Leblond, 1959) and proline (Leblond,

[1]Present address: Department of Zoology, University of Toronto, Toronto, Ontario, Canada M5S 1A1.

1963) and secrete them into a portion of prebone (osteoid), which, within about a day, becomes bone tissue. That the glycine and proline labels are incorporated into collagenous material is demonstrated by treatment with collagenase; the label is then extracted from the sections (Carneiro and Leblond, 1966).

In the meantime, much has been learned about collagen and the chemical events in its formation. Five types of collagen have been identified; the most common, type I, which is present in dentin and bone, will be examined in this presentation. The production of type I collagen by a producer cell starts on ribosomes with the synthesis of two types of randomly coiled polypeptides, the pro $\alpha1(I)$ and pro $\alpha2$ chains (Bornstein, 1974; Diegelmann *et al.*, 1973; Harwood *et al.*, 1974). The pro $\alpha$ chains consist of three parts: a central region made of repetitive units composed of three amino acids (glycine, proline or hydroxyproline, and a third, variable one), and two different groups of amino acids at each end, referred to as the $NH_2$- and COOH-terminal propeptides. When the synthesis of a pro $\alpha$ chain is completed, it is released into the lumen of a rough endoplasmic reticulum (RER) cisterna (Harwood *et al.*, 1974). At a later stage, the COOH terminals of two pro $\alpha1(I)$ chains and one pro $\alpha2$ chain bind to one another through disulfide bonds; the central region then begins to coil into a helix and continues doing so until the three $NH_2$ terminals become associated (Bornstein, 1974). The result, a rodlike molecule known as type I procollagen, has been recently visualized in the electron microscope by Bruns *et al.* (1979). The procollagen molecule is then released from the producer cell, and enzymes remove the two terminal propeptides. The remaining central portion constitutes the type I collagen molecule (Bornstein, 1974). This is the unit making up the bulk of dentin and bone matrix as well as the collagen fibrils of connective tissue. Our presentation will review how various morphological methods may be used to clarify the steps in the elaboration of the procollagen precursor of collagen.

## II.   Methods

### A.   Routine Electron Microscopy

Details of ultrastructure have been provided in collaboration with Weinstock for the odontoblasts of rat incisor teeth (Weinstock and Leblond, 1974a,b) and by

---

FIG. 1.   A portion of the Golgi apparatus in an odontoblast fixed in glutaraldehyde. Profiles of rough endoplasmic reticulum cisternae (RER) may be observed along the left side and lower right corner of the figure. Smooth endoplasmic reticulum (SER) continuous with a portion displaying ribosomes is seen in close association with some elements of the Golgi apparatus; it has been referred to as "transitional element." Large spherical distensions (Sd) of the Golgi saccules contain finely entangled threads. Cylindrical distensions (Cd) contain threads parallel to one another. Other elongated profiles (psg) contain parallel threads and regularly arranged, electron-dense spherical particles. These structures are considered to be prosecretory granules. Intermediate vesicles (V) are found throughout the region. Magnification 50,000×. From Weinstock and Leblond (1974b).

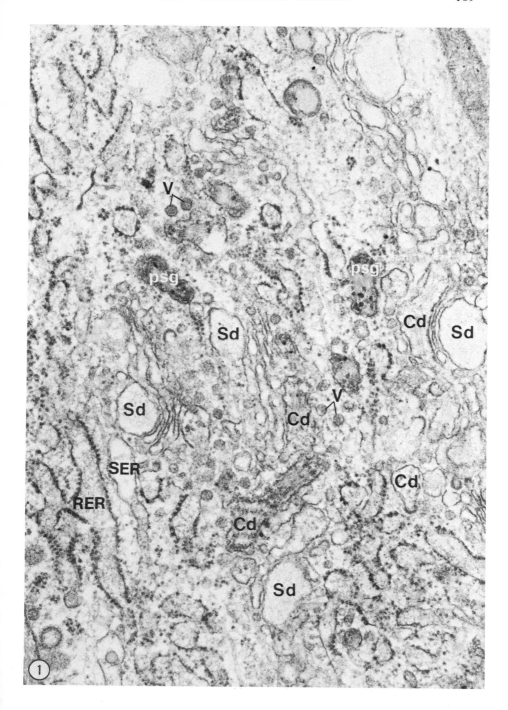

Weinstock (1975) for the osteoblasts in the alveolar bone surrounding these teeth (see also Gartner et al., 1979).

The tissues were obtained from rats (35–40 gm) sacrificed by perfusion with 2.5% glutaraldehyde in 0.05 $M$ Sörensen's phosphate buffer (pH 7.3) with the addition of 0.1% sucrose and 0.5% dextrose (400 mOsm). Teeth and bones were demineralized for 2 weeks at 4°C in EDTA, followed by an overnight rinse in 0.15 $M$ Sörensen's phosphate buffer (pH 7.3). After osmication and acetone dehydration, the tissues were embedded in Epon. The Golgi apparatus of odontoblasts is depicted in Fig. 1; the diagram in Fig. 2 summarizes the salient features of this organelle in odontoblasts and osteoblasts.

## B.   Radioautography

Radioautographs of incisor teeth and bone were prepared in collaboration with Weinstock from rats (35–40 gm) sacrificed at various times after intravenous injection of 0.2 ml of 0.05 $N$ HCl containing 2.5 mCi of L-[2,3-$^3$H]proline (45.7 Ci/mmol). The animals were perfused with a 3% formaldehyde solution (prepared from paraformaldehyde, containing 0.1 $M$ Sörensen's phosphate buffer and 0.1% sucrose at pH 7.3).

The radioautographs were prepared by routine methods for light microscopy (Kopriwa and Leblond, 1962) and electron microscopy (Kopriwa, 1973). Counts of silver grains over the organelles of odontoblasts are summarized in Fig. 3 (Weinstock and Leblond, 1974a).

## C.   Immunohistochemistry

Antibodies were prepared against type I procollagen obtained from acetic acid extracts of rat skin (Smith et al., 1977). Rabbits were given intramuscular and subcutaneous injections of the solution of procollagen in 0.15 $M$ sodium chloride, mixed with an equal volume of Freund's complete adjuvant. A booster shot consisting of procollagen in Freund's incomplete adjuvant was given 1 or 2 weeks later. The rabbits were bled periodically over several months or completely after 4–6 weeks. The anti-procollagen I (anti-pro I) serum was separated from the blood and stored at −70°C. The antibodies were extracted from the serum by affinity chromatography using type I procollagen bound to Sepharose beads.

The anti-pro I antibodies were coupled to horseradish peroxidase by the two-step method of Avrameas and Ternynck (1971). Ten milligrams of horseradish peroxidase (Sigma type VI, Rz = 3.03) were first reacted with glutaraldehyde, which, in a second step, was linked to 5 mg of affinity-purified antibodies. The complex was then precipitated by ammonium sulfate.

Sherman rats aged 20 ± 2 days for the study of odontoblasts or 24 ± 6 hours for the study of osteoblasts were anesthetized with ether and perfused via the left

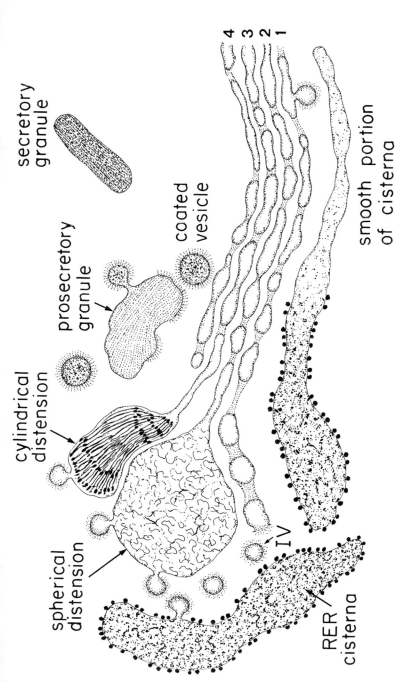

Fig. 2. A diagrammatic representation of a portion of the Golgi apparatus in odontoblasts depicting the significant features of the various organelles. The RER cisternae are coated with ribosomes and contain indistinct material (dots and short, hazy threads). Next to the first Golgi saccule, they are often continued by a smooth, ribosome-free portion, the so-called "transitional element." Intermediate vesicles (IV) are also present in this region. The flat portion of the first and second saccules often ends in a "spherical distension" containing convoluted threads with a diameter varying from 0.5 to 3 nm. The flat portion of the third and fourth saccules often ends in a "cylindrical distension" containing aggregates of parallel threads 2–3 nm thick and about 340 nm long. Distinct structures containing parallel threads of the same length as well as bristle-coated patches and buds along the surface are referred to as "prosecretory granules." Other structures with a denser content in which the threads are barely visible are "secretory granules." From Weinstock and Leblond (1974a).

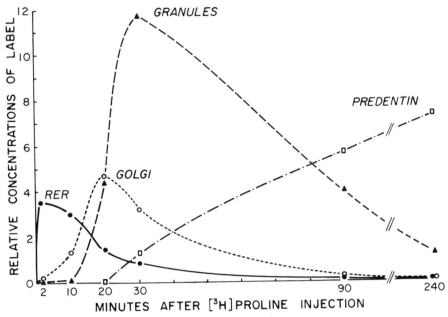

FIG. 3.    Graphic representation of silver grain counts over odontoblast organelles at various times after an intravenous injection of [³H]proline. The counts are expressed as an index of the relative concentration of label over the rough endoplasmic reticulum (RER), Golgi apparatus (Golgi), prosecretory and secretory granules (granules), and predentin, determined by dividing the percentage of silver grains over each structure by the percentage volume of that structure (Weinstock and Leblond, 1974a). Peaks of labeling occur successively in RER (2 minutes), Golgi (20 minutes), granules (30 minutes), and predentin (240 minutes), indicating successive steps in the intracellular migration of an exportable protein labeled with [³H]proline. The high concentration of label within secretory granules at 30 minutes suggests that the exportable protein is carried by these granules to predentin. From Weinstock and Leblond (1974a).

ventricle, first for 30 seconds with ice-cold Ringer's solution and then for 15 minutes with ice-cold 5% formaldehyde freshly prepared from paraformaldehyde in 0.088 $M$ phosphate buffer at pH 6.0 containing 2% sucrose. They were then decapitated, and the incisor teeth or tibiae were dissected out and washed for 1.5 to 2 hours in the same buffer used to prepare the fixative.

The tissues were then placed on cold plastic disks covered with filter paper; a 7% agar solution at about 47°C was poured over the tissue, and the disks were placed at 4°C to allow the agar to solidify. The excess agar was removed, and the tissue was chopped into slices approximately 60 $\mu$m thick on a Sorvall TC-2 tissue sectioner. The slices were collected in ice-cold phosphate-buffered saline (PBS), and the best ones were selected for immunostaining.

All reagents were passed through Millipore filters before use. First, the slices were placed in a 30% concentration of normal goat serum in PBS for 20 minutes at room temperature. The slices were then rinsed for 3 minutes in PBS to which 1% normal goat serum has been added (1% PBSS), then incubated overnight at 4°C in peroxidase-linked anti-pro I antibodies. Control slices were incubated with immunoglobulin from nonimmune rabbit serum linked to peroxidase. Experimental and control slices were then washed for 10 minutes in 1% PBSS at room temperature and washed again for 10 minutes in PBS, followed by a brief fixation in 2.5% glutaraldehyde in PBS for 15 minutes. The slices were then washed in 0.05 $M$ Tris–HCl at pH 7.6 for 10 minutes and treated with a mixture of 0.5 mg/ml of 3,3'-diaminobenzidine (DAB) and 0.02% hydrogen peroxide ($H_2O_2$) in 0.05 $M$ Tris–HCl at pH 7.6 for 10 minutes. This was followed by a wash of 0.05 Tris–HCl for 15 minutes, postfixation in 1% osmium tetroxide for 15 minutes, and a final washing in distilled water for 10 minutes.

The slices were dehydrated in acetone and infiltrated in acetone–Epon mixtures of increasing Epon concentration and, finally, into pure Epon overnight. Each slice was then placed into a fresh drop of Epon on a Teflon plate, and a polymerized block of Epon was positioned over it and gently depressed. The Teflon plate and blocks were kept in a 60°C oven for 2 days for polymerization of the liquid Epon. Thin sections were cut from the surface of the block with a diamond knife, and the first 100 sections were placed on 300-mesh naked copper grids and used for electron microscopy.

Control sections of material exposed to peroxidase-linked immunoglobulin from nonimmune rabbits and treated with DAB–$H_2O_2$ followed by osmium tetroxide showed only a moderate overall contrast as a result of the uptake of osmium (see Fig. 5). But, after immunostaining with peroxidase-linked anti-pro I antibodies, electron-dense material in the form of minute dots (see Figs. 8 and 9) was observed in some organelles of odontoblasts and osteoblasts, a result of the oxidation of DAB by the peroxidase moiety associated with the antibodies.

## D.   Other Reactions

For the detection of glycoprotein, glutaraldehyde-fixed incisor teeth were embedded in glycol methacrylate, and thin sections were floated for 30–60 minutes on a freshly prepared solution of 1% phosphotungstic acid in 1 $N$ HCl (Weinstock and Leblond, 1974a), rinsed, and transferred to grids (Fig. 4).

## III.   Results and Discussion

## A.   Role of the Cisternae of Rough Endoplasmic Reticulum

The odontoblasts are elongated cells, polarized in the direction of a long apical process associated with predentin, whereas the osteoblasts are more massive cells

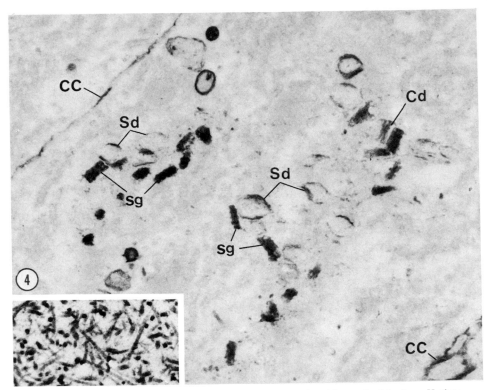

FIG. 4.    Electron micrograph of the Golgi region of an odontoblast stained by low-pH phosphotungstic acid for the detection of glycoprotein (Rambourg, 1971). The RER cisternae scattered throughout the figure are not significantly stained. The spherical distensions (Sd) show some stain lining part or most of the wall, whereas their center remains unstained. Bundles of parallel threads in cylindrical distensions (Cd) are strongly stained along their length. Prosecretory and secretory granules (sg) are similarly stained. Stain is also found in the cell coat (CC) at the surfaces of adjacent cells. Magnification 21,600×. The inset depicts a portion of the predentin matrix where collagen fibrils display intense staining but the interfibrillar substance is little or not stained. Magnification 21,600×.

displaying a similar ultrastructure and polarized toward the prebone (osteoid). A common feature of the two cell types is the abundance of cisternae of rough endoplasmic reticulum.

*Radioautographic examination* of odontoblasts (Weinstock and Leblond, 1974a) and osteoblasts (Weinstock, 1975) was carried out after intravenous injection of [³H]proline in the hope of tracing the migration of collagen precursors within these cells. Two minutes after injection, radioautographs show the label almost exclusively in the RER cisternae of the cells (Fig. 3). The amount of label in other organelles, such as mitochondria and nucleus, is very low at 2 minutes and later times; in the Golgi apparatus, label is also negligible at 2 minutes, but it appears later, as will be shown below.

By 10 minutes after [³H]proline injection, the proportion of label in RER cisternae is decreased, and by 20 minutes it is low, even though the total count of silver grains over the radioautographed odontoblasts has increased. Hence, the label incorporated by the RER must have migrated to other cell organelles. Biochemical analysis of rough microsome fractions of tendon and cartilage cells subjected to a [¹⁴C]proline pulse has also shown rapid entry of label and departure within 10–20 minutes (Harwood *et al.*, 1976).

These results demonstrate that protein material rich in proline is processed in RER cisternae. Further information on this material has been provided by other methods.

*Immunostaining* has been carried out with antibodies against type I procollagen coupled to horseradish peroxidase. The sites of binding of these antibodies in sections of rat incisor teeth or tibia have been detected by the deposition of a dense reaction product resulting from the interaction of the antibodies' peroxidase with diaminobenzidine in the presence of hydrogen peroxide (Karim *et al.*, 1979). Control sections exposed to nonimmune rabbit immunoglobulin linked to peroxidase show no density in any of the cytoplasmic organelles (Fig. 5).

In odontoblasts (Fig. 6) or osteoblasts (see Fig. 9), the reaction product is not found within nuclei and mitochondria but is present in RER cisternae as well as in other structures to be mentioned below. To be reactive, an organelle must be cut open during preparation of the tissue slices and thus allow entry of the antibodies. Moderately dense reactions are then observed in both the narrow RER cisternae oriented along the length of the odontoblasts (Fig. 6) and the dilated cisternae often present below their nucleus (see Fig. 8). In the osteoblasts, reactions occur in narrow and dilated portions, which alternate along the length of the cisternae (see Fig. 9). There is a tendency for the reactive material to line the surface of the wall in narrow cisternae, with the central lumen being less reactive, as seen at lower center in Fig. 6, whereas the material is distributed uniformly throughout the dilated cisternae (see Figs. 8 and 9).

The antigens capable of reacting with anti-procollagen I (anti-pro I) antibodies include not only type I procollagen itself, but also closely related substances, particularly the pro $\alpha$ chains (but not type I collagen). Hence, the reaction observed in RER cisternae may be produced by either type I procollagen or pro $\alpha$ chains or both. Since pro $\alpha$ chains are formed in RER cisternae, it is likely that some, if not all, of the antigenicity observed is due to these chains.

It is known that collagen and procollagen have short carbohydrate side chains composed of galactose with or without terminal glucose residues and may, therefore, be considered to be glycoprotein (Neutra and Leblond, 1966; Clark *et al.*, 1975; Oikarinen *et al.*, 1976). It was of interest to learn whether the incorpora-

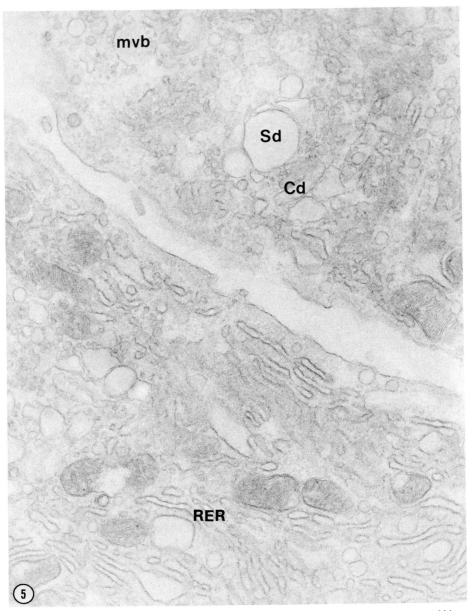

Fig. 5.   Control section of a portion of two osteoblasts after exposure to nonimmune rabbit immunoglobulin linked to peroxidase and staining with DAB–H$_2$O$_2$ and osmium tetroxide. Structures are distinguishable, but no reaction product is seen within the rough endoplasmic reticulum (RER), the spherical distensions (Sd) and cylindrical distensions (Cd) of Golgi saccules, the multivesicular body (mvb), or any other structure. Magnification 30,000×.

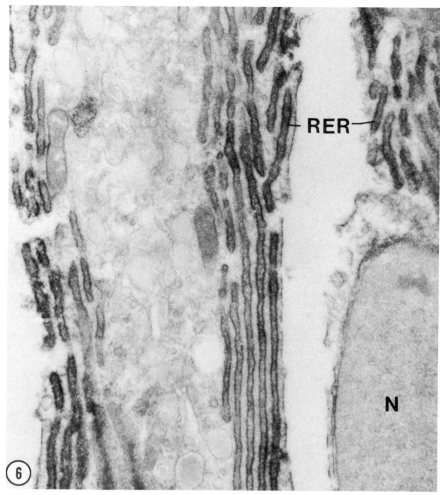

FIG. 6.   Longitudinal section of odontoblasts immunostained by exposure to peroxidase-labeled anti-pro I antibodies followed by DAB–H$_2$O$_2$ and osmium tetroxide. Dense reaction product representing DAB oxidized by the peroxidase component of the antibodies is observed within the cisternae of rough endoplasmic reticulum (RER), which are presumed to have been cut open during the chopping of the tissue prior to staining. The nucleus (N) at lower right shows some reactivity of the nuclear envelope. At lower center, several long RER cisternae show predominance of the reaction along the wall rather than in their center (preparation of Dr. A. Karim). Magnification 20,000×.

tion of these sugars takes place in the RER. However, preliminary radioautographic tests by Weinstock 5 minutes after [$^3$H]galactose injection indicate that little label is detected outside the Golgi apparatus. Moreover, when histochemical detection of glycoproteins has been attempted by a brief treatment with low-pH

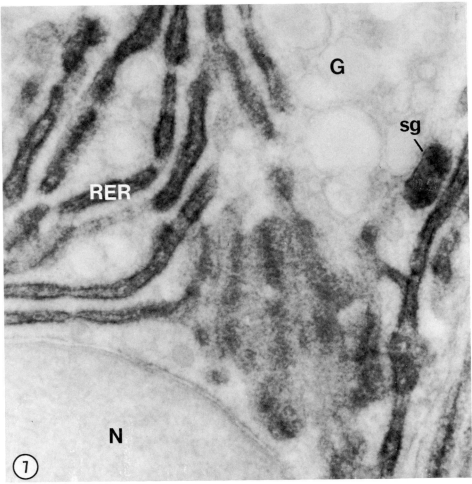

FIGS. 7 and 8.    Portions of odontoblasts immunostained by exposure to peroxidase-labeled anti-pro I antibodies (preparations of Dr. A. Karim). In Fig. 7, strong staining is seen within the profiles of RER cisternae in the supranuclear region of the cell. Portions of the cisternae in the lower central portion of the figure appear to be cut obliquely. At the edge of the Golgi region (G), a secretory granule (sg) is cut obliquely. Even so, the staining of the two poles may be distinguished. N, nucleus. Magnification 44,000×. In Fig. 8, strong staining in the form of fine dots is distributed uniformly throughout a swollen portion of the rough endoplasmic reticulum in the infranuclear area of the cell. This dilated cisterna is continuous at left with flattened cisternae, in which reaction predominates at the periphery. N, nucleus. Magnification 18,000×.

phosphotungstic acid, as proposed by Rambourg (1971)—a method that stains collagen fibrils (Fig. 4, inset), presumably by reaction with the glycosylated side chains—the reactivity of RER cisternae does not exceed background (Fig. 4)

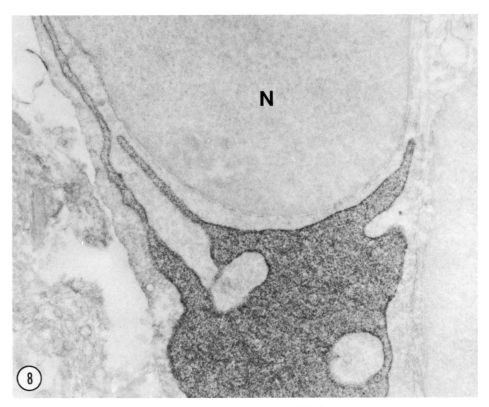

FIG. 8. See legend on p. 178.

(Weinstock and Leblond, 1974a). It thus seems that no significant glycosylation of collagen precursors takes place in RER cisternae. Moreover, since the galactosyl transferase of collagen-producing cells is believed to donate galactose before completion of the procollagen helix (Risteli *et al.*, 1976), it appears that the RER cisternae are not the site of helix formation.

If the procollagen helix does not arise in RER cisternae, then the rodlike elements about 340 nm long described as procollagen by Weinstock (1977) and Bruns *et al.* (1979) should not be present. Ultrastructural examination of the narrow cisternae that constitute the bulk of the RER in odontoblasts and osteoblasts has shown dots of various sizes and short hazy threads, but no rodlike elements. However, in the dilated cisternae, an occasional rodlike element has been seen. Hence, it is possible that some procollagen arises in this particular location.

In *conclusion,* the proline-labeled material initially appearing in RER cisternae consists of pro $\alpha$ chains, but it is not known with certainty where these chains combine to give rise to procollagen. Current biochemical thinking is that helix formation and the associated glycosylation take place in the cisternae of the RER (Brownell and Veis, 1976; Harwood *et al.*, 1975, 1976, 1977; Risteli *et al.*, 1976). Rough microsomes separated by fractionation of tendon cells are said to contain procollagen along with the pro $\alpha1(I)$ and pro $\alpha2$ chains (the respective proportions, calculated from Fig. 6a in Harwood *et al.*, 1977, being about 3, 2, and 1). However, the publications of this group (Harwood *et al.*, 1974, 1975, 1976, 1977) do not mention whether the content of the cell fractions has been checked by quantitative electron microscopy and estimation of marker enzymes. Hence, the rough microsomes may have been contaminated by other cell fractions. Nevertheless, the conclusions of these authors indicate that formation of the procollagen helix in RER cisternae is a possibility.

Yet the search for rodlike molecules has been negative in the narrow cisternae making up most of the RER network of odontoblasts as well as in the various cisternae of osteoblasts. Glycoprotein staining, which is supposed to precede helix formation, is negligible in RER cisternae. It is tentatively concluded that, with the possible exception of minimal helix formation in dilated cisternae, the collagenous material in most RER cisternae consists of pro $\alpha$ chains. Thus, the immunostained material that lines the walls of the narrow cisternae is likely to be composed of nascent pro $\alpha$ chains, whereas the lesser immunostaining of their lumen indicates that few or no released pro $\alpha$ chains accumulate at this site, but instead they migrate away in the direction of other organelles, in accord with the short time spent by labeled material in the RER.

## B.   Role of the Golgi Apparatus

The *ultrastructure* of this organelle is fairly similar in odontoblasts and osteoblasts (Weinstock, 1975; Weinstock and Leblond, 1974a). In both, it consists of stacks of saccules, usually four per stack. Along the edges of the saccules, there are sac-like dilations of two main types referred to as spherical and cylindrical distensions (Fig. 2). The spherical distensions are related to the first or second saccule (counting from the cis face); they contain entangled threads (Sd, Fig. 1), which are thin and irregular, with a diameter of 0.5–2 (rarely 3) nm and show no definite orientation. Hence, unlike RER cisternae, the spherical distensions are packed with filamentous elements.

The cylindrical distensions (Cd, Fig. 1) are related to the third or fourth saccule (being therefore located next to the trans face of the Golgi stacks); they contain thread aggregates 2–3 nm thick, which are grouped into about ten small bundles and measure around 340 nm in length (Weinstock, 1977). The threads within a cylindrical distension are straight, parallel, and in register with one another (Fig. 2).

Radioautography of odontoblasts or osteoblasts shows no silver grains over the Golgi apparatus 2 minutes after intravenous injection of [$^3$H]proline. By 10

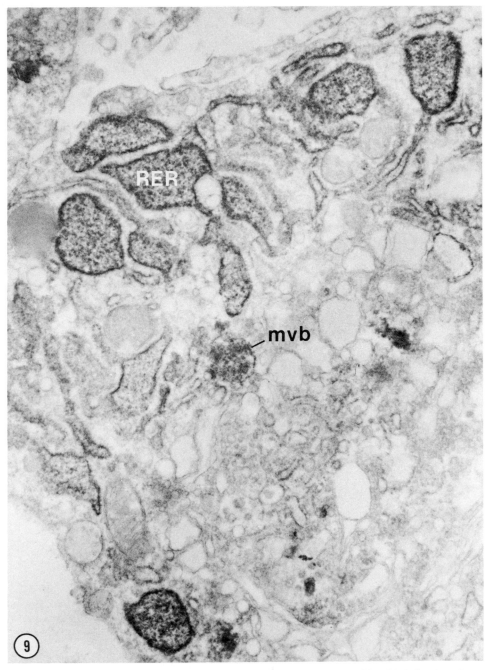

FIG. 9.    A portion of an osteoblast immunostained by exposure to peroxidase-labeled anti-pro I antibodies. A dotlike reaction product can be seen within flat and dilated cisternae of rough endoplasmic reticulum (RER). A multivesicular body (mvb) is stained. Magnification 23,000×.

minutes, silver grains appear over the Golgi region where, by 20 minutes, their number reaches a peak (Fig. 3). Since the Golgi label is lacking at 2 minutes when the blood level of [$^3$H]proline is high, and since it increases thereafter while the blood level declines sharply, it is likely to come from the RER pool rather than from the blood.

At the 10-minute interval, much of the label is localized in the spherical distensions (Weinstock, 1975; Weinstock and Leblond, 1974a,b) as well as on the cis side of the Golgi apparatus. The label on the cis side is presumed to be migrating from the RER through the local organelles, intermediate vesicles, and/or transitional elements (Fig. 2), to reach the spherical distensions. By 20 minutes, the label predominates in the cylindrical distensions. At either time, few silver grains are over the flat portions of the saccules. The successive labeling of spherical and cylindrical distensions, as well as the presence of intermediate forms of these two structures, indicates that the material present in the spherical ones gives rise to the material present in the cylindrical ones (Weinstock and Leblond, 1974a).

The *immunostaining* indicative of the presence of type I procollagen antigenicity (attributable to either pro $\alpha$ chains or procollagen itself) is lacking or rather weak within the flattened portion of the saccules, but is strong in both types of distensions, stronger in fact than in RER cisternae (Fig. 10). In the spherical distensions, the reaction product is uniformly distributed and may predominate as a dense band at the periphery (Fig. 13 in Karim *et al.*, 1979). In the cylindrical distensions, reaction density is even greater than in spherical distensions.

The *glycoprotein reactions,* negative in RER cisternae, are positive in the Golgi region (Fig. 4). A glycoprotein reaction is observed along the membrane of spherical distensions, whereas in cylindrical distensions it affects the bundles of threads, which are stained along their length (Fig. 4).

In conclusion, the proline label rapidly migrates through organelles containing material similar in antigenicity but different in ultrastructure. At 2 minutes, the label is over the indistinct material of RER cisternae; by 10 minutes, some label is over the entangled fine threads of spherical distensions; and by 20 minutes, some label overlies the neatly packed rodlets in cylindrical distensions. Hence, the sequence of events indicates a change in the structures supporting the label as it migrates from organelle to organelle.

Let us first consider the cylindrical distensions. Their content possesses properties—strong antigenicity, glycosylation, arrangement in 340-nm rodlets—that are those of procollagen. Presumably, the material present at this step along the biogenetic pathway consists of bundles of procollagen molecules.

The nature of the entangled threads in spherical distensions is less clear. They probably represent stages intermediate between pro $\alpha$ chains and procollagen— that is, the stages when the chains link through disulfide bonds and then coil into a helix. These events may take place in relation to the membrane of spherical

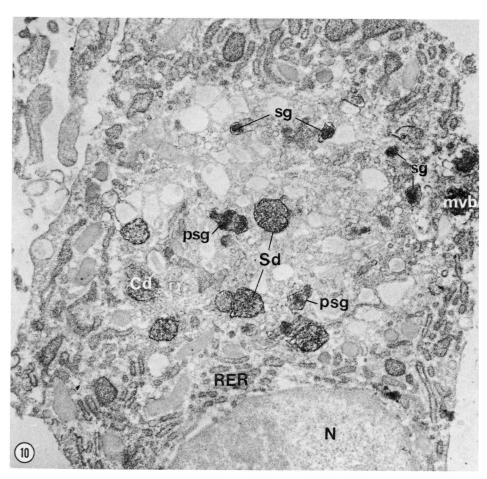

FIG. 10.   A section of an osteoblast immunostained with peroxidase-labeled anti-pro I antibodies. Reaction product is observed within the lumen of the cisternae of rough endoplasmic reticulum (RER) that are found above the nucleus (N) and around the periphery of the cell. A strong reaction is present in some elements of the Golgi region seen in the center of the figure. Large spherical distensions (Sd) and a few cylindrical distensions (Cd) demonstrate the dotlike reaction product. At lower right center, a structure labeled psg is a prosecretory granule arising from the transformation of a cylindrical distension; the stained poles are distinguishable. In center, a prosecretory granule is cut longitudinally, demonstrating an intense staining at the two poles (psg); a coated bud protrudes on its left. Intensely stained secretory granules in the upper half of the figure (sg) are cut transversely through their polar region. Multivesicular bodies (mvb) at the cell periphery are strongly stained. Magnification 21,600×.

distensions, since some density may be seen along their wall when staining for glycoprotein. Whatever the case may be, formation of the procollagen helix would cause the entangled threads to straighten into parallel bundles and thus reshape the spherical distensions into cylindrical ones.

Since the spherical distensions are usually associated with the first or second saccule of the Golgi stack (cis side), and the cylindrical distensions with the third or fourth saccule (trans side), the transformation of the entangled threads into rodlets implies a concomitant displacement in the cis to trans direction. Such a movement could be the result of a wholesale migration of the saccules, in which case their membrane as well as their content would undergo renewal, as proposed in 1966 for the Golgi saccules of goblet cells (Neutra and Leblond, 1966). In pancreatic acinar cells, however, Meldolesi has shown that Golgi membranes are renewed far more slowly than the secretory material that they enclose (Meldolesi, 1974). If these conclusions apply to odontoblasts and osteoblasts, the membranes of the flat and distended portions of the saccules would be relatively static, while their content would migrate toward the trans face, presumably through lateral connections of the saccules as described by Rambourg et al. (1979) in Sertoli cells and spermatids (Y. Clermont et al., unpublished). However, the low responsiveness of the flat portion of saccules to immunostaining and to [$^3$H]proline radioautography suggests another possibility; that is, the flat portion of the saccules would be static, while the distended portion, including membrane and content, would be transferred from cis to trans saccules.

## C.   Secretory Granules

In the past, the formation of collagen fibrils was explained in various ways. Some believed that collagen in a "monomeric" or partly aggregated form arose within the cytosol, from which it was shed outside the cell (Porter and Pappas, 1959; Salpeter, 1968; Wassermann, 1954). When it became clear that the cell elaborated collagen precursors rather than collagen itself, some authors proposed that the precursors were released from RER cisternae to the outside by direct fusion of their membrane with the plasmalemma (Ross and Benditt, 1965) or by transport in smooth vesicles (Goldberg and Green, 1964). However, the investigation of odontoblasts and osteoblasts revealed the presence of structures responsible for the transport of collagen precursors from Golgi apparatus to cell surface (Weinstock, 1972, 1975; Weinstock and Leblond, 1974a). These were referred to as "secretory granules."

*Ultrastructural* examination of the trans face of Golgi stacks shows elongated profiles averaging 150 by 440 nm, which are not attached to saccules. Some of these contain distinguishable thread aggregates and are enclosed by an irregular membrane, which displays bristle-coated "blebs," referred to as "prosecretory granules." Others are narrower and longer; they have a smooth membrane and a fairly homogeneous content in which threads are only faintly distinguishable; they are termed "secretory granules." These granules are seen not only at the periphery of the Golgi region, but also at various points between this region and the apical membrane, along which they frequently accumulate. Release of the thread bundles by exocytosis from these granules has been observed at the apical surface of odontoblasts (Weinstock and Leblond, 1974a) and osteoblasts (Weinstock, 1975).

*Radioautography* reveals that the prosecretory and secretory granules are unlabeled at 2 and 10 minutes after [³H]proline injections, but are labeled occasionally at 20 minutes in the Golgi region and frequently at 30 minutes in the Golgi and apical regions (Fig. 3). Later, the number of labeled granules decreases markedly, while label appears in the adjacent matrices—predentin or prebone (Weinstock, 1975; Weinstock and Leblond, 1974a). This finding confirms that the labeled material is carried by secretory granules from Golgi apparatus to extracellular matrix.

*Glycoprotein reactions* are intense in both types of granules (Fig. 4). The rectangular appearance of the stained granules suggests that the reaction is localized along the length of the bundled threads (Weinstock and Leblond, 1974a).

*Immunostaining* with anti-pro I antibodies is intense in prosecretory and maximal in secretory granules (Fig. 10). There is a tendency for the reaction to predominate at the two poles of the granules (Figs. 7 and 10).

In conclusion of these various observations, the threads that may be distinguished within prosecretory granules, and, less clearly, in secretory granules, stain for glycoprotein along their length, whereas immunostaining predominates at the poles. The glycoprotein reactivity along the threads may be attributed to the helical part of the procollagen where the galactose and galactosyl-glucose groups are located, whereas the polar arrangement of the immunostaining is explained by the much stronger immunogenicity of the terminal peptides than of the central helical region (Timpl *et al.*, 1977). Since the respective sizes of the NH₂ terminal, the helical region, and the COOH terminal are in approximate ratios of 1:5:2, one might have expected smaller poles and a longer central region than has been observed. To explain the large size of the polar densities, it is necessary to assume a substantial overlap of the antigenic terminals within each pole, in contrast to the cylindrical distension aggregates, which are in register with one another. The transformation from these distensions to the secretory granules would thus involve some sliding of procollagen molecules toward one or the other pole. The resulting overlap would also account for the granules being longer (400 to over 500 nm, with a mean of 440 nm) than the procollagen molecules (about 340 nm).

## D.   The Extracellular Matrices: Predentin and Prebone

The predentin is a regular, sharply outlined band of nonmineralized matrix separating odontoblasts from dentin; its thickness is maintained at around 30 nm. Similarly, the prebone is a nonmineralized matrix layer lying between osteoblasts and bone tissue, but its thickness and regularity are variable.

The *ultrastructure* of these matrices consists of collagen fibrils embedded within material that is poorly stained by routine uranyl-lead but shows fine granularity. In predentin, collagen fibrils are scarce in the proximal region—that is, next to the cells—but they gradually increase in number to reach a maximum in the distal region—that is, toward the predentin–dentin junction. Conversely, the interstitial material decreases from the proximal to the distal region.

*Radioautographic* examination has shown a discrete reaction 30–35 minutes after injection of [$^3$H]proline or other precursors within predentin and prebone. The labeling rapidly increases at 90 minutes and especially at 4 hours. After one to several days, the labeled material of predentin is found in dentin, where it is maintained indefinitely (Greulich and Leblond, 1954).

*Glycoprotein reactions* are intense along the length of collagen fibrils, but weak or absent on the interstitial material (inset in Fig. 4).

The *immunostaining* shows strong pro I antigenicity in proximal predentin (Cournil *et al.*, 1979; Karim *et al.*, 1979) and prebone, but follows a rapidly decreasing gradient of intensity away from the cells, so that only a faint reaction is present in the distal third.

In conclusion, the proline label ends in predentin or prebone, where pro I antigenicity testifies to the presence of procollagen (although some antigenicity may be due to the $NH_2$- and COOH-terminal peptides excised by enzymes in the course of procollagen transformation to collagen). Away from the cell border, the antigenicity decreases and disappears, presumably owing to the progressive disappearance of procollagen, as it transforms into unreactive collagen; and indeed collagen fibrils increase in number and size.

## E.   Lysosomal Structures

Secondary lysosomes and multivesicular bodies are commonly found in odontoblasts and osteoblasts. *Immunostaining* has revealed that, unlike most of the organelles located outside the biosynthetic pathway of procollagen, such as the nucleus and mitochondria, secondary lysosomes and multivesicular bodies (Fig. 10) may be immunostained by anti-pro I antibodies in odontoblasts (Karim *et al.*, 1979) and osteoblasts (Figs. 9 and 10) and, therefore, are endowed with pro I antigenicity. In multivesicular bodies, the matrix but not the vesicle lumen is reactive.

After this property of lysosomes had been recognized, the *radioautographs* made after [$^3$H]proline injection as previously described (Weinstock and Leblond, 1974a) were re-examined. Melvyn Weinstock also examined radioautographs exposed for several years. A discrete but significant labeling of lysosomal structures was observed, mainly at the 90-minute and 4-hour intervals. These results indicate that a small amount of label enters lysosomal structures. One possibility is that, although most of the collagen precursors reach secretory granules, a small proportion of them end in lysosomes. Another possibility is that the $NH_2$ and COOH propeptides released in the course of procollagen transformation to collagen are endocytosed and taken into lysosomes. The hydrolytic enzymes known to be present in lysosomes, such as acid phosphatase and aryl sulfatase, have been detected in odontoblasts by C. E. Smith (unpublished). It is, therefore, expected that the material taken up by lysosomal structures is broken down. Such destruction has also been observed after addition of [³H]proline to cultures of lung fibroblasts, since free [³H]hydroxyproline appears in the medium (Bienkowski *et al.*, 1978).

# IV. General Conclusion

Odontoblasts and osteoblasts have been immunostained by using antibodies that recognize pro $\alpha$(I) chains and type I procollagen, but not type I collagen. Immunostaining by these antibodies is moderate in RER cisternae, strong in the spherical distensions and even stronger in the cylindrical distensions of Golgi saccules, and maximal at the poles of secretory granules. This progressive increase in antigen concentration occurs along the biosynthetic pathway revealed by the successive arrival of [³H]proline label in RER at 2 minutes, spherical and cylindrical distensions at 10 and 20 minutes, respectively, and secretory granules at 30 minutes. However, it is not until the cylindrical distensions are reached that there are bundles of rodlets showing the glycosylation and ultrastructural features characteristic of procollagen. It is proposed that, with the possible exception of happenings in dilated RER cisternae, the disulfide coupling of pro $\alpha$ chains and the maturation into a procollagen helix occur within the distensions of Golgi saccules. After these processes are completed, the distensions free themselves of the saccules to become prosecretory granules, and these in turn condense into secretory granules, which migrate to the cell apex.

Substances with procollagen antigenicity but so far unidentified find their way into lysosomal structures, where they are presumably broken down.

Finally, the secretory granules of odontoblasts and osteoblasts release their procollagen content into predentin and prebone, respectively. The concentration of the released procollagen is high in these matrices along the cell border, but gradually decreases in a distal direction reflecting its transformation into collagen. The latter appears to be immediately built into fibrils.

ACKNOWLEDGMENTS

This work was carried out with the support of Grant DE-04547 from the National Institutes of Health and Grant MT-906 from the Medical Research Council of Canada. The assistance of Drs. George R. Martin, Isabelle Cournil, and Melvyn Weinstock is acknowledged.

REFERENCES

Avrameas, S., and Ternynck, T. (1971). *Immunochemistry* **8**, 1175-1179.
Bienkowski, R. S., Cowan, M. J., McDonald, J. A., and Crystal, R. G. (1978). *J. Biol. Chem.* **253**, 4356-4363.
Bornstein, P. (1974). *Annu. Rev. Biochem.* **43**, 567-603.
Brownell, A. G., and Veis, A. (1976). *J. Biol. Chem.* **251**, 7137-7143.
Bruns, R. R., Hulmes, D. J. S., Therrien, S. F., and Gross, J. (1979). *Proc. Natl. Acad. Sci. U.S.A.* **76**, 313-317.
Carneiro, J., and Leblond, C. P. (1959). *Exp. Cell Res.* **18**, 291-300.
Carneiro, J., and Leblond, C. P. (1966). *J. Histochem. Cytochem.* **14**, 334-344.
Clark, C. C., Fietzer, P. P., and Bornstein, P. (1975). *Eur. J. Biochem.* **56**, 327-333.
Cournil, I., Leblond, C. P., Pomponio, J., Hand, A. R., Sederlof, L., and Martin, G. R. (1979). *J. Histochem. Cytochem.* **27**, 1059-1069.
Diegelmann, R. F., Bernstein, L., and Peterkofsky, B. (1973). *J. Biol. Chem.* **248**, 6514-6521.
Gartner, L. P., Seibel, W., Hiatt, J. L., and Provenza, D. V. (1979). *Acta Anat.* **103**, 16-33.
Goldberg, B., and Green, H. (1964). *J. Cell Biol.* **22**, 227-258.
Greulich, R. C., and Leblond, C. P. (1964). *J. Dent. Res.* **33**, 859-872.
Harwood, R., Grant, M. E., and Jackson, D. S. (1974). *Biochem. Biophys. Res. Commun.* **59**, 947-954.
Harwood, R., Grant, M. E., and Jackson, D. S. (1975). *Biochem. J.* **152**, 291-302.
Harwood, R., Grant, M. E., and Jackson, D. S. (1976). *Biochem. J.* **156**, 81-90.
Harwood, R., Merry, A. H., Woolley, D. E., Grant, M. E., and Jackson, D. S. (1977). *Biochem. J.* **161**, 405-418.
Karim, A., Cournil, I., and Leblond, C. P. (1979). *J. Histochem. Cytochem.* **27**, 1070-1083.
Kopriwa, B. M. (1973). *Histochemie* **37**, 1-17.
Kopriwa, B. M., and Leblond, C. P. (1962). *J. Histochem. Cytochem.* **10**, 269-284.
LeBlond, C. P. (1963). *Ann. Histochim.* **8**, 43-50.
Meldolesi, J. (1974). *Philos. Trans. R. Soc. London Ser. B* **268**, 39-53.
Neutra, M., and Leblond, C. P. (1966). *J. Cell Biol.* **30**, 119-136.
Oikarinen, A., Anttinen, H., and Kivirikko, K. I. (1976). *Biochem. J.* **156**, 545-551.
Porter, K. R., and Pappas, G. O. (1969). *J. Biophys. Biochem. Cytol.* **5**, 153-166.
Rambourg, A. (1971). *Int. Rev. Cytol.* **31**, 57-114.
Rambourg, A., Clermont, Y., and Hermo, L. (1979). *Am. J. Anat.* **154**, 455-476.
Risteli, L., Myllyla, R., and Kivirikko, K. I. (1976). *Biochem. J.* **155**, 145-153.
Ross, R., and Benditt, E. P. (1965). *J. Cell Biol.* **27**, 83-106.
Salpeter, M. M. (1968). *J. Morphol.* **124**, 387-421.
Smith, B. D., McKenney, K. H., and Lustberg, T. L. (1977). *Biochemistry* **16**, 2980-2985.
Timpl, R., Wick, G., and Gay, S. (1977). *J. Immunol. Methods* **18**, 165-182.
Wassermann, R. (1954). *Am. J. Anat.* **94**, 399-437.
Weinstock, M. (1972). *Z. Zellforsch. Mikrosk. Anat.* **129**, 455-470.

Weinstock, M. (1975). *In* "Extracellular Matrix Influences on Gene Expression" (H. C. Slavkin and R. C. Greulich, eds.), pp. 119–128. Academic Press, New York.
Weinstock, M. (1977). *J. Ultrastruct. Res.* **61,** 219–229.
Weinstock, M., and Leblond, C. P. (1974a). *J. Cell Biol.* **60,** 92–127.
Weinstock, M., and Leblond, C. P. (1974b). *Fed. Proc., Fed. Am. Soc. Exp. Biol.* **33,** 1205–1218.

# Chapter 13

# Role of the 6-Phosphomannosyl-Enzyme Receptor in Intracellular Transport and Adsorptive Pinocytosis of Lysosomal Enzymes

WILLIAM S. SLY, H. DAVID FISCHER,
ALFONSO GONZALEZ-NORIEGA, JEFFREY H. GRUBB,
AND MARVIN NATOWICZ

*Departments of Pediatrics and Genetics, Washington University School of Medicine, St. Louis, Missouri, and Division of Medical Genetics, St. Louis Children's Hospital, St. Louis, Missouri*

## I. Introduction

This chapter reviews the evidence that the recognition marker on lysosomal enzymes contains mannose 6-phosphate (Man 6-P) and describes experiments

more potent than other hexosephosphates tested and proved to be over 1000 times more potent than mannose as an inhibitor of enzyme pinocytosis (Kaplan *et al.*, 1977a). If phosphate were present on the enzyme as a phosphomonoester, we reasoned that it might be sensitive to phosphatases. This proved to be the case. Alkaline phosphatase treatment of ''high-uptake'' β-glucuronidase from platelets converted the enzyme to less acidic, low-uptake enzyme forms that were no longer susceptible to pinocytosis (Kaplan *et al.*, 1977a). These observations led us to propose (Kaplan *et al.*, 1977a; Sly *et al.*, 1978) that Man 6-P was present in the recognition marker for uptake of human β-glucuronidase by fibroblasts. Until this report (Kaplan *et al.*, 1977a), there was no precedent for Man 6-P in a mammalian glycoprotein. Very soon, however, the observations providing indirect evidence for Man-6-P recognition in pinocytosis of human platelet β-glucuronidase were extended to several other enzymes (Kaplan *et al.*, 1977b; Sando and Neufeld, 1977; Ullrich *et al.*, 1978). These studies led us to propose (Kaplan *et al.*, 1977b; Sly *et al.*, 1978) that Man 6-P was an essential component of the ''common recognition marker'' for uptake of acid hydrolases by human fibroblasts, and to predict (Kaplan *et al.*, 1977b; Sly *et al.*, 1978) that Man 6-P would be missing or masked in enzyme secreted by I-cell disease fibroblasts. This prediction appears to have been confirmed by the recent studies of Hasilik *et al.* (1979) showing $^{32}$P incorporation into newly synthesized acid hydrolases from normal fibroblasts, and the failure to see $^{32}$P incorporation into newly synthesized acid hydrolases from I-cell fibroblasts. Related findings were reported by Bach *et al.* (1979).

# III.  Direct Evidence for Mannose 6-Phosphate on High-Uptake Lysosomal Enzymes

Natowicz *et al.* (1979) recently reported direct evidence for Man 6-P in the recognition marker for human β-glucuronidase purified from spleen. Using an enzymatic assay for Man 6-P, they showed that Man 6-P was released from high-uptake enzyme on acid hydrolysis. In addition, they found that the Man-6-P content of β-glucuronidase varied directly with its susceptibility to pinocytosis by fibroblasts. These experiments are presented in Figs. 1 and 2. Figure 1 shows the CM–Sephadex elution profile from the final step in the purification of spleen β-glucuronidase to homogeneity. This figure shows that when the uptake properties of the enzyme across the peak were analyzed, there was approximately 20-fold greater uptake activity in the acidic high-uptake enzyme eluting at low salt concentrations than in the less-acidic lower-uptake fractions eluting at higher salt concentrations. The correlation between the uptake activity of the enzyme from fractions across the profile shown in Fig. 1, and the Man 6-P released on

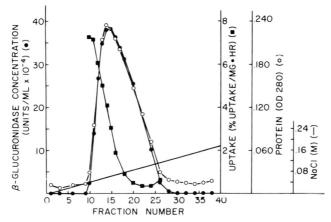

FIG. 1.   Uptake properties of forms of pure human spleen $\beta$-glucuronidase resolved by CM–Sephadex chromatography. This profile was taken from the second ion-exchange step described in Method B by Natowicz *et al.* (1979). Enzyme was taken from fractions across the peak and assessed for susceptibility to pinocytosis in a standard uptake assay in which 2000 units of $\beta$-glucuronidase was added to 35-mm dishes of $\beta$-glucuronidase-deficient fibroblasts in 1 ml of medium, and the dishes were incubated for 2 hours at 37°C. Then cells were washed thoroughly and lysed in deoxycholate, and the lysates were assayed for cell-associated enzyme and protein. Uptake was expressed as the percentage of added enzyme internalized per milligram of protein per hour. After Natowicz *et al.* (1979).

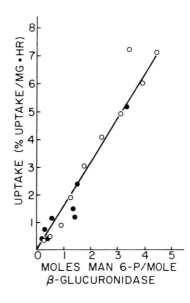

FIG. 2.   Correlation between the Man-6-P content and the uptake of pure human spleen $\beta$-glucuronidase. Enzyme from fractions from the CM–Sephadex profile in Fig. 1 (open circles) and from other purifications (solid circles) was assayed for susceptibility to pinocytosis, as described in Fig. 1, and for Man-6-P content. Mannose 6-phosphate was measured enzymatically as described in detail by Natowicz *et al.* (1979) on acid-hydrolyzed aliquots of $\beta$-glucuronidase (1–4 $\mu$g) using the enzymatic cycling assays developed by Lowry. For details, see Natowicz *et al.* (1979).

acid hydrolysis of enzyme from the same fractions, is presented in Fig. 2 (open circles). The solid circles in Fig. 2 represent data from fractions from several other enzyme preparations. There was clearly a correlation between the Man-6-P content of the enzyme from these preparations and the susceptibility of the enzyme to pinocytosis.

Natowicz *et al.* (1979) showed further than the Man 6-P on the enzyme was present in oligosaccharides released by treatment of the enzyme with endoglycosidase H. As is seen in Table I, this treatment produced a concomitant loss in susceptibility of the enzyme to pinocytosis, although it produced no loss in catalytic activity of the enzyme. Sahagian *et al.* (1979) also reported Man 6-P on $\beta$-galactosidase from beef testes, and Distler *et al.* (1979a) demonstrated Man 6-P in glycopeptides from bovine testicular glycoproteins that had been initially purified as inhibitors of enzyme pinocytosis. Von Figura and Klein (1979) recently reported that endoglycosidase H converted high-uptake $\alpha$-N-acetylglucosaminidase into an enzyme form no longer susceptible to adsorptive pinocytosis. This treatment was reported to liberate acidic oligosaccharides, which were susceptible to degradation by alkaline phosphatase, $\alpha$-mannosidase, and $\beta$-N-acetylglucosaminidase. The authors concluded that the recognition marker was present on high-mannose oligosaccharide chains bearing phosphory-

TABLE I

EFFECT OF ENDOGLYCOSIDASE H ON UPTAKE PROPERTIES AND MAN-6-P CONTENT OF "HIGH-UPTAKE" $\beta$-GLUCURONIDASE[a]

| $\beta$-Glucuronidase | Uptake activity (%/mg/hr) | Reduction of uptake (%) activity | Man 6-P (mol/mol enzyme) | | Man 6-P released (%) |
|---|---|---|---|---|---|
| | | | Ethanol (ppt) | Supernatant | |
| Untreated | 3.8 | — | 3.5 | 0.0 | — |
| Endoglycosidase H-treated | 0.1 | 97 | 0.3 | 3.9 | 93 |

[a] Aliquots of pure spleen $\beta$-glucuronidase (10,000 units) were incubated with or without 9 milliunits of endoglycosidase H in a final volume of 15 ml containing 0.024 $M$ sodium acetate, 0.012% sodium azide, pH 5.5, and 0.02% bovine serum albumin for 24 hours at 37°C. No $\beta$-glucuronidase catalytic activity was lost during this incubation. Following the incubation, some samples were diluted into minimum essential media containing 15% fetal calf serum and assayed for enzymatic activity and susceptibility to pinocytosis. Other samples were precipitated with ice-cold 80% ethanol, and Man 6-P was measured on the precipitated $\beta$-glucuronidase and on the oligosaccharides present in the supernatant. Ninety percent of the initial $\beta$-glucuronidase catalytic activity was recovered in the ethanol precipitate; the supernatant contained no measurable enzymatic activity. From Natowicz *et al.* (1979).

lated mannose and/or $N$-acetylglucosamine residues at the nonreducing termini. Thus, three different laboratories have provided evidence for Man 6-P in the recognition marker on three different enzymes by three different approaches. Thus, the prediction based on indirect evidence (Kaplan *et al.*, 1977b; Sly *et al.*, 1978) that Man 6-P would be found in the "common recognition marker" for acid hydrolases has been confirmed.

## IV.    Are Other Structural Features Involved in Enzyme Recognition?

High-uptake enzymes had much higher affinity for the receptor ($K_m = 1 \times 10^{-9}M$) than Man 6-P ($K_i = 2 \times 10^{-5}$), suggesting that there were additional structural features of the recognition marker that contributed to high-affinity binding, or that binding of enzyme to receptors involved more than one recognition marker. Distler *et al.* (1979b) reported that Man $\alpha 1 \rightarrow 2$ Man was 100-fold more potent as a pinocytosis inhibitor than mannose. The same group (Distler *et al.*, 1979a) later reported finding Man 6-P linked $\alpha 1 \rightarrow 2$ to mannose in a disaccharide isolated from hydrolyzates of bovine testicular inhibitory glycopeptides. This raised the possibility that Man 6-P in $1 \rightarrow 2$ linkage to mannose is a structural feature of the recognition marker, which contributes to high-affinity binding and potent inhibition, since Man 6-P in $1 \rightarrow 3$ linkage to mannose at the reducing terminus of the phosphomannan fragment pentamannosyl monophosphate was not significantly more inhibitory than Man 6-P as a monosaccharide (Fischer *et al.*, 1980a). Several laboratories are presently working out the structure of the recognition marker on high-uptake enzymes. In addition, rapid progress is being made in isolating the phosphomannosyl-enzyme receptor (G. Sahagian and G. W. Jourdian, personal communication). Complementary studies on the structure of the recognition marker and on structural features of inhibitors of binding to the isolated receptor should lead to a precise definition of the requirements for high-affinity binding to the pinocytosis receptor on fibroblasts. High-affinity binding might also be conferred by a multivalent interaction involving more than one Man 6-P on high-uptake enzymes with one or more cell-surface receptors. Kaplan *et al.* (1977) reported that high-molecular-weight multivalent phosphomannan fragments that contained many Man-6-P groups were very potent inhibitors of enzyme pinocytosis compared with free Man 6-P. Fischer *et al.* (1979, 1980a) extended these studies and reported that a large-molecular-weight phosphomonoester fragment (PPME) of *H. holstii* phosphomannan (MW = $1 \times 10^6$), which contains about 1000 Man-6-P groups per molecule, was 100 times more potent as an inhibitor of pinocytosis, per mole of phosphate, than the monovalent fragment pentamannosyl monophosphate (MW 1000). This means that the PPME was 100,000 times more potent per molecule. Furthermore, the large-molecular-weight multivalent fragment was found to be subject to adsorptive endocytosis by the same pinocytosis receptors that mediate uptake of lysosomal enzymes (Fischer *et al.*, 1979, 1980a). The pinocytosis of PPME was subject to substrate inhibition; i.e., the PPME inhibited its own pinocytosis at high concentrations, as one would predict if it were acting as a multivalent ligand. In contrast to the PPME, the monovalent low-molecular-weight pentamannosyl monophosphate was not taken up at a rate faster than one could explain by the rate of nonspecific fluid endocytosis. These studies suggested that a multivalent interaction could be important for high-affinity binding to the acid hydrolase pinocytosis receptors, at least for artificial ligands, and could also contribute to the high-uptake property—i.e., the susceptibility to pinocytosis. By analogy, we suggested that acid hydrolases themselves might act as multivalent ligands for these receptors. They could do so if more than one Man 6-P were present on a single enzyme recognition marker, or if multiple Man 6-P-containing recognition markers were to interact with one or more pinocytosis receptors to confer high-affinity binding and susceptibility to pinocytosis. The report by Natowicz *et al.* (1979) that high-uptake fractions of human $\beta$-glucuronidase contained over 4 moles of Man 6-P per mole of enzyme made a multivalent interaction for this enzyme seem possible. Since many acid hydrolases are multimeric enzymes, many enzymes may be capable of multivalent interactions with multiple phosphomannosyl pinocytosis receptors.

## V.    Alternatives to the Secretion–Recapture Hypothesis for Enzyme Transport

The findings in I-cell disease fibroblasts led Neufeld and co-workers (Hickman and Neufeld, 1972; Neufeld *et al.*, 1977) to propose that all enzymes are normally "secreted" into the extracellular medium and delivered to lysosomes by receptor-mediated "recapture." This proposal became known as the "secretion–recapture hypothesis." The discovery that Man 6-P inhibited uptake of exogenous enzymes made it possible to estimate the fraction of enzymes secreted and recaptured by normal fibroblasts, and thus to test the secretion–recapture hypothesis quantitatively. If enzymes must first be secreted into the medium and then recaptured to reach lysosomes, growth of cells in the presence of a competitive inhibitor such as Man 6-P that blocks enzyme pinocytosis should trap secreted enzyme outside and lead to depletion of intracellular enzyme levels (Sly *et al.*, 1978; Sly and Stahl, 1978). This kind of experiment was done in several laboratories in different ways (von Figura and Weber, 1978; Sly and Stahl, 1978; Vladutiu and Rattazzi, 1979). Table II presents an experiment

TABLE II

EFFECT OF 10 mM MAN 6-P ON THE DISTRIBUTION OF "NEWLY SYNTHESIZED" HEXOSAMINIDASE B IN GROWING FIBROBLASTS

| | Intracellular enzyme | | | Extracellular enzyme | |
|---|---|---|---|---|---|
| | A<br>Final<br>units | B<br>Initial<br>units | C<br>Newly<br>synthesized<br>units | D<br>Units in<br>medium | D/(C + D)<br>(% of newly<br>synthesized<br>extracellular) |
| Control | 1433 | 498 | 935 | 233 | 20 |
| +Man 6-P | 1409 | 498 | 911 | 301 | 24.8 |

[a] Tay-Sachs disease fibroblasts were split 1 : 5, allowed to attach for 24 hours, and then grown for 9 more days with and without 10 mM Man 6-P. Medium was changed on days 4 and 7. Initial hexosaminidase levels (at 24 hours) were 498 U/plate. Cells were harvested on day 10 by freeze-thawing, and hexosaminidase was measured in cell lysates and in the collected media. Newly synthesized intracellular enzyme (C) was calculated by subtracting the initial enzyme level (B) from the final intracellular enzyme level (A). Total newly synthesized enzyme was assumed to be C + D.

showing the effect of Man 6-P in the growth medium on enzyme distribution in Tay-Sachs disease fibroblasts that secrete mainly heat-stable hexosaminidase B. In this experiment, like many others, growth of cells in media containing Man 6-P did not reduce intracellular enzyme levels and produced only marginal changes in extracellular enzyme levels. This and similar experiments suggested that secretion of lysosomal enzymes into the extracellular medium and receptor-mediated recapture is not the major pathway for acid hydrolases to reach lysosomes. These observations convinced us that most lysosomal enzymes reach lysosomes without leaving the cell and led us to propose an intracellular pathway for the majority of receptor-mediated enzyme delivery to lysosomes (Gonzalez-Noriega *et al.*, 1980; Sly, 1979; Sly and Stahl, 1978), which is discussed further in the next section. Lloyd (1977) published an alternative to the secretion–recapture hypothesis to explain the findings in I-cell disease fibroblasts. He suggested that acid hydrolases enter lysosomes as they form by budding off the Golgi or GERL, and that the recognition marker is not needed for lysosomal enzymes to enter lysosomes but is needed, instead, for enzymes to be retained in lysosomes. Because endocytic vesicles continually join lysosomes and exocytic vesicles presumably pinch off from lysosomes and return membrane components to the plasma membrane, he reasoned that the recognition marker was important to anchor lysosomal enzymes to lysosomal membranes, and to prevent the hydrolases from being lost to the extracellular medium in exocytic vesicles that bud off of lysosomes. This proposal viewed secretion of lysosomal enzymes into the extracellular medium in I-cell disease, not as a result of failure to deliver them to lysosomes, but rather as a result of failure to retain them in lysosomes because they lacked their anchor. Two observations made Lloyd's suggestion improbable. First, Glaser *et al.* (1975) presented results indicating that the recognition marker was removed soon after enzyme pinocytosis. When acidic high-uptake β-glucuronidase was taken up by enzyme-deficient fibroblasts, it was rapidly (within 24 hours) converted into less-acidic, low-uptake enzyme forms. Yet, the half disappearance time of pinocytosed β-glucuronidase in enzyme-deficient fibroblasts following uptake was at least 2 weeks (Nicol *et al.*, 1974; Sly *et al.*, 1975). Thus, enzyme was retained long after the recognition marker that was important for its delivery to lysosomes had been removed. These observations convinced us that the recognition marker, though important for delivery to lysosomes, was not important for retention in lysosomes following pinocytosis.

To explain the failure of Man 6-P in the growth medium to deplete intracellular enzyme levels, von Figura and Weber (1978) proposed another alternative to the secretion–recapture hypothesis. They suggested that lysosomal enzymes are normally delivered to lysosomes indirectly by vesicles that bring receptor-bound acid hydrolases first to the plasma membrane, after which receptor-bound enzyme is internalized without ever dissociating from the cell-surface receptors. The proposal that the "major part of the lysosomal enzyme cycles via the cell surface in a receptor-bound form" would explain why only a minor fraction of lysosomal enzymes are normally released into the extracellular space except in I-cell disease, where all the enzymes are released. I-cell enzymes would be released as vesicles fuse with the plasma membrane because the I-cell enzymes lack the recognition marker and are not receptor-bound. Evidence for enzyme on the cell surface (von Figura and Weber, 1978; von Figura and Voss, 1979) was of two types. First, four different enzymes were demonstrated by immunofluorescence on the surface of normal cells and found not to be present on the surface of I-cell disease fibroblasts. Second, gentle trypsinization of normal fibroblasts released measurable enzyme from the cell surface.

## VI. The Intracellular Pathway for Receptor-Mediated Segregation of Lysosomal Enzymes

We proposed (Sly and Stahl, 1978; Sly *et al.*, 1979) another alternative to the secretion–recapture hypothesis for enzyme delivery to lysosomes that recognizes the existence of a pinocytic pathway for enzymes to reach lysosomes in fibroblasts, but postulated that the primary route for delivery of acid hydrolases to lysosomes involves an intracellular pathway. We suggested that acid hydrolases produced in the endoplasmic reticulum receive the phosphomannosyl recognition marker that enables them to bind to intracellular membrane receptors, which collect into specialized vesicles in the Golgi or GERL and bud off as primary lysosomes. The key components of this intracellular delivery system are the phosphomannosyl recognition marker on the enzymes, and its receptor on intracellular membranes. We proposed (Fischer *et al.*, 1980b; Gonzalez-Noriega *et al.*, 1980; Sly *et al.*, 1979; Sly and Stahl, 1978) that these two elements provide a system that segregates most newly synthesized lysosomal enzymes from other products of the endoplasmic reticulum that are destined for secretion into the extracellular medium. The pinocytic pathway, which was identified by the uptake of corrective factors by enzyme-deficient fibroblasts, clearly exists, but we suggest that it is a secondary, quantitatively less important pathway, even in fibroblasts. In this view, enzyme that appears in the medium of normal fibroblasts is not considered an obligatory intermediate in enzyme transport, but rather enzyme that has escaped the normal intracellular segregation process. If the recognition marker is not added to newly synthesized enzyme, as in I-cell disease (Hickman and Neufeld, 1972; Hasilik *et al.*, 1979), the enzymes lacking the recognition marker would fail to bind receptors, would fail to be segregated into lysosomes, and would be secreted instead. This proposal (Sly, 1979; Sly *et al.*, 1979; Sly and Stahl, 1978) explains why nearly all newly synthesized enzymes in I-cell disease fibroblasts are secreted into the extracellular medium. We have found that the normal pathway for intracellular transport can also be disrupted by chloroquine and ammonium chloride (Gonzalez-Noriega *et al.*, 1980), which divert newly synthesized enzyme to the extracellular medium (enhance secretion) and also block pinocytosis of exogenous enzymes. We interpret the effects of amines to result, at least in part, from elevation of intralysosomal pH, one consequence of which is the failure of reutilization of the phosphomannosyl-enzyme receptors following delivery of enzyme to lysosomes (Gonzalez-Noriega *et al.*, 1980). This aspect will be discussed further below.

The observations summarized above led us to propose (Gonzalez-Noriega *et al.*, 1980; Sly, 1979; Sly *et al.*, 1979; Sly and Stahl, 1978) that, although enzyme can reach lysosomes in fibroblasts by two pathways, the major pathway is the intracellular one. The secretion (or escape) of enzyme into the medium and return by adsorptive endocytosis (secretion–recapture) is felt to be a much less important pathway. However, we suggest that both pathways depend on the phosphomannosyl-enzyme receptor for delivery of enzyme to lysosomes, and that both pathways depend on receptor reutilization. Both pathways appeared to be disrupted by lysosomotropic amines (Gonzalez-Noriega *et al.*, 1980), which suggested to us that amines disrupt both pathways by interfering with receptor reutilization.

The proposal by von Figura and Weber (1978) and our proposal (Gonzalez-Noriega *et al.*, 1980; Sly, 1979; Sly *et al.*, 1979; Sly and Stahl, 1978) that most

enzyme is segregated at the Golgi or GERL by collecting into specialized vesicles that bud off as primary lysosomes have certain similarities. The major
difference between these proposals is that in one case vesicles containing
receptor-bound enzymes go directly to lysosomes, and in the other the vesicles
go first to the plasma membrane and the receptor-bound enzymes are subsequently internalized by endocytosis without dissociating from the receptors that
delivered them to the plasma membrane. We suspected that the fraction of
enzyme that reaches the cell surface was not large because growth of cells in 10
m*M* Man 6-P, which displaced over 90% of the bound enzyme from the cell
surface in 15 minutes at 37°C (Gonzalez-Noriega *et al.,* 1980), did not detect-
ably deplete cells of intracellular enzyme (Table II). However, if Man 6-P were
significantly less effective in displacing the receptor-bound biosynthetic intermediate than in displacing added high-uptake enzyme, or if the time that the
receptor-bound intermediate spends on the cell surface en route to lysosomes is
too brief to permit displacement by Man 6-P, the conclusions drawn from failure
to deplete cells by growth in the presence of competitive inhibitors could be
incorrect.

## VII.  Evidence That Cell-Surface Receptors Must Recycle

The kinetics of binding of enzyme to membranes from broken fibroblasts
(Fischer *et al.,* 1980b,c) and the kinetics of enzyme pinocytosis (Gonzalez-
Noriega *et al.,* 1980) provided additional insight into the role of the recognition
marker and its receptor in enzyme transport. We calculated (Gonzalez-Noriega *et
al.,* 1980) from the amount of enzyme bound to the cell surface at saturation (73
U/mg) and the maximum rate of enzyme internalization (806 U/mg × hr) that
cell-surface receptors must be replaced or reutilized every 5 minutes to explain
the rates of enzyme pinocytosis by the amount of cell-surface binding activity
(Gonzalez-Noriega *et al.,* 1980). Yet, enzyme pinocytosis was observed to
continue at maximal rates in the presence of inhibitors or protein synthesis such
as cycloheximide for up to 3 hours (Gonzalez-Noriega *et al.,* 1980). This observation would suggest either a pool of receptors containing 36 times the number of
receptors present on the cell surface, or that some of the cell-surface receptors are
reutilized for additional rounds of pinocytosis following internalization. Studies
of enzyme binding to membranes from broken fibroblasts (Fischer *et al.,*
1980b,c) revealed that over 80% of the specific enzyme binding (i.e., the Man-
6-P-inhibitable enzyme binding) was to intracellular membranes. Only 20% of
the total enzyme binding to membranes from homogenates was present on the
cell surface prior to breaking the cells. The 20% of the total binding activity that
was present on the cell surface was destroyed by trypsinizing the fibroblasts. The

80% of binding activity inside the cells became accessible to trypsin degradation only when the cells were disrupted. The finding that the number of internal receptors was at least four times as large as the number measurable on the cell surface is consistent with the idea that intracellular phosphomannosyl-enzyme receptors play a role in segregating newly synthesized lysosomal enzymes. However, it is not clear how the internal pool of receptors is related to those on the cell surface. Even if the entire intracellular pool of receptors were in equilibrium with the cell-surface receptors, the total number of cell receptors that would be available for enzyme pinocytosis would still be too small to explain the observed rates of pinocytosis of exogenous enzyme by fibroblasts without invoking reutilization or recycling of cell-surface receptors (Fischer *et al.*, 1980b; Gonzalez-Noriega *et al.*, 1980).

## VIII.   Chloroquine and Other Amines Impair Receptor Reutilization

Lysosomotropic amines disrupt enzyme transport; therefore, studies with these agents give some additional insight into the role of the recognition marker and its receptor in enzyme transport. Chloroquine has been found to inhibit pinocytosis of exogenous enzyme by several laboratories (Gonzalez-Noriega *et al.*, 1980; Sando *et al.*, 1979; Wicsmann *et al.*, 1975). We recently reported experiments (Gonzalez-Noriega *et al.*, 1980) suggesting that this inhibition of enzyme pinocytosis was due to inhibition of reutilization of cell-surface receptors following internalization of enzyme–receptor complexes. We reported that amines not only inhibited pinocytosis of exogenous enzymes, but also caused normal fibroblasts to secrete large amounts of acid hydrolases. In fact, secretion was enhanced to approximately the same level of secretion as that seen in I-cell disease fibroblasts. The lysosomotropic amines appeared to cause normal fibroblast lines (Gonzalez-Noriega, *et al.*, 1980; Wilcox and Rattray, 1979) and any non-I-cell disease fibroblast line tested (Gonzalez-Noriega *et al.*, 1980) to divert newly synthesized enzymes to the extracellular medium. By contrast, these agents did not enhance the already high level of secretion seen in I-cell fibroblasts or those from patients with mucolipidosis III, who are thought to have a similar defect. Although amine treatment of normal cells produced the I-cell phenotype, both in terms of the amount of enzyme secreted into the extracellular medium and in terms of depressing the intracellular level of lysosomal enzymes in normal cells, the enzyme secreted by amine-treated normal cells was not I-cell-like enzyme; i.e., it was not recognition-defective. In fact, it was greatly enriched for high-uptake enzyme (Gonzalez-Noriega *et al.*, 1980). These observations suggested that amines blocked the normal segregation of newly synthesized enzymes, even though the enzymes had normal recognition markers, and also that the intracellular traffic pathway for receptor-mediated delivery of enzymes like the pinocytic pathway that amines disrupted was dependent on receptor recycling. If this interpretation is correct, amines might be thought of as producing the equivalent of a receptor-negative phenotype, with newly synthesized enzyme failing to be segregated and being secreted, not because they lack the recognition marker as was true for I-cell disease enzymes, but rather because they found no free receptors to bind. A simple mechanism to explain the failure of receptors to discharge their ligand and recycle as free receptors was suggested by the known effects (Ohkuma and Poole, 1978) of these amines on intralysosomal pH (raising the pH from 4.5 to above 6.0) and the findings (Fischer *et al.,* 1980b; Gonzalez-Noriega *et al.,* 1980) that the dissociation of lysosomal enzymes from cell-surface receptors was very slow at pHs above 6.0 but very rapid at pHs below 6.0. On the basis of these observations, we suggested that the delivery system for lysosomal enzymes may depend on the pH-dependent release of enzyme from receptor in lysosomes to permit free receptors to be reutilized. Thus, by raising the intralysosomal pH, amines could block receptor reutilization by interfering with receptor-ligand dissociation. Another possibility for the action of amines was suggested by the experiments of Helenius and co-workers (1980) showing that certain enveloped virus–vesicle fusion processes, which are important to viral multiplication in mammalian cells and normally take place in lysosomes, are pH-dependent and are blocked by amines. This at least raised the possibility that amines might disrupt traffic of lysosomal enzymes by interfering with some vesicle–vesicle fusion process on which enzyme delivery and receptor reutilization depends. In any case, it was clear that amines diverted most newly synthesized enzymes to the extracellular medium. These effects of amines on enzyme transport strengthened our view that most enzyme segregation normally occurs via an intracellular receptor-mediated process.

## IX.   Mannose 6-Phosphate May Affect Processing of Oligosaccharide Chains on Lysosomal Hydrolases

The presence of the phosphomannosyl recognition marker clearly influences the fate of the acid hydrolases—i.e., into which intracellular compartment they are delivered. It may also influence the processing of the oligosaccharide chains on lysosomal enzymes. Many secretory glycoproteins contain "complex-type" oligosaccharide chains. The complex-type oligosaccharide chains are the end result of a series of processing steps on oligosaccharide chains that were initially high-mannose-type chains when transferred cotranslationally from lipid-linked intermediates to nascent glycoproteins entering the cisternal space of the endoplasmic reticulum (Kornfeld *et al.*, 1978). These high-mannose-type oligosaccharide chains are first trimmed to smaller mannose-containing core oligosaccharides, and then built back up into complex-type chains by the action of several glycosyl transferases, which transfer *N*-acetylglucosamine, galactose, and sialic acid to the oligosaccharide chains in the Golgi region (Kornfeld *et al.*, 1978). The experiment summarized in Table I showed that the Man-6-P recognition marker was present on endoglycosidase H-sensitive oligosaccharides. All the Man-6-P-containing oligosaccharides were released from the lysosomal enzyme by treating with endoglycosidase H, which produced a concomitant loss of susceptibility of enzyme to pinocytosis. This result indicated that the Man 6-P on high-uptake acid hydrolases was present in either high-mannose or hybrid-type oligosaccharides. In other words, the Man-6-P-bearing oligosaccharides had not been processed to complex-type oligosaccharides, which are common in secretory glycoproteins. However, I-cell enzymes, which have been reported to lack the recognition marker, have also been reported to have oligosaccharide side chains with features that suggest further processing to complex type. Two pieces of evidence support this conclusion. First, excess sialic acid (the terminal sugar in complex-type oligosaccharide chains) has been reported in I-cell secretion enzymes (Vladutiu and Rattazzi, 1975). Second, we have found that I-cell secretion hexosaminidase was quantitatively retained on *Ricinus communus*–Sepharose columns. This suggests that the oligosaccharide chains contain galactose, a monosaccharide also found in chains that have been processed further. Only 10% of hexosaminidase secreted by normal fibroblasts (up to 30% of hexosaminidase secreted in the presence of amines) was specifically adsorbed to this galactose-recognizing lectin (Sly *et al.*, 1979). When the ricin-adsorbed enzyme secreted by normal fibroblasts was eluted from ricin–Sepharose columns and tested for susceptibility to pinocytosis by fibroblasts, Man-6-P-inhibitable uptake was observed. Treatment of the eluted enzyme with endoglycosidase H destroyed the susceptibility to pinocytosis of the enzyme, without reducing its ability to be readsorbed to ricin–Sepharose columns (A. Gonzalez-Noriega and W. S. Sly, unpublished observations). We interpreted these observations to mean that hexosaminidase secreted by normal fibroblasts can contain Man-6-P-bearing oligosaccharide side chains (which are endoglycosidase H-sensitive) and galactose-containing oligosaccharide side chains (which have been further processed to hybrid-type or complex-type oligosaccharides). This evidence could be explained if the presence of the Man-6-P recognition marker on the oligosaccharide chain of normal lysosomal enzymes either directly, or by binding to the Man-6-P receptor, prevents further processing of the Man-6-P-bearing oligosaccharide chains into complex-type oligosaccharide chains on the enzyme (Sly *et al.*, 1979). In this regard, it is known that a mannosidase is necessary for pruning the high-mannose-type oligosaccharide chains down to the core before building them back up into complex chains (Kornfeld *et al.*, 1978), and that von Figura and Weber (1978) found that some of the oligosaccharides released by endo-glycosidase H from a high-uptake enzyme were not sensitive to mannosidase until first converted from acidic to neutral oligosaccharide by a phosphatase. Thus, we have suggested (Sly *et al.*, 1979; Sly, 1979) that the Man 6-P on the enzyme protects the enzyme from processing and that, in the absence of Man 6-P on the enzyme as in I-cell disease, the oligosaccharide chains are processed to either complex-type or hybrid-type oligosaccharide chains that are more characteristic of the long circulating secretory glycoproteins (Kornfeld *et al.*, 1978) than of lysosomal enzymes. That would explain why the sialic acid content and the ricin-binding properties of I-cell enzymes differ from those of high-uptake enzymes.

## X.    Generality of the Receptor-Mediated Intracellular Pathway for 6-Phosphomannosyl Enzymes

If the Man-6-P recognition marker and its receptor provide a general mechanism for delivery of acid hydrolases from the endoplasmic reticulum to lysosomes, one might expect that the enzyme transport pathway should not be restricted to fibroblasts. In that case, we should be able to find high-uptake enzyme forms produced by other cell types, and we would predict the presence of the receptor on intracellular membranes from all cell types that use this intracellular pathway for enzyme delivery. We recently reported an enzyme binding assay that made it possible to measure the receptor in membranes from human fibroblasts. This assay permitted us to demonstrate that in fibroblasts at least 80% of the total membrane receptor activity was on intracellular membranes (Fischer *et al.*, 1980b,c). Table III presents the results of experiments using a similar assay for binding of high-uptake $\beta$-hexosaminidase to membranes from fibroblasts and from other human tissues. Enzyme binding activity with the properties of the phosphomannosyl-enzyme receptor on human fibroblasts was found in membranes from testes, brain, kidney, liver, adipose tissue, and lung. These studies suggested that the phosphomannosyl-enzyme receptor is widely distributed in human tissues, as one would predict if it has a general role in delivery of acid hydrolases to lysosomes.

## XI.    The Relationship of the Phosphomannosyl Recognition Marker to the "Secretory Pathway" and the "Signal Hypothesis"

The synthesis of secretory proteins, their passage to various compartments, and their exocytosis has been analyzed by Palade and co-workers (Jamieson and Palade, 1977; Palade, 1975). The "secretory pathway" (Jamieson and Palade,

TABLE III

SPECIFIC BINDING OF $\beta$-HEXOSAMINIDASE B TO MEMBRANES
FROM HUMAN TISSUES[a]

| Tissue | Total binding (units/mg) | +Man 6-P | Specific binding (units/mg) |
|---|---|---|---|
| Fibroblasts | 375 | 63 | 312 |
| Testis | 590 | 81 | 509 |
| Spleen | 590 | 78 | 512 |
| Brain | 555 | 70 | 485 |
| Lung | 533 | 74 | 459 |
| Liver | 467 | 64 | 403 |
| Kidney | 305 | 60 | 245 |
| Adipose | 220 | 40 | 180 |
| Heart | 205 | 40 | 165 |
| Muscle | 185 | 33 | 152 |

[a] Approximately 1 gm of each tissue was homogenized by three 10-second pulses with a Brinkman Polytron (setting #10). Binding of $\beta$-hexosaminidase was done with membranes sedimenting at between 500 and 12,500 $g$. One thousand units of high-uptake $\beta$-hexosaminidase was incubated with 50 $\mu$g of membrane protein at 20°C for 1 hour. The membranes were sedimented for 5 minutes at 10,000 $g$, washed twice in binding buffer (25 m$M$ Tris, 5 m$M$ EDTA, 0.5% saponin, pH 7.0), and assayed for $\beta$-hexosaminidase and protein. The $\beta$-hexosaminidase used was enzyme purified from hexosaminidase secreted into serum-free Waymouth medium containing 10 m$M$ NH$_4$Cl by Tay-Sachs disease fibroblasts (Gonzalez-Noriega et al., 1980).

1977) is thought to involve (1) synthesis of exportable proteins on the rough endoplasmic reticulum (RER), (2) segregation of exportable proteins in the cisternal space of the RER, (3) intracellular transport from the RER to the Golgi complex in some form of transporting vesicles, (4) concentration of secretory proteins in condensing vacuoles, (5) storage in secretory vesicles; and/or (6) exocytosis, which achieves delivery of the secretory products of the endoplasmic reticulum to the extracellular milieu via fusion of secretory granules or vesicles with the plasma membrane.

To explain how secretory proteins are transferred from the cytosol across the membrane of the RER to the cisternal space, the "signal hypothesis" (Blobel, 1977) was formulated. This hypothesis proposes that mRNAs for secretory proteins contain signal codons that are translated into a sequence that triggers ribosome attachment to the membranes of the endoplasmic reticulum and leads to the formation of a tunnel through the membrane through which the protein is segregated as it is translated. Soon after the signal peptide section has been transferred to the lumen of the cisternal space, the signal sequence may be cleaved from the nascent chain by a peptidase. The "processed" nascent chain grows until chain termination, when the secretory protein is released into the cisternal space. Considerable experimental support has accumulated for the involvement of "signal sequences" in transfer of proteins across intracellular membranes.

We have suggested (Sly, 1979; Sly *et al.*, 1979) that the phosphomannosyl recognition marker on acid hydrolases is also a form of signal or address marker for intracellular transport of acid hydrolases. It is not suggested as a substitute for the signal sequence. Presumably lysosomal enzymes would require such a signal sequence to gain access to the cisternal space of the endoplasmic reticulum. Rather, we propose that the Man 6-P on the enzymes acts as a "second signal" for the next level of sorting, allowing the cell to target receptor-bound acid hydrolases for lysosomes and to segregate them from other products of the endoplasmic reticulum that have other intracellular or extracellular destinations. We propose that segregation may be effected by collection of the receptor-bound hydrolases into specialized regions that bud off as primary lysosomes. Novikoff (1976) has argued from histochemical evidence that lysosomes arise in a smooth-membrane tubular network that was closely associated with the endoplasmic reticulum and the concave (trans) face of the Golgi apparatus that he named GERL. Friend and Farquhar (1967) and Novikoff (1976) identified coated vesicles that stained for acid phosphatase and that appeared to arise from some component of the Golgi complex or GERL. These were proposed as transport vesicles for acid hydrolases (primary lysosomes). If that interpretation is correct, these would be the structures that we would suggest are segregating receptor-bound enzymes for delivery to lysosomes. Specifically, we suggest (Sly, 1979) that these structures arise after newly synthesized acid hydrolases bearing phosphomannosyl recognition markers bind to receptors in the endoplasmic reticulum and collect in the specialized regions of the Golgi or GERL to bud off as coated vesicles. In our view, extracellular lysosomal enzymes are not obligatory intermediates in this pathway to lysosomes, but rather reflect the failure of some enzymes to be segregated by this pathway.

It should be mentioned that many acid hydrolases have recently been reported to be synthesized as precursors that have molecular weights much higher than the average molecular weights of these hydrolases isolated from lysosomes (Hasilik *et al.*, 1979; Sahagian *et al.*, 1979). To date, there is no evidence to suggest that the excess polypeptide sequence, although large, has any transport function. There is, however, considerable evidence (Kaplan *et al.*, 1977a,b; Sando and Neufeld, 1977; Sando *et al.*, 1979; Sly, 1979; Sly *et al.*, 1975, 1978, 1979; Sly and Stahl, 1978; Ullrich *et al.*, 1978) to implicate the phosphomannosyl recognition marker on acid hydrolases in enzyme transport.

We view I-cell disease as a disorder resulting from failure of the normal receptor-mediated segregation of acid hydrolases because the recognition marker is not put on the enzymes. In the absence of segregation, most of the enzymes are secreted. In addition, in the absence of Man 6-P on the enzyme, most of the oligosaccharide side chains are processed into complex type. A small amount of enzyme may go through this "secretory pathway" in normal fibroblasts. A small amount of sialated lysosomal enzyme is found in normal serum (Wilcox and Renwick, 1977) that may also be made in this way. In addition, failure to add the recognition marker may provide a secretory pathway for lysosomal enzymes that is actually an important pathway in some cell types (Vladutiu and Rattazzi, 1975, 1979), such as those that produce seminal fluid where sialated enzymes appear to form a significant fraction of the total enzyme. Although we have suggested that all cell types utilize the Man-6-P recognition marker and its receptor for the intracellular pathway for delivery of newly synthesized lysosomal enzymes to lysosomes, and have found the phosphomannosyl-enzyme receptor activity in membranes from homogenates of every human organ we have examined, it is not yet clear how many human cell types express the receptor on their cell surface. Nor is it clear how many cell types utilize the pinocytic pathway for uptake of extracellular enzyme that has been demonstrated in fibroblasts. Furthermore, although we have proposed that there are two pathways for lysosomal enzymes to reach lysosomes in fibroblasts and that both pathways utilize phosphomannosyl-enzyme receptors, it is not presently known whether the receptors involved in adsorptive endocytosis are in equilibrium with the receptors mediating intracellular transport. The relationship of the two pathways is still be be defined.

## XII.   Alternative Routes for Acid Hydrolases in Mammalian Cells

Not all lysosomal enzymes go to lysosomes, and not all lysosomal enzymes that are taken up from the extracellular milieu are taken up by the phosphomannosyl-enzyme recognition system. Several alternative routes for lysosomal enzymes have been recognized. Some are specific for specific enzymes, and others are specific for specific differentiated mammalian cell types.

First, two enzymes may have an alternative intracellular pathway to reach lysosomes in fibroblasts. These two enzymes, acid phosphatase and $\beta$-glucocerebrosidase, have been reported to be exceptional enzymes in that they are present in normal amounts in I-cell disease fibroblasts, which are deficient for most other hydrolases (Neufeld et al., 1975). If it can be shown that they are not only present in normal amounts, but are present in lysosomes, it would suggest that these two enzymes may not require the phosphomannosyl-enzyme recognition marker to reach lysosomal enzymes in fibroblasts, as appears to be the case for most other acid hydrolases. Both acid phosphatase and $\beta$-glucocerebrosidase are very hydrophobic and are tightly bound to membranes in homogenates. This tight association with membranes could mean that these two enzymes rely on a different type of interaction with membranes than do most other acid hydrolases for delivering them to lysosomes or for preventing them from being secreted by I-cell disease fibroblasts. One can imagine how it might be to the cell's advantage to segregate acid phosphatase from other phosphomannosyl-acid hydrolases until they reach lysosomes.

Second, $\beta$-glucuronidase provides a well-studied example of "dual localization" of a lysosomal enzyme, at least in mouse liver (Paigen *et al.*, 1975). One form is in lysosomes, and the other is in microsomes. Both forms are products of the same structural gene. The phosphomannosyl recognition marker can provide a mechanism for delivery of the enzyme to lysosomes in mouse liver. The mechanism of segregation and transport of the microsomal form of $\beta$-glucuronidase is still a mystery, though specific $\beta$-glucuronidase-binding proteins have been identified (Paigen *et al.*, 1975) that may play a role either in delivery of $\beta$-glucuronidase or in retention of $\beta$-glucuronidase in microsomes of mouse liver.

Third, there are two other receptors that mediate adsorptive pinocytosis of acid hydrolases that may have physiological significance in "recapture" of acid hydrolases secreted into the extracellular medium or the plasma. One of these is the mannosyl/$N$-acetylglucosamine-glycoprotein receptor that is present on mononuclear phagocytes (Kupffer cells, alveolar macrophages, and other fixed tissue macrophages in spleen and bone). This is the receptor that mediates uptake of many acid hydrolases by macrophages *in vitro* (Achord *et al.*, 1978; Stahl *et al.*, 1978) and also mediates clearance of acid hydrolases from plasma following intravenous infusion (Achord *et al.*, 1977, 1978; Sly *et al.*, 1979; Sly and Stahl, 1978; Stahl *et al.*, 1976; Steer *et al.*, 1978). The other receptor that probably mediates uptake of acid hydrolases that have predominantly complex-type oligosaccharide chains (Furbish *et al.*, 1978; Steer *et al.*, 1978) is the galactosyl-glycoprotein receptor that has been so well characterized by Ashwell and co-workers (Ashwell and Morell, 1974).

Thus, although we visualize the phosphomannosyl-enzyme recognition marker and its receptor as the major delivery system for most lysosomal enzymes to lysosomes in fibroblasts and probably in most tissues, there are alternative routes by which lysosomal enzymes can reach other intracellular localizations and alternative receptors by which lysosomal enzymes can be taken up from the extracellular milieu in certain situations. For a more detailed review of the carbohydrate recognition systems involved in these alternative traffic pathways, see Sly (1979).

## XIII. Summary Statement and Model for Receptor-Mediated Transport of Lysosomal Enzymes

Adsorptive pinocytosis of lysosomal enzymes by fibroblasts depends on Man-6-P containing recognition markers on lysosomal enzymes and on high-affinity receptors on the cell surface of fibroblasts. Fischer *et al.* (1980d) found by subcellular fractionation that most (90%) of the phosphomannosyl-enzyme receptors in rat liver were intracellular and faced the interior of endoplasmic reticulum, Golgi apparatus, and lysosomes, and that the highest specific activity of enzyme receptors was present in these fractions. Intracellular receptors were found to be occupied by endogenous enzymes which could be displaced by added Man-6-P and presented an occupancy gradient which ran downhill from endoplasmic reticulum to lysosomes. Moreover, endogenous, receptor-bound $\beta$-hexosaminidase which was displaced by Man 6-P had properties of high uptake phosphomannosyl enzymes. These results suggested that proteins which receive phosphomannose moieties in the endoplasmic reticulum are bound and transferred to primary lysosomes by an essentially intracellular pathway. Presumably lysosomal enzymes share a common component for recognition at another level, possibly in their primary sequence, which allows the cell to distinguish the proper candidates to receive the common recognition marker that targets them to lysosomes.

Recent studies have also clarified many questions concerning the biosynthesis of the phosphomannosyl recognition marker on lysosomal enzymes. Tabas and Kornfeld (1980) have shown that oligosaccharides on biosynthetic intermediates of $\beta$-glucuronidase from mouse lymphoma cells consist of high mannose-type units in which mannose residues located at or near the nonreducing termini are linked by phosphodiester bonds to another moiety, $\alpha$-$N$-acetylglucosamine. These results suggested that the mechanism by which mannose residues of lysosomal enzymes are phosphorylated is through transfer of $N$-acetylglucosamine-1-phosphate to mannose residues of high mannose-type oligosaccharides. Using dephosphorylated $\beta$-hexosaminidase as acceptor and [$\beta$-$^{32}$P] UDP-$N$-acetylglucosamine as donor for the phosphate group, Hasilik *et al.* (1981) demonstrated phosphorylation of $\beta$-hexosaminidase by microsomes from rat liver, human placenta, and human skin fibroblasts. The transferase activity was deficient in fibroblasts from patients affected with I-cell disease, and this deficiency is proposed to be the primary enzyme defect in I-cell disease.

Previous results outlined above showed that alkaline phosphatase treatment of acid hydrolases abolished their high affinity cellular uptake and indicated that the phosphate in high uptake enzyme forms is present as a phosphomonoester group in these molecules. Thus, Varki and Kornfeld (1980) postulated that during the maturation of acid hydrolases the covering $N$-acetylglucosamine residues are removed to unmask the targeting function of the phosphate residues and demonstrated that smooth membrane preparations of rat liver contain an $\alpha$-$N$-acetylglucosaminyl phosphodiesterase capable of uncovering the blocked phosphate groups of high mannose oligosaccharides of lysosomal enzymes.

The addition of these recent results to those obtained previously now suggest the following steps in the life cycle of lysosomal enzymes.

1. Transcription of 40–50 unlinked genes for acid hydrolases in the nucleus.

2. Translation of mRNAs on membrane bound ribosomes.

3. Transfer of $(Glc)_3$ $(Man)_9$ $(GlcNAC)_2$ from lipid-linked intermediates to Asn residues of nascent acid hydrolases.

4. Transfer of GlcNAc-1-P from UDP-GlcNAc to the 6 position of mannoses.

5. Release of GlcNAc by GlcNAc (1) P (6) Man-phosphodiesterase which exposes Man 6-P groups to produce a "high-uptake" enzyme.

6. Enzymes bind to phosphomannosyl enzyme receptors and collect into vesicles which bud off of Golgi or GERL to form primary lysosomes. If they fail to bind, the are secreted.

7. Some cell types, for example, fibroblasts, express similar receptors on the cell surface and take up secreted high-uptake enzymes by pinocytosis (i.e., secretion-recapture provides an alternate route to lysosomes). The physiological role of this adsorptive pinocytosis system on fibroblasts is still unclear as is its distribution on other cell types and its relationship to the intracellular pathway.

8. As pH falls below 6, enzymes dissociate from receptors and free receptors can be reutilized (recycled). Chloroquine and other basic amines disrupt both adsorptive pinocytosis of acid hydrolases and segregation of newly synthesized enzymes inside cells, at least partly, by blocking receptor recycling. As a consequence, these agents divert newly synthesized enzymes to the extracellular medium.

9. Enzymes undergo "post-lysosomal" processing by (a) acid phosphatase, which inactivates the phosphomannosyl recognition marker and (b) acid proteases, which trim off excess polypeptide.

10. Enzymes participate in degradation activities of the lysosomes.

11. Enzymes die after prolonged exposure to proteolytic neighbors.

## REFERENCES

Achord, D. T., Brot, F. E., Gonzalez-Noriega, A., Sly, W. S., and Stahl, P. (1977). *Pediatr. Res.* **11,** 816–822.

Achord, D. T., Brot, F. E., Bell, C. E., and Sly, W. S. (1978). *Cell* **15,** 269–278.

Ashwell, G., and Morell, A. G. (1974). *Adv. Enzymol.* **41,** 99–128.

Bach, G., Bargel, R., and Cantz, M. (1979). *Biochem. Biophys. Res. Commun.* **91,** 976–981.

Ballou, C. E. (1974). *Adv. Enzymol.* **40,** 239–265.

Blobel, G. (1977). *In* "International Cell Biology 1976–1977" (B. R. Brinkley and K. R. Porter, eds.), pp. 318–315. Rockefeller Univ. Press, New York.

Brot, F. E., Glaser, J. H., Roozen, K. J., Sly, W. S., and Stahl, P. D. (1974). *Biochem. Biophys. Res. Commun.* **57,** 1–8.

Distler, J., Hieber, V., Sahagian, G., Schmickel, R., and Jourdian, G. W. (1979a). *Proc. Natl. Acad. Sci. U.S.A.* **76,** 4235–4239.

Distler, J., Hieber, V., Schmickel, R., and Jourdian, G. W. (1979b). *ACS Symp. Ser.* **74,** 163–180.

Fischer, H. D., Natowicz, M., Sly, W. S., and Bretthauer, R. K. (1979). *Fed. Proc., Fed. Am. Soc. Exp. Biol.* **38,** 467 (abstr.).

Fischer, H. D., Natowicz, M., Sly, W. S., and Bretthauer, R. K. (1980a). *J. Cell Biol.* **84,** 77–86.
Fischer, H. D., Gonzalez-Noriega, A., and Sly, W. S. (1980b). *J. Biol. Chem.* **255,** 5069–5074.
Fischer, H. D., Gonzalez-Noriega, A., and Sly, W. S. (1980c). *Fed. Proc., Fed. Am. Soc. Exp. Biol.* **39,** 1797 (abstr.).
Fischer, H. D., Gonzalez-Noriega, A., Sly, W. S., and Morré, D. J. (1980d). *J. Biol. Chem.* **255,** 9608–9615.
Fratantoni, J. C., Hall, C. W., and Neufeld, E. F. (1968). *Science* **162,** 570–572.
Friend, D. S., and Farquhar, M. G. (1967). *J. Cell Biol.* **35,** 357–376.
Furbish, F. S., Steer, C. J., Barringer, J. A., Jones, E. A., and Brady, R. O. (1978). *Biochem. Biophys. Res. Commun.* **81,** 1047–1053.
Glaser, J. H., Roozen, K. J., Brot, F. E., and Sly, W. S. (1975). *Arch. Biochem. Biophys.* **166,** 536–542.
Gonzalez-Noriega, A., Grubb, J. H., Talkad, V., and Sly, W. S. (1980). *J. Cell. Biol.* **85,** 839–852.
Hasilik, A., Rome, L. N., and Neufeld, E. F. (1979). *Fed. Proc., Fed. Am. Soc. Exp. Biol.* **38,** 467 (abstr.)
Hasilik, A., Waheed, A., and von Figura, K. (1980). *Biochem. Biophys. Res. Commun.* **98,** 761–767.
Helenius, A., Kartenbeck, J., Simons, K., and Fries, E. (1980). *J. Cell Biol.* **84,** 404–420.
Hickman, S., and Neufeld, E. F. (1972). *Biochem. Biophys. Res. Commun.* **49,** 992–999.
Hickman, S., Shapiro, L. J., and Neufeld, E. F. (1974). *Biochem. Biophys. Res. Commun.* **57,** 55–61.
Hieber, V., Distler, J., Myerowitz, R., Schmickel, R. D., and Jourdian, G. W. (1976). *Biochem. Biophys. Res. Commun.* **73,** 710–717.
Jamieson, J. D., and Palade, G. E. (1977). *In* "International Cell Biology 1976–1977" (B. R. Brinkley and K. R. Porter, eds.), pp. 308–317. Rockefeller Univ. Press, New York.
Kaplan, A., Achord, D. T., and Sly, W. S. (1977a). *Proc. Natl. Acad. Sci. U.S.A.* **74,** 2026–2030.
Kaplan, A., Fischer, D., Achord, D. T., and Sly, W. S. (1977b). *J. Clin. Invest.* **60,** 1088–1093.
Kaplan, A., Fischer, D., and Sly, W. S. (1978). *J. Biol. Chem.* **253,** 647–650.
Kornfeld, S., Li, E., and Tabas, I. (1978). *J. Biol. Chem.* **253,** 7771–7778.
Lloyd, J. B. (1977). *Biochem. J.* **164,** 281–282.
Natowicz, M. R., Chi, M.-M. Y., Lowry, O. H., and Sly, W. S. (1979). *Proc. Natl. Acad. Sci. U.S.A.* **76,** 4322–4326.
Neufeld, E. F., and Cantz, M. J. (1971). *Ann. N. Y. Acad. Sci.* **179,** 580–587.
Neufeld, E. F., Lim, T. W., and Shapiro, L. J. (1975). *Annu. Rev. Biochem.* **44,** 357–376.
Neufeld, E. F., Sando, G. N., Garvin, A. J., and Rome, L. H. (1977). *J. Supramol. Struct.* **6,** 95–101.
Nicol, D. M., Lagunoff, D., and Pritzl, P. (1974). *Biochem. Biophys. Res. Commun.* **59.** 941–946.
Novikoff, A. B. (1976). *Proc. Natl. Acad. Sci. U.S.A.* **73,** 2781–2787.
Ohkuma, S., and Poole, B. (1978). *Proc. Natl. Acad. Sci. U.S.A.* **75,** 3327–3331.
Paigen, K., Swank, R. T., Tomino, S., and Ganschow, R. E. (1975). *J. Cell. Physiol.* **85,** 379–392.
Palade, G. E. (1975). *Science* **189,** 347–358.
Rome, L. H., Weissman, B., and Neufeld, E. F. (1979). *Proc. Natl. Acad. Sci. U.S.A.* **76,** 2331–2334.
Sahagian, G., Distler, J., Hieber, V., Schmickel, R., and Jourdian, G. W. (1979). *Fed. Proc., Fed. Am. Soc. Exp. Biol.* **38,** 467 (abstr.).
Sando, G. N., and Neufeld, E. F. (1977). *Cell* **12,** 619–627.
Sando, G. N., Titus-Dillon, P., Hall, C. W., and Neufeld, E. F. (1979). *Exp. Cell Res.* **119,** 359–364.
Skudlarek, M. D., and Swank, R. T. (1979). *J. Biol. Chem.* **254,** 9939–9942.
Sly, W. S. (1979). *In* "Structure and Function of Gangliosides" (L. Svennerholm, ed.), pp. 433–451. Plenum, New York.

Sly, W. S., and Stahl, P. (1978). *In* "Transport of Macromolecules in Cellular Systems" (S. C. Silverstein, ed.), pp. 229–244. Dahlem Konferenzen, Berlin.

Sly, W. S., Glaser, J. H., Roozen, K., Brot, F., and Stahl, P. (1975). *In* "Enzyme Therapy in Lysosomal Storage Diseases" (J. M. Tager, G. J. M. Hooghwinkel, and W. T. Daems, eds.), pp. 288–291. North-Holland Publ., Amsterdam.

Sly, W. S., Achord, D. T., and Kaplan, A. (1978). *In* "Protein Turnover and Lysosome Function" (H. L. Segal and D. J. Doyle, eds.), pp. 497–519. Academic Press, New York.

Sly, W. S., Gonzalez-Noriega, A., Natowicz, M., Fischer, H. D., and Chambers, J. P. (1979). *Fed. Proc., Fed. Am. Soc. Exp. Biol.* **38,** 467 (abstr.).

Stahl, P. D., Six, H., Rodman, J. S., Schlesinger, P., Tulsiani, D. R. P., and Touster, O. (1976). *Proc. Natl. Acad. Sci. U.S.A.* **73,** 4045–4049.

Stahl, P. D., Rodman, J. S., Miller, M. J., and Schlesinger, P. H. (1978). *Proc. Natl. Acad. Sci. U.S.A.* **75,** 1399–1403.

Steer, C. J., Furbish, F. S., Barringer, J. A., Brady, R. O., and Jones, E. A. (1978). *FEBS Lett.* **91,** 202–205.

Tabas, I., and Kornfeld, S. (1980). *J. Biol. Chem.* **255,** 6633–6639.

Ullrich, K., Mersmann, G., Weber, E., and von Figura, K. (1978). *Biochem. J.* **170,** 643–650.

Varki, A., and Kornfeld, S. (1980). *J. Biol. Chem.* **255,** 8398–8401.

Vladutiu, G. D., and Rattazzi, M. C. (1975). *Biochem. Biophys. Res. Commun.* **67,** 956–964.

Vladutiu, G. D., and Rattazzi, M. (1979). *J. Clin. Invest.* **63,** 595–601.

von Figura, K., and Klein, U. (1979). *Eur. J. Biochem.* **94,** 347–354.

von Figura, K., and Kresse, H. (1974). *J. Clin. Invest.* **53,** 85–90.

von Figura, K., and Voss, B. (1979). *Exp. Cell Res.* **121,** 267–276.

von Figura, K., and Weber, E. (1978). *Biochem. J.* **176,** 943–956.

Wiesmann, U. N., DiDonato, S., and Herschkowitz, N. N. (1975). *Biochem. Biophys. Res. Commun.* **66,** 1338–1343.

Wilcox, P., and Rattray, P. (1979). *Biochim. Biophys. Acta* **586,** 442–452.

Wilcox, P., and Renwick, A. G. C. (1977). *Eur. J. Biochem.* **13,** 579.

# Part III.  Translocation of Secretory Granules

# Chapter 14

# *Actin Filaments and Secretion: The Macrophage Model*

THOMAS P. STOSSEL

*Hematology–Oncology Unit,
Massachusetts General Hospital,
and Department of Medicine,
Harvard Medical School,
Boston, Massachusetts*

## I.  Introduction

Secretion is definable as the vectorial movement of molecules across the plasma membrane from the inside of the cell to the extracellular fluid. Often it is first necessary to move these molecules to the plasma membrane from some other location in the cell where synthesis and processing of these molecules occurred. In view of the requirements for movements in secretion, cellular systems for

however, the actin filaments in the macrophages seem to have some kind of firm association with the plasma membrane and therefore would attempt to draw it toward the center of the cell with the cortical actin. Presumably an equal and opposite hydrostatic force would oppose this inward movement because this system is closed. Therefore, no net movement could occur. The same argument is applicable to the dynamics in the plane of the membrane. If the force generation by myosin throughout the cortex and the mass upon which the force acts (actin and membranes) are equal throughout the cell periphery, no net movement occurs, according to Newton's first law. But local alterations of either the force or the mass in this cortical "tug-of-war" would result in a net acceleration and could produce net movement in some direction.

Knowledge about any changes in the power of the force-generating mechanism, the myosin–actin interaction, in the cytoplasm of macrophages is very incomplete at the present time. Assuming that the $Mg^{2+}$-ATPase activity of myosin in the presence of F-actin correlates with force generation, the contractile activity of mixtures of actin and myosin purified from macrophages is very weak compared with the force-generating capacity of actin and myosin from skeletal muscle (Stossel and Hartwig, 1975; Trotter and Adelstein, 1979). The poor activity of macrophage proteins is ascribable to the purified myosins and not the actins, which are functionally nearly equivalent to skeletal muscle actin. A crude protein fraction, designated "cofactor," from rabbit macrophages activates the $Mg^{2+}$ATPase activity and superprecipitation of actin–myosin mixtures to a level comparable to that of the skeletal muscle proteins (Stossel and Hartwig, 1975, 1976). Since cofactor has not been purified, we do not know if it is a regulator or modulator of macrophage contractile force. Calcium does not influence its activity. The idea that kinase enzymes phosphorylate the 15,000-dalton light chains of smooth muscle and certain nonmuscle cell myosins and thereby activate the myosins' contractile activity is currently popular (Adelstein, 1978) although controversial (Hirata et al., 1977). Since kinases that phosphorylate proteins and phosphatases that remove the phosphates from proteins establish a reciprocal control mechanism in many biological systems, the activity of myosin in macrophages is potentially regulated in this fashion. A crude protein preparation isolated from macrophages with cofactor activity phosphorylates this 15,000-dalton light chain of macrophage myosin; and the $Mg^{2+}$-ATPase activity of different macrophage myosin preparations in the presence of F-actin correlated with the extent of light-chain phosphorylation (Trotter and Adelstein, 1979). Like cofactor activity, phosphorylation activity was not dependent on the presence of calcium ions (Trotter and Adelstein, 1979). Resolution of the mechanism of activation of macrophage myosin by cofactor awaits it purification, but that such a factor is required is clear.

A different concept is that the control of orientation arises from changes in the actin filament mass and that gelatin or crosslinking of actin filaments controls these changes. Crude soluble extracts of macrophages solidify slowly (hours) in the cold or rapidly (minutes) when warmed (Stossel and Hartwig, 1976). Different rearrangements of molecular structures can cause liquid macromolecular solutions to solidfy, and all the arrangements involve interlocking of the macromolecules into a network to the extent that they become relatively insoluble. The insoluble network structure must have sufficient rigidity to counteract the ambient forces to remain solid. The molecular rearrangement occurring in the cytoplasmic extracts of macrophages is the polymerization of actin and the crosslinking of the actin fibers to form such a network. Solidification of cytoplasmic extracts was reported many years ago (Hultin, 1950), but the role of actin was recognized only recently (reviewed by Taylor and Coudeelis, 1979).

Two properties are useful for characterizing gels: (1) critical conditions and (2) mechanical properties (Flory, 1942, 1944). The gel state occurs when a critical number of molecules become linked into a continuous network. Obviously, long fibers have a greater chance of contacting one another than do small round molecules. Therefore, a much smaller concentration of long fibers, such as actin filaments, is required to create a gel than of globular molecules. F-actin filaments are highly entangled. However, since the interaction between the actin filaments is very weak, only high-actin-filament concentrations yield networks with much rigidity. Even then, sufficient force can cause the filaments to slide past one another so that interfilament contact and the network structure diminish. The fibers may flow away from the direction of force. When the force is removed, molecular motion tends to return the filaments to their original positions, restoring the network. This behavior defines a thixotropic gel–sol transformation (Goodeve, 1939). If exogenous crosslinking molecules tie the filaments together with some force, a true gel exists. This exogenous factor in macrophages is a high-molecular-weight protein called actin-binding protein (ABP), which noncovalently binds actin filaments together (Stossel and Hartwig, 1975, 1976; Brotschi et al., 1978). At a critical concentration of crosslinking molecules, a continuous network of the polymers abruptly occurs. A very small change in the crosslinker concentration near this critical point creates a sol–gel transformation. With addition of crosslinking molecules above the critical crosslinker concentration, the mechanical rigidity of the polymer solution rises sharply as the lattice propagates. The degree to which a crosslinker increases the rigidity of a polymer gel depends on the number of crosslinks introduced and on the uniformity of the distribution of crosslinkers within the network. The sol–gel transformation is therefore analogous to condensation of a vapor, although the dispersed phase (sol) is mixed in with the condensed phase (gel) (Flory, 1942). The potency of a gelling agent therefore depends on the capacity of as few crosslinking molecules to link as many fibers as possible. An inefficient agent may form many bridges between adjacent filaments. The redundant crosslinking can create a very stable and rigid aggregate of the involved fibers, but at the expense of failing to recruit more filaments into the network and to stabilize the network as a whole. Measurement of the critical crosslinker concentration is a good way, therefore, to compare the efficiency of different gelling agents.

The theoretical minimal concentration in moles per liter of crosslinker required for incipient gelation of isotropically oriented long-actin polymers of heterogeneous length is given by the expression

$$M_{cl} = \frac{(\text{moles of actin monomers in filaments/liter})}{\bar{x}_w} \tag{1}$$

where $\bar{x}_w$ is the weight average degree of polymerization (Flory, 1942; Stockmayer, 1944). The value of $M_{cl}$ can be computed by using dimensions of actin filaments determined from electron micrographic measurements of negatively stained actin filaments that were polymerized under defined conditions. The experimentally determined critical concentration of ABP is quite close to the theoretical and indicates that ABP is a potent crosslinking agent (Hartwig and Stossel, 1981). The critical concentration of macrophage ABP for crosslinking actin filaments into a continuous network or gel is much lower than that of other actin-crosslinking proteins such as myosin or ABP from smooth muscle (Brotschi et al., 1978) [smooth muscle ABP is also called filamin (Wang, 1977)]. Above the critical crosslinker concentration a small additional amount of ABP increases the rigidity of actin much more than equivalent quantities of myosin or filamin (Brotschi et al., 1978). The differences in the crosslinking ability of these three proteins are not ascribable to differences in their binding affinity to F-actin, since ABP, filamin, and myosin all bind to actin with high affinity. The simplest explanation, therefore, for the relative potency of ABP as an actin-crosslinking agent is that it is a highly flexible dimer which can easily locate specific domains for binding on randomly overlapping actin filaments (Hartwig and Stossel, 1981). Furthermore, it preferentially crosslinks actin filaments at right angles. Such perpendicular binding would minimize the formation of redundant crosslinks and preserve the isotropy of actin solutions (Hartwig et al., 1980).

ABP constitutes about 1% of the total protein of macrophages (Hartwig and Stossel, 1975). The bulk of actin-crosslinking activity in macrophage extracts copurifies with ABP (Brotschi et al., 1978). Evidence for other actin-crosslinking proteins in other cells exists, but these proteins are insufficiently characterized to be compared with ABP (Bryan and Kane, 1978; Maruta and Korn, 1977; Schloss and Goldman, 1979; Schollmeyer et al., 1978). The slow development of definite information is due to the difficulty in purifying proteins present in low concentrations in various cells and to problems with proteolysis. Degradation of ABP is disrupted macrophages is preventable only with potent combinations of proteolysis inhibitors (Brotschi *et al.*, 1978). If unimpeded, proteolysis can convert actin-crosslinking activity in macrophage extracts from high- to low-molecular-weight components (Brotschi *et al.*, 1978). Proteolysis can also cause protein solutions to gel by nonspecifically generating associations between molecules (Tombs, 1974). The gelation of cell extracts following trypsinization (Moore and Carraway, 1978) probably occurs by this process.

The existence of ABP, a protein that is a highly potent crosslinker of F-actin, strongly supports the idea that controlled lattice formation in cortical cytoplasm could provide for amplification and directionality of the force-generating mechanism. It could also account for local changes in stability of cortex underlining the lipid bilayer. If these notions are correct, then the lattice state of actin must be under some kind of control.

## C.   Control

The solidification of crude macrophage extracts is prevented by calcium (Stossel and Hartwig, 1976), and calcium dissolves solidified extract gels. The crosslinking of F-actin by purified ABP is unaffected by calcium (Brotschi *et al.*, 1978), implicating some other factor in conferring calcium control in the consistency of macrophage cytoplasmic extracts. Up to 60% of the activity that sensitizes ABP F-actin interaction to calcium in cytoplasmic extracts purifies with a heat-labile calcium-binding protein called gelsolin (Yin and Stossel, 1979, 1980; Yin *et al.*, 1980). Gelsolin purified from macrophages dissolves gels of actin crosslinked by ABP or filamin when the calcium concentration exceeds $2 \times 10^{-7}$ $M$. When the calcium concentration falls below this level, the sol partially gels again. Gelsolin binds to actin filaments when the calcium concentration is above the threshold level. The activity of gelsolin is maximal at a calcium concentration of about $10^{-5}$ $M$. Our present understanding of how activated gelsolin operates is that in the presence of calcium it reversibly breaks actin filaments into shorter pieces. The evidence for this conclusion is that gelsolin lowers the reduced viscosity and flow birefringence of F-actin and decreases the contour length of actin filaments seen in the electron microscope. However, calcium-activated gelsolin does not increase the concentration of monomeric actin in proportion to the degree of shortening of the filaments. The effect of gelsolin on actin is rapid and stoichiometric, one molecule of activated gelsolin apparently causing one break in an actin filament.

Since severing of the actin filaments obligatorily increases their number, the ratio of crosslinker to actin filaments can fall abruptly below the critical value, and solvation of gels then takes place. Therefore, if the calcium concentration rises in one domain of an actin network containing uniformly dispersed myosin filaments, the myosin filaments in that domain will act on a smaller mass of actin and therefore draw it toward the domains in which the calcium concentration remains lower, in which the network is more intact. Alternatively, if crosslinking increases the efficiency of contraction, the contractile force in the dissolved domain of the gel will be lower than where the gel is intact. In either or both cases, fluctuations in the calcium concentrations could rapidly control directionality of movement. In the cell cortex, if the lattice beneath the plasma membrane has a supportive role in strengthening the lipid bilayer, local decreases in the lattice strength could permit the hydrostatic pressure opposing the centripetal force to create bulges or blebs in the periphery. Since the swelling of gels is inversely proportional to the degree of crosslinking (Flory and Rehner, 1943), solvent might move from more to less crosslinked areas. Such movements could amplify the shape changes created by the alterations in the stability of the lattice and have some function in the translocation of organelles.

The hypothesis stated here predicts that a mass of cytoplasmic lattice will flow from high to low calcium concentration. This prediction is experimentally verifiable in horizontal capillary tubes containing muscle actin, muscle myosin, macrophage ABP, and macrophage gelsolin. If the calcium concentration rises in one side of the capillary, the actin gel flows to the other side (Stendahl and Stossel, 1980). In the absence of calcium and gelsolin, ABP reduces by a factor of 4 the concentration of myosin required for a maximal contraction of actin in the capillary, although it has no affect on the $Mg^{2+}$-ATPase activity of the myosin. Therefore, ABP appears to act by rendering the actin lattice more susceptible to movement by interspersed myosin filaments, presumably by recruitment of crosslinked actin filaments. Activation of gelsolin by calcium is equivalent to removing ABP from the filaments.

This discussion has dealt with movement of cortical cytoplasm, but the lipid bilayer must also move. As stated previously, there appears to be some connection between actin filaments and the membrane. However, at present the nature of the connection between these proteins and the bilayer of macrophages is obscure.

In summary, the force-generating mechanism is a superprecipitation of actin and myosin filaments, a process requiring hydrolysis of ATP and presumably based on the sliding-filament interaction characterized in striated muscle. This energy-dependent mechanism may be a major consumer of the macrophage's metabolic activity and can account for the susceptibility of secretion to inhibition by metabolic poisons. Directionality and amplification of the force generated arise from controlled focal changes in the crosslinking of actin filaments. Gelsolin, a calcium-activated protein, controls lattice rigidity of actin by severing actin filaments between points of crosslinking by ABP. The free calcium concentrations that regulate the activity of this protein are levels found in living cells. We call this mechanism for directional cytoplasmic movement control by calcium the "tug-of-war" hypothesis.

## D. Evidence for the Existence of the "Tug-of-War" Mechanism in Macrophages

The evidence that the interactions of proteins described in the foregoing sections exist in macrophages is cytochemical, biochemical, and pharmacological. If directional movement of cortical microfilaments power secretion in macrophages by the mechanism proposed, ABP and myosin should reside there. Studies employing biochemical and immunocytochemical techniques reveal an enrichment of ABP and myosin in the cell periphery of macrophages and document the presence of these proteins in pseudopod extensions of the cortical cytoplasm (Davies and Stossel, 1977; Hartwig et al., 1977; Stendahl et al., 1980).

Cytochalasin B very effectively dissolves actin-ABP gels (Hartwig and Stossel, 1976). Like the inhibition of cell motility, the solvation of actin gels by cytochalasin B is reversible following removal of the drug. Cytochalasin B acts by binding to F-actin and by severing the actin filaments, thereby having a similar effect on actin gels as those of calcium-activated gelsolin. Cytochalasin B does not bind to ABP nor prevent the crosslinking of F-actin by ABP per se, although it interferes with the ability of ABP to confer network structure on F-actin. Cytochalasin B binds to one high-affinity site ($K_a = 10^8\ M^{-1}$) and more than one lower affinity ($K_a$ app. $= <2 \times 10^6\ M^{-1}$) sites in actin filaments. Cytochalasin B, in concentrations greater than $10^{-8}\ M$, effectively dissolves actin gels estimated to be crosslinked by ABP at distances on each filament corresponding to one site per filament length (Hartwig and Stossel, 1979). Therefore, binding of cytochalasin B to the high-affinity sites leads to mechanical rupture of the filaments. Because of the important relationship between crosslink density and filament length in gelation, higher cytochalasin B concentrations ($>10^{-6}\ M$) are required for dissolution of heavily crosslinked actin networks or for lowering the weak rigidity of actin in the absence of an exogenous crosslinker. The "crosslinker" of F-actin by itself is the overlap points of the filaments. These increase exponentially as the number of filaments increases secondary to addition of more actin (Brotschi et al., 1978) or even to breakage of filaments by cytochalasin B (Hartwig and Stossel, 1979). Hence, the lower affinity sites are also important for the mechanical effect of cytochalasin B.

Since cytochalasin B shortens and divides actin filaments, an increase in $M_{cl}$ for ABP is predictable from Eq. (1). By measuring the length distribution of actin filaments after exposure to cytochalasin B, the increase in $M_{cl}$ can be calculated and shown to agree within experimental error with that measured (Hartwig and Stossel, 1979). Appreciation of the structure of ABP-actin gels brings to order the apparent paradox that concentrations of cytochalasin B that have profound effects on cell motility have little or no effect on actin viscosity or its interaction with myosin (Spudich and Lin, 1972).

In contrast to the proposed action of gelsolin, cytochalasin B unevenly penetrates cells, and, by weakening the actin lattice, it promotes the focal contraction of the broken cortical actin network by myosin, thereby causing widespread local superprecipitation rather then directional contraction characteristic of the cell's own control mechanism. Therefore, the effect of cytochalasin B depends on active contractions, explaining why the expression of the effects of cytochalasin B requires that cellular energy metabolism be unimpaired. Disorganized cortical contraction prevents the assembly of pseudopodia involved in various movements. On the other hand, the depletion of cortex from certain areas into focal aggregates could permit cytoplasmic secretory granules to approach the plasmalemma, explaining the enhancement of exocytosis in secretagogue-stimulated cells by cytochalasin B. The mechanism of cytochalasin B action on actin predicts that mainly microfilament functions involving the crosslinked actin lattice under tension will be susceptible to inhibition by the drug (Hartwig and Stossel, 1979). Therefore, failure of cytochalasin B to inhibit secretion does not rule out a role for actin in the process.

Considering the central role of calcium in the hypothesized control of microfilament movement in macrophages, the effects of calcium on movement in macrophages might seem interesting. Although the extracellular calcium concentration influences some activities of macrophages (Stossel, 1973), these results are not readily interpretable in terms of intracellular actions of calcium. The stimulation of secretion in macrophages by a calcium ionophore (Schneider *et al.*, 1978) is possibly ascribable to activation of gelsolin by the ionophore-induced rise in intracellular calcium concentration.

## E.    Actin Filaments and Secretion in the Macrophage

The information is still insufficient to explain in detail how actin might function to drive secretion in macrophages. However, for discussion purposes it is possible to synthesize a scheme for secretion of cytoplasmic granules into developing phagocytic vacuoles during phagocytosis by macrophages (Fig. 1). Binding of recognition-eliciting molecules on the surfaces of an object to receptors on the macrophage plasma membrane hypothetically results in activation of a membrane calcium pump that extrudes calcium out of the cortical cytoplasm across the membrane. Evidence for the existence of such a pump has been obtained (Lew and Stossel, 1980). A local fall in calcium decreases the activity of gelsolin in that region, resulting in recruitment by myosin of more actin filaments crosslinked by ABP into the region beneath the object, pleating of membrane against the object, which creates more receptor–object contacts, and further activation of calcium pumps. This sequence propagates around the object, and actin filaments and membrane are drawn from the adjacent cortex. Actin filaments are also taken from the base of the developing vacuole. The supply of

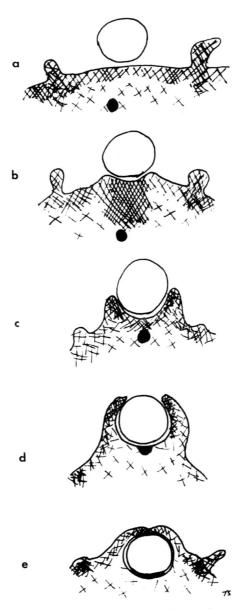

FIG. 1.   (a) The macrophage cortical cytoplasm and membrane is in a "tug-of-war" with local movements resulting from gradations in cortical calcium concentrations that influence the activity of gelsolin. (b) An ingestible particle activates a calcium efflux mechanism locally, switching off the subjacent gelsolin, and causing thereby increased gelation and contraction of actin filaments. (c) The reaction forms pseudopodia and propagates around the objects (see text), (d) recruiting cortical actin from the adjacent cell periphery and the base of the forming phagocytic vacuole. The latter results in attenuation of subplasmalemmel actin and permits granules to reach the membrane for fusion. (e) After fusion of the pseudopodia, the condition in (a) resumes. Cortical tension drives the phagolysosome to the cytocenter.

filaments is limited in this region; therefore, actin filament attenuation permits granules to approach and eventually fuse with the plasma membrane.

This scheme does not take into account any change in distribution between monomeric and filamentous actin, because no evidence for acute redistribution of this nature currently exists for this cell type. The general model is adaptable with slight modification to other cells (e.g., mast cells) in which perturbation of membrane receptors by ligands is believed to cause depolarization and local influxes of calcium (Douglas, 1968; Foreman *et al.*, 1973). If the motor network is similar to that of macrophages, gelsolin would be activated at the points of calcium entry, the cortical gel would dissolve, and secretion would take place at those points.

## ACKNOWLEDGMENTS

This work was supported by grants from the U.S. Public Health Service (HL19429), The Council for Tobacco Research USA (1116), The Edwin S. Webster Foundation, and Edwin W. Hiam.

## REFERENCES

Ackerman, N. R., and Beebe, J. R. (1974). *Nature (London)* **247**, 475–477.
Adelstein, R. S. (1978). *Trends Biochem Sci.* **6**, 27–30.
Adelstein, R. S., and Pollard, T. D. (1978). *Prog. Hemostasis Thromb.* **4**, 37–58.
Allison, A. C. (1973). *Ciba Found. Symp.* [N.S.] **14**, 109–148.
Begg, D. A., Rodewald, R., and Rebhun, L. I. (1978). *J. Cell Biol.* **79**, 846–852.
Brandon, D. L. (1976). *Eur. J. Biochem.* **65**, 139–146.
Brotschi, E. A., Hartwig, J. H., and Stossel, T. P. (1978). *J. Biol. Chem.* **253**, 8988–8993.
Bryan, J., and Kane, R. E. (1978). *J. Mol. Biol.* **125**, 207–224.
Cohn, Z. A., and Wiener, E. (1963). *J. Exp. Med.* **118**, 91–117.
Colten, H. R., and Gabbay, K. H. (1972). *J. Clin. Invest.* **51**, 1927–1931.
Davies, W. A., and Stossel, T. P. (1977). *J. Cell Biol.* **75**, 941–955.
D'Haese, J., and Komnick, H. (1972). *Z. Zellforsch. Mikrosk. Anat.* **134**, 427–434.
Dickson, J. G., Malan, P. G., and Ekins, R. P. (1979). *Eur. J. Biochem.* **97**, 471–479.
Douglas, W. W. (1968). *Br. J. Pharmacol.* **34**, 451–474.
Fine, R. E., and Blitz, A. L. (1975). *J. Mol. Biol.* **95**, 447–454.
Flory, P. J. (1942). *J. Phys. Chem.* **46**, 132–140.
Flory, P. J. (1944). *Chem. Rev.* **35**, 51–75.
Flory, P. J., and Rehner, J., Jr. (1943). *J. Chem. Phys.* **11**, 523–530.
Foreman, J. C., Mongar, J. L., and Gomperts, B. D. (1973). *Nature (London)* **245**, 249–251.
Gabbiani, G., Malaisse-Lagae, F., Blondel, B., and Orci, L. (1974). *Endocrinology* **95**, 1630–1635.
Gabbiani, G., Chaponnier, C., and Lüscher, E. F. (1975). *Proc. Soc. Exp. Biol. Med.* **149**, 618–621.
Goodeve, C. F. (1939). *Trans. Faraday Soc.* **35**, 342–358.
Govindan, V. M., and Wieland, T. H. (1975). *FEBS Lett.* **59**, 117–119.
Hall, P. F., Chapponier, C., Nakamura, M., and Gabbiani, G. (1979). *J. Biol. Chem.* **254**, 9080–9084.

Hartwig, J. H., and Stossel, T. P. (1975). *J. Biol. Chem.* **250**, 5696-5705.

Hartwig, J. H., and Stossel, T. P. (1976). *J. Cell Biol.* **71**, 295-303.

Hartwig, J. H., and Stossel, T. P. (1979). *J. Mol. Biol.* **134**, 539-553.

Hartwig, J. H., and Stossel, T. P. (1981). *J. Mol. Biol.* **145**, 563-581.

Hartwig, J. H., Davies, W. A., and Stossel, T. P. (1977). *J. Cell Biol.* **75**, 956-967.

Hartwig, J. H., Tyler, J., and Stossel, T. P. (1980). *J. Cell Biol.* **87**, 841-848.

Hirata, M., Mikawa, T., Nonomura, Y., and Ebashi, S. (1977). *J. Biochem. (Tokyo)* **82**, 1793-1796.

Hultin, T. (1950). *Exp. Cell Res.* **11**, 272-283.

Huxley, H. E. (1963). *J. Mol. Biol.* **3**, 281-308.

Huxley, H. E., and Hanson, J. (1954). *Nature (London)* **173**, 973-976.

Jockush, B. M., Burger, M. M., Da Prada, M., Richards, J. G., Chaponnier, C., and Gabbiani, G. (1978). *Nature (London)* **270**, 628-629.

Johnson, D. H., McCubbin, W. D., and Kay, C. M. (1977). *FEBS Lett.* **77**, 69-74.

Kane, R. E. (1975). *J. Cell Biol.* **66**, 305-315.

Korn, E. D. (1978). *Proc. Natl. Acad. Sci. U.S.A.* **75**, 588-599.

Kuo, I. C. Y., and Coffee, C. J. (1976). *J. Biol. Chem.* 251, 1603-1609.

Lew, D., and Stossel, T. P. (1979). *J. Biol. Chem.* **255**, 5841-5846.

Maruta, H., and Korn, E. D. (1977). *J. Biol. Chem.* **252**, 399-402.

Meyer, D. I., and Burger, M. M. (1979). *FEBS Lett.* **101**, 129-133.

Miranda, A. F., Godman, G. C., and Tannenbaum, S. W. (1974). *J. Cell Biol.* **62**, 406-423.

Moore, P. B., and Carraway, K. L. (1978). *Biochem. Biophys. Res. Commun.* **80**, 560-567.

Moore, P. L., Bank, H. L., Brissie, T., and Spicer, S. S. (1976) *J. Cell Biol.* **71**, 659-666.

Myrvik, Q. N., Leake, E. S., and Fariss, B. (1961). *J. Immunol.* **86**, 128-132.

Oosawa, F., and Kasai, M. (1971). *In* "Subunits in Biological Systems" (S. N. Timasheff and G. D. Fasman, eds.), Part A, pp. 261-322. Dekker, New York.

Orci, L., Gabbay, K. H., and Malaisse, W. J. (1972). *Science,* **175**, 1128-1130.

Prentki, M., Chaponnier, C., Jeanrenaud, B., and Gabbiani, G. (1979). *J. Cell Biol.* **81**, 592-607.

Puszkin, S., and Kochwa, S. (1974). *J. Biol. Chem.* **249**, 7711-7714.

Reaven, E. P., and Axline, S. G. (1973). *J. Cell Biol.* **59**, 12-27.

Schloss, J. A., and Goldman, R. D. (1979). *Proc. Natl. Acad. Sci. U.S.A.* **76**, 4484-4488.

Schneider, C., Gennaro, R., de Nicola, G., and Romeo, D. (1978). *Exp. Cell Res.* **112**, 249-256.

Schollmeyer, J. V., Rao, G. H. R., and White, J. G. (1978). *Am. J. Pathol.* **93**, 433-446.

Sorber, W. A. (1978). *Infect. Immun.* **19**, 799-806.

Spooner, B. S., Yamada, K. M., and Wessells, N. K. (1971). *J. Cell Biol.* **49**, 595-613.

Spudich, J. A., and Lin, S. (1972). *Proc. Natl. Acad. Sci. U.S.A.* **69**, 442-446.

Stendahl, O. I., and Stossel, T. P. (1980). *Biochem. Biophys. Res. Commun.* **92**, 675-681.

Stendahl, O. I., Hartwig, J. H., Brotschi, E. A., and Stossel, T. P. (1980). *J. Cell Biol.* **84**, 215-224.

Stock, C., Launay, J. F., Grenier, J. F., and Bauduin, H. (1978). *Lab. Invest.* **38**, 157-164.

Stockmayer, W. H. (1944). *J. Chem. Phys.* **12**, 125-131.

Stossel, T. P. (1973). *J. Cell Biol.* **58**, 346-356.

Stossel, T. P., and Hartwig, J. H. (1975). *J. Biol. Chem.* **250**, 5706-5712.

Stossel, T. P., and Hartwig, J. H. (1976). *J. Cell Biol.* **68**, 602-619.

Taylor, E. W. (1979). *CRC Crit. Rev. Biochem.* **6**, 103-164.

Taylor, D. L., and Coudeelis, J. (1979). *Int. Rev. Cytol.* **56**, 57-144.

Tilney, L. G., and Mooseker, M. (1971). *Proc. Natl. Acad. Sci. U.S.A.* **68**, 2611-2615.

Tombs, M. P. (1974). *Faraday Discuss, Chem. Soc.* **57**, 158-163.

Trotter, J. A., and Adelstein, R. S. (1979). *J. Biol. Chem.* **254**, 8781-8785.

Vanderkerkhoeve, J., and Weber, K. (1978). *J. Mol. Biol.* **126**, 783-802.

Vial, J. D., and Garrido, J. (1976). *Proc. Natl. Acad. Sci. U.S.A.* **73**, 4032-4236.

Wang, K. (1977). *Biochemistry* **16,** 1857–1865.

Watterson, D. M., Harrelson, W. G., Jr., Keller, P. M., Sharief, R., and Vanaman, T. C. (1976). *J. Biol. Chem.* **251,** 4501–4513.

Weihing, R. R. (1976). *In* "Biological Handbooks, Cell Biology" (P. L. Altman and D. D. Katz, eds.), vol. 1, pp. 341–356. FASEB Publications, Bethesda, Maryland.

Wessells, N. K., Spooner, B. S., Yamada, K. M., Ash, J. F., and Ludueña, E. (1971). *Science* **171,** 135–143.

Wolosewick, J. J., and Porter, K. R. (1979). *J. Cell Biol.* **82,** 114–139.

Yin, H. L., and Stossel, T. P. (1979). *Nature (London)* **281,** 583–586.

Yin, H. L., and Stossel, T. P. (1980). *J. Biol. Chem.* **255,** 9490–9493.

Yin, H. L., Zaner, K. S., and Stossel, T. P. (1980). *J. Biol. Chem.* **255,** 9494–9500.

Zurier, R. B., Hoffstein, S., and Weissman, G. (1973). *Proc. Natl. Acad. Sci. U.S.A.* **70,** 844–848.

# Chapter 15

# The Effect of Colchicine on the Synthesis and Secretion of Rat Serum Albumin

## C. M. REDMAN, D. BANERJEE, AND S. YU

*The Lindsley F. Kimball Research Institute
of the New York Blood Center, New York, New York*

## I. Introduction

Albumin is the major secretory protein produced by the liver. Its site of synthesis, intracellular route, kinetics of secretion, and the processing steps it undergoes prior to discharge have been established (Peters, 1977). This basic information makes albumin a useful model with which to study the mechanisms of hepatic secretion. Colchicine is an alkaloid that interferes with microtubular function (Wilson *et al.*, 1974) and inhibits secretion in a variety of systems. Thus, studying the effect of colchicine on the synthesis and secretion of albumin may prove useful in determining at which steps of hepatic secretion colchicine exerts its inhibitory action; and this may lead to a further understanding of the molecular mechanisms that govern the secretory steps affected by colchicine.

Rat albumin is synthesized on polysomes attached to the endoplasmic reticulum (ER) membrane, is vectorially discharged into the lumen of the rough ER, and then is transported stepwise to the smooth ER, to the Golgi complex, and to secretory vesicles, which fuse with the plasma membrane and secrete albumin into the Space of Disse (Peters, 1977). Initially albumin mRNA is translated as a larger molecule (pre-proalbumin) with an octadecapeptide extension on the amino-terminal portion of proalbumin (Strauss *et al.*, 1977; Yu and Redman, 1977). This extension is thought to act as a signal that aids in attaching the albumin-synthesizing polysomes to the ER membranes (Blobel, 1977). This "signal" portion of nascent albumin is cleaved during the vectorial transport of pre-proalbumin into the lumen of the ER. The major intracellular nascent form of albumin is proalbumin, which contains the following additional amino-terminal sequence to serum albumin; Arg-Gly-Val-Phe-Arg-Arg- (Urban *et al.*, 1974; Russel and Geller, 1975; Quinn *et al.*, 1975). *In vivo,* this basic hexapeptide is removed from proalbumin late in the secretory process, producing serum albumin with an amino-terminal glutamic acid. The exact intracellular sites of conversion of proalbumin to serum albumin are not clear, since conversion may occur in both the smooth ER and the Golgi complex in isolated hepatocytes (Edwards *et al.*, 1976), but *in vivo* it appears to occur mostly in the Golgi complex, and predominantly in secretion vesicles rather than in the cisternal elements of the Golgi complex (Ikehara *et al.*, 1976; Redman *et al.*, 1978).

Colchicine's major pharmacological role is to inhibit the polymerization of tubulin into microtubules. Since polymerization and deploymerization is a constantly ongoing process, treatment of tissues with colchicine leads to a decrease in number or to the eventual disappearance of microtubules (Inoué and Sato, 1967; Olmsted and Borisy, 1973). In liver, colchicine binds to soluble tubulin at concentrations similar to those needed to inhibit hepatic secretion, but colchicine also demonstrates afinity for other hepatic proteins (Patzelt *et al.*, 1977; Reaven, 1977). Colchicine has other cellular effects that possibly may not be linked to its antimicrotubular action. For example, it affects nucleoside transport (Mizel and Wilson, 1972), it modifies the shape of erythrocytes (Seeman *et al.*, 1973), and, as will be discussed later, it also inhibits protein synthesis. It is not known whether these cellular effects of colchicine are mediated through a primary antimicrotubular action.

## II.   Plasma Protein (Albumin) Secretion by Liver Slices

## A.   Effect of Colchicine on the Secretion and Intracellular Levels of Nascent Albumin

Rat liver slices were pulse-labeled with radioactive L-leucine for 7–9 minutes and then "chase"-incubated for varying periods of time in a medium containing nonradioactive L-leucine. Colchicine, at various concentrations, was added to the chase medium, and the release of radioactive albumin into the incubation medium was measured. Colchicine inhibited the release of albumin at concentrations as low as $10^{-6}$ $M$, with optimal inhibition (60–70%) at $10^{-5}$ $M$. Concomitant with the inhibition of albumin secretion, the amount of radioactive albumin retained within the slices was increased (Fig. 1). These initial observations indicated that colchicine inhibited secretion and that the nonsecreted albumin accumulated within the intracellular compartments.

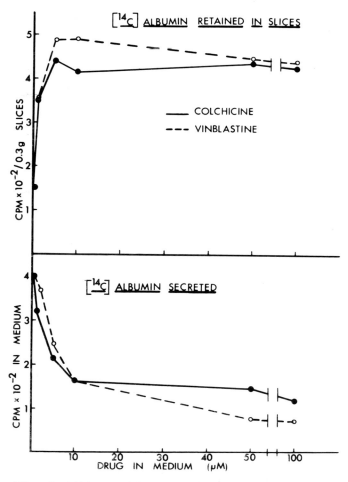

FIG. 1.   Effect of colchicine and vinblastine on albumin secretion by rat liver slices. Rat liver slices were pulse-labeled for 9 minutes at 37°C with L-[$^{14}$C]leucine and then "chase"-incubated for 90 minutes. Radioactive albumin retained in the slices and secreted into the incubation medium was determined. From Redman *et al.* (1975), with the permission of the *Journal of Cell Biology.*

## B.  Effect of Related Drugs on Secretion

The inhibitory effect of colchicine on secretion is specific in that lumicolchicine, an inactive isomer of colchicine, at comparable concentrations was one-fifth as effective as colchicine in inhibiting albumin secretion by rat liver slices. Also, griseofulvin, an antimitotic agent that does not affect microtubules, and cytocholasin B, which is thought to react with microfilaments, had little or no effect on albumin secretion by rat liver slices (Redman et al., 1975). Vinblastine inhibits albumin secretion similarly to colchicine (Fig. 1).

## III.   In Vivo Effects of Colchicine on Secretion

## A.   Concentrations of Colchicine Used and Time Course of Inhibitory Effect

Colchicine (5–25 μmol per 100 gm body weight) given intravenously to rats causes inhibition of plasma protein secretion and an accumulation of the nonsecreted proteins within the liver. At these doses the inhibitory effect of colchicine is prompt and may be noticed within 2 minutes of colchicine administration (Fig. 2). It is important to note that colchicine acts quickly in inhibiting in vivo secretion. By contrast, most studies on the effect of colchicine on secretion involve exposing the tissues to colchicine for relatively long periods of time. Lengthy exposure to the drug is necessary because these studies measure the accumulation of nonsecreted materials at specific hepatic organelles. In order to measure accumulation of exportable protein at a specific blocked site, sufficient time must elapse to allow for transport of proteins to that site and for sufficient accumulation of protein to occur. Therefore, many of the hepatic changes that are

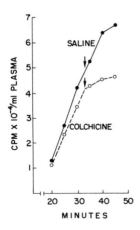

FIG. 2.    Effect of colchicine on the in vivo secretion of albumin. Rats were injected intravenously with L-[14C]leucine at zero time. Saline or colchicine (25 μmol per 100 gm body weight) were injected 33 minutes after the administration of L-[14C]leucine (arrows). Blood samples were taken from the tail at times shown, and the amount of radioactive albumin in the blood samples was determined. From Redman et al. (1975) with the permission of the Journal of Cell Biology.

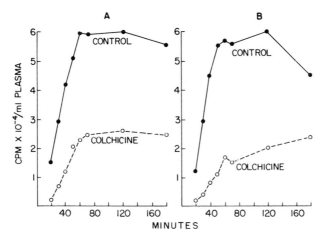

FIG. 3.   Effect of colchicine on the *in vivo* secretion of albumin and other plasma protein. Rats received L-[$^{14}$C]leucine via the femoral vein; this was followed by a second injection with either colchicine (25 $\mu$mol per 100 gm body weight) or an equal volume of saline. The rats were bled from the tail at the times shown above, and albumin and other plasma proteins were determined by immunoprecipitation. To determine the radioactivity in albumin a rabbit antibody specific to rat albumin was used, and to determine the radioactivity in other plasma proteins a multivalent antiserum, from which antialbumin had been removed by absorption with rat albumin, was employed. (A) describes the appearance of radioactive albumin in plasma and (B) describes that of the other proteins. From Redman *et al.* (1975), with permission from the *Journal of Cell Biology.*

measured in response to colchicine are caused by these relatively lengthy treatments, and they may not be a true index of the primary or earliest effect of colchicine on secretion.

Colchicine appears to be a general inhibitor of *in vivo* secretion, since it inhibits the secretion of albumin and also that of other plasma proteins produced by the liver (Fig. 3). The inhibitory effect of colchicine on the secretion of hepatic lipoprotein has been studied in detail (Stein and Stein, 1973; Stein *et al.*, 1974; Le Marchand *et al.*, 1973). The effect of colchicine, however, is not indiscriminate, as a variety of hepatic physiological parameters has been studied and found not to be affected (Le Marchand *et al.*, 1973; Redman *et al.*, 1975).

The inhibitory effect of colchicine, as mentioned before, is prompt, but it is also long-lasting; in mice the effect is still apparent after 8 hours but is reversed after 24 hours (Le Marchand *et al.*, 1975).

## B.  Intracellular Localization of the Colchicine Effect

When rats are pulse-labeled *in vivo* by the intravenous administration of radioactive L-leucine, it is possible to follow, with time, the intracellular movement of radioactive albumin. At 10 minutes after the administration of radioactive L-leucine to the rat, there is peak albumin radioactivity in the rough ER, and at 30 minutes it is highest in the smooth ER and the Golgi complex. Colchicine given to rats just prior to the administration of radioactive L-leucine does not affect the movement of radioactive albumin from the rough to the smooth ER, nor does it affect the entry of nascent albumin into the Golgi complex (see 10 and 20 minutes, Fig. 4). Colchicine appears to affect the release of radioactive albumin from the Golgi complex in that in the control rats there was a disappearance of radioactive albumin from these cell fractions with increasing time (after 30 minutes), while in the colchicine-treated rats the radioactive albumin was retained for up to 90 minutes (Fig. 4).

These experiments indicate that colchicine does not affect the transport of albumin in the rough and smooth ER, nor does it affect the filling of the Golgi vesicles. It appears that colchicine affects a stage in albumin secretion after the secretory vesicles have been filled, but before they fuse with the plasma membrane. This effect of colchicine, on a late stage in secretion, is in agreement with the observation that, *in vivo,* colchicine can block albumin secretion within 2 minutes after its administration to the rat. It has been calculated by Morgan and Peters (1971) that the intracellular transit time for albumin is 16 minutes, and if colchicine had blocked an early step in secretion then a longer delay in the time taken to elicit an inhibition of secretion would have been observed.

Albumin does not undergo glycosylation, but like the plasma glycoproteins, whose secretion is also inhibited by colchicine, it does pass through the Golgi complex. In the Golgi complex the secretory glycoproteins are modified in that the penultimate and terminal sugars, galactose and sialic acid, are added to nascent glycoproteins (Leblond and Bennett, 1977; Schacter, 1974). Since colchicine appears to inhibit secretion after the secretory proteins have reached the Golgi complex, it was of interest to determine whether the colchicine-induced block occurred prior to or after the addition of these sugars to nascent secretory glycoproteins.

D-[$^{14}$C]Glucosamine or D-[G-$^3$H]galactose was injected into rats, and at various times after the administration of these sugars the plasma glycoproteins were isolated by immunoprecipitation from various hepatic cell fractions and the incorporation of radioactivity into these proteins was measured. The incorporation of radioactivity from D-glucosamine into the sialic acid moiety of plasma proteins was also measured by acid hydrolysis of the isolated glycoproteins and by recovering the released radioactive sialic acid by ion-exchange chromatography. These studies showed that radioactive glucosamine was incorporated into nascent plasma glycoproteins in the rough and smooth ER and in the Golgi complex, but that incorporation of sialic acid radioactivity into protein occurred only in the Golgi complex. D-[$^{14}$C]Galactose incorporation into plasma glycoproteins was also limited to the Golgi cell fractions. This is in keeping with studies by others who have shown that the Golgi complex is the site of terminal glycosylation of

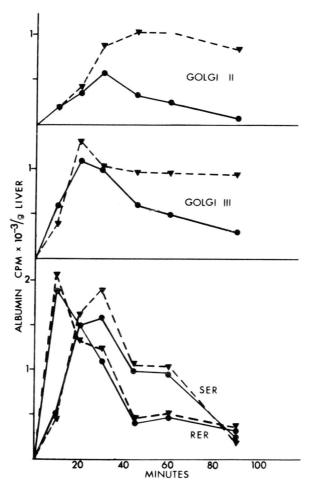

FIG. 4. Effect of colchicine on the distribution of nascent albumin in various liver cell fractions at different times after the administration of L-[¹⁴C]leucine to rats. Rough (RER) and smooth (SER) endoplasmic reticulum and two Golgi cell fractions were prepared at various times after the intravenous administration of L-[¹⁴C]leucine to rats. The dotted lines represent the amount of radioactive albumin in colchicine-treated rats, and the solid lines that in the control animals. For a description of the cell fractions, see Ehrenreich *et al.* (1973) and Redman *et al.* (1975). Taken from Redman *et al.* (1975), with the permission of the *Journal of Cell Biology.*

secretory proteins. (Leblond and Bennett, 1977; Schacter, 1974). The administration of colchicine to rats caused an increased incorporation of radioactive galactose and sialic acid into the plasma glycoproteins, indicating that colchicine inhibits secretion after terminal glycosylation had taken place, and this then results in retention of radioactive glycoprotein in the Golgi cell fractions (Banerjee *et al.*, 1976). If colchicine had limited the filling of the Golgi complex by nascent glycoproteins, then an inhibition of terminal glycosylation would have been noticed.

In addition to terminal glycosylation, another step in the processing of proteins that occurs in the Golgi complex is the proteolytic conversion of proalbumin to serum albumin. The Golgi complex, in rat liver, may be subfractionated into three subfractions by sucrose density gradient centrifugation. The least dense of these subfractions, Golgi fraction 1, consists of rounded secretion vesicles filled with VLDL particles and contains very few Golgi cisternae. Golgi fraction 2, which is of intermediate density, is also predominantly composed of secretion vesicles but has more cisternal elements than Golgi fraction 1, and Golgi fraction 3, the most dense of the fractions, contains mostly flattened Golgi cisternae showing distended rims. These cell fractions have been well characterized by morphological, enzymatic, and histochemical procedures (Ehrenreich *et al.*, 1973; Farquhar *et al.*, 1974).

In the rat, the *in vivo* conversion of proalbumin to serum albumin appears to occur in the rounded, filled secretion vesicles represented by Golgi fraction 1 and 2. Analyses of proalbumin and serum albumin in the three Golgi subfractions show that only nascent proalbumin occurs in Golgi fraction 3, while Golgi fractions 1 and 2 contain a mixture of both proalbumin and serum albumin (Ikehara *et al.*, 1976; Redman *et al.*, 1978). Administration of radioactive L-leucine to rats and isolation of proalbumin and serum albumin (from the three Golgi cell fractions) at various times shows that radioactive proalbumin reaches a peak level in all three Golgi fractions within 20–30 minutes. Radioactive serum albumin is not seen in the cisternal elements of the Golgi complex (fraction 3), but it appears within 20–30 minutes in the less dense Golgi fractions 1 and 2 (Fig. 5). This indicates that the conversion of proalbumin to serum albumin is more prominent in the secretion vesicles than in the cisternal elements of the Golgi and that conversion is rapid, since it is not possible to observe a precursor product time relation between proalbumin and serum albumin in the secretion vesicles.

Administration of colchicine to rats, at the same time that radioactive L-leucine is given, shows that after 30 minutes both proalbumin and serum albumin accumulate in secretion vesicles (Golgi fractions 1 and 2). The accumulation of nascent albumin appears to be less pronounced in the cisternal elements of the Golgi (Golgi fraction 3), where only radioactive proalbumin is seen at early times (Fig. 5). Colchicine appears therefore to have a mixed effect in that inhibiting secretion causes both species of albumin (proalbumin and serum albumin) to accumulate. If colchicine had specifically blocked secretion prior to the conversion of proalbumin to serum albumin, then we would expect only radioactive proalbumin in the Golgi fractions, and if inhibition of secretion occurs after the production of serum albumin, then the Golgi cell fractions should accumulate

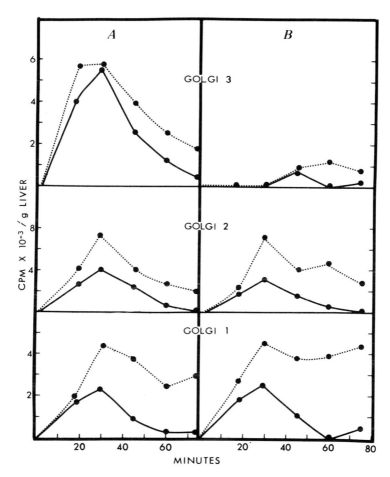

FIG. 5.    Time course of appearance of proalbumin and serum albumin in Golgi cell fractions of saline- and colchicine-treated rats. Rats were fed alcohol (0.6 gm per 100 gm body weight) 1 hour before the intravenous administration of L-[¹⁴C]leucine and colchicine (25 μmol per 100 gm body weight). At the times shown, the Golgi cell fractions were prepared and radioactive proalbumin and serum albumin were isolated by immunoprecipitation followed by DEAE–cellulose chromatography. The dotted lines show the amount of radioactive proalbumin (left column) and serum albumin (right column) in the colchicine-treated rats, and the solid lines that in the control animals. From Redman *et al.* (1978), with permission from the *Journal of Cell Biology*.

only serum albumin, in response to colchicine. The fact that colchicine causes both proalbumin and serum albumin to accumulate in the secretion vesicles indicates that colchicine affects secretion during the span of time at which conversion of proalbumin to serum albumin occurs.

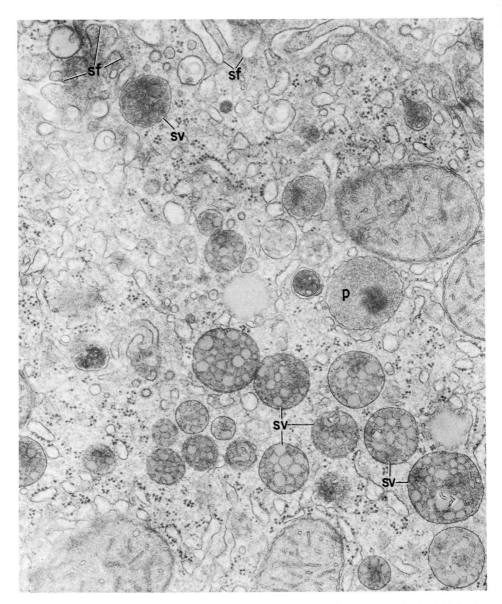

FIG. 6.   The cytoplasm of a hepatocyte from a colchicine-treated rat. Rats received colchicine intravenously (25 $\mu$mol per 100 gm body weight) 1–3 hours before liver samples were taken for electron microscopy. All specimens were fixed with glutaraldehyde-formaldehyde and postfixed in osmium tetroxide. The field shows secretion vesicles (sv), filled with VLDL particles, throughout the cytoplasm and at the sinusoidal front (sf) of the cell. Peroxisome (p). Magnification 32,000×. From Redman *et al.* (1975), with the permission of *Journal of Cell Biology*.

# IV. Hepatocyte Morphology

The liver of colchicine-treated rats, in comparison with their corresponding controls, showed a larger number of secretion vesicles, loaded with VLDL. These secretion vesicles were not only accumulated at the sinusoidal front of the hepatocyte but were also more numerous throughout the cytoplasm (see Fig. 6). In rats that were fed alcohol (a procedure used to increase the yield of Golgi cell fraction), the VLDL content of the Golgi elements was more pronounced in colchicine-treated rats.

In hepatocytes of control rats, microtubules are few in number but are usually seen in the Golgi regions and in the peripheral cytoplasm. The distribution of microtubules is random, and they do not appear to be concentrated near the secretion vesicles. At 1 hour after colchicine treatment (25 $\mu$mol per 100 gm body weight), a time at which inhibition of albumin secretion is maximal, microtubules could still be detected in the Golgi regions, along the bile canaliculli, and on the sinusoidal front. Similar results were seen 3 hours after colchicine treatment (Redman *et al.*, 1975). Recent morphometric studies by Reaven and Reaven (1980) show, however, that there is a close association between loss of microtubules, alterations to the Golgi complex induced by colchicine, and decreases in hepatic VLDL secretion.

# V. Effect of Colchicine on Protein Synthesis

Colchicine studies have been performed with perfused livers, in liver slices, and with intact animals. It has been generally observed in all these systems that colchicine inhibits the secretion of all plasma proteins, without markedly affecting total hepatic protein synthesis. There are conditions, however, when colchicine affects both secretion and protein synthesis (Dorling *et al.*, 1975); this usually occurs after intact animals have been exposed to prolonged treatment with colchicine. This has led to the proposal that the delayed effect of colchicine on protein synthesis is a feedback mechanism, which shuts off further synthesis of plasma proteins in response to the initial inhibition of secretion.

Colchicine administered to rats inhibits the synthesis of TCA-precipitable proteins, and of albumin, after the animals have been exposed to colchicine for 1 hour or longer. The delayed inhibition of protein synthesis can be elicited at concentrations as low as 0.5 $\mu$mol of colchicine per 100 gm of body weight (Figs. 7 and 8).

This delayed effect of colchicine on protein synthesis is not specific to any class of hepatic protein. It affects proteins synthesized by both free and

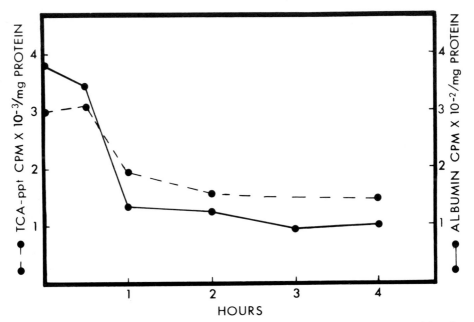

F<small>IG</small>. 7.    Delayed effect of colchicine in inhibiting hepatic protein synthesis. Rats were injected intravenously with colchicine (25 $\mu$mol per 100 gm body weight), and at the indicated times a 10-minute incorporation of L-[$^{14}$C]leucine into TCA-precipitable protein (solid lines) and into albumin (dotted lines) was measured as described by Redman *et al.* (1978). Published with the permission of the *Journal of Cell Biology*.

membrane-attached polysomes, and it inhibits to a similar degree both secretory and nonsecretory proteins (Redman *et al.*, 1978). This indicates that the delayed effect of colchicine on protein synthesis is a general effect and is probably not involved in a feedback inhibition of secretory proteins.

## VI.    Discussion

These experiments suggest that microtubules are involved in hepatic secretion and that they may function at some secretory steps after the exportable proteins have reached the Golgi complex. Unfortunately, we cannot, at present, deduce the function of microtubules in these late stages of secretion. The evidence that microtubules are involved in hepatic secretion is based on experiments that demonstrate that colchicine and vinblastine (but not lumicolchicine) inhibit secretion and that these drugs are known to interact with hepatic tubulin at concentrations that inhibit secretion (Le Marchand *et al.*, 1973, 1974; Patzelt *et al.*,

1977). In the liver, only a small percentage of tubulin is in the form of microtubules (Reaven, 1977), and we do not see an obvious decrease in the number of microtubules at conditions of colchicine treatment that inhibit secretion. It should be noted, however, that others, in studies with mice, have seen marked hepatic microtubular disruption (Le Marchand *et al.*, 1973), and quantitative studies on the number of rat hepatic microtubules in response to colchicine indicate a coupling between the decrease in microtubules and VLDL secretion (Reaven and Reaven, 1980). Also, colchicine may have subtle effects on microtubular function that are not easily detectable but may influence albumin secretion.

Our studies localize the effect of colchicine on secretion at a post-Golgi stage after nascent albumin has entered the secretion vesicles, probably after plasma glycoproteins have been fully glycosylated and during the conversion of proalbumin to albumin. During the late stages of secretion, after secretory proteins have entered the secretion vesicles, the following steps may be identified. (1) The secretion vesicles mature. This is often seen (by electron microscopy) as vesicles containing different densities of material, which probably reflects differing concentrations of protein within the vesicles. (2) The secretion vesicles travel from the region of the Golgi complex to the site of discharge, which in

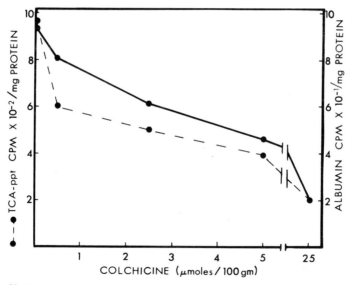

FIG. 8.   Various concentrations of colchicine that elicit a delayed inhibition of hepatic protein synthesis. Rats were treated with various amounts of colchicine 3 hours prior to the administration of L-[¹⁴C]leucine. Then, 10 minutes after the administration of radioactive L-leucine, the radioactivity in TCA-precipitable proteins and into albumin was determined. Taken from Redman *et al.* (1978), with the permission of the *Journal of Cell Biology*.

the case of albumin and other hepatic plasma proteins is the plasmalemma that fronts with the Space of Disse. (3) Exocytosis takes place. This involves recognition of mature secretory vesicles with specific areas of the plasma membrane, and attachment and fusion of the membranes of the secretion vesicle to the plasma membrane, leading to the emptying of the contents of the secretion vesicles to the extracellular space.

Colchicine probably affects one or more of these processes. It appears unlikely that colchicine specifically inhibits the final stages of exocytosis, since we do not notice an accumulation of secretion vesicles at the sinusoidal front, but rather there is an accumulation of secretion vesicles throughout the cytoplasm. Also, studies with other secretory tissues, such as the pancreas and the parotid in which exocytosis is triggered by hormones, indicates that colchicine does not affect the discharge of stored zymogen granules, but rather that this drug inhibits the secretion of newly formed secretion vesicles (Patzelt *et al.*, 1977; Williams and Lee, 1976; Boyd *et al.*, 1979). Taken together, these observations suggest that colchicine affects either the maturation of the secretory vesicles or their translocation from the Golgi complex to the site of discharge.

## References

Banerjee, D., Manning, C. P., and Redman, C. M. (1976). *J. Biol. Chem.* **251**, 3887–3892.

Blobel, G. (1977). *In* "International Cell Biology 1976–1977" (B. R. Brinkley and K. R. Porter eds.), pp. 318–325. Rockefeller Univ. Press, New York.

Boyd, A. E., Bolton, W., and Brinkley, B. R. (1979). *J. Cell Biol.* **83**, 333a (abstr.).

Dorling, P. R., Quinn, P. S., and Judah, J. D. (1975). *Biochem. J.* **152**, 341–348.

Edwards, K., Fleisher, B., Dryburgh, H., Fleisher, S., and Schreiber, G. (1976). *Biochem. Biophys. Res. Commun.* **72**, 310–318.

Ehrenreich, J. A., Bergeron, J. J. M., Siekevitz, P., and Palade, G. E. (1973). *J. Cell Biol.* **59**, 45–72.

Farquhar, M. G., Bergeron, J., and Palade, G. E. (1974). *J. Cell Biol.* **60**, 8–25.

Ikehara, Y., Oda, K., and Kato, K. (1976). *Biochem. Biophys. Res. Commun.* **72**, 319–326.

Inoué, S., and Sato, H. (1967). *J. Gen. Physiol.* **50**, Suppl., 259–288.

Leblond, C. P., and Bennett, G. (1977). *In* "International Cell Biology 1976–77" (B. R. Brinkley and K. R. Porter, eds.), pp. 326–336. Rockefeller Univ. Press, New York.

Le Marchand, Y., Singh, A., Assimacopoulos-Jeannet, F., Orci, L., Rouiller, C., and Jeanrenaud, B. (1973). *J. Biol. Chem.* **248**, 6862–6870.

Le Marchand, Y., Patzelt, C., Assimacopoulos-Jeannet, F., Loten, E. G., and Jeanrenaud, B. (1974). *J. Clin. Invest.* **53**, 1512–1517.

Le Marchand, Y., Singh, A., Paltzelt, C., Orci, L., and Jeanrenaud, B. (1975). *In* "Microtubules and Microtubule Inhibitors" (M. Borgers and M. de Brabander, eds.), pp. 153–164. North-Holland Publ., Amsterdam.

Mizel, S. B., and Wilson, L. (1972). *Biochemistry* **11**, 2573–2578.

Morgan, E. H., and Peters, T., Jr. (1971). *J. Biol. Chem.* **246**, 3500–3507.

Olmsted, J. B., and Borisy, G. G. (1973). *Annu. Rev. Biochem.* **42**, 507–540.

Patzelt, C., Brown, D., and Jeanrenaud, B. (1977). *J. Cell Biol.* **73**, 578–593.

Peters, T. (1977). *In* "Albumin Structure, Function and Uses" (V. M. Rosenoer, M. Oratz, and M. A. Rothschild, eds.), pp. 305–332. Pergamon, Oxford.

Quinn, P. S., Gamble, M., and Judah, J. D. (1975). *Biochem. J.* **146,** 389–393.

Reaven, E. P. (1977). *J. Cell Biol.* **75,** 731–742.

Reaven, E. P., and Reaven, G. M. (1980). *J. Cell Biol.* **84,** 28–39.

Redman, C. M., Banerjee, D., Howell, K., and Palade, G. E. (1975). *J. Cell Biol.* **66,** 42–59.

Redman, C. M., Banerjee, D., Manning, C., Huang, C. Y., and Green, K. (1978). *J. Cell Biol.* **77,** 400–416.

Russell, J. H., and Geller, D. M. (1975). *J. Biol. Chem.* **250,** 3409–3413.

Schacter, H. (1974). *Adv. Cytopharmacol.* **2B,** 207–218.

Seeman, P., Chau-Wong, M., and Moyen, S. (1973). *Nature (London), New Biol.* **241,** 22.

Stein, D., and Stein, Y. (1973). *Biochim. Biophys. Acta* **306,** 142–147.

Stein, D., Sanger, L., and Stein, Y. (1974). *J. Cell Biol.* **62,** 90–103.

Strauss, A. W., Donohue, A. M., Bennett, C. D., Rodney, J. A., and Alberts, A. W. (1977). *Proc. Natl. Acad. Sci. U.S.A.* **74,** 1358–1362.

Urban, J., Inglis, A. S., Edwards, K., and Schreiber, G. (1974). *Biochem. Biophys. Res. Commun.* **61,** 494–501.

Williams, J. A., and Lee, M. (1976). *J. Cell Biol.* **71,** 795–806.

Wilson, L., Bamburg, J. R., Mizel, S. B., Brisham, L. M., and Creswell, K. M. (1974). *Fed. Proc., Fed. Am. Soc. Exp. Biol.* **33,** 158–166.

Yu, S., and Redman, C. M. (1977). *Biochem. Biophys. Res. Commun.* **76,** 469–476.

# Chapter 16

# Effects of Antimitotic Agents on Ultrastructure and Intracellular Transport of Protein in Pancreatic Acini

JOHN A. WILLIAMS

*Department of Physiology,*
*University of California,*
*San Francisco, California*

## I. Introduction

A role for microtubules in the secretory process has been studied in the pancreas and other exocrine cells (Temple *et al.*, 1972; Butcher and Goldman, 1972; Seybold *et al.*, 1975; Williams and Lee, 1976; Patzelt *et al.*, 1977; Stock *et al.*, 1978; Chambaut-Guérin *et al.*, 1978). In exocrine cells the structural and functional polarization of the cell suggests the possibility of microtubular participation in intracellular packaging, transport, and release steps of secretion. Earlier studies of pancreatic secretion employed pancreatic slices or fragments, and these studies have indicated that, after prolonged exposure (1–3 hours) to antimitotic agents such as vinblastine or colchicine, cellular microtubules disappeared or decreased and enzyme secretion was reduced (Williams and Lee, 1976; Stock *et al.*, 1978). Seybold *et al.* (1975), in a study using guinea pig pancreatic lobules, concluded that intracellular transport of newly synthesized protein was

more sensitive to inhibition by antimitotic agents but that these drugs also inter-
fered with the release of digestive enzymes. These studies indicated that antimitotic agents do not compromise cellular energy reserves as deduced from
measurement of ATP levels or cellular ionic concentrations (Williams and Lee,
1976; Patzelt *et al.*, 1977; Chambaut-Guérin *et al.*, 1978). Since vinblastine
does not interfere with secretagogue-induced mobilization of cellular calcium
(Williams and Lee, 1976), it can also be concluded that the initial steps in the
stimulus–secretion coupling mechanism are not dependent on intact microtubules.

Recently, we have developed a new preparation of isolated pancreatic acini,
prepared by limited enzymatic digestion with collagenase (Williams *et al.*,
1978). These acini retain both structure and function, are exquisitely sensitive to
hormones and neurotransmitters, and thus have obviated the problems of tissue
penetration, present in the studies using either fragments or slices. We therefore
reinvestigated the effects of antimitotic agents in this preparation, carrying out
combined studies of ultrastructure, enzyme secretion, and intracellular transport
of newly synthesized protein.

## II.  Methods

Isolated pancreatic acini were prepared as previously described (Williams *et
at.*, 1978). Briefly, 0.9–1.2 g of pancreatic tissue was obtained from male
Swiss mice (20–25 gm), which had been fasted overnight. The basic medium
used for dissociation and incubation of acini was Krebs–Henseleit bicarbonate
solution (KHB) enriched with minimal Eagles medium essential amino acid
supplement, and 0.1 mg/ml soybean trypsin inhibitor. The dissociation medium
had the $Ca^{2+}$ content reduced to 0.1 m$M$ and contained 60–75 U/ml purified
collagenase, 15–20 $\mu$g/ml chymotrypsin, and 1.8 mg/ml hyaluronidase. This
medium was injected with a 27-gauge needle into the pancreatic parenchyma,
and the tissue was incubated for a total of 45–50 minutes at 37°C with shaking at
120 cycles/min. Acini were then dissociated by repeated suction through poly-
propylene pipettes with decreasing orifices, filtered through 150-$\mu$m nylon cloth,
and then purified by centrifugation through KHB containing 4% bovine serum
albumin. The acini were finally suspended in KHB containing 1.28 m$M$ CaCl$_2$
and 1% bovine serum albumin and preincubated for 1 hour prior to use.[1]

Intracellular transport and secretion of newly synthesized protein was studied

---

[1]We have found that slightly more precise results in secretory studies can be obtained by using a
HEPES-or Tris-buffered Ringer,which obviates the need for repeated gassing with 95% $O_2$–5% $CO_2$.
However, these buffers (HEPES less than Tris) both cause swelling and vacuolization of the Golgi
area, making morphological evaluation difficult. As we were particularly interested in this part of the
cell for this study, we chose to use bicarbonate-buffered Ringer.

by means of a pulse–chase protocol with [³H]leucine (New England Nuclear, 57 Ci/mmol). Acini were incubated for 10 minutes in plastic flasks at a protein density of 10 mg/ml with 5 $\mu$Ci/ml of [³H]leucine, centrifuged (50 $g$ for 2 minutes), washed, and resuspended in medium without radioactivity at a density of 1–2 mg/ml protein for an 80-minute chase period. This time was chosen on the basis of studies of both pancreatic slices and isolated acinar cells showing that after this time newly synthesized protein had been transported to a pool whose release could be stimulated by secretagogues (Jamieson and Palade, 1971; Amsterdam and Jamieson, 1974). Acini were then resuspended in fresh medium and incubated as 2-ml aliquots with or without secretagogues for 30 minutes. Acini were then pelleted in an Eppendorf microcentrifuge tube, the surface of the pellet was washed once with 0.9% NaCl, and the acini were dispersed in 1 ml $H_2O$ by sonication. Aliquots of both the supernatant and dispersed pellet were precipitated by addition of trichloracetic acid (TCA) to a final concentration of 10% at 4°C. The pellets were washed once with 10% TCA and resuspended in 0.1 $M$ NaOH; radioactivity was measured by liquid scintillation counting using a Triton X-100–toluene cocktail.

Amylase release was measured as previously described (Williams *et al.*, 1978); the method of Rinderknecht *et al.* (1967), utilizing amylose azure as substrate, was used to assay amylase activity.

The $^{45}Ca^{2+}$ efflux was measured as described previously (Williams *et al.*, 1978). Acini were preincubated for 1 hour with $^{45}CaCl_2$ (2 $\mu$Ci/ml), centrifuged, washed with ice-cold KHB medium, and resuspended in nonradioactive medium at 37°C. Duplicate samples were taken immediately (time 0), secretagogue was added, and further samples were taken at various times thereafter. Acini were separated from the extracellular medium by dilution into 20 volumes of ice-cold 0.9% NaCl with immediate suction through a Nuclepore filter (3 $\mu$m pore size), followed by a single wash with a further 10 volumes of ice-cold saline. The filters were placed in plastic tubes, 2 ml $H_2O$ was added, and the acini were removed from the filter and disrupted by sonication. Aliquots were then assayed for $^{45}Ca^{2+}$ by liquid scintillation counting and protein content by the method of Lowry *et al.* (1951).

For morphological evaluation the acinar suspension was added to an equal volume of 1.5% glutaraldehyde and 1% paraformaldehyde in 0.08 $M$ cacodylate buffer (pH 7.4), centrifuged (100 $g$ for 3 minutes), and resuspended in the above fixative. Acini were then immediately pelleted at 10,000 $g$ for 1 minute in an Eppendorf microcentrifuge and after 1 hour at room temperature were stored overnight in fixative at 4°C. Pellets were postfixed with 2% $OsO_4$ in cacodylate buffer and block-stained with 1% uranyl acetate in 50 m$M$ acetate buffer (pH 5.0), all at 4°C, and embedded in Araldite resin (Grade Cy-212 [British]). Thin sections were doubly stained with uranyl acetate and lead citrate and examined at 60 kV in a JEM 100-B electron microscope.

# III.   Results

## A.   Release of Amylase and Newly Synthesized Protein

Initially isolated mouse pancreatic acini were prepared and preincubated for 90 minutes with vinblastine (30 $\mu M$) or podophyllotoxin (30 $\mu M$); secretion was then stimulated with 1 $\mu M$ carbachol, the optimal dose for stimulating enzyme release (Williams *et al.*, 1978). As shown in Fig. 1, neither antimitotic agent significantly affected basal amylase release; vinblastine reduced carbachol-stimulated release by about 50%, and podophyllotoxin was without effect. These results are almost identical to our previous results with mouse pancreatic fragments similarly pretreated with vinblastine and colchicine (Williams and Lee, 1976). Podophyllotoxin was employed in the present study because its rate of binding to tubulin is faster than that of colchicine (Cortese *et al.*, 1977). The lack of an effect of podophyllotoxin on amylase release was surprising, since our structural studies discussed below showed that both antimitotic agents were affecting cellular microtubules. We therefore looked at the release of newly synthesized protein to determine if this process was more sensitive to antimitotic agents.

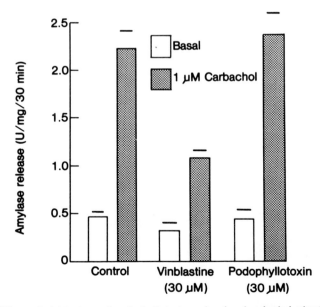

FIG. 1.   Effects of vinblastine and podophyllotoxin on basal and carbachol-stimulated amylase release by isolated mouse pancreatic acini. Acini were prepared, preincubated for 90 minutes with the specified agent, and then incubated for 30 minutes with or without carbachol. Bars show the mean ± S.E. of four flasks.

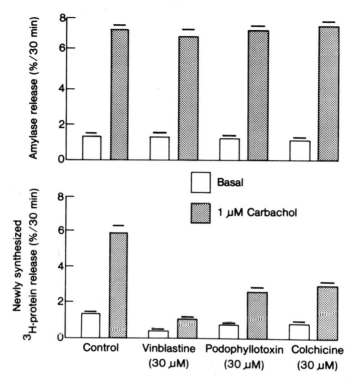

Fig. 2. Effect of antimitotic agents on release of amylase and newly synthesized protein by isolated pancreatic acini. Acini were preincubated for 40 minutes in KHB and resuspended in medium containing drugs for 90 minutes, during which a 10-minute pulse with [³H]leucine and an 80-minute chase with nonradioactive medium were carried out. After this, acini were incubated for 30 minutes with or without carbachol, and release of amylase and newly synthesized protein was measured. Bars show the mean ± S.E. for four flasks.

Following dissociation, acini were preincubated for 40 minutes to recover after their preparation, and then the antimitotic drugs were added for 90 minutes; during this time the acini were allowed to take up [³H]leucine during the initial 10 minutes of drug exposure and then chased for 80 minutes. When the release of newly synthesized protein and that of amylase were measured simultaneously, neither vinblastine, colchicine, nor podophyllotoxin had any effect on amylase release, although all three significantly inhibited release of newly synthesized protein (Fig. 2). As shown in Fig. 3, the effect of vinblastine was primarily on carbachol-stimulated release, with 50% inhibition at about 5 $\mu M$ drug.

To exclude the possibility that the effect of vinblastine was on one of the earlier steps in stimulus–secretion coupling, the carbachol-stimulated mobilization of cellular $Ca^{2+}$ prelabeled with $^{45}Ca^{2+}$ was studied. As shown in Fig. 4, carbachol-stimulated $Ca^{2+}$ efflux was not inhibited by vinblastine pretreatment.

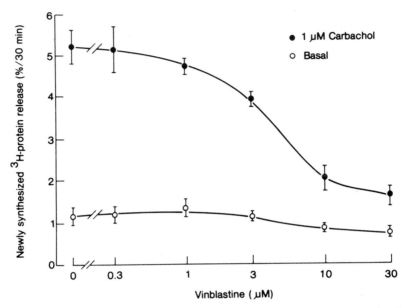

FIG. 3. Concentration dependence of vinblastine inhibition of newly synthesized protein release by isolated pancreatic acini. Acini were preincubated and incubated as in Fig. 2. All points are the mean ± S.E. of three flasks.

## B. Ultrastructural Studies

The ultrastructure of isolated pancreatic acini is similar to that of acini in intact pancreas (Williams *et al.*, 1978). As shown in Fig. 5, microtubules are present in the apical portion of the cell, along with a relatively homogeneous population of zymogen granules. Golgi complexes consisting of condensing vacuoles, cisternae, and small vesicles are present between the nucleus and the apical surface. The presence of any one of the three antimitotic agents during preincubation (90 minutes) and incubation (30 minutes) was sufficient to eliminate all the microtubules. Although the overall polarization of the cell was preserved, the zymogen granules were now grouped in clusters that contained a large number of small granules (Fig. 6). In acini treated with vinblastine, paracrystals were present along with an increased number of autophagic vacuoles (Fig. 6). Mitochondria, by contrast, appeared basically normal. The most striking observation however, was that in acini treated with any one of these agents there were large masses of small vesicles next to the enlarged Golgi cisternae (Fig. 7). Collections of these vesicles filling an area up to 2–3 $\mu$m across were present in every cell. The ultrastructural appearance of acini incubated directly in medium containing the antimitotic agents was identical to those first preincubated in KHB and then incubated in medium containing the drugs.

## IV. Discussion

The use of isolated pancreatic acini has allowed a clearer definition of the action of antimitotic agents on pancreatic acinar cell secretion than was previously possible with tissue fragments or slices. This is probably due in part to the reduced problem of tissue penetration, but it may be also due to the better structural and functional preservation of isolated acini.

The primary aim of this study was to distinguish the relative importance of microtubules in intracellular transport and release of secretory materials. In the pancreas, preformed secretory product exists as zymogen granules, and release of digestive enzymes such as amylase is not dependent on new protein synthesis (Jamieson and Palade, 1968). In contrast, when release of newly synthesized protein is studied, intracellular transport as well as the final release process is

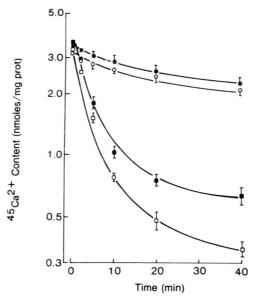

FIG. 4. Control and carbachol-stimulated efflux of $^{45}Ca^{2+}$ from acini preincubated in KHB (closed symbols) or KHB containing 30 $\mu M$ vinblastine (open symbols). Acini were preincubated for 90 minutes with $^{45}Ca^{2+}$ present for the last 60 minutes, then washed and resuspended in similar medium but without radioisotope, either with ($\square$, $\blacksquare$) or without ($\bigcirc$, $\bullet$) 1 $\mu M$ carbachol.

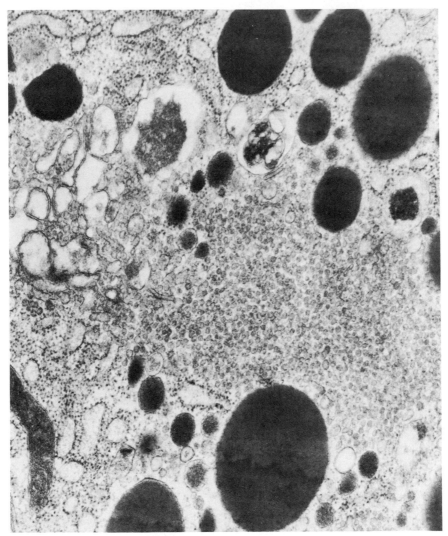

Fɪɢ. 7.   Golgi complex of isolated mouse pancreatic acini incubated for 120 minutes in 30 $\mu M$ podophyllotoxin. Magnification 32,600×.

measured. Thus, if an inhibitor simultaneously blocks the release of both amylase and newly synthesized protein, it can be concluded that the release process is being affected, whereas if amylase secretion is not reduced but that of newly synthesized protein is, it can be concluded that intracellular transport (in the broadest sense) is being affected. In the present study, when acini were allowed to recover and then treated for 90 minutes with antimitotic agents at a concentration sufficient to abolish cellular microtubules, subsequent carbachol-stimulated release of amylase was not affected, while the simultaneous release of newly synthesized protein was greatly reduced. This finding indicates that under these conditions the release process (exocytosis) is intact, while intracellular transport is disrupted. Since the antimitotic agents were applied at the same time as the [³H]leucine pulse (in order to minimize inhibition of protein synthesis), the maximal effects of the agents were not observed. In fact, when acini were first preincubated for 90 minutes with drugs, the subsequent release of newly synthesized protein was even further reduced. Even so, under the conditions where there was no inhibition of amylase release, vinblastine at a concentration as low as 3 $\mu M$ significantly reduced the release of protein when stimulated by carbachol.

The reason why exposure to vinblastine for 90 minutes immediately after preparation of the acini inhibits subsequent amylase release, whereas no effect is seen if the acini are initially preincubated in control medium, is not clear at present. Possibly, granules in a "releasable pool" were depleted during preparation of the acini, and refilling of this pool is sensitive to vinblastine. Alternatively, the newly prepared acini may be metabolically compromised and hence more susceptible to vinblastine.

The morphological results of treatment with all three antimitotic agents are also consistent with a primary effect on intracellular transport. Ultrastructural changes include the absence or swelling of Golgi cisternae, the accumulation of small vesicles normally found at the cis Golgi face, and the appearance of a number of small zymogen granules. As these changes are associated with a loss of microtubules, and since no other known effects of the antimitotic agents can explain this effect, it seems reasonable to assume that microtubules are necessary to maintain a normal configuration of these subcellular structures. Although ATP levels in isolated acini were not measured, both amylase release and $^{45}Ca^{2+}$ movements, shown here to be normal, are very dependent on an intact energy supply. Thus there is no reason to believe the agents studied affected the cellular energy production. The lack of effect of vinblastine on $^{45}Ca^{2+}$ efflux also indicates that the early steps in stimulus–secretion coupling including receptor occupancy and $Ca^{2+}$ mobilization are not being inhibited.

The dependence of pancreatic intracellular transport on intact microtubules is consistent with studies in other exocrine glands including the parotid (Patzelt *et al.*, 1977) and the lacrimal gland (Chambaut-Guérin *et al.*, 1978). Although it is not unlikely that during prolonged stimulation the release of amylase could be inhibited by treatment with antimitotic agents, it seems that the primary effect of these agents, and by inference the dependence on intact microtubules, is on intracellular transport and specifically during packaging in the Golgi complex.

## ACKNOWLEDGMENTS

This research was supported by National Institutes of Health Grant GM19998. Technical assistance provided by E. Roach and A. Bailey is gratefully acknowledged.

## REFERENCES

Amsterdam, A., and Jamieson, J. D. (1974). *J. Cell Biol.* **63,** 1057–1073.

Butcher, F., and Goldman, R. H. (1972). *Biochem Biophys. Res. Commun.* **48,** 23–29.

Chambaut-Guérin, A. M., Muller, P., and Rossignol, B. (1978). *J. Biol. Chem.* **253,** 3870–3876.

Cortese, F., Bhattacharyya, B., and Wolff, J. (1977). *J. Biol. Chem.* **252,** 1134–1140.

Jamieson, J. D., and Palade, G. E. (1968). *J. Cell Biol.* **39,** 580–588.

Jamieson, J. D., and Palade, G. E. (1971). *J. Cell Biol.* **48,** 503–522.

Lowry, O. H., Rosebrough, N. J., Farr, A. L., and Randall, R. J. (1951). *J. Biol. Chem.* **193,** 265–275.

Patzelt, C., Brown, D., and Jeanrenaud, B. (1977). *J. Cell Biol.* **73,** 578–593.

Rinderknecht, J., Wilding, P., and Haverback, B. J. (1967). *Experientia* **23,** 805.

Seybold, J., Bieger, W., and Kern, H. (1975). *Virchows Arch. A: Pathol. Anat. Histol.* **368,** 309–329.

Stock, C., Launay, J. F., Grenier, J. F., and Bauduin, H. (1978). *Lab. Invest.* **38,** 157–164.

Temple, R., Williams, J. A., Wilber, J. F., and Wolff, J. (1972). *Biochem. Biophys. Res. Commun.* **46,** 1454–1461.

Williams, J. A., and Lee, M. (1976). *J. Cell Biol.* **71,** 795–806.

Williams, J. A., Korc, M., and Dormer, R. L. (1978). *Am. J. Physiol.* **235,** E517–E524.

# Chapter 17

# The Role of Microtubules and Microfilaments in Lysosomal Enzyme Release from Polymorphonuclear Leukocytes

## SYLVIA T. HOFFSTEIN

*Division of Rheumatology,*
*Department of Medicine,*
*New York University*
*School of Medicine,*
*New York, New York*

## I.  Introduction

Secretion in human neutrophils is initiated by ligand binding to surface receptors. A variety of ligands are capable of inducing secretion. These include the Fc portions of IgG molecules, particularly when these have been clustered by heat aggregation or by formation of immune complexes with specific antigens. Complement components also have specific receptors on neutrophils. Thus, C3b when bound to surfaces induces secretion, and soluble C5a, normally a chemotactic agent, also induces secretion in the presence of cytochalasin B

259

(Goldstein *et al.*, 1973). Subsequently, it was found that virtually all soluble chemotactic stimuli, which by themselves do not cause degranulation, can, in the presence of cytochalasin B, induce exocytosis from neutrophils (Showell *et al.*, 1976).

## II.   Morphology of Secretion

Normal, resting polymorphonuclear leukocytes (PMN) prior to exposure to phagocytic or chemotactic stimuli are rounded, with nuclear lobes arranged around a centrally placed centriole (Fig. 1). Granules are distributed randomly throughout the cytoplasm except for a thin hyaline border around the cell that excludes organelles. This resting morphology is drastically altered when resting cells are exposed to chemotactic stimuli in the absence of cytochalasin B. Seconds after exposure to such stimuli, the PMN becomes polarized. The cytocenter, in which are all of the cell's visible microtubules and the bulk of its granule population, appears to constrict, moving the granules closer to each other and to the centrioles. Simultaneously, agranular cytoplasm is protruded to form lamellipodia and ruffles from one side of the cell (Fig. 2). Since stimulus-induced exocytosis or degranulation requires that plasma and granule membranes become closely approximated rather than separated, we studied by ultrastructural means the role of cytoskeletal elements and calcium in this process.

Ultrastructural studies of neutrophils engaged in phagocytosis show that binding of a phagocytosable particle to the cell membrane leads to a localized contraction under the area of contact, forming a cup-shaped depression into which the particle fits. The contraction is associated with the formation of a web of visible filaments oriented parallel to the plane of the membrane (Fig. 3). The particle is drawn deeper into the cytocenter, and granule fusion begins whenever vacuoles and granules are in close proximity. Contents of discharged granules can diffuse into the extracellular space until the lips of the vacuole close (inset, Fig. 3). Release of granule enzymes from such phagocytosing cells was partially inhibited by colchicine and enhanced by heavy water (Zurier *et al.*, 1974), suggesting that microtubules played a role in this process, but the nature of that role remained obscure.

Neutrophils that are induced to secrete by soluble chemotactic factors in the presence of cytochalasin B do so very rapidly. Secretion is initiated after a 15- to 30-second lag period and is complete within a minute or two. Examination of these cells at early time points showed that secretion was associated with an array of elongated invaginations oriented toward the centrioles. The invaginations appeared to parallel the microtubules radiating from the centrioles, and many more microtubules were visible in these cells than in corresponding control cells treated with cytochalasin B only.

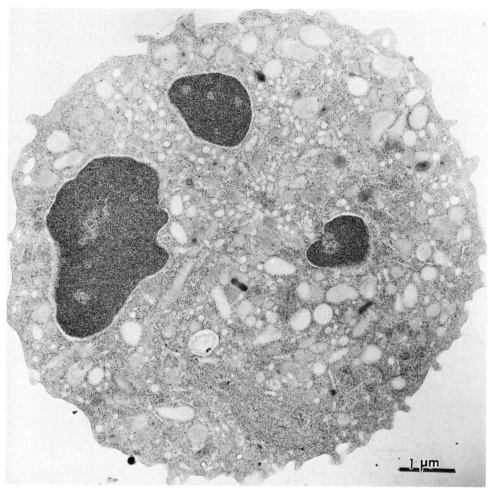

Fig. 1. A human peripheral blood neutrophil fixed while still unstimulated. The cell is rounded, with only a few small protrusions from the cell surface. Numerous granules are distributed throughout the cytoplasm but are always at least 100 nm from the plasma membrane.

## III.   Morphometry of Microtubule Assembly *in Vivo*

The ability to use soluble secretory stimuli and affect all cells in a population equally, simultaneously, and symmetrically permitted us to quantitate a phenomenon that had previously been only a subjective impression. Using morphometric methods of analysis, we were able to confirm that microtubules actually are present in greater numbers in cells stimulated to secrete than in resting cells. By this technique, electron micrographs of cells or tissues are analyzed quantita-

tively, and the number per unit area or the fractional cytoplasmic volume of microtubules or any other organelle can be determined.

Ultrastructural stereology (Weibel and Bolender, 1973) is a powerful tool, but the validity of a statistical comparison of treated and untreated tissue is basically dependent on the technique for drawing representative samples. The individual cells in a dilute suspension of cells and particles will vary considerably in the amount of material ingested. When a pellet of such cells is sectioned and examined in the electron microscope, a single grid square will show cells that have ingested much material and cells that have no phagolysosomes at all. Thus, statistical analysis of treatment effects on degranulation requires selection of micrographs that show equivalent amounts of internalized material. When cytochalasin B-treated leukocytes are exposed to phagocytic stimuli, with and without pretreatment with drugs, phagocytosis is inhibited, although the stimulating particles can be seen adhering to the plasma membrane and degranulation occurs at the plasma membrane as if it were a phagosome membrane. However, the individual cells will still vary in their degree of exposure to particulate stimuli.

These difficulties were overcome by the use of chemotaxins such as zymosan-treated serum in which complement has been activated by the alternative pathway, and the synthetic n-formylmethionyl peptides. Both the low-molecular-weight product of complement activation, C5a, and the synthetic chemotactic peptides interact with cytochalasin B-treated human PMN and provoke the extracellular release of lysosomal enzymes (Showell et al., 1976). Enzyme release occurs in the absence of particulates; it is not accompanied by release of cytoplasmic enzymes, and cell viability is not altered. Thus, a population of cells can be uniformly stimulated, providing a model system in which stereological methods of analysis can best be applied to the study of degranulation and microtubule assembly.

In order to correlate these morphological changes with enzyme release as measured biochemically, duplicate reaction mixtures were prepared, and alternate tubes were either assayed for released enzymes or were rapidly fixed in suspension and prepared for electron microscopy. The fixative used was an aldehyde and osmium tetroxide mixture that acts extremely rapidly and gives excellent contrast. It is composed of two stock solutions, equal volumes of which are mixed together immediately before being added to the cell suspension. The first solution consists of 5% glutaraldehyde and 3% acrolein in 0.1 $M$ cacodylate buffer, pH 7.4. The second is merely 2% aqueous osmium tetroxide. Eight milliliters of combined fixative is used for each milliliter of cell suspension, and fixation must be followed by a water or saline wash before en bloc staining and alcohol dehydration. The use of buffered sucrose in the washing step drastically reduces the amount of contrast obtained.

PMN fixed in this mixture at room temperature are as well fixed after 15 seconds as they are after 15 minutes. The extreme rapidity with which neutrophils are fixed by this method may account for the fact that cytoskeletal architecture is better preserved than when other, more standard fixation methods are used. Luftig *et al.* (1977) have shown that the buffers normally used in electron microscopy (cacodylate and phosphate) depolymerize microtubules. Microtubule dissolution can therefore occur before fixation is complete when a slowly penetrating glutaraldehyde solution is used.

To quantitate centriole-associated microtubule assembly, all centrioles visible on at least six sections from each of the experiments were photographed at 17,000× and printed at 50,000× on high-contrast paper. Microtubules were counted from all the centrioles photographed. Profiles were considered to be microtubules if they had straight parallel sides, 240–280 Å apart, were at least 550 Å long, and were more electron-dense than the ground cytoplasm. Only those microtubules were counted that were within a $2 \mu m \times 2 \mu m$ square centered upon the centriole.

Microtubule assembly in the cell periphery was quantitated by photographing the right-hand side of nine consecutive cells per treatment from each of four separate experiments. To minimize observer bias the areas to be photographed were selected at a magnification too low to distinguish microtubules and subsequently focused at higher magnification. Negatives were printed on high-contrast paper to a final magnification of 50,000×. Each print was examined carefully, and microtubules were counted, measured, and circled in ink. A coherent double-lattice test system (Weibel and Bolender, 1973) was then superimposed on each print. The coarse points 2.5 cm apart were used to measure cytoplasmic areas, and the fine points 0.25 cm apart were used to measure microtubule density. The ratio of test points on microtubule profiles to total points on cytoplasm is an estimate of microtubule volume density, and the number of microtubules per square micron estimates microtubule frequency.

## A. Effects of Cyclic Nucleotides

By means of electron microscopic stereology it was possible to demonstrate that lysosomal enzyme release in response to receptor–ligand interactions is associated with assembly of microtubules. This appears to hold true whether both specific and azurophil granules are discharged (C5a: Goldstein *et al.*, 1973, 1975a,b; FMLP: S. T. Hoffstein, unpublished) or specific granules alone (phorbol myristate acetate: Goldstein *et al.*, 1975a; Con A: Hoffstein *et al.*, 1976; Oliver, 1976). Moreover, cyclic nucleotide-modulated increments and decrements in enzyme release (cGMP ↑ cAMP ↓) correlate closely with increments and decrements in microtubule numbers (Goldstein *et al.*, 1973, 1975a; Weissmann *et al.*, 1975a,b). However, cAMP, while preventing any increase in microtubule numbers in response to stimulation, does not decrease microtubule numbers in resting cells (Table I). Cyclic GMP can induce assembly of microtubules in the absence of any other stimulus but is incapable of inducing enzyme release by itself. Although it enhances enzyme release in response to other stimuli such as C5a (Weissmann *et al.*, 1975b) or zymosan (Zurier *et al.*, 1974), it does not enhance Con A-induced enzyme release. This may be because Con A induces maximal microtubule assembly by itself, and thus cGMP cannot enhance. In other systems where cGMP has an enhancing effect on enzyme release, it also can enhance microtubule assembly.

## B.   Effects of Colchicine

Lysosomal enzyme release is not, however, completely dependent on microtubule function. Although modulation of microtubule assembly by various agents is concordant with modulation of lysosomal enzyme release, the effects on microtubule assembly are often greater than the corresponding effect on enzyme release. Colchicine concentrations that affect microtubule assembly also inhibit lysosomal enzyme release in a dose-related fashion, but concentrations that cause the virtual disappearance of microtubules inhibit enzyme release by no more than 40% (Hoffstein *et al.*, 1977; Weissmann *et al.*, 1971; Wright and Malawista, 1973; Zurier *et al.*, 1973a). Thus, lysosomal enzyme release from PMN is influenced by the state of assembly of microtubules, but not in a simple, direct way.

TABLE I

Effect of Cyclic Nucleotides and Con A upon Microtubule Assembly in, and Lysozyme Release from, Human PMN

| Compounds added | Mean microtubule numbers[a] | p vs. control | Lysozyme release[b] | p vs. control |
|---|---|---|---|---|
| None | $24.2 \pm 1.8$ | Control | $4.0 \pm 0.8$ | Control |
| Con A (100 $\mu$g/ml) | $39.5 \pm 2.4$ | <0.001 | $19.4 \pm 1.4$ | <0.001 |
| cAMP + theophylline ($10^{-3}$ $M$) | $22.2 \pm 2.1$ | NS | $2.4 \pm 0.9$ | NS |
| cAMP + theophylline + Con A | $27.5 \pm 1.7$ | NS | $10.1 \pm 0.5$ | <0.001 |
| cGMP ($10^{-6}$ $M$) | $37.8 \pm 2.3$ | <0.001 | $3.1 \pm 0.9$ | NS |
| cGMP + Con A | $36.7 \pm 3.0$ | <0.001 | $19.0 \pm 1.2$ | <0.001 |

[a] Mean microtubule number obtained from electron micrographs by counting microtubules in a $4$-$\mu$m$^2$ area centered upon a visible centriole ($n = 17$, two experiments). Specimens fixed for morphology after 3 minutes of incubation with Con A.

[b] Expressed as percentage of total activity released by 0.2% Triton X-100. Pretreatments with cAMP + theophylline and cGMP were 30 minutes and 3 minutes, respectively. Enzyme release was measured 60 minutes after addition of Con A ($n = 4$).

TABLE II

EFFECT OF COLCHICINE ON MICROTUBULE FREQUENCY[a] IN PERIPHERAL CYTOPLASM OF
CYTOCHALASIN B-TREATED HUMAN PMN STIMULATED BY C5a

| Experiment | Resting cells (no colchicine) | Stimulated cells, colchicine concentration | | |
|---|---|---|---|---|
| | | None | $10^{-5}$ M | $10^{-6}$ M |
| 1 | 0.23 | 0.60 | 0.11 | 0.15 |
| 2 | 0.33 | 0.43 | 0.06 | 0.09 |
| 3 | 0.21 | 0.49 | 0.07 | 0.19 |
| 4 | 0.14 | 0.33 | 0.12 | 0.15 |
| M ± S.E.M. | 0.23 ± 0.04 | 0.46 ± 0.06 | 0.09 ± 0.01 | 0.15 ± 0.02 |

[a] Mean number of microtubules per square micron of peripheral cytoplasm, determined from nine micrographs per treatment per experiment.

[b] Cells were preincubated with cytochalasin B (5.0 μg/ml) for 10 minutes and exposed to C5a for 1 minute at 37°C.

Fluid phase and phagocytic stimuli cause an increase in microtubule numbers throughout the cytoplasm (Table II). Microtubules are most conspicuous in a radial array centered on centriole–associate microtubule organizing sites (Fig. 4). However, microtubule assembly in response to stimuli, as well as disassembly in response to colchicine, occurs in parallel, in peripheral, and in pericentriolar regions (Hoffstein *et al.*, 1976, 1977). Invaginations of plasma membrane are prominent along the radiating microtubules in the pericentriolar cytoplasm of such stimulated cells (Fig. 5). This occurs whether or not the PMN are cytochalasin B-treated and whether or not the stimulus presented is soluble (C5a, FMLP, or Con A) or is composed of small particulates (immune precipitates) (Goldstein *et al.*, 1973; Hoffstein *et al.*, 1977).

In strong contrast, cells that are prevented by colchicine from assembling microtubules in response to stimuli have a random pattern of invaginations (Fig. 6). In the absence of microtubule assembly, there appears to be no coordinated translocation of stimulated plasma membrane or phagosomes to the centriolar region. The morphological correlate of reduced numbers of microtubules in unstimulated PMN is an apparent disorganization of the central region of the cell, with the centrioles and Golgi apparatus displaced from their usual position between the nuclear lobes. The agranular cytoplasm appears not to be adversely affected by colchicine and remains capable of forming pseudopods. Such a change in cell form is not seen when microtubule assembly in response to a stimulus is inhibited by cAMP. Although there are fewer microtubules than in the

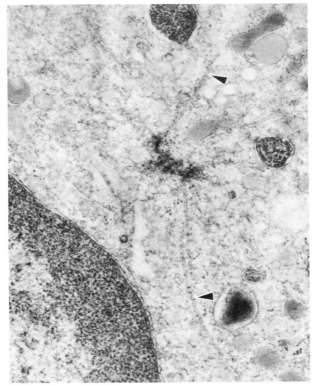

FIG. 4. A section through a centriole and the tips of two microtubule organizing sites from a PMN stimulated with zymosan-treated serum. Many microtubules (arrowheads) radiate in all directions from each of the organizing sites but are rarely seen projecting from the centriole itself.

absence of exogenous cAMP, those microtubules that are present maintain the organization of the cell and serve as guides to direct membrane invaginations toward the interior of the cell.

The data suggest that translocation of phagosomes rather than fusion is modulated by microtubules. Microtubules appear to be involved in maintaining the internal organization of PMN and the topologic relationships between their organelles and the plasma membrane. Correlations between tubule assembly and disassembly and enhanced or diminished enzyme release probably reflect events earlier in the secretory process of PMN. Assembly may enhance, or disassembly diminish, the chances for contact between PMN granules and stimulated areas of the plasma membrane. Other structures, perhaps contractile proteins, probably play a more direct role in permitting fusion between granules and phagosomes or the membrane of the cell itself.

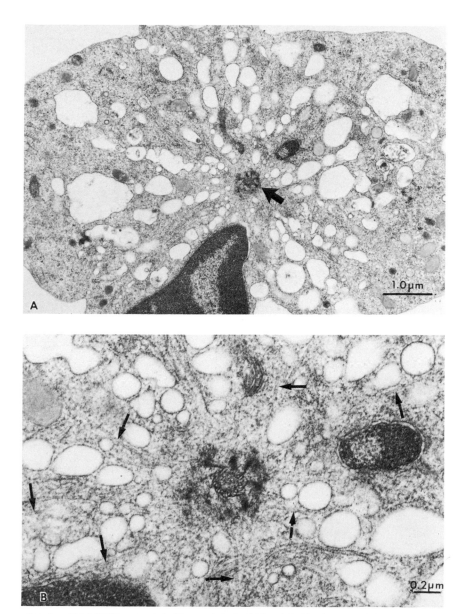

Fig. 5. (A) A PMN pretreated with cytochalasin B (5 μg/ml, 10 minutes) and then exposed for 1 minute to serum in which complement had been activated and C5a generated. Microtubules have been assembled and radially symmetrical membrane invaginations have formed along microtubule defined tracts to the centriolar region. (B) This high-magnification view of the centriolar region of the same cell shows many microtubules radiating from the cell center. Massive degranulation has occurred, as evidenced by the presence of numerous empty vacuoles in the pericentriolar region and very few granules with intact contents.

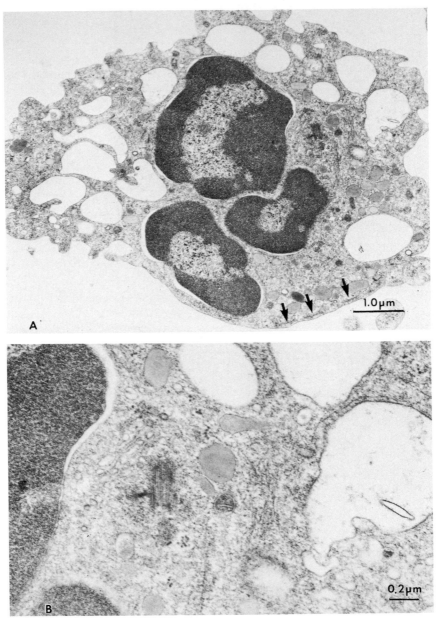

FIG. 6.   (A)The cell in this figure was pretreated with colchicine ($10^{-5}$ $M$, 60 minutes) before the addition of cytochalasin B and activated serum as seen in Fig. 5. Massive enzyme release has also occurred from this cell, but the membrane invaginations take the form of vacuoles distributed around the perimeter of the cell rather than elongated invaginations oriented toward the pericentriolar cytoplasm. The arrows indicate microfilament bundles, which are observed only in stimulated cells. (B) This view of the centriolar region of the cell in Fig. 6A at higher magnification shows the absence both of microtubules and of membrane invaginations.

# IV.  The Role of Calcium

Bundles of microfilaments were commonly seen in PMN stimulated to secrete enzymes either in the presence or in the absence of colchicine or of cytochalasin B. They were frequently seen adjacent to vacuoles where degranulation was occurring (Fig. 7) or parallel to the plasma membrane (Fig. 6, arrows). These figures suggest that a calcium-sensitive contractile microfilament network might play a role in plasma membrane movement and degranulation and that changes in intracellular calcium might play an important role.

## A.  Effects of the Calcium Ionophore A23187

We then undertook a series of experiments in which the calcium ionophore A23187 was used to bypass receptor–ligand interaction and by elevating intracellular calcium induce enzyme release. The morphological concomitant of such calcium ionophore-induced enzyme release was an alteration of neutrophil shape suggestive of movement in two planes. One series of movements occurred parallel to the plane of the membrane, drawing the membrane up into a series of blebs and folds. A subplasmalemmal web of microfilaments was visible just subjacent to the blebbed region. The other movement occurred at right angles to the cell surface, and part of the membrane was drawn into the interior of such cells. Degranulation occurred into the invaginations so formed (Fig. 8). Simultaneously, a centripetal movement of granules occurred (Fig. 9), and granules became more crowded in the centriolar regions of the cell. This centripetal movement was similar to that seen when PMN were exposed to secretory or chemotactic stimuli. Microtubule numbers were not affected by either the ionophore alone or ionophore plus calcium. Colchicine, as expected, reduced microtubule numbers but had no effect on ionophore-induced secretion.

Thus, lysosomal enzyme release from PMN, whether induced by particle ingestion or by elevating intracellular calcium with the ionophore, resulted in similar morphological configurations. Membrane is moved centripetally to form invaginations, and membrane is moved laterally to form closed vacuoles in normal cells. In ionophore-treated cells lateral movement is uncoordinated with centripetal movement and thus surface blebs are formed. The effect of cytochalasin B in both systems is to inhibit lateral membrane movement without affecting centripetal movement.

Although the extent of ionophore-induced secretion depended on the concentration of extracellular calcium, many of the morphological changes seen in this study could be elicited by A23187 in the absence of added calcium, although to a lesser extent than in its presence. Lateral movement of the plasma membrane to form blebs and folds occurred even when A23187 alone was added to a cell

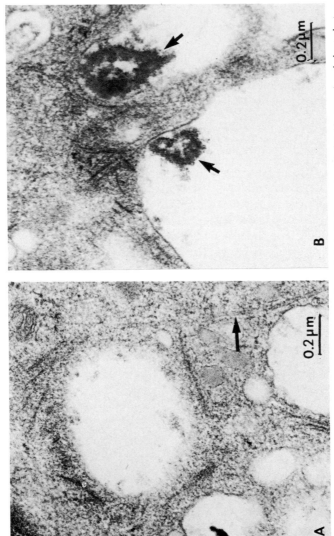

Fig. 7. PMN stimulated to secrete in the presence of cytochalasin B show microtubules (A, arrow) and short electron-opaque filaments up to 250 nm long by 10–20 nm wide oriented tangentially to the plasma membrane of membrane invaginations. In B, they are conspicuous immediately subjacent to a region where granules have recently fused with the vacuolar membrane. Fixation occurred so rapidly after the discharge event that the granule contents (arrows) did not have time to dissolve and diffuse away.

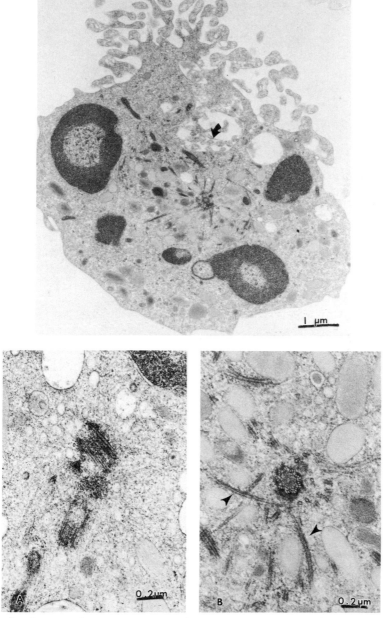

FIG. 8.    A neutrophil treated with A23187 ($10^{-5}$ $M$) and 1 m$M$ calcium. The cell margin shows numerous blebs and folds. Two vacuoles containing evidence of recent degranulation are visible in the interior. The vacuole indicated with an arrow also contains blebs, suggesting that it actually represents an invagination of a blebbed portion of the plasma membrane.

FIG. 9.    A higher magnification of the centriolar region of a control neutrophil (A) and of the ionophore-treated neutrophil shown in Fig. 8 (B). Microtubules decorated with a filamentous material (arrowheads) radiate from the centriole, and granules are closer together in the cell center than they are in unstimulated cells.

suspension in buffered saline. Similarly, centripetal membrane movement and granule discharge occurred in cytochalasin B-treated PMN exposed to A23187 without calcium.

## B.   Calcium Localization with Pyroantimonate-Osmium

Data obtained in this and other laboratories point to the fact that many PMN responses to stimuli do not require calcium in the medium but can be enhanced by exogenous calcium. For example, granule discharge and $O_2^-$ production can be induced by a variety of soluble secretagogues in the absence of extracellular calcium (Korchak and Weissmann, 1978; Goldstein *et al.*, 1975a), but the effect of these agents is enhanced with added calcium. Likewise, ingestion of aggregated immune complex does not require calcium and is only slightly decreased by EDTA in the medium; however, the process becomes calcium-dependent after prolonged preincubation with EDTA (Henson and Oades, 1975). Furthermore, both C5a (Gallin and Rosenthal, 1974) and phagocytic stimuli (Barthelemy *et al.*, 1977) can induce a calcium efflux. These data suggest that cell-associated calcium can be dissociated by receptor–ligand interactions to create localized, physiologically active concentrations of ionic calcium.

Whereas the sites at which calcium is sequestered and from which it can be mobilized have been identified in skeletal muscle, its sites of sequestration have not been so well defined in nonmuscle cells. It has recently been shown, using chlortetracycline as a fluorescent probe, that membrane-associated calcium is displaced when PMN are stimulated by chemotactic factors, but the specific membranes involved were not identified (Naccache *et al.*, 1979). Mitochondria and microsomes (Borle, 1973; Chandler and Williams, 1978; Lehninger, 1970), and plasma membrane and storage granules (Moore *et al.*, 1975; Howell, 1977; Clemente and Meldolesi, 1975; Cardasis *et al.*, 1978), have all been suggested as the predominant sequestration sites in various cell types. Pyroantimonate was therefore used as a tool to investigate the localization of cell-associated calcium within human PMN by cytochemical means and to identify those sites from which calcium can be mobilized by surface stimuli.

The pyroantimonate-osmium fixative (Komnick and Komnick, 1963) was prepared by mixing 4% $K_2H_2SB_2O_7 \cdot 4H_2O$ (Fisher Scientific Co., Fairlawn, NJ) 1 : 1 with 2% $OsO_4$ in $H_2O$, adjusting the pH to 7.3–7.8 with acetic acid. The cells were fixed in the above reagent, at 0°–4°C for 1 hour, rinsed in 70% ethanol, dehydrated in ethanol, and embedded in epoxy the same day. Two batches of potassium pyroantimonate were used and gave identical localization. One was from Lot 730550, described by Simson and Spicer (1975) as a "good" batch, and when that was used up, Lot 771302 was used. The latter differed from the former in that the pH did not require adjusting but gave the same results.

The amount of reaction product generated by pyroantimonate-osmium precipitation of cations was variable, and the procedure did not always give good staining. In some experiments there was little or no precipitate in the large granule population, although precipitate was present on the plasma membrane. In other experiments cations in the buffer were precipitated by the pyroantimonate, resulting in an unsectionable mass of antimony salts in the cell pellet. In both of these instances the entire experiment was discarded.

Some of the sources of variation were eventually identified and could be correlated with the results obtained by Simson and Spicer (1975). Thus, membrane-associated pyroantimonate precipitation without accompanying intracellular localization occurred when approximately equal volumes of fixative and cell suspension were used and correlated with the results obtained by them with slightly lower pyroantimonate concentrations. Larger volumes of fixative precipitated sodium from the cell medium. The best and most consistent results were obtained when the cells were allowed to settle and most of the medium was drawn off before the stimuli were added and the cells fixed.

Two approaches were made toward the problem of determining the nature of the cations associated with the pyroantimonate precipitate. Energy-dispersive X-ray spectra were obtained from cells fixed in both fashions with an Ortec X-ray microanalyzer attached to a Zeiss EM 10A (accelerating voltage 40 kV). One-hundred second counts were obtained from unstained semithick specimens mounted on copper grids. The energy spectra obtained from membrane-associated pyroantimonate precipitates revealed a large peak between 3.6 and 4.2 kV representing calcium $K\alpha_{12}$ (3.69 kV) and $K\beta$ (4.01 kV) and antimony $L\alpha$ (3.69 kV) and $L\beta_{12}$ (3.84 kV). Because of the large overlap between the calcium and antimony peaks, the data only suggest the presence of calcium pyroantimonate in the precipitate; they do not prove it. Since the tissue was osmicated, osmium peaks [$M\alpha$ 1, 2 (1.01 kV) and $L\alpha_1$ (8.9 kV)] were consistently present. Neither magnesium nor sodium could be detected in the membrane pyroantimonate deposits (Fig. 10). Potassium was detected only in membrane precipitates of specimens that had been prepared with potassium phosphate buffer.

The other approach was to selectively dissolve the precipitate with chelators. When freshly cut sections were incubated in EGTA (either 100 m$M$ for 10 minutes of 3 m$M$ for 3 hours), finely divided pyroantimonate precipitates were removed from heterochromatin and cytoplasm and partially removed from the granules. The large particles, found sometimes in cytoplasm and consistently on the plasma membrane, were loosened, and small and medium-sized particles completely removed. Incubation with EDTA and $Mg^{2+}$ had a similar effect; however, sections incubated in $H_2O$ at the same pH for the same length of time were unaltered. Only freshly cut sections were extractable with chelators. Sections cut weeks before treatment with chelators retained their pyroantimonate deposits at all sites.

A consistent pattern of pyroantimonate precipitation was observed in control,

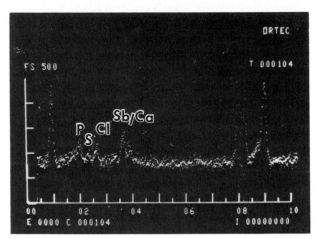

FIG. 10. X-ray spectrum from the plasma membrane of the PMN fixed with pyroantimonate and osmium as in Figs. 11–16. The broad peak between 3.6 and 4.2 kV represents the combined emission of calcium and antimony. Phosphorus, sulfur, and chlorine were also present, but magnesium and sodium were not detected.

unstimulated cells (Fig. 11). The nucleus was pyroantimonate-positive with relatively more, finely divided precipitate deposited in areas of heterochromatin than in areas of euchromatin. Approximately one-fourth to one-third of the granule population, including most of the larger granules, were pyroantimonate-positive, and the remainder, including dumbbell-shaped granules, were negative, as were the mitochondria. Course granular deposits were also present at the plasma membrane of the PMN. These membrane-associated precipitates varied somewhat from preparation to preparation and consisted either of numerous small particles or less numerous larger particles. It was not possible to determine precisely where the particles were located with respect to the plasma membrane. Some appeared to project outward, toward the extracellular space; some projected equally inward and outward.

In order to determine whether the subcellular sites of pyroantimonate deposition represented a mobilizable pool of cations, PMN were exposed to a variety of soluble and particulate stimuli before fixation with pyroantimonate and osmium. Soluble stimuli of enzyme release including Con A (30 $\mu$g/ml), A23187 ($10^{-6}$ $M$), zymosan-activated serum (10%) plus cytochalasin B (5 $\mu$g/ml), and F-Met-Leu-Phe ($10^{-7}$ $M$) plus cytochalasin B, but not cytochalasin B alone, caused the loss of membrane-associated pyroantimonate-precipitable cations from the entire cell surface (Fig. 12). Pyroantimonate was still present intracellularly in the same locations as in control cells, nuclear heterochromatin, and some of the granules. When Con A, which induces discharge of specific but not azurophil granules (Hoffstein et al., 1976), was used as the stimulus, a greater proportion

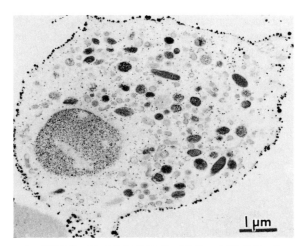

FIG. 11. A resting PMN fixed in 2% $K_2H_2Sb_2O_7 \cdot 4H_2O$ and 1% $OsO_4$. Pyroantimonate precipitates are conspicuous in the heterochromatin of the nuclear lobes and in a portion of the granule population, staining these intensely black. Small particles of precipitate are also present along the plasma membrane.

of the remaining granules were pyroantimonate-positive, suggesting that these were azurophils.

When particulate stimuli were incubated with PMN before fixation with pyroantimonate-osmium, reaction product was visible on the plasma membrane

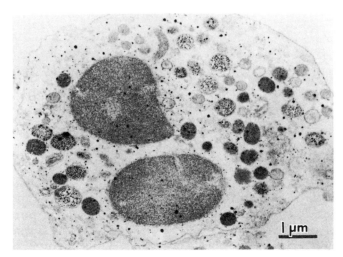

FIG. 12. A PMN exposed to Con A (30 μg/ml), then fixed as described in Fig. 11. Pyroantimonate precipitates are present in all the usual locations except the plasma membrane. Similar results were obtained with any of the soluble secretagogues.

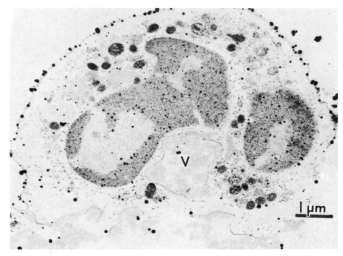

FIG. 15.   This figure is similar to Fig. 14, except that BSA/anti-BSA immune precipitates were used as phagocytic stimuli. The immune complexes are visible as gray amorphous material. The plasma membrane in contact with the immune precipitates is free of precipitates, as is the vacuole (v) that has formed in the center of the cell. The upper surface of the cell to which immune complexes have not adhered has pyroantimonate precipitates.

incubating the cells with EGTA before fixing with pyroantimonate leaves sufficient membrane-associated calcium to form small amounts of precipitate. Whatever its distribution across the membrane, the data reported here indicate that membrane-associated cations can be mobilized by physiological stimuli. This displaced free calcium can, in turn, act on the plasma membrane, affecting its charge, permeability, or enzymatic complement. Depending upon the experimental design, some of this calcium can be detected as an efflux (Barthelemy *et al.*, 1977; Gallin and Rosenthal, 1974).

## V.   Summary

Our results have shown that a pool of rapidly (less than 10 seconds) mobilizable calcium is located at the plasma membrane of human neutrophils. This calcium is displaced subsequent to receptor–ligand interaction. Displacement of calcium occurs at the site of receptor–ligand interaction and is not generalized to the entire cell surface. Microfilament bundles assemble whenever secretion is induced and are located subjacent to regions of plasma membrane from which calcium has been displaced. These bundles are associated with both lateral and centripetal movement of the plasma membrane. Whereas the former is inhibited by cytochalasin B, the latter is not.

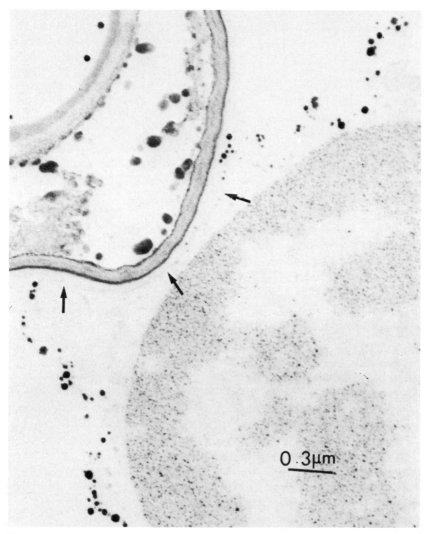

FIG. 16. PMN fixed in pyroantimonate and osmium immediately (less than 10 seconds) after the addition of opsonized zymosan. The cell shown has attached to the particle of zymosan (Z) but has not yet internalized it. Pyroantimonate deposits are present over the surface of the cell except in the region immediately subjacent to the particle (arrows).

Microtubules are assembled rapidly in response to secretagogues that act via cell surface receptors but not when secretion is induced by the calcium ionophore A23187, a secretagogue that bypasses receptors. Modulation of secretion by cyclic nucleotides and colchicine correlates with increments and decrements in microtubule assembly, but induction and modulation of secretion by calcium does not. Microtubules appear to serve a topologic function, maximizing stimulus-induced association of granules and plasma membrane. However, their complete absence reduces receptor-mediated secretion by only 40% and has no effect on calcium-induced secretion. Our data, therefore, indicate that calcium-sensitive, cytochalasin B-insensitive microfilaments, not microtubules, mediate stimulus–secretion coupling in human PMN.

## REFERENCES

Barthelemy, A., Paridaens, R., and Schell-Frederick, E. (1977). *FEBS Lett.* **82**, 283–287.
Borle, A. (1973). *Fed. Proc., Fed. Am. Soc. Exp. Biol.* **2**, 1944–1950.
Cardasis, C. A., Scheul, H., and Herman, L. (1978). *J. Cell Sci.* **31**, 101–115.
Chandler, D. E., and Williams, J. A. (1978). *J. Cell Biol.* **76**, 386–399.
Clemente, F., and Meldolesi, J. (1975). *J. Cell Biol.* **15**, 88–102.
Gallin, J. I., and Rosenthal, A. S. (1974). *J. Cell Biol.* **62**, 594–609.
Goldstein, I. M., Hoffstein, S. T., Gallin, J., and Weissmann, G. (1973). *Proc. Natl. Acad. Sci. U.S.A.* **70**, 2916–2920.
Goldstein, I. M., Hoffstein, S. T., and Weissmann, G. (1975a). *J. Cell Biol.* **66**, 647–652.
Goldstein, I. M., Hoffstein, S. T., and Weissmann, G. (1975b). *J. Immunol.* **155**, 665–670.
Henson, P. M., and Oades, Z. (1975). *J. Clin. Invest.* **56**, 1053–1061.
Hoffstein, S., Soberman, R., Goldstein, I., and Weissmann, G. (1976). *J. Cell Biol.* **68**, 781–787.
Hoffstein, S., Goldstein, I., and Weissmann, G. (1977). *J. Cell Biol.* **73**, 242–256.
Howell, S. L. (1977). *Biochem. Soc. Trans.* **5**, 875–879.
Komnick, H., and Komnick, V. (1963). *Z. Zellforsch. Mikrosk. Anat.* **60**, 136–203.
Korchak, H. M., and Weissmann, G. (1978). *Proc. Natl. Acad. Sci. U.S.A.* **74**, 3818–3822.
Lehninger, A. L. (1970). *Biochem. J.* **19**, 129–134.
Luftig, R. B., McMillan, P. M., Weatherbee, J. A., and Weihing, R. R. (1977). *J. Histochem. Cytochem.* **25**, 175–187.
Moore, L., Chen, T., Knapp, H. R., Jr., and Landon, E. J. (1975). *J. Biol. Chem.* **250**, 4563–4567.
Naccache, P. H., Volpi, M., Showell, H. J., Becker, E. L., and Sha'afi, R. I. (1979). *Science* **203**, 461–463.
Oliver, J. M. (1976). *Am. J. Pathol.* **85**, 395–418.
Showell, H. J., Freer, R. J., Zigmond, S. H., Shiffmann, E., Aswanikumar, S., Corcoran, B., and Becker, E. L. (1976). *J. Exp. Med.* **143**, 1154–1169.
Simson, J. A. V., and Spicer, S. S. (1975). *J. Histochem. Cytochem.* **23**, 575–598.
Weibel, E. K., and Bolender, R. P. (1973). *In* "Principles and Techniques of Electron Microscopy" (M. A. Hayat, ed.), Vol. 3, pp. 239–310. Van Nostrand-Reinhold, Princeton, New Jersey.
Weissmann, G., Zurier, R. B., Spieler, P. J., and Goldstein, I. M. (1971). *J. Exp. Med.* **134**, 149s–165s.
Weissmann, G., Goldstein, I., Hoffstein, S., and Tsung, P.-K. (1975a). *Ann. N.Y. Acad. Sci.* **253**, 750–762.
Weissmann, G., Goldstein, I., Hoffstein, S., Chauvet, G., and Robineaux, R. (1975b). *Ann. N.Y. Acad. Sci.* **256**, 222–232.
Wright, D. G., and Malawista, S. E. (1973). *Arthritis Rheum.* **16**, 749–758.
Zurier, R. B., Hoffstein, S., and Weissmann, G. (1973a). *J. Cell Biol.* **58**, 27–41.
Zurier, R. B., Hoffstein, S., and Weissmann, G. (1973b). *Proc. Natl. Acad. Sci. U.S.A.* **70**, 844–848.
Zurier, R. B., Weissmann, G., Hoffstein, S., Kammerman, S., and Tai, H.-H. (1974). *J. Clin. Invest.* **53**, 297–309.

# Part IV.   Exocytosis

# Chapter 18

# Exocytosis–Endocytosis as Seen with Morphological Probes of Membrane Organization

## L. ORCI, R. MONTESANO, AND A. PERRELET

*Institute of Histology and Embryology, University of Geneva Medical School, Geneva, Switzerland*

## I.   Introduction

To our knowledge, the story of exocytosis started 22 years ago with a picture published by G. E. Palade showing an acinar lumen of the exocrine pancreas receiving the content of a zymogen granule through continuity of the granule membrane with the cell membrane (Palade, 1959). In 1959, Palade had thus identified the mechanism by which a polypeptide-containing secretory granule can be discharged at the cell exterior without any break in the cell membrane continuity; he had also envisaged the consequence of this process: "It is highly improbable that the membrane moves only from the centrosphere to the surface. A unidirectional movement would soon result in considerable enlargement of the lumen and in exhaustion of the intracellular membranous material. It is reasonable to assume that a concomitant movement, from the surface to the centrosphere region, takes place, and the presence of small, 'empty' vesicles below the luminal membrane is compatible with this assumption.'' Twenty-two years ago, therefore, the obligatory coupling between exocytosis and endocytosis was suspected, and the title of this article already justified. As will be seen in the data to follow, a fair amount of new information has been collected since then. An equally fair—for not saying much larger—amount of data is still needed, however, to understand all features governing exocytosis and the reverse process, endocytosis.

The morphological aspects of exocytosis have been the object of a recent review (Orci and Perrelet, 1978); we shall summarize here only the salient features obtained with the thin-sectioning and freeze-fracture approaches.

## II.   Exocytosis—Thin Sections

The two partners in exocytosis—namely, the secretory granule membrane and the plasma membrane (Fig. 1)—both show the trilaminar unit structure

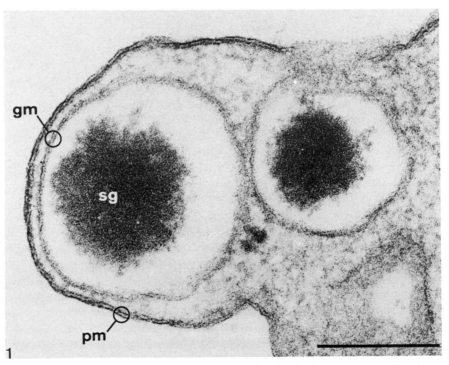

FIG. 1.   Thin section of the periphery of a B cell showing the trilaminar appearance of the plasma membrane (pm) underlined by the trilaminar membrane (gm) of a closely apposed secretory granule (sg). Bar, 0.2 μm.

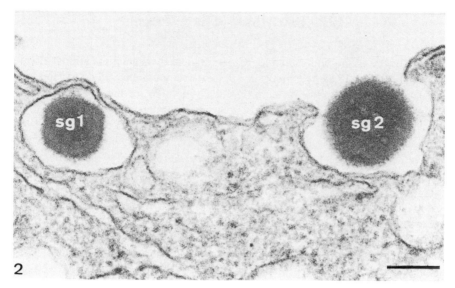

FIG. 2.   Periphery of a B cell showing two secretory granules, one of which is undergoing exocytosis. While the core of the granule to the left (sg1) is still separated from the outside of the cell by intact secretory granule and plasma membranes, the core of the granule to the right (sg2) is exposed to the extracellular space following the incorporation of the secretory granule membrane into the plasma membrane; this represents the characteristic thin-section image of exocytosis. Bar, 0.2 μm.

(Robertson, 1959), and the first detectable step in exocytosis is the merging of the outer leaflet of the secretory granule membrane with the inner leaflet of the plasma membrane, resulting in a so-called pentalaminar fusion (Lagunoff, 1973). Leaflets of both membranes then fuse and become continuous, a process that incorporates the granule membrane into the plasma membrane and exposes the granule content to the extracellular space (Fig. 2). Although thin sections of conventionally prepared material (i.e., aldehyde and osmium fixation followed by lead and uranyl staining) do not allow the detection of any change in the appearance of the membranes involved in exocytosis, the use of electron-dense probes for carbohydrate residues of the cell coat (residues attached to either proteins and/or lipids of the membrane on its external side) show that such residues, among them lectin-binding sites, are absent from the segment of the plasma membrane to be involved or already involved in a pentalaminar fusion with the membrane of a secretory granule (Lawson *et al.*, 1977; Orci *et al.*, 1979). Thin sections have also revealed electron-dense bridges between the two interacting membranes (Orci *et al.*, 1979). These bridges, detected in freeze-fracture (Orci and Perrelet, 1976), are interpreted as microfilamentous elements of the cell web (McNutt, 1978; Orci *et al.*, 1979).

with corresponding particle-free patches of the plasma membrane: the process of exocytotic fusion has thus been interpreted as involving pure (by freeze-fracture criteria) lipid domains of the respective membranes (Ahkong et al., 1975; Lucy, 1978). An exception to this scheme is represented by the exocytotic granule discharge in protozoans (Plattner, 1973; Satir et al., 1973; Hausmann and Allen, 1976; Allen, 1978). In these cells, regions of the membranes about to fuse appear instead decorated by a specific particle arrangement, the "rosette." Moreover, the existence of particle-free patches in aldehyde-fixed, glycerol-impregnated mammalian cells has been recently ascribed to an artifactual process, since they were not found in unfixed, unimpregnated cells freeze-fractured by the rapid freezing method (Chandler and Heuser, 1979). This point is not definitely settled.[1] At present we hold the view that, if the particle-free patches in fixed and glycerinated tissue represent an "artifact," it is at least a valuable one to detect exocytotic sites in the membranes. The reproducibility of this "artifact" and its precise topographical occurrence also favor the view that it is underlined by a distinct change in the interacting membranes, previously undetectable with thin sectioning alone.

Until the last few years, particles represented the only available probe for the visualization of a specific membrane component: protein. Recently, a probe for the localization of cholesterol in freeze-fractured membranes has been introduced (Elias et al., 1979; Robinson and Karnovsky, 1980). This probe, the polyene antibiotic filipin, by specifically interacting with cholesterol (for a recent review, see Norman et al., 1976), induces the formation of large (25–30 nm) protuberances in the plane of the plasma membrane. These protuberances, representing filipin–sterol complexes, are revealed upon splitting of the membrane bilayer during freeze-fracture (Tillack and Kinsky, 1973; Verkleij et al., 1973; Elias et al., 1979) and are easily distinguishable from the intramembrane particles by their much larger size. The filipin probe has revealed interesting features of cholesterol distribution at the sites of exocytosis. In addition to the pancreatic B cell, the rat peritoneal mast cell was studied.

When isolated islets of Langerhans were fixed in a glutaraldehyde–filipin mixture, it appeared that the degree of labeling of the plasma membrane by filipin–sterol complexes was not the same in all B cells. Whereas the membrane of some cells was heavily labeled (see Fig. 12), others had only a sparse labeling (Fig. 5). At the present time, we are unable to establish whether these variations reflect an heterogeneity of B-cell membranes or are due to the uneven penetration of filipin within the islet. In sparsely labeled membranes, filipin–sterol com-

---

[1]Although there is no obligatory link between movement of intramembrane particles and movement of surface binding sites (Pinto da Silva et al., 1975), the fact that the latter are also undetectable in thin sections of fusing plasma membrane (Lawson et al., 1977; Orci et al., 1979) may be viewed against the purely artifactual nature of particle removal at exocytotic sites.

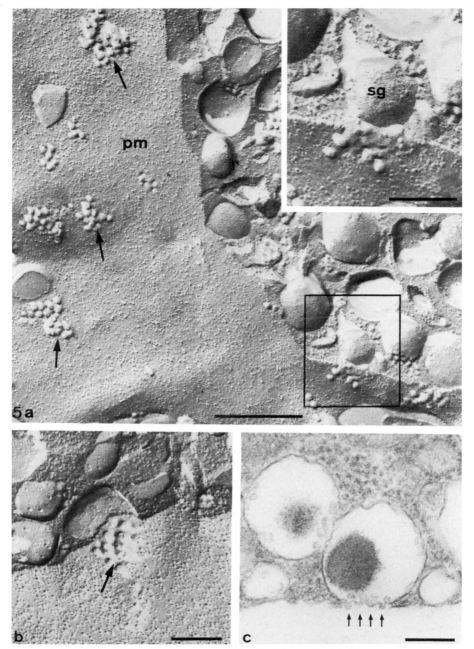

FIG. 5. (a) Islet cells fixed in the presence of filipin. In this case the labeling appears very sparse (compare with Fig. 12 for the appearance of heavy labeling), with only a few clusters of filipin–sterol complexes (arrows) on the plasma membrane (pm) and on granule membrane (sg). The enlargement of the outlined area in the inset in (a) and Fig. 5b reveal a cluster of filipin–sterol complexes at a site of close apposition between secretory granule and plasma membrane (arrow in b); the complexes are located in a particle-poor area of the plasma membrane, indicating a prospective site of exocytosis. (c) Thin section of a filipin-treated B cell showing the characteristic filipin–sterol complexes at a site of secretory granule and plasma membrane interaction (arrows). (a) Bar, 0.5 μm. Inset: Bar, 0.2 μm. (b) Bar, 0.2 μm. (c) Bar, 0.2 μm.

plexes were clustered in discrete regions corresponding to sites of close interaction between secretory granule and plasma membrane (Fig. 5). Similar observations were made in granular cells of the toad bladder (Orci et al., 1980). This suggests that sites of exocytotic membrane fusion might be enriched in cholesterol with respect to the surrounding plasma membrane. Such interpretation is strengthened by another interesting observation made in filipin-treated B cells. When the fracture exposed the membrane of secretory granules, one could see that filipin-sterol complexes were mostly clustered in patches on the granule membrane (Fig. 6); the patch in one granule was often in register with a patch in the neighboring granule. This was also evident in thin sections, where regions of close contact between adjacent granule membranes showed a characteristic scalloped appearance due to the presence of filipin-sterol complexes (Fig. 7). Since compound exocytosis (Douglas, 1974)—that is, the sequential fusion of several secretory granules with one another and finally with the plasma membrane—is a common event in degranulating B cells (Orci et al., 1973), the preferential localization of filipin-sterol complexes at sites of granule interaction further suggests that exocytotic membrane fusion may occur in cholesterol-rich membrane areas. It is also interesting to note that the membranes involved in fusion—namely, the secretory granule and plasma membrane—are richly labeled by filipin, whereas other intracellular membranes not committed to fusion (rough endoplasmic reticulum, mitochondria) are poorly or not labeled. These morphological findings fit well with biochemical data concerning the cholesterol content of these respective membranes (Keenan and Morré, 1970; Colbron et al., 1971; Meldolesi et al., 1971; Morin et al., 1972; Zambrano et al., 1975). Taking into account the membrane flow hypothesis (for a recent review, see Morré et al., 1979), we can anticipate that an enrichment in cholesterol should take place somewhere between the rough endoplasmic reticulum and secretory granule membranes. In this respect, recent observations in B cells, as well as in pancreatic acinar cells, showed an increase in filipin labeling from proximal to distal Golgi cisternae, suggesting that this may be the site of cholesterol enrichment during membrane flow (Orci et al., 1981). The distribution of filipin-sterol complexes at sites of exocytosis has been studied in another model system: the peritoneal mast cell (Montesano et al., 1980). The extensive membrane interactions occur-

---

FIG. 6.    Freeze-fracture replica of an islet cell fixed in the presence of filipin. This cytoplasmic fracture exposes numerous secretory granules (sg) whose membrane appears labeled to a variable extent by filipin–sterol complexes. Note that the complexes form patches on the secretory granule membrane. Bar, 0.5 μm.

FIG. 7.    Thin section of a filipin-treated B cell showing a cytoplasmic field containing numerous secretory granules. Many granules are apposed to one another, and the site of membrane contact is decorated with filipin–sterol complexes (arrows). The inset shows a similar labeling of the region of contact between two secretory granule membranes (sg) as seen in freeze-fracture. Bar, 0.2 μm. Inset: Bar, 0.2 μm.

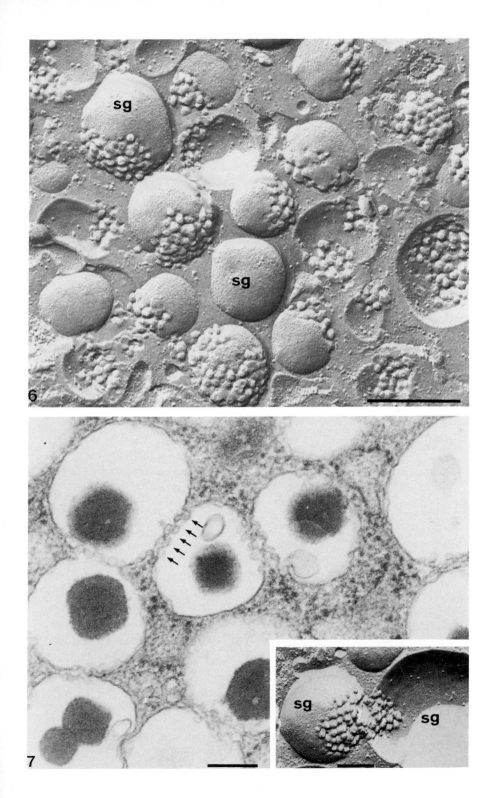

6

7

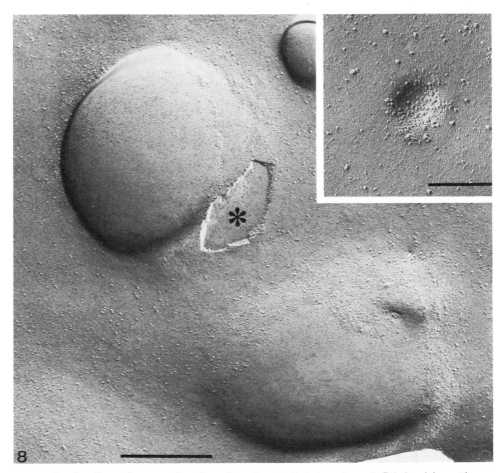

FIG. 8.   Freeze-fracture replica of a peritoneal mast cell during polymyxin B-induced degranulation. Incipient exocytosis takes the form of large bulges in the plasma membrane face, which are devoid of intramembrane particles. A "window" (*) opened in the plasma membrane near the upper bulge allows the particle-free membrane of the underlying secretory granule to be visualized. The inset shows by contrast that invaginating segments of the plasma membrane (presumptive endocytosis) are *not* deprived of particles. Bar, 0.5 μm. Inset: Bar, 0.2 μm.

ring during the massive degranulation (Röhlich *et al.*, 1971; Lagunoff, 1973) that can be induced in this cell type render mast cells particularly attractive for studying the distribution of cholesterol during exocytotic membrane fusion. An additional advantage over other secretory cells is that problems of filipin penetration are circumvented by the use of single-cell suspensions. When mast cells obtained by peritoneal lavage and centrifugation through a discontinuous density gradient (Lagunoff, 1975) are exposed to polymyxin B, rapid and extensive degranulation occurs (Chi *et al.*, 1976; Burwen and Satir, 1977). Freeze-fracture replicas of aldehyde-fixed and glycerinated mast cells show a clearing of intramembrane particles from plasma membrane bulges overlying secretory granules (Chi *et al.*, 1976; Lawson *et al.*, 1977) (Fig. 8). When mast cells fixed in the same conditions were exposed to filipin, one observed a dense labeling of the entire plasma membrane and of the membrane of all secretory granules (Figs. 9 and 10). This seems to indicate that, in mast cells undergoing exocytosis, cholesterol redistribution does not occur, or occurs on such a small scale that it is not detectable with the filipin technique. Although we cannot definitely conclude from these experiments that sites of exocytosis in mast cells are enriched in cholesterol, we can at least say that, morphologically, they do not appear cholesterol-deprived. The general conclusion that can be drawn from the application of the filipin probe to secreting pancreatic B cells and peritoneal mast cells is that exocytosis seems to take place in cholesterol-containing areas of the plasma membrane.

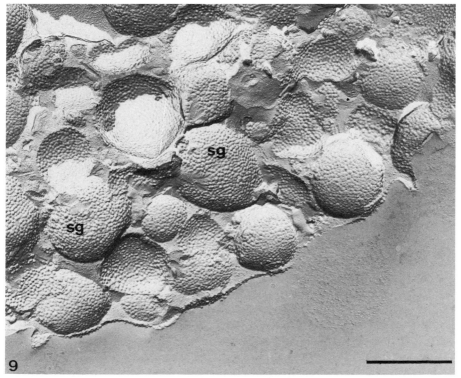

9

FIG. 9. Freeze-fracture replica of a degranulating peritoneal mast cell treated with filipin. This cross-fracture of the cytoplasm reveals numerous secretory granules (sg) homogeneously and densely labeled with filipin–sterol complexes. Bar, 1 μm.

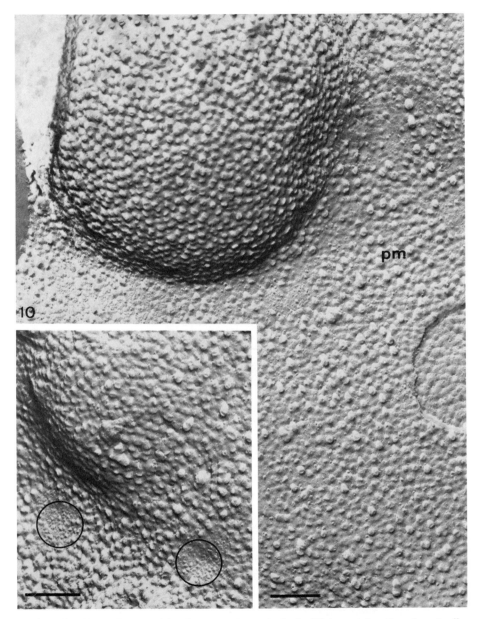

FIG. 10.   Freeze-fracture of the plasma membrane (pm) of a filipin-treated peritoneal mast cell undergoing exocytosis. The filipin–sterol complexes decorate the entire plasma membrane face exposed (pm), including the bulge signaling prospective exocytosis. By contrast, the inset shows two depressed areas of the membrane (encircled) at the base of a labeled bulge, which are free of filipin–sterol complexes (compare with the inset of Fig. 8). Bar, 0.2 $\mu$m. Inset: Bar, 0.2 $\mu$m.

## IV. Endocytosis

Conceptually, endocytosis is the reverse process of exocytosis—namely, the recapturing of plasma membrane in the cytoplasm as membrane-bound vesicles. Exocytosis-coupled endocytosis has been shown to occur in a variety of polypeptide hormone-containing cells after stimulation of secretion, and a reasonable question is to enquire, as was done for exocytosis, into the possible changes of the plasma membrane associated with endocytosis. Although studies with electron-dense tracers (for review, see Herzog and Miller, 1979; Silverstein *et al.*, 1977) have proved essential to follow the intracellular pathway of internalized plasma membrane (see Chapter 25 by M. G. Farquhar), one must turn to freeze-fracture in order to visualize the organization of the membrane involved in endocytosis. However, even with this approach, exocytosis-coupled endocytosis does not represent an ideal model. Indeed, this type of endocytosis occurs mostly via small (about 50 nm) vesicles (micropinocytosis). The small radius of curvature of such vesicles makes them rarely exposed by the fracture process, and this prevents the detailed study of their membrane, since most frequently only the fractured neck of the vesicular opening is available for examination. Another model of endocytosis was thus sought, and we turned to the study of cultured fibroblasts, which have much larger segments of their plasma membrane involved in endocytosis. These invaginating segments or pits have a characteristic bristle coat on the cytoplasmic side of the membrane inner leaflet (Fig. 11a), which has been shown to contain clathrin (Anderson *et al.*, 1978). Coated pits are involved in receptor-mediated uptake of macromolecules (adsorptive endocytosis) (for review, see Goldstein *et al.*, 1979) and may give rise to coated cytoplasmic vesicles upon pinching off from the plasma membrane. In freeze-fracture replicas of cultured human fibroblasts, coated pits were identified as large, shallow membrane depressions differing from the remainder of the plasma membrane by the presence of a higher density of intramembrane particles, which were also of larger mean diameter (Orci *et al.*, 1978) (Fig. 11b). In fibroblasts exposed to filipin during fixation, filipin–sterol complexes were missing from localized regions of the plasma membrane which corresponded in shape, size, and freeze-fracture appearance to the coated pits observed in untreated cells (Montesano *et al.*, 1979) (Fig. 11c). In addition to the large pits, fibroblasts also display invaginations with smaller diameters (microinvaginations). As explained previously, these small invaginations tend to fracture at their neck without showing the entire invaginating membrane, and the presence or absence of filipin–sterol complexes at these sites is more difficult to ascertain. In the rare instances in which the limiting membrane was exposed, it appeared free of filipin–sterol complexes. In thin sections of fibroblasts fixed in the aldehyde-filipin mixture, the plasma membrane appeared corrugated by filipin–sterol complexes except at the level of the coated pits (see Fig. 11a).

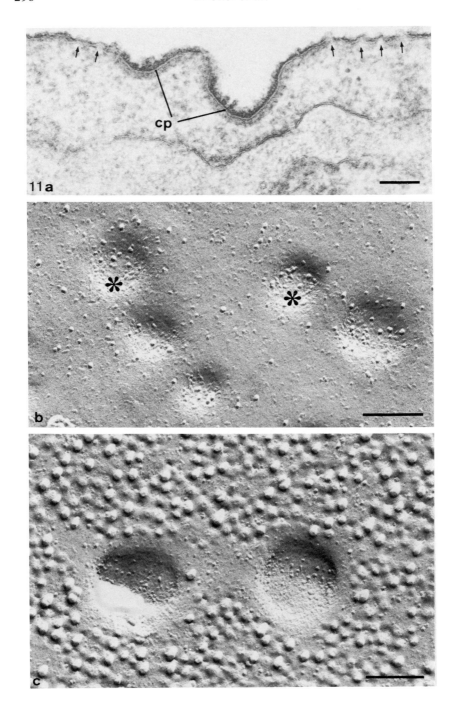

Viewed with the filipin probe, therefore, the membrane architecture at sites of endocytosis in fibroblasts appears the reverse of the one shown at sites of exocytosis. Invaginating segments of the plasma membrane (at least those underlined by a bristle coat) appear protein (particle)-rich and cholesterol (filipin–sterol complexes)-poor (see Addendum). Careful examination of thin sections of filipin–treated pancreatic B cells indicated a similar absence of filipin–sterol complexes from coated endocytotic membrane segments (Fig. 12, inset). Likewise, in freeze-fracture replicas of heavily labeled B cells, patches lacking filipin–sterol complexes appeared in invaginated areas of the plasma membrane, presumably corresponding to coated pits (Fig. 12; cf. also the mast cell, Fig. 10, inset).

# V.   Summary and Conclusions

Thin-section and freeze-fracture data of membrane organization in pancreatic B cells, peritoneal mast cells, and cultured fibroblasts during exocytosis and/or endocytosis, with intramembrane particles as protein markers and filipin–sterol complexes as cholesterol markers, have been reviewed. These data suggest a difference in membrane organization (as revealed with such morphological markers) between exocytosis and endocytosis. Aldehyde-fixed, glycerol-impregnated cells show that secretory granule and plasma membrane segments interacting during exocytosis are particle (protein)-free but display filipin–sterol complexes. The reverse appears true for coated endocytotic segments of the plasma membrane, which appear to be particle (protein)-rich, but devoid of filipin–sterol complexes. The functional significance of this difference is presently unknown.

### ACKNOWLEDGMENT

This work was supported by a grant from the Swiss National Science Foundation, Berne, Switzerland.

FIG. 11.   (a) Thin section of a cultured fibroblast fixed in the presence of filipin. Coated pits (cp) appear as membrane invaginations underlined on their cytoplasmic side by a thick bristle coat. Filipin–sterol complexes (arrows) are missing at the level of the coated invaginations. Bar, 0.2 μm. (b) Freeze-fracture replica of the plasma membrane of a cultured fibroblast. Coated pits (two of them are marked by asterisks) appear as shallow membrane depressions containing a high density of intramembrane particles of large size (for quantitative evaluation, see Orci *et al.*, 1978). Bar, 0.2 μm. (c) Freeze-fracture replica of the plasma membrane of a cultured fibroblast fixed in the presence of filipin. Filipin–sterol complexes (clearly distinguishable by their large size from the much smaller particles) homogeneously label the membrane face except at the level of two depressions corresponding to coated pits. Bar, 0.2 μm.

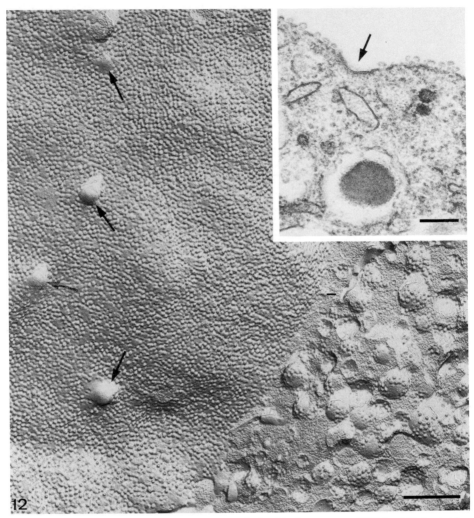

FIG. 12. Freeze-fracture replica of an islet cell fixed in the presence of filipin. Filipin–sterol complexes appear as small protuberances dotting the entire plasma membrane face (to the left) and parts of the secretory granule membrane in the cytoplasmic fracture (to the right). In the plasma membrane (E face) heavily labeled with filipin–sterol complexes (see Fig. 5 for sparse labeling), four circular patches (arrows) are free of complexes. These patches, which appear here as elevations, since they are viewed from the inside of the cell, probably correspond to the coated invaginations of the plasma membrane seen in thin section (inset). In thin section (inset), filipin–sterol complexes induce deformation of the plasma membrane, which appears scalloped, except at the site of the coated invagination (arrow) in which the membrane is not modified. Bar, 0.5 μm. Inset: Bar, 0.2 μm.

# VI. Addendum

Recent data obtained in this laboratory with exocrine pancreatic cells (Orci *et al.*, unpublished data) allow us to modulate the conclusion that exocytosis-stimulated endocytosis occurs mainly via invaginating, cholesterol-poor, protein-rich coated membrane segments. Following stimulated exocytotic release, a population of intracellular, subplasmalemmal vesicles appears that show a density of filipin-cholesterol complexes comparable to that of secretory granule membrane. Since the luminal plasma membrane is relatively poor in filipin-cholesterol complexes, this suggests that cholesterol-rich vesicles represent recaptured granule membrane following release of the granule content; only discrete areas of this recaptured membrane may appear coated (i.e., cholesterol-poor in filipin-treated material). Thus, according to these data, plasma membrane retrieval after exocytosis would include both cholesterol-poor (coated vesicles) and cholesterol-rich (former granule membrane) segments. That the membrane internalized during endocytosis may include both cholesterol-poor and cholesterol-rich areas also emerges from recent studies on macrophages (Montesano *et al.*, 1981).

## REFERENCES

Ahkong, Q. F., Fisher, D., Tampion, W., and Lucy, J. A. (1975). *Nature (London)* **253**, 194–195.
Allen, R. D. (1978). *Cell Surf. Rev.* **5**, 657–763.
Anderson, R. G. W., Vasile, E., Mello, R. J., Brown, M. S., and Goldstein, J. L. (1978). *Cell* **15**, 919–933.
Burwen, S. J., and Satir, B. H. (1977). *J. Cell Biol.* **73**, 660–671.
Chandler, D. E., and Heuser, J. (1979). *J. Cell Biol.* **83**, 91–108.
Chi, E. Y., Lagunoff, D., and Koehler, J. K. (1976). *Proc. Natl. Acad. Sci. U.S.A.* **73**, 2823–2827.
Colbron, A., Nachbaur, J., and Vignais, P. M. (1971). *Biochim. Biophys. Acta* **249**, 462–492.
Deamer, D. W. (1977). *In* "Mammalian Cell Membranes" (G. A. Jamieson and D. M. Robinson, eds.), Vol. 4, pp. 1–31. Butterworth, London.
Douglas, W. W. (1974). *Biochem. Soc. Symp.* **39**, 1–28.
Elias, P. M., Friend, D. S., and Goerke, J. (1979). *J. Histochem. Cytochem.* **27**, 1247–1260.
Goldstein, J. L., Anderson, R. G. W., and Brown, M. S. (1979). *Nature (London)* **279**, 679–685.
Hausmann, K., and Allen, R. D. (1976). *J. Cell Biol.* **69**, 313–326.
Herzog, V., and Miller, F. (1979). *Symp. Soc. Exp. Biol.* **33**, 101–115.
Keenan, T. W., and Morré, J. D. (1970). *Biochemistry* **9**, 19–25.
Lagunoff, D. (1973). *J. Cell Biol.* **57**, 252–259.
Lagunoff, D. (1975). *Tech. Biochem. Biophys. Morphol.* **2**, 283–305.
Lawson, D., Raff, M. C., Gomperts, B., Fewtrell, C., and Gilula, N. B. (1977). *J. Cell Biol.* **72**, 242–259.
Lucy, J. A. (1978). *Cell Surf. Rev.* **5**, 267–304.
McNutt, N. S. (1978). *J. Cell Biol.* **79**, 774–787.
Meldolesi, J., Jamieson, J. D., and Palade, G. E. (1971). *J. Cell Biol.* **49**, 130–149.

Montesano, R., Vassalli, P., and Orci, L. (1981). J. Cell Sci. (in press).

Montesano, R., Perrelet, A., Vassalli, P., and Orci, L. (1979). Proc. Natl. Acad. Sci. U.S.A. **76**, 6391–6395.

Montesano, R., Vassalli, P., Perrelet, A., and Orci, L. (1980). Cell Biol. Intern. Rep. **4**, 975–984.

Morin, F., Tay, S., and Simpkins, H. (1972). Biochem. J. **129**, 781–788.

Morré, D. J., Kartenbeck, J., and Franke, W. W. (1979). Biochim. Biophys. Acta **559**, 71–152.

Norman, A. W., Spievogel, A. M., and Wong, R. G. (1976). Adv. Lipid Res. **14**, 127–170.

Orci, L. (1974). Diabetologia **10**, 163–187.

Orci, L., and Perrelet, A. (1976). In "Endocrine Gut and Pancreas" (T. Fujita, ed.), pp. 295–299. Elsevier, Amsterdam.

Orci, L., and Perrelet, A. (1978). Cell Surf. Rev. **5**, 630–656.

Orci, L., Montesano, R., and Brown, D. (1980). Biochim. Biophys. Acta **601**, 443–452.

Orci, L., Amherdt, M., Roth, J., and Perrelet, A. (1979). Diabetologia **16**, 135–138.

Orci. L., Amherdt, M., Malaisse-Lagae, F., Rouiller, C., and Renold, A. E. (1973). Science **179**, 82–84.

Orci. L., Carpentier, J.-L., Perrelet, A., Anderson, R. G. W., Goldstein, J. L., and Brown, M. S. (1978). Exp. Cell Res. **113**, 1–13.

Orci, L., Montesano, R., Meda, P., Malaisse-Lagae, F., Brown, D., Perrelet, A., and Vassalli, P. (1981). Proc. Natl. Acad. Sci. U.S.A. **78**, 293–297.

Palade, G. E. (1959). In "Subcellular Particles" (T. Hayashi, ed.), pp. 64–83. Ronald Press, New York.

Pinto da Silva, P., Martinez-Palomo, A., and Gonzalez-Robles, A. (1975). J. Cell Biol. **64**, 538–550.

Plattner, H. (1973). J. Cell Sci. **13**, 687–719.

Robertson, J. D. (1959). Biochem. Soc. Symp. **16**, 3–43.

Robinson, J. M., and Karnovsky, M. J. (1980). J. Histochem. Cytochem. **28**, 161–168.

Röhlich, P., Anderson, P., and Uvnäs, B. (1971). J. Cell Biol. **51**, 465–483.

Satir, B., Schooley, C., and Satir, P. (1973). J. Cell Biol. **56**, 153–176.

Tillack, T. W., and Kinsky, S. C. (1973). Biochim. Biophys. Acta **323**, 43–54.

Verkleij, A. J., De Kruijff, B., Gerritsen, W. J., Demel, R. A., Van Deenen, L. L. M., and Ververgaert, P. H. J. (1973). Biochim. Biophys. Acta **291**, 577–581.

Zambrano, F., Fleischer, S., and Fleischer, B. (1975). Biochim. Biophys. Acta **380**, 357–369.

# Chapter 19

# Comparison of Compound with Plasmalemmal Exocytosis in Limulus Amebocytes

## RICHARD L. ORNBERG AND
## THOMAS S. REESE

*Section on Functional Neuroanatomy,*
*Laboratory of Neuropathology and*
*Neuroanatomical Sciences,*
*National Institutes of Health,*
*Bethesda, Maryland,*
*and*
*Marine Biological Laboratory*
*Woods Hole, Massachusetts*

## I. Introduction

A previous study of *Limulus* amebocytes established the sequence of events leading to exocytosis of secretory granules at the plasmalemma (Ornberg and Reese, 1981). Within seconds of stimulation, the plasmalemma buckles inward to form broad appositions with secretory granules lying near the cell surface. The distribution of intramembrane particles at these appositions is not different from that over the rest of the plasmalemma. This approach step is somehow related to the filaments that extend from some peripheral granules to the plasmalemma, even in resting cells. Numerous five-layered contacts then form along each apposition, and an exocytotic opening begins as a minute pore at one of these five-layered contacts. We inferred that pores widen very rapidly because they are rare compared with the numerous wide exocytotic openings. A summary of these steps is shown in Fig. 1; exocytosis at the plasmalemma is divided into an *approach* step, a *contact* step, and a *pore formation* step. The present study undertakes a comparison of this scheme of exocytosis at the plasmalemma to the subsequent exocytosis between granules deeper within the amebocyte, a sequential type of release known as compound exocytosis (Röhlich *et al.,* 1971).

Amebocytes from *Limulus* have several advantages in studies of exocytosis. Horseshoe crabs are readily available the year around and are hardy enough to be shipped inland. Amebocytes are virtually the only blood cell type; cell counts are high, particularly in the late summer and fall, and they are uniformly ready to secrete. Therefore, cells can be withdrawn and used directly for studies of exocytosis without concentration or purification. Since amebocytes are so sensitive to bacterial endotoxins, addition of micrograms per milliliter will suffice to excite exocytosis in the vast majority of amebocytes within a few seconds. Therefore, good synchronization of exocytotic events between different cells can be achieved. This sensitivity to endotoxin does introduce the constraint that careful methods of drawing blood and handling cells are necessary in order not to introduce extraneous endotoxin (see Cohen, 1979, for reviews of the biology of amebocytes).

We have used the technique of rapid freezing (Ornberg and Reese, 1981) followed by freeze-substitution or freeze-fracture to study exocytosis in the amebocyte. Rapid freezing offers the advantage that cells are immobilized within milliseconds of contact with the freezing surface without the addition of chemical agents (Heuser *et al.,* 1979). This speed permits estimations of the rates of different steps in a rapid process to be made on the basis of their frequency or, if the initial synchronization is sufficient, on the basis of the time elapsed after stimulation (Ornberg and Reese, 1981). These methods would also be expected to avoid any of the physiological changes incident to the interaction of chemical fixatives with secretory cells (Smith and Reese, 1980), and to avoid entirely the use of the aldehyde fixatives that have been linked to actual distortions in the exocytotic process (Chandler and Heuser, 1980).

## II.   Methods

Blood drawn by puncturing joint membranes of *Limulus polyphemus* was either frozen immediately or mixed with bacterial endotoxin to a final concentration approaching 10 $\mu$g/ml while it was being ejected from the withdrawal syringe onto the specimen mount of the freezing machine. This dose of endotoxin is orders of magnitude above threshold, so practically all the amebocytes in the

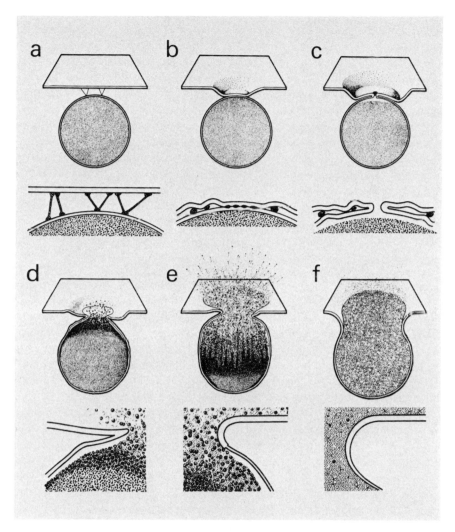

FIG. 1.   Sequence of exocytosis between secretory granules and the plasmalemma of *Limulus* amebocytes diagrammed for comparison with the exocytosis between granules deeper in the same cell, the subject of the present article. Interactions of the membrane around the granule with the plasmalemma is shown in more detail below each granule. In resting cells (a), granules are separated from the plasmalemma by a gap of 50 nm or more, and filamentous densities bridge the intervening space. The approach step results in the formation of punctate five-layered contacts between the granule and plasma membranes (b). A pore then forms at one of these contacts (c). The pore progressively widens as the granule contents expand and diffuse into the space around the amebocyte (d–e). The joint between the plasmalemma and the thinner granule membrane persists as the granule widens (e, below). The sequence shown here is estimated to last less than a second. Reproduced from Ornberg and Reese (1981).

sample began to secrete within 2–5 seconds of being placed on the specimen mount. By these means, exocytosis was synchronized in most of the cells within a preparation. This made it possible to use different intervals of endotoxin exposure of up to 60 seconds to put most of the cells in a preparation into a similar stage of exocytosis at the time of freezing. Freezing was accomplished by pressing the drop of amebocytes against a copper block cooled with liquid helium by means of a procedure described elsewhere (Ornberg and Reese, 1981).

Frozen drops of amebocyte blood were prepared by either freeze-substitution or freeze-fracture. Freeze-substituted cells were stained en bloc with either hafnium chloride or uranyl acetate; freeze-fracture was performed in a conventional Balzers apparatus equipped with a microtome (see Ornberg and Reese, 1981, for details of preparative methods).

# III.  Results

Exocytosis evoked by endotoxin treatment became compound or sequential after 5 seconds or more of endotoxin treatment. Granules deep in the cytoplasm formed exocytotic openings on the deep or lateral sides of more superficial granules, which had already opened and released their content. No instance was found where neighboring granules fused without one having already opened to the outside of the cell. In fact, a gradient of opening times seemed to exist between the outside and inside of a cell, judging from the degree to which granule contents had dispersed at the time a cell was frozen (Fig. 2).

How granule–granule exocytosis begins proved to be more difficult to determine than the sequence of granule–plasmalemma exocytosis (Fig. 1). This difficulty may be due to there being fewer granule–granule events in a cell at any single time because granule–granule exocytosis is less synchronized. Appositions between the outer dense laminae of apposed granule membranes were frequent, even in unstimulated amebocytes. These appositions consisted of five layers with the middle dense layer as thick or thicker than the two dense layers next to the granule contents (Fig. 3). Following endotoxin treatment and peripheral granule release, granule–granule appositions were frequent between the membranes of released and unreleased granules. In many instances, these

---

FIG. 2.    Amebocyte 15 seconds after stimulation with endotoxin, showing different degrees of emptying of granule contents. Several peripheral granules have contributed their membranes to the large cavity on the right, which remains filled with matrix. A granule (below) opening into this cavity has a denser content, and opening into it is another granule with an even denser content, suggesting that it opened very recently. These granules illustrate sequential compound exocytosis between secretory granules. Magnification 30,000×.

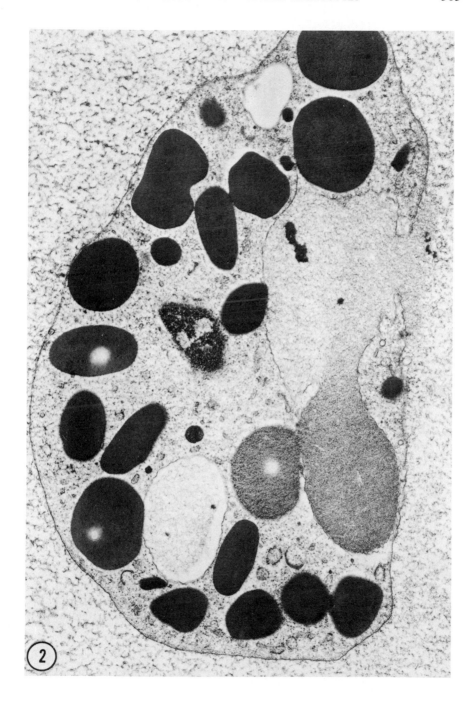

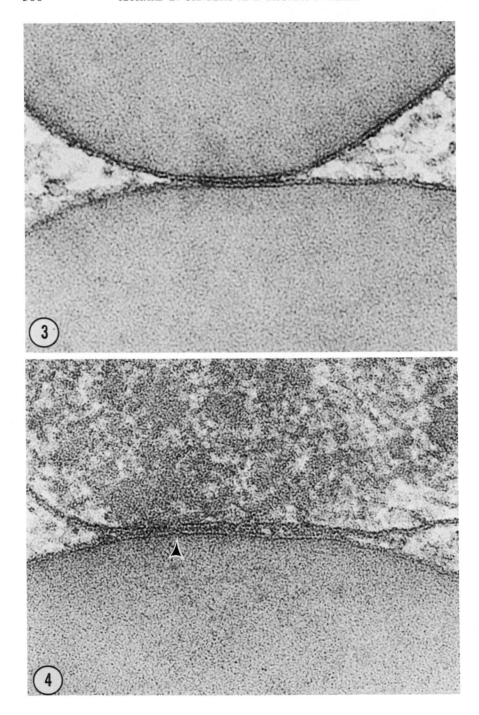

contacts were similar to the appositions seen in control cells. However, in other instances, small punctate regions of close membrane contact formed along the region of apposition (Fig. 4).

A transition from punctate contacts to exocytotic openings proved difficult to see. We did find frequent examples of granules connected by large exocytotic necks with their contents in an advanced state of dissolution (Fig. 2). The lumens of these openings had a smooth contour, and, unlike exocytotic lumens of peripheral granules, no discontinuity in membrane thickness or contour was found that might represent the original site of membrane fusion.

Membrane appositions were also recognized in freeze-fractured amebocytes both at rest (Fig. 5) and after stimulation with endotoxin (Fig. 6). However, no changes in intrinsic membrane structure, such as redistribution of intramembrane particles or sharp changes in membrane contour, were found. It also proved difficult to see in replicas how exocytosis begins, even with the large sampling of membrane appositions provided by the freeze-fracture technique.

Filaments connect peripheral granules with the adjacent plasmalemma and have been thought to be involved in initiating the contact between the granule and surface membranes that precedes exocytosis (Ornberg and Reese, 1981). We were unable to find comparable filaments arrayed at present or future sites of granule–granule exocytosis.


## IV.   Discussion


The sequence of membrane relationships underlying exocytosis at the surface of *Limulus* amebocytes is summarized in Fig. 1 (Ornberg and Reese, 1981). We use this scheme for peripheral granule exocytosis as a basis of comparison with the compound exocytosis of more internal granules reported here.

One striking difference between exocytosis at the surface of amebocytes and internal, compound exocytosis is in the formation of an apposition between the fusing membranes. Appositions of secretory granules with the plasmalemma are rare in unstimulated cells but common between secretory granules in the cell interior. Also, an array of filaments connects the plasmalemma with some peripheral granules and may be responsible for pulling the plasmalemma into

---

FIGS. 3 and 4.   Apposition between granules in a resting cell (Fig. 3) and a stimulated cell (Fig. 4) prepared by freeze-substitution. The upper granule in Fig. 4 has already opened to the surface; dispersion of its contents is apparent. The arrowhead indicates one of the punctate contacts that characterize pentalaminar appositions between pairs of granules, where one of the pair has undergone exocytosis. Some of these contacts may be sites of exocytotic pore formation. Magnification 150,000×.

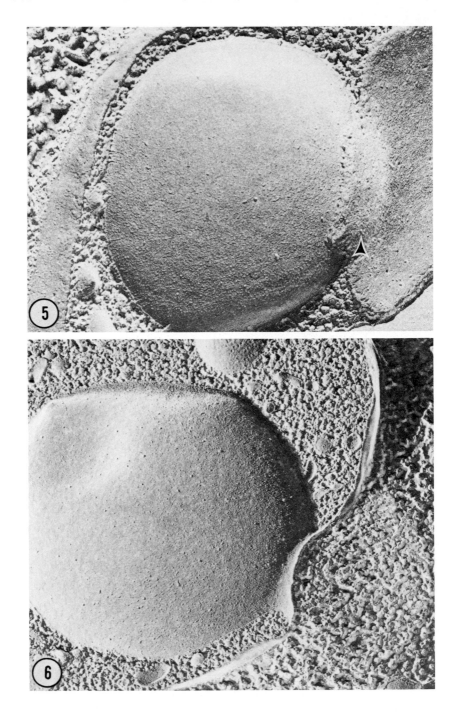

contact with the underlying granule (Ornberg and Reese, 1981). A similar concentration of filaments interconnecting adjacent secretory granules has not been observed.

The typical granule–granule apposition seen in unstimulated cells has a uniformly spaced, five-layered structure, which resembles early membrane fusion sites reported for other secretory cells prepared by chemical fixation (Palade and Bruns, 1968; Orci and Perrelet, 1978; Neutra and Schaeffer, 1977). The frequency of these appositions at rest, and the apparent closeness of membrane contact, might be taken as evidence that these are stable, long-lived structures. However, these contacts must be transient, since we have observed granules moving with respect to one another in amebocytes crawling on glass slides (Cohen, 1979). Although we have no proof that these resting appositions are the precursors of exocytosis, the internal granules, unlike the surface granule, may be regarded as having undergone the approach step of exocytosis prior to stimulation.

The uniformly pentalaminar granule–granule apposition seen at rest is different from the punctate pentalaminar contacts seen at plasmalemma–granule appositions, which we previously concluded to be the sites where exocytosis begins as a minute pore (Ornberg and Reese, 1981). Although we did not capture comparable pores at granule–granule appositions, the observation of numerous punctate pentalaminar contacts at some granule–granule appositions, where one granule of the granule pair has undergone exocytosis, suggests these may be precursors of exocytotic pore formation between granules. The frequency of punctate membrane contacts and the abundance of large exocytotic lumens compared with the rarity of small pores suggests that any small pores that form at a punctate pentalaminar contact must widen rapidly into exocytotic lumens.

It has been reported and generally accepted that intramembrane particles are laterally displaced from appositions between secretory granules and from contacts between isolated secretory granules forced together by centrifugation (Tanaka *et al.*, 1980). Particle-free blebs and vesicles are also thought to be part of exocytosis in most cells (Lawson *et al.*, 1977) and of membrane fusion in myoblasts (Kalderon and Gilula, 1979), although other recent studies have suggested that they may be artifacts of aldehyde fixation (Tanaka *et al.*, 1980; Chandler and Heuser, 1980; Rees and Reese, 1981). Our studies with amebocytes have never demonstrated any particle-free blebs and vesicles or particle-

---

FIGS. 5 and 6. Secretory granules from endotoxin-stimulated amebocytes prepared by freeze-fracture. Intramembrane particles (arrowhead) persist on the split membrane half (P-face) of a granule indented by a neighboring granule (Fig. 5). Figure 6 shows a granule that has opened onto the surface (right) and has been dented (left), presumably by a neighboring granule. The distribution of intramembrane particles within the dent is not different from that over the rest of the split granule membrane (E-face). Magnification: Fig. 5, 60,000×; Fig. 6, 50,000×.

free regions at appositions between granules or between granules and the plasmalemma. We do not know whether amebocytes are fundamentally different from other cells or whether preparation by rapid freezing prevents particle clearing. The former possibility is a real one, since we have seen no particle-free regions at plasmalemma–granule appositions in amebocytes fixed in aldehydes (Ornberg and Reese, 1981).

The fact that granule–granule contacts of the punctate pentalaminar type are more frequently observed than are small exocytotic pores or even larger exocytotic openings suggests that some type of energy barrier to exocytosis exists even at membrane contacts. Recent evidence from experiments with model systems has suggested that osmotic gradients may be an important force in initiating exocytosis (Cohen *et al.*, 1980; Zimmerberg *et al.*, 1980); our previous evidence on exocytosis of peripheral granules in amebocytes is consistent with this idea (Ornberg and Reese, 1981). Although an osmotic gradient is likely to exist between secretory granules and the outside of the cell, it would not be expected between secretory granules. Under these circumstances the interposition of an approach step before exocytosis at peripheral granules can begin could serve to prevent exocytosis in resting cells. Once exocytosis begins, however, deeper granules in contact with exocytosing surface granules would immediately experience these osmotic forces and form granule–granule pores. By this means, exocytosis would be constrained to proceed sequentially inward. Such a conjecture fits the available information, but proving it seems beyond the reach of current morphological techniques, since the critical issue is whether swelling of granule contents indicative of water uptake precedes or follows early pore formation. The initial swelling of the granule contents might be so small that it would escape detection as well as being statistically indeterminable because of the difficulties in finding examples. So far, we have not been able even to find small pores between granules in stimulated amebocytes, and they are very rare between peripheral granules and the surface membrane of exocytosing cells. Nevertheless, *Limulus* amebocytes offer the necessary synchronous secretion, and rapid freezing offers the necessary time resolution eventually to determine the role of osmotic forces in initiating compound exocytosis.

## REFERENCES

Chandler, D. E., and Heuser, J. E. (1980). *J. Cell Biol.* **83,** 91–108.
Cohen, E., ed. (1979). "Biomedical Applications of the Horseshoe Crab (Limulidae)," Prog. Clin. Biol. Res., Vol. 29. PCBR, New York.
Cohen, F. S., Zimmerberg, J., and Finkelstein, A. (1980). *J. Gen. Physiol.* **75,** 251–270.
Heuser, J. E., Reese, T. S., Dennis, M. J., Jan, Y., Jan, L., and Evans, L. (1979). *J. Cell Biol.* **81,** 275–300.
Kalderon, N., and Gilula, N. B. (1979). *J. Cell Biol.* **81,** 411–425.

Lawson, D., Raff, M. C., Gomperts, B., Fewtrell, C., and Gilula, N. B. (1977). *J. Cell Biol.* **72,** 242–259.

Neutra, M. R., and Schaeffer, S. F. (1977). *J. Cell Biol.* **74,** 983–991.

Orci, L., and Perrelet, A. (1978). *In* "Membrane Fusion" (G. Poste and G. L. Nicholson, eds.), pp. 629–656. Elsevier North-Holland Publ., Amsterdam and New York.

Ornberg, R. L., and Reese, T. S. (1981). *J. Cell Biol.* (in press).

Palade, G. E., and Bruns, R. R. (1968). *J. Cell Biol.* **37,** 633–648.

Rees, R., and Reese, T. S. (1981). *Neuroscience* (in press).

Röhlich, P., Anderson, P., and Unvas, B. (1971). *J. Cell Biol.* **51,** 465–483.

Smith, J., and Reese, T. S. (1980). *J. Exp. Biol.* **87,** 1–11.

Tanaka, Y., DeCammili, P., and Meldolesi, J. (1980). *J. Cell Biol.* **84,** 438–453.

Zimmerberg, J., Cohen, F. S., and Finkelstein, A. (1980). *J. Gen. Physiol.* **7,** 241–250.

# Chapter 20

# Role of Ions and Intracellular Proteins in Exocytosis

HARVEY B. POLLARD, CARL E. CREUTZ, AND
CHRISTOPHER J. PAZOLES[1]

*Section on Cell Biology and Biochemistry of the Clinical
Hematology Branch,
National Institute of Arthritis, Metabolism and Digestive Diseases,
National Institute of Health,
Bethesda, Maryland*

## I. Introduction

The term "exocytosis" represents a general hypothesis for secretion of substances such as transmitters, hormones, and enzymes that are initially sequestered in intracellular storage organelles. This hypothesis, based primarily on ultrastructural data summarized by Palade (1975), is that the storage organelle membrane attaches to the inner aspect of the plasma membrane of the secretory cell. Subsequently, the two membranes "fuse" by forming a "pentalaminar structure," which may eventually become a single bilayer separating vesicle contents from the external medium. Finally, this bilayer undergoes "fission" or breakage, and the vesicle contents are released.

---

[1]Present address: Pfizer Central Research, Eastern Point Road, Groton, Connecticut 06340.

The biochemical data supporting the exocytosis hypothesis are simply that (1) storage organelles from different cells can be isolated and therefore shown to be discrete objects and (2) when secretion occurs the entire organelle contents, ranging from small molecules to large proteins, can be found in the extracellular medium. The data are certainly circumstantial, yet apparently compelling in the support they lend for the exocytosis hypothesis (Smith, 1968; Viveros, 1974; Winkler, 1977; Pollard *et al.*, 1979a).

It has also been appreciated for many years that calcium plays a critical role in regulating exocytosis in a number of systems. In particular, an increase in intracellular free calcium concentration often seems to be a prerequisite for secretion, and it is presently thought that calcium might mediate that initial interaction between the storage organelle and the plasma membrane just prior to secretion. Among the best evidence for this concept are kinetic data from studies by Llinas and his colleagues (1976) on the squid stellate ganglion showing that calcium penetrates the presynaptic nerve terminal and elicits secretion over a time interval that is too short to allow calcium to diffuse more than a short distance from the plasma membrane. Parsegian (1977) has computed this distance to be less than 10 nm.

How calcium might act, once free on the cytoplasmic side of the plasma membrane, has been the focus of numerous proposals ranging from simple charge screening processes to mediation by specific proteins (Pollard *et al.*, 1979a). In this article we shall describe in some detail a new protein called "synexin," which induces $Ca^{2+}$-dependent formation of pentalaminar complexes between secretory granule membranes. In this sense synexin has proved distinct from other, more conventional candidates for $Ca^{2+}$-effect mediators such as calmodulin, actomyosin, and tubulin. We shall also describe our investigation into the process of membrane breakage or "fission," which finally results in release of secretory vesicle contents. This remains a poorly understood problem, which we have approached experimentally over the last several years by supposing that it might occur by local osmotic rupture at the fusion site, a process analogous to the chemiosmotic lysis of isolated chromaffin granules. We propose that synexin-like molecules may mediate calcium action during exocytosis, and we shall describe studies comparing the granule lysis reaction with predicted behavior of secreting chromaffin and other cells.

## II.  Synexin and Calcium Action

In spite of the long history and generality of the observation that $Ca^{2+}$ is required for exocytosis, the molecular details of the regulatory role of $Ca^{2+}$ have not been elucidated. A major difficulty has been the lack of biochemical assays for the morphological events that define exocytosis, particularly the fusion of the secretory vesicle to the plasma membrane. However, in many secretory systems, secretory vesicles fuse not only with the plasma membrane but with the membranes of other vesicles, forming channels to the exterior of the cell in a process called "compound exocytosis." It seemed likely to us that the interaction between vesicles in this process might be controlled by the same molecular factors that regulate the interaction between the secretory vesicle and the plasma membrane. Therefore, in order to facilitate a biochemical approach to examining the role of $Ca^{2+}$, we studied the *in vitro* interactions between purified secretory vesicles as monitored by turbidity measurements of vesicle suspensions (Creutz *et al.*, 1978, 1980).

Chromaffin granules, the secretory vesicles from the adrenal medulla, were particularly suitable for this study, inasmuch as the $Ca^{2+}$ requirement for exocytosis in this system had been well documented, compound exocytosis had been observed in chromaffin cells (Fenwick *et al.*, 1978), and the vesicles themselves could be readily isolated in large quantities. For our studies chromaffin granules were isolated by either of two procedures: differential centrifugation in isotonic sucrose solution (Pollard *et al.*, 1976), or equilibrium density gradient centrifugation in isotonic mixtures of sucrose and metrizamide (Pollard *et al.*, 1979b). The latter procedure gave greater yields of more highly purified chromaffin granules.

It had previously been reported that isolated chromaffin granules would attach to one another when exposed to high concentrations of divalent cations (Schober *et al.*, 1977). However, in contrast to secretion *in vivo*, this type of interaction showed no specificity for $Ca^{2+}$ as opposed to $Mg^{2+}$, and the concentrations of cation required (several millimolar) seemed too high to be relevant to an intracellular process. We found, however, that a very sensitive $Ca^{2+}$-specific aggregation process could be induced if granules were incubated at 37°C in the presence of the postmicrosomal supernatant from homogenized chromaffin tissue. The soluble factor responsible for this $Ca^{2+}$-dependent process was heat- and trypsin-sensitive, suggesting that it was protein in nature. We named the factor "synexin," from the Greek word *synexis,* which means "meeting," because of its ability to bring chromaffin granules together.

We were able to devise an isolation scheme for synexin, using as an assay for the protein its ability to cause an increase in the turbidity of a chromaffin granule suspension. The scheme, which is outlined in Fig. 1, consists in the following steps: (1) preparation of the postmicrosomal supernatant; precipitation of the factor in 20% $(NH_4)_2SO_4$; (2) repetition of this precipitation step, which proved important for removing serum albumin; (3) gel filtration on Ultrogel AcA34; and (4) hydroxylapatite chromatography. (Complete details are given in Creutz *et al.*, 1978.) After the final chromatographic step the preparation appeared as a single band on an SDS gel with an apparent molecular weight of 47,000. However, the losses in this step were high, and for many experimental purposes the preparation after gel filtration was found to be suitable. This material was

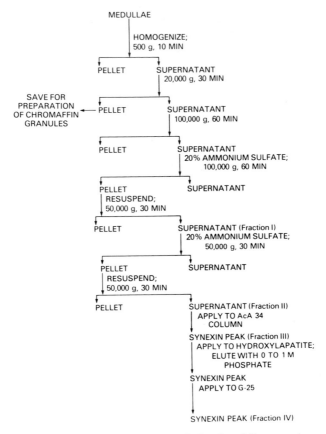

FIG. 1.    Outline for the purification of adrenal medullary synexin.

80–90% pure and behaved in an identical fashion to the more highly purified material in its granule aggregation and self-association (see below) assays. The yield of synexin, after gel filtration, was about 500 $\mu$g from 50 gm of tissue.

The influence of synexin on the turbidity of a granule preparation is illustrated in Fig. 2. This assay was conducted at 37°C in the presence of 240 m$M$ sucrose, 30 m$M$ KCl, 28 m$M$ histidine–HCl (pH 6.0), 2.5 m$M$ EGTA, and sufficient CaCl$_2$ to obtain the desired free Ca$^{2+}$ concentration. The turbidity increase began immediately and was complete in 8–10 minutes. Barium, strontium, or magnesium could not be substituted for Ca$^{2+}$. The calcium appeared to be titrating a site with a dissociation constant of 200 $\mu M$, thereby stimulating the reaction. The reaction occurred as well at 25°C, but below 25°C it declined in activity and it did not occur at 0°C. The reaction was only partly reversible in that the removal of Ca$^{2+}$ by chelation with EGTA after the reaction was complete resulted in a

10–20% reversal of the turbidity increase. During the reaction synexin appeared to bind to the granules in a $Ca^{2+}$-dependent manner, as defined by the fact that the protein could be removed from solution by sedimenting aggregating granules. From such binding studies we calculated that there appeared to be approximately 90 synexin-binding sites on each granule.

When chromaffin granules that had been exposed to $Ca^{2+}$ and synexin were examined in thin sections, it was seen that nearly all the granules became attached to at least one other granule at $Ca^{2+}$ concentrations as low as 6 $\mu M$. At higher $Ca^{2+}$ concentration larger aggregates of granules formed. The granules appeared to be separated at their regions of mutual contact by a pentalaminar structure formed by the close apposition of their trilaminar limiting membranes. This pentalaminar structure appeared similar to those observed in electron microscopic studies of intact cells undergoing exocytosis. Under the conditions of our assay the pentalaminar structures appeared stable, as vesicles did not fuse to

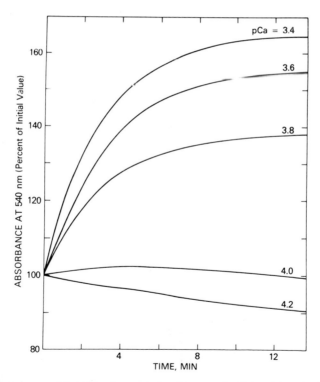

FIG. 2. Time course of turbidity changes induced by synexin (3.4 $\mu g$) in a 1-ml chromaffin granule suspension in the presence of different concentrations of free calcium. Initial absorbance $\simeq$ 0.3 (130 $\mu g$ of granule protein). pCa represents the negative logarithm of the free calcium concentration.

give larger structures with common lumina. Therefore it appeared that synexin could be responsible only for an initial step in exocytosis, although it might be the critical step at which $Ca^{2+}$ regulated the entire process.

In order to investigate the site of action of $Ca^{2+}$, we examined the behavior of purified synexin in the presence of varying concentrations of free $Ca^{2+}$. Using 90° light-scattering measurements we determined that $Ca^{2+}$ caused synexin to self-associate (Creutz *et al.*, 1979). The calcium dependence and cation specificity of the self-association reaction was identical to that of the synexin-induced chromaffin granule aggregation, strongly suggesting that the mechanism of granule aggregation included a step in which $Ca^{2+}$ became bound to synexin and caused the protein in turn to self-associate. By negative staining of purified synexin with sodium phosphotungstate, we determine that $Ca^{2+}$ caused synexin to self-associate into rodlike particles of dimensions 50 by 150 Å. Subsequently these rods rapidly associated with one another to form bundles of parallel rods. The native molecular weight of synexin in the absence of $Ca^{2+}$ is such that the rods may consist of four or five synexin monomers. Whether the rods actually form or line up side by side when chromaffin granules are present, or whether synexin monomers interact individually with the membranes, is not known at present. Figure 3 schematically outlines one possible interpretation of the steps that occur when $Ca^{2+}$ activates synexin to cause membrane interaction.

In order to further examine the morphological relationship of synexin in the intact chromaffin cell to the chromaffin granules and the plasma membrane, and to assist our search for synexin-like proteins in other secretory tissues, we have prepared an antibody to synexin. To prepare the antibody, synexin was partially purified by precipitation in ammonium sulfate and gel filtration. This material was then run on a preparative SDS slab gel, and the major band on the gel,

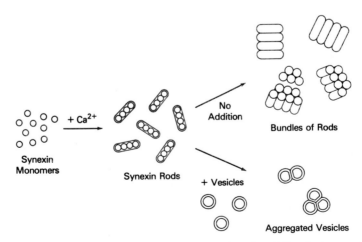

Fig. 3.    A schematic interpretation of the events occurring when $Ca^{2+}$ activates synexin to cause the aggregation of secretory vesicles.

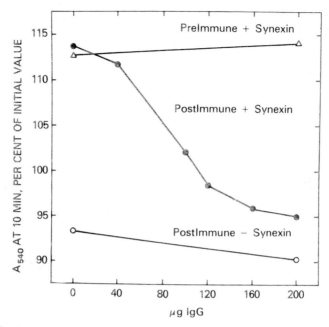

FIG. 4.   Inhibition of synexin-dependent chromaffin granule aggregation by rabbit anti-synexin IgG. Approximately 1 μg of synexin was incubated with 130 μg of chromaffin granules in 1 ml in the presence of various amounts of preimmune or anti-synexin IgG. The anti-synexin has blocked the aggregation of the granules, which is monitored as a turbidity change in 10 minutes.

migrating with an apparent molecular weight of 47,000 and constituting 80% of the protein, was cut from the gel; the protein was removed from this slice by electrophoresis. The eluted protein was emulsified with complete Freund's adjuvant and injected into rabbits. The presence of anti-synexin IgG in the serum of a rabbit was detected by the ability of the IgG to inhibit the aggregation of chromaffin granules by synexin (Fig. 4). We have also used the anti-synexin IgG to determine the localization of synexin in chromaffin cells by indirect immunofluorescence. Chromaffin cells, isolated from medullary tissue by collagenase digestion, were fixed in 2% formaldehyde for 20 minutes at room temperature, then spread on glass slides, air-dried, and further fixed and permeabilized in 95% ethanol. The cells were then incubated with the anti-synexin antibody and, subsequently, with a fluorescein-tagged goat anti-rabbit-IgG immunoglobulin. An excess of normal goat IgG was included in the immune reagent to saturate nonspecific binding sites. As illustrated in Fig. 5, specific anti-synexin staining occurred in a diffuse manner throughout the cytoplasm of the chromaffin cell. In some cells the degree of staining was noticeably enhanced near the cell surface.

To summarize our results with synexin, we have found that it is a $Ca^{2+}$-binding protein that, *in vitro,* causes a $Ca^{2+}$-specific association of chromaffin granule membranes at levels of $Ca^{2+}$ that may occur near the plasma membrane when the

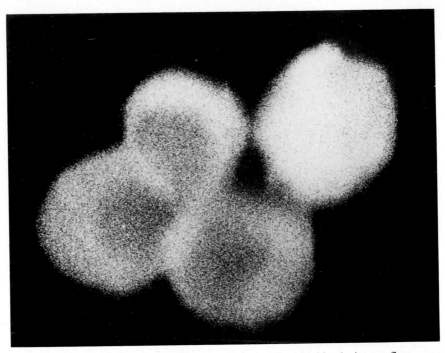

Fig. 5.   A small cluster of bovine adrenal chromaffin cells stained by the immunofluorescence technique for the presence of synexin. Specific staining is seen throughout the cytoplasm, leaving darker areas at the positions of cell nuclei.

chromaffin cell is stimulated. Furthermore, synexin is present in the cytoplasm of the cell. These facts suggest that the *in vivo* role of synexin may be to act as an intracellular receptor for $Ca^{2+}$, which initiates the process of exocytosis. In further work on this protein it will be of particular importance to determine whether synexin will indeed attach secretory vesicles to the inner surface of the plasma membrane. It will also be of interest to characterize synexin-like proteins in other tissues, as we have already determined that $Ca^{2+}$-dependent chromaffin granule aggregation activity exists in crude preparations obtained from the parotid gland, whole brain, and blood platelets; it has been purified to homogeneity from liver.

## III.   Chemistry of Chemiosmotic Lysis of Isolated Chromaffin Granules

As discussed earlier, the ultimate event in exocytosis is release of the vesicle contents into the extracellular space subsequent to breakage of the membrane(s) separating the two compartments. It was apparent, from both chemical and ultrastructural analysis, that, although synexin might induce formation of pentalaminar complexes, it did not cause breakage or "fission" of the adjacent membranes. A plausible alternative candidate for a process related to fission was the ATP-driven lysis of isolated chromaffin granules in an isoomotic medium, a reaction first described in 1965 by Oka *et al.* Such lysis results in the release of granule contents including catecholamines, dopamine $\beta$-hydroxylase, ATP, and chromogranins, and is greatly potentiated by the presence of MgATP and permeant anions such as chloride. In recent years it has become evident that the loss of granule integrity is due to osmotic lysis subsequent to the coupled influx of protons and permeant anions via two granule membrane activities: a proton pumping MgATPase, and a selective anion transport site (summarized in Pollard *et al.*, 1979a). Our understanding of this granule lysis process led us to postulate, by analogy, that the mechanism of fission in exocytosis might involve local osmotic lysis of the fused secretory granule. We supposed the process might utilize intracellular ATP and the concentration gradient of chloride ions, which exists across most cell membranes. We therefore studied the characteristics of the chromaffin granule lysis reaction in detail in order to generate specific, testable predictions regarding the secretory behavior of chromaffin and other cells.

In our experiments, granule lysis was monitored by measuring the release of epinephrine from suspensions of bovine adrenal chromaffin granules. The procedure involved the addition of 250 $\mu$l of a suspension of granules (250–500 $\mu$g protein) in cold 0.3 $M$ sucrose (4°C), prepared as described earlier, to 1.5 ml prewarmed (37°C) medium containing various substances of interest (salts, drugs, ATP, etc.). The mixture was buffered with 50 m$M$ MES–NaOH (pH 6.0). After a 10-minute incubation at 37°C, release was terminated by the addition of 1 ml ice-cold 335 m$M$ sucrose. After sedimentation of the granules by centrifugation (20 minutes at 35,000 $g$, 4°C), released epinephrine was assayed in the supernatant by a fluorometric method (Pazoles and Pollard, 1978).

Using this protocol, we examined the effects of $Mg^{2+}$-ATP and various salts on granule lysis (Table I). We found that epinephrine release was potentiated by $Mg^{2+}$-ATP and was greatly influenced by the nature of the anion, but not the cation present. Permeant anions such as $Cl^-$, $Br^-$, and $I^-$ supported release, whereas impermeant anions such as isethionate and $PO_4^{3-}$ did not. In addition, the ATP-dependent release rate was a saturable function of $[Cl^-]$ and was competitively inhibited by isethionate ions, suggesting that an anion-selective membrane transport site might be involved (Pazoles and Pollard, 1978). To pursue this possibility, we made use of a number of compounds that had been previously shown to block anion transport in erythrocytes. These compounds included SITS (a disulfonic stilbene), probenecid, and pyridoxal phosphate. These drugs contained both anionic and hydrophobic moieties in keeping with their inhibitory function and were impermeant under our conditions. Figure 6 shows that these

### TABLE I

INFLUENCE OF $Mg^{2+}$-ATP AND VARIOUS SALTS
ON EPINEPHRINE RELEASE
FROM CHROMAFFIN GRANULES[a]

| | Epinephrine released (%/10 min) | |
|---|---|---|
| Salt (80 m$M$) | $-Mg^{2+}$-ATP | $+Mg^{2+}$-ATP |
| None (335 m$M$ sucrose) | $10.6 \pm 0.8$[b] | $11.4 \pm 0.1$ |
| KCl | $19.5 \pm 2.1$ | $65.9 \pm 5.2$ |
| NaCl | $17.3 \pm 1.8$ | $66.4 \pm 6.6$ |
| Choline Cl | $14.2 \pm 0.7$ | $63.3 \pm 2.9$ |
| Tetraethylammonium Cl | $15.1 \pm 1.3$ | $60.4 \pm 4.2$ |
| KBr | $19.9 \pm 2.3$ | $86.1 \pm 4.4$ |
| K isethionate | $10.4 \pm 0.2$ | $14.6 \pm 1.0$ |
| KHPO$_4$ | $10.9 \pm 1.1$ | $11.6 \pm 0.4$ |

[a] Incubations were at 37°C in isotonic media buffered with
50 m$M$ MES, pH 6.0.
[b] Values $\pm$ S.D. ($n \geq 4$).

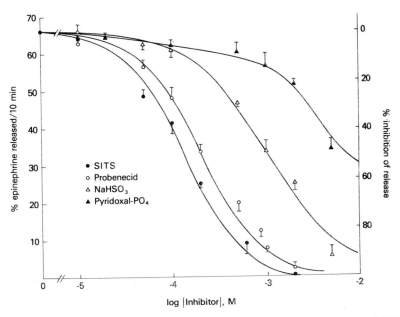

FIG. 6. Inhibition of $Mg^{2+}$-ATP-Cl$^-$-mediated epinephrine release by anion transport-inhibiting
drugs. The medium contained 1 m$M$ MgSO$_4$, 1 m$M$ Na$_2$ATP, 90 m$M$ KCl, 50 m$M$ HEPES–NaOH,
pH 6.0, and sufficient sucrose to adjust the final osmolarity to 335 mOsm.

compounds all inhibited $Mg^{2+}$-ATP-$Cl^-$-mediated epinephrine release, and their potencies were very similar to those reported for inhibition of anion transport in erythrocytes (Cabantchik and Rothstein, 1974). Using kinetic analysis, we were also able to show that the mechanism of inhibition in each case was competitive with respect to [$Cl^-$] (Pazoles and Pollard, 1978). Finally, we demonstrated that SITS directly inhibited entry of $^{36}Cl^-$ into granules in the presence of MgATP (Pazoles and Pollard, 1978). We concluded from these experiments that the mechanism of MgATP–anion-induced granule lysis involved anion transport through a selective and saturable membrane site. We are currently attempting to identify the specific granule membrane components involved in anion transport.

We, and others, have also studied the role of ATP in the granule lysis process. It had been known for some time that the granule membrane contained an exo-$Mg^{2+}$-ATPase (Smith, 1968; Viveros, 1974), and subsequent studies indicated that it might function to pump protons into the granule interior (Casey *et al.*, 1976, 1977; Njus and Radda, 1978). If this were the case, then granule lysis might involve entry of both protons and anions into the granule. This possibility was especially attractive, since some cation entry was required to maintain electroneutrality upon anion entry, and since we had shown that uptake of the cation of the chloride salt did not occur (Pazoles and Pollard, 1978). Our approach to the study of this proton-pumping ATPase was to examine the effects of MgATP addition on the electrical and pH gradients across the granule membrane. The transmembrane electrical potential was determined by first measuring the distribution of a lipophilic anion, [$^{14}C$]thiocyanate, across the granule membrane as described in detail elsewhere (Pollard *et al.*, 1976; Pazoles and Pollard, 1978). The potential ($\Delta E$) was then calculated from this distribution via the Nernst equation. Chromaffin granules in an isoosmotic sucrose medium (including 50 m$M$ MES-NaOH, pH 6.0) exhibited a $\Delta E$ of $+28.6 \pm 1.0$ mV, positive inside. Upon addition of 1 m$M$ MgATP, the $\Delta E$ increased to $+66.4 \pm 6.3$ mV. These data were obtained at 0°C to minimize granule lysis, but qualitatively similar results were found at 37°C. These findings were consistent with the suggestion that the granule ATPase acted to move protons electrogenically into the granule, and supporting this conclusion was the additional finding that FCCP, a proton ionophore, strongly inhibited $Mg^{2+}$-ATP-$Cl^-$-induced granule lysis, presumably by counteracting the inward proton pump.

The transmembrane proton concentration gradient ($\Delta pH$) was determined by measuring the intragranular pH under conditions of known external pH. To do this, we (Pollard *et al.*, 1979b) and others (Njus *et al.*, 1978) have made use of the facts that ATP was an endogenous component of the granule core and that the $^{31}P$ NMR resonance signal from the $\gamma$-phosphate of ATP was sensitive to its local pH. By artificially adjusting the internal pH to known values, we were able to calibrate the chemical shift of this resonance with the known value of intragranular pH (Pollard *et al.*, 1979b). These studies showed that the intragranular pH was buffered to approximately 5.7, a value consistent with data from methylamine distribution (Pollard *et al.*, 1976, 1979b; Njus *et al.*, 1978; Johnson and Scarpa, 1976). Addition of MgATP had no effect on this value because of substantial intragranular buffering. If, however, a permeant anion were also present, the granule interior was progressively acidified (Fig. 7); permeant anions alone did not support this change. These results were obtained at external pHs of both 7.1 and 6.3, suggesting that the granule ATPase was capable of sustaining proton movement into the granule against the $H^+$ concentration gradient only in the form of electroneutral coupled transport with anions.

To further test this hypothesis we repeated the above experiment but replaced the ATPase as the driving force for proton entry with an inward proton concentration gradient created by incubating the granules at pH 5.2. Once again, intragranular acidification depended on the presence of permeant anions (Fig. 8). In an isoosmotic medium, the combination of low pH and permeant anions resulted in granule lysis, and, therefore, experiments designed to measure changes in internal granule compartments were performed in hypertonic media. This lysis, like that in the presence of MgATP and permeant anions at higher pH, was inhibited by anion transport blockers (Fig. 9).

From these experiments it was apparent that, in the presence of permeant anions and either MgATP or low external pH, there was net accumulation of ions within the granule leading to osmotic lysis. This was confirmed experimentally by demonstrating that MgATP-induced epinephrine release was progressively inhibited by increasing the osmotic strength of the medium (Fig. 10). Similar osmotic suppression was also found for release occurring at low pH.

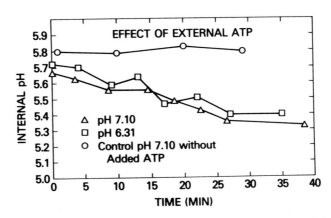

Fig. 7.   Influence of external $Mg^{2+}$-ATP and chloride on the acidification of the chromaffin granule interior. The reaction mixture contained 100 m*M* KCl, 7.5 m*M* MgSO$_4$, 7.5 m*M* Na$_2$ATP, 50 m*M* PIPES–NaOH buffer pH 7.10, or 50 m*M* MES–NaOH buffer, pH 6.31, and sufficient sucrose to bring the osmotic strength to 970 mOsM.

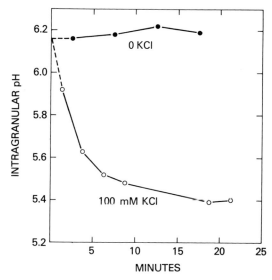

FIG. 8.   Intragranular acidification induced by incubation in a chloride medium with a pH less than the intragranular pH. Media contained either zero or 100 m$M$ KCl, 50 m$M$ MES–NaOH buffer, pH 5.2, and enough sucrose to adjust the osmotic strength to 970 mOsM.

Taken together, our studies and those of others have led to a chemiosmotic model of the granule lysis process. In the presence of an inward concentration gradient of permeant anions, such as chloride, the activity of a granule membrane proton-pumping MgATPase results in the electroneutral, bulk uptake of protons and anions. The resulting increase in granule osmotic content induces water entry and subsequent lysis.

We have recently described the successful construction of a mathematical model for the release process that depends on ATP and Cl$^-$ (Creutz and Pollard, 1980). This model incorporates the basic concepts set forth above of an ATP-dependent proton pump and the concurrent influx of Cl$^-$ into the granule via a specific transport site. Since the release event is hypothesized to be due to osmotic lysis, the model relates the influx of osmotically active particles to a degree of lysis within a granule preparation by referring to an experimentally determined osmotic fragility curve for that same population. Such a fragility curve, as determined by incubating granules in hypotonic media and monitoring lysis as a drop in turbidity or a release of epinephrine, is illustrated in Fig. 11. Based on this fragility curve, the model was able to accurately describe the kinetics of the release reaction with different levels of Cl$^-$ present, as illustrated in Fig. 12. In addition, the model was able to make accurate predictions about the ATP dependence of the reaction, and the suppression of the reaction by increased osmotic strength. The predictions concerning osmotic strength suppression were

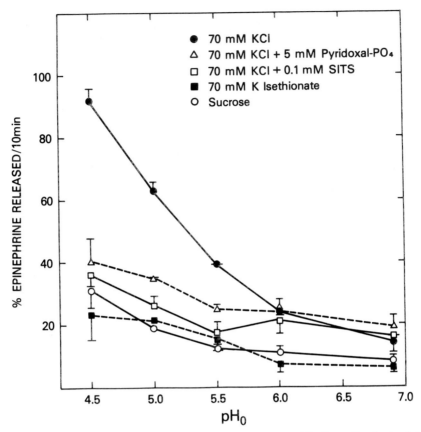

FIG. 9.   Influence of anions and anion transport-blocking drugs on $H^+$–$Cl^-$-mediated epinephrine release from chromaffin granules. The medium was buffered with 50 m$M$ MES–NaOH and brought to 335 mOsm with sucrose.

also accurate when applied to observations on exocytosis from whole cells (see subsequent sections). The success of this modeling effort provided dramatic support for the conception of the roles of ATP, anions, and osmotic lysis in the chromaffin granule release reaction and also supported the notion that the same events might be involved in the process of exocytosis from cells.

## IV.   Exocytosis from Chromaffin and Other Cells

We have considered the mechanism of chemiosmotic lysis of isolated chromaffin granules, described in detail in the preceding section, to be a possible model for the release event during exocytosis (Pollard *et al.*, 1979a). In fact, this was our primary reason for characterizing this process so extensively over the last few years, and we anticipated being able to test our hypothesis with chromaffin and other cell types if we understood the chemistry of granule lysis in some detail.

As a possible scenario of secretion from the viewpoint of the chromaffin granule, we imagined that the average granule within the chromaffin cell might be immersed in ~1 m$M$ Mg$^{2+}$-ATP, so that the proton pump might be functioning at all times. Should the local calcium concentration rise, the granule might become closely juxtaposed to the plasma membrane, thereby bringing it into proximity with the greatly elevated chloride concentration outside the cell. If the anion transport site on the granule membrane were in some way introduced into the membrane region separating granule interior and external milieu, then net H$^+$

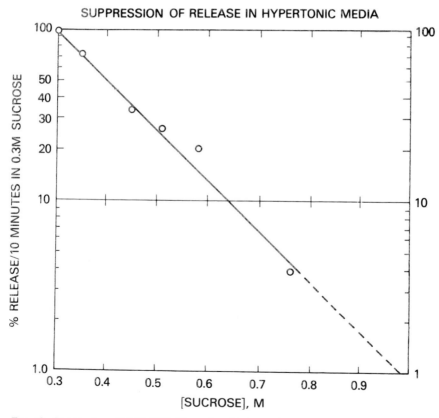

**SUPPRESSION OF RELEASE IN HYPERTONIC MEDIA**

FIG. 10. Suppression of Mg$^{2+}$-ATP-Cl$^-$-mediated epinephrine release from chromaffin granules by elevation of the extragranular osmotic strength.

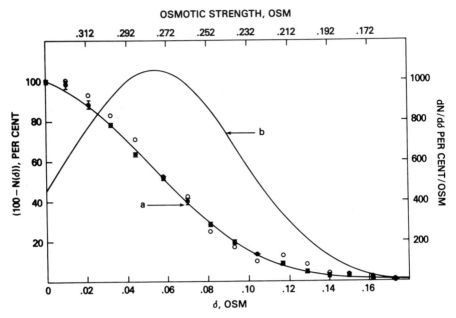

FIG. 11.   Osmotic fragility curve of chromaffin granules prepared in 0.3 *M* sucrose. Curve a is
$100 - N(\delta)$, the normalized turbidity of the suspension in percent, where 100% corresponds to the
turbidity of 0.332 Osm medium (A540 $\simeq$ 0.09). The curve is plotted as a function of the total
osmolarity of the incubation medium, or of $\delta$, the deviation from 0.332 Osm. The points are
measurements of turbidity in duplicate $\pm$ S.D.; the open circles are the percentage of total epinephrine remaining in the granules; the line is 100 − the integral of curve b. Curve b is $dN/d\delta$, a Gaussian
of the form $Ce^{-(\delta-A)^2/B}$, with A, B, and C chosen so that the integral data are $A = 0.0540$ Osm;
$B = 3.51 \times 10^{-3}$ Osm$^2$; and $C = 1052\%/$Osm. This curve is referred to as the osmotic fragility
curve (Creutz and Pollard, 1980).

and Cl$^-$ transport could ensue, and osmotic rupture, as in the *in vitro* granule
system, might occur.

If this scenario were correct, then several experimental predictions could be
made for chromaffin cells. For example, increasing the osmotic strength of the
medium might suppress exocytosis. Perhaps replacing chloride by an impermeant anion such as isethionate or adding an anion transport blocker might
suppress exocytosis. Finally, drugs that interfere with ATP synthesis or deplete
proton gradients might be expected to suppress exocytosis.

To test these predictions we prepared chromaffin cells by treating adrenal
medullary tissue with collagenase, by a variation of the method described by
Hochman and Perlman (1976) and others (Fenwick *et al.*, 1978; Schneider *et
al.*, 1977; Brooks, 1977). These cells, in our hands, were found to secrete
epinephrine and dopamine $\beta$-hydroxylase when incubated with cholinergic

agonists (acetylcholine or carbachol), the calcium ionophore A23187, or with veratridine, a drug that activates action potential sodium channels. Tetrodotoxin, a drug that specifically blocks veratridine action on nerves, was also found to block epinephrine secretion by veratridine. This cell preparation thus appeared to be secreting as expected, and we proceeded to test our chemiosmotic hypothesis.

As expected, progressive elevation of the osmotic strength of the medium with either NaCl or sucrose led to suppression of epinephrine secretion induced with

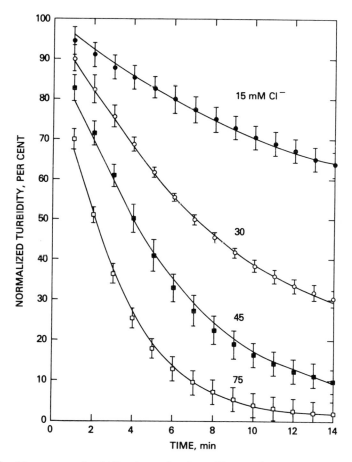

FIG. 12.   Time course of turbidity change for chromaffin granule suspensions undergoing the MgATP- and Cl⁻-dependent release reaction at 37°C with different levels of Cl⁻. The turbidity axis is in percent, with 100% corresponding to the initial turbidity of A540 ≃ 0.3 and 0% corresponding to the turbidity at 15 minutes with 90 m$M$ Cl⁻. The MgATP concentration is 5 m$M$. The error bars represent the standard deviation for triplicate determinations. The solid lines represent the best simultaneous fit to the data of the model for the release reaction described in detail in Creutz and Pollard (1980).

either veratridine (75 $\mu M$) or acetylcholine (500 $\mu M$). The suppression curve was essentially very similar to that observed with isolated chromaffin granules in the presence of ATP and chloride (cf. Fig. 10). A23187 could not be studied under hypertonic conditions because it induced virtually complete cell lysis above 450 mOsM.

We also studied the relative levels of epinephrine release when different monovalent anions were substituted for chloride in the medium. When both veratridine and A23187 were used as secretagogues, NaBr and NaI were found to support higher levels of epinephrine secretion than NaCl, whole Na acetate supported a lower level and Na isethionate essentially supported very little release (see Table II). This relationship was quite similar to that observed for relative rates of ATP-driven lysis of isolated chromaffin granules, and we concluded that the predictions about anions made on the basis of chromaffin granule studies also appeared to be borne out in studied on cells.

Finally, we examined the influences of various drugs affecting both anion and proton transport in granules on secretion by intact cells. Proton ionophores such as FCCP (carbonylcyanide-$p$-trifluoromethoxyphenylhydrazone) and TTFB (4,5,6,7-tetrachloro-$\alpha$-trifluoromethylbenzimidazole) inhibited veratridine-activated release by almost 90% at a 5 $\mu M$ concentration, and the dose-response curves were quite similar to that observed for inhibition of chromaffin granule lysis. However, profound inhibition was observed with other metabolic inhibitors that did not affect the granule system. For example, these included rotenone, $CN^-$, and azide, which affect mitochondrial function, and 2-deoxyglucose, which affects glucose metabolism. We concluded that ATP production was probably important for secretion, but that a specific interpretation in terms of a direct action of ATP on the fused granule prior to secretion seemed necessarily precluded. A wide variety of anion transport inhibitors were also tested; stilbene disulfonates (such as SITS and DIDS) were inactive, whereas probenecid and pyridoxal phosphate inhibited secretion at higher concentrations.

TABLE II

INFLUENCE OF DIFFERENT ANIONS ON RELEASE OF
EPINEPHRINE FROM CHROMAFFIN CELLS

| | Release relative to that in chloride media | |
| --- | --- | --- |
| $Na^+$-anion$^-$ | Veratridine | A23187 |
| $Br^-$ | 178 | 311 |
| $I^-$ | 158 | 279 |
| $Cl^-$ | 100 | 100 |
| Acetate | 78 | 83 |
| Isethionate | 6 | 21 |

These findings may reflect the specific geometry of granules in a fused state or, perhaps, a heterogeneity of anion transport sites, which we have begun to detect and study in isolated chromaffin granules (Pazoles *et al.*, 1980).

These studied appeared to indicate that it was possible to make predictions about secretion from chromaffin cells based on what was known about chemiosmotic lysis of isolated chromaffin granules. An important question to consider then was how general this effect might be. During and preceding our studies on chromaffin cells and granules, we did have the opportunity to study serotonin secretion from human platelets (Pollard *et al.*, 1977) and parathormone (PTH) secretion from bovine parathyroid cells (Brown *et al.*, 1978). Both cell types had general similarities to the chromaffin cell system but quite a few specific differences.

Similarities, for example, included the fact that secretion of both serotonin and PTH were inhibited by hyperosmotic media. The osmotic suppression curves for both cell types were quite similar to the curve described earlier for the chromaffin granule and the chromaffin cell. Furthermore, secretion from both cell types was inhibited by proton ionophores such as FCCP, indicating that metabolic energy was required for secretion to occur. Finally, both cell types were profoundly inhibited by anion transport inhibitors of several types (e.g., stilbene disulfonates and probenecid). However, although $Cl^-$ appeared to be the critical anion for PTH release, $OH^-$ (or $HCO_3^-$) appeared to be important for platelet secretion. A few other cell types appear to express some secretory behavior resembling the predictions of the chromaffin granule-based model. These include lysosomal enzyme secretion from the human polymorphonuclear leukocyte (Korchak *et al.*, 1980) and the dogfish blood phagocyte (Weissmann *et al.*, 1978); release of oxytosin–vasopressin from posterior pituitary nerve terminals (Poisner and Hong, 1974); and release of clotting protein from *Limulus* amebocytes (Armstrong and Rickles, 1980).

The main conclusion from studies on the chromaffin system was that when secretion occurred the chromaffin granule appeared to confer some of its chemical properties on the whole cell. This seems plausible when one considers that when secretion occurs the secretory granule becomes closely juxtaposed with the plasma membrane. Whether this is true of other cells such as the parathyroid or platelet will depend on the results of future studies on the homologous secretory granules. At present, the work on these other cells, although suggestive, is still only phenomenological.

## V.   Conclusion

One of the problems confronting those studying exocytosis from a biochemical viewpoint is the "black-box" character of the various biological preparations available for study. The cell is observed to secrete material known to be otherwise stored in secretory vesicles, and electron microscopic views consistent with fusion and fission of secretory vesicles and plasma membranes are regularly captured. But our understanding of secretion based on such studies has not been profound enough to permit us to devise successful "secretion in a test tube" experiments.

Indeed, many who have had the good fortune to study the convenient chromaffin system have mixed granules, plasma membranes, and calcium together, hoping for "secretion in a test tube." But the event has proved depressingly difficult to detect on any regular basis. Attempts to add calcium-dependent enzymes or proteins such as actomyosin–troponin or calmodulin or even tubulin have not met with success. Furthermore, these particular proteins do not affect or induce membrane fusion even in the most complex model systems.

It is for these reasons that we chose to pay so much attention to synexin as a possible mediator of membrane fusion during exocytosis, and have seriously considered the chemiosmotic lysis of chromaffin granules as a possible mechanism for fission during exocytosis. Synexin was in fact discovered while we were looking for something that would support aggregation of granules in the presence of calcium. The formation of "pentalaminar" fusion complexes especially in fairly low ($\sim$5 $\mu M$) calcium concentrations but concomitantly in a veritable sea of free magnesium (1 m$M$ or more) convinced us that the protein had properties to be expected of an intracellular calcium-sensitive mediator of membrane interaction. Only time and more experiments will reveal whether synexin is only an interesting biological byway or a main road toward a fundamentally important principle in cell biology.

The chemiosmotic lysis reaction of chromaffin granules seems to us an intrinsically interesting model for the fission step in exocytosis for a number of reasons. Synexin itself does not cause granule fission or breakage *in vitro,* and no other protein or membrane preparation seems to do so either. It also seems to us unlikely that Nature would go to such immense troubles to construct an otherwise irrelevant chemiosmotic lysis mechanism for granules whose primary role in life, in any event, is to eventually break. Finally, the evidence from microscopy indicates that when secretion occurs the granule could indeed come into close association with the high-chloride extracellular medium, and that it might there confer its chemiosmotic properties on a localized portion of the cell surface. The evidence at hand testing this hypothesis on intact chromaffin cells seems to lend positive support to the concept that the fission process may indeed be based on directed osmotic lysis of the fused granule. Very recently studies have been reported by Zimmerberg *et al.* (1980) and by Cohen *et al.* (1980) from Finkelstein's laboratory showing that fluorescein-filled multilamellar liposomes could fuse with and transfer liposome contents across a black lipid membrane, provided an osmotic gradient existed across the black lipid membrane. It is possible that this physical system may model aspects of the chemiosmotic hypothesis for exocytosis described for chromaffin and other cells in this article.

## REFERENCES

Armstrong, P. B., and Rickles, F. R. (1980). *J. Cell Biol.* **87**, 295a.

Brooks, J. C. (1977). *Endocrinology* **101**, 1360–1378.

Brown, E. M., Pazoles, C. J., Creutz, C. E., Aurbach, G. D., and Pollard, H. B. (1978). *Proc. Natl. Acad. Sci. U.S.A.* **75**, 876–880.

Cabantchik, Z. I., and Rothstein, A. (1974). *J. Membr. Biol.* **15**, 207–226.

Casey, R. P., Njus, D., Radda, G. K., and Sehr, P. A. (1976). *Biochem. J.* **158**, 583–588.

Casey, R. P., Njus, D., Radda, G. K., and Sehr, P. A. (1977). *Biochemistry* **16**, 972–978.

Cohen, F. S., Zimmerberg, J., and Finkelstein, A. (1980). *J. Gen. Physiol.* **75**, 251–270.

Creutz, C. E., and Pollard, H. B. (1980). *Biophys. J.* **31**, 255–270.

Creutz, C. E., Pazoles, C. J., and Pollard, H. B. (1978). *J. Biol. Chem.* **253**, 2858–2866.

Creutz, C. E., Pazoles, C. J., and Pollard, H. B. (1979). *J. Biol. Chem.* **254**, 553–558.

Creutz, C. E., Pazoles, C. J., and Pollard, H. B. (1980). In "Catecholamines: Basic and Clinical Frontiers" (I. Kopin and E. Usdin, eds.), Vol. 1, pp. 346–349. Pergamon, Oxford.

Fenwick, E. M., Fajdiga, P. B., Howe, N. B. S., and Livett, B. G. (1978). *J. Cell Biol.* **76**, 12–30.

Hochman, J., and Perlman, R. L. (1976). *Biochim. Biophys. Acta* **421**, 168–175.

Johnson, R. G., and Scarpa, A. (1976). *J. Biol. Chem.* **241**, 2189–2191.

Korchak, H. M., Eisenstat, B. A., Hoffstein, S. T., and Weissmann, G. (1980). *Proc. Natl. Acad. Sci. U.S.A.* (in press).

Llinas, R., Steinberg, I. Z., and Walton, K. (1976). *Proc. Natl. Acad. Sci. U.S.A.* **73**, 2918–2922.

Njus, D., and Radda, G. K. (1978). *Biochim. Biophys. Acta* **463**, 219–244.

Njus, D., Sehr, P. A., Radda, G. K., Ritchie, G. A., and Seeley, P. J. (1978). *Biochemistry,* **17**, 4337–4343.

Oka, M., Ohuchi, T., Yoshida, H., and Imaizumi, R. (1965). *Biochim. Biophys. Acta* **97**, 170–171.

Palade, G. E. (1975). *Science* **189**, 347–357.

Parsegian, V. A. (1977). *Soc. Neurosci. Symp.* **2**, 161–194.

Pazoles, C. J., and Pollard, H. B. (1978). *J. Biol. Chem.* **253**, 3962–3969.

Pazoles, C. J., Creutz, C. E., Ramu, A., and Pollard, H. B. (1980). *J. Biol. Chem.* **255**, 7863–7869.

Poisner, A. M., and Hong, J. S. (1974). *Adv. Cytopharmacol.* **2**, 303–310.

Pollard, H. B., Zinder, O., Hoffman, P. G., and Nikodijevik, O. (1976). *J. Biol. Chem.* **251**, 4544–4550.

Pollard, H. B., Tack-Goldman, K. M., Pazoles, C. J., Creutz, C. E., and Shulman, N. R. (1977). *Proc. Natl. Acad. Sci. U.S.A.* **74**, 5295–5299.

Pollard, H. B., Pazoles, C. J., Creutz, C. E., and Zinder, O. (1979a). *Int. Rev. Cytol.* **58**, 159–197.

Pollard, H. B., Shindo, H., Creutz, C. E., Pazoles, C. J., and Cohen, J. S. (1979b). *J. Biol. Chem.* **254**, 1170–1177.

Schneider, A. S., Herz, R., and Rosenheck, K. (1977). *Proc. Natl. Acad. Sci. U.S.A.* **74**, 5036–5040.

Schober, R., Nitsch, C., Rinne, U., and Morris, S. J. (1977). *Science* **195**, 495–497.

Smith, A. D. (1968). In "The Interaction of Drugs and Subcellular Components in Animal Cells" (P. N. Campbell, ed.), pp. 239–292. Churchill, London.

Viveros, O. H. (1974). In "Handbook of Physiology" (S. R. Geiger, ed.), Sect. 7, Vol. VI, pp. 389–426. Am. Physiol. Soc., Washington, D.C.

Weissmann, G., Finkelstein, M. C., Csernonsky, J., Quigley, J. P., Quinn, R. S., Techner, L., Troll, W., and Dunham, R. B. (1978). *Proc. Natl. Acad. Sci. U.S.A.* **75,** 1825–1829.
Winkler, H. (1977). *Neuroscience* **2,** 657–683.
Zimmerberg, J., Cohen, F. S., and Finkelstein, A. (1980). *J. Gen. Physiol.* **75,** 241–250.

NOTE ADDED IN PROOF: Since this article was completed, a number of new studies by ourselves and others have been reported that bear directly on concepts discussed here. These are summarized in *Cold Spring Harbor Symp. Quant. Biol.* **47,** by Pollard *et al.*

# Chapter 21

# Studies of Isolated Secretion Granules of the Rabbit Parotid Gland: Ion-Induced Changes in Membrane Integrity

## J. DAVID CASTLE AND ANNA M. CASTLE

*Section of Cell Biology*
*Yale University School of Medicine,*
*New Haven, Connecticut*

## WAYNE L. HUBBELL

*Department of Chemistry,*
*University of California, Berkeley, California*

## I.  Introduction

The mechanism of exocytosis entails the reorganization of membrane components of the secretion granule and plasmalemma in the region of their mutual contact leading to the establishment of open communication between the storage compartment and the extracellular space. As yet there is no insight, beyond the suggestion of an induced bilayer instability, into the actual events that constitute the rearrangement process. The ability to use lysolipids as fusogenic agents in model systems has prompted attempts to obtain evidence for the involvement of phospholipid degradation (via phospholipases) in exocytosis. In spite of suggestions to the contrary (Rutten *et al.*, 1975), it remains possible that a putative phospholipase activity within cells stimulated to discharge secretion could be restricted to action in contacting regions representing only a few percent of the granule or plasma membrane surface areas. In such a case degradation of phospholipids might go unnoticed, especially if the cell provided some sort of repair (reacylation) mechanism that could mask any attempts to detect and quantitate bulk changes in intact phospholipids. The results presented below as well as the fact that isolated secretion granules from pancreas, parotid, pituitary, and adrenal medulla contain substantial quantities of lysophosphatidylcholine (Meldolesi *et al.*, 1975; Blaschko *et al.*, 1967) have raised the question of whether displacement of cellular organelles from their intracellular environment has removed restraints on normally tightly controlled phospholipid hydrolysis mechanisms. We have been prompted to reconsider the possible involvement of phospholipase activities in membrane–membrane interactions once certain technical problems concerning organelle studies *in vitro* have been overcome.

The tissue used for the present study is the parotid salivary gland of the rabbit. Its acinar cells comprise an apparently homogeneous population of secretory cells (Castle *et al.*, 1972), each containing numerous apically located secretion granules (diameter 1.0–1.2 $\mu$m) that collectively occupy $\sim$35% of the cellular volume (Bedi *et al.*, 1974). The secretion granules have been isolated according to the procedure of Castle *et al.* (1975), the purified fraction being obtained, after discontinuous sucrose gradient centrifugation, in 2 $M$ sucrose buffered with 40 m$M$ potassium phosphate and containing 2 m$M$ EDTA,[1] pH 7.2. The sucrose concentration was subsequently reduced to 0.4 $M$ by the continuous addition (for $\geqslant$ 1 hour at 4°C using a peristaltic pump) of 0.25 $M$ buffered sucrose. The diluted suspension was subjected to low-speed centrifugation at 2000 $g_{av}$ for 25 minutes, and the pelleted granules were gently resuspended for further studies in fresh 0.4 $M$ buffered sucrose. The intactness of the granule suspension at this point and in all subsequent studies was determined by measuring the amount (expressed as percentage of total) of $\alpha$-amylase, a secretory protein normally packaged in the granule content, which is sedimentable during centrifugation of $\leqslant$60 $\mu$l suspension for 1.5 minutes at 8000 $g_{av}$ (Eppendorf 3200 centrifuge). The latter conditions ensure complete granule sedimentation, and suspensions were found to be 82–92% intact at this stage.

[1]Abbreviations used: EDTA, ethylenediamine tetraacetic acid; buffered sucrose, sucrose containing 40 m$M$ potassium phosphate + 2 m$M$ EDTA (pH 7.2); PC, phosphatidylcholine; PE, phosphatidylethanolamine; PS, phosphatidylserine; LPC, lysophosphatidylcholine; LPE, lysophosphatidylethanolamine; LPS, lysophosphatidylserine; S, sphingomyelin; NL, neutral lipid; FA, fatty acid; Chol., cholesterol; MG, monoglyceride; DG, diglyceride; TG, triglyceride; UnK, unknown.

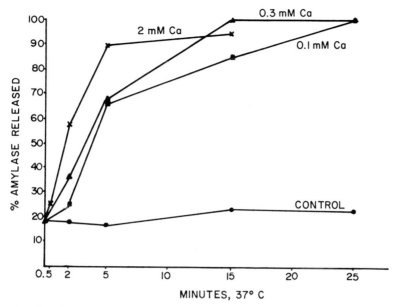

FIG. 1. Secretion granule lysis as a function of time in the presence of $Ca^{2+}$ at 37°C. Granule suspensions were incubated for 1 minute at 37°C before 0.1 volume of a stock $CaCl_2$ solution (in 0.4 $M$ sucrose) was added to give the final concentration indicated in the figure. Lysis is denoted as percentage of amylase released determined according to the centrifugation assay described in the text. (●) Control; (X) 2 m$M$ $CaCl_2$; (▲) 0.3 m$M$ $CaCl_2$; (■) 0.1 m$M$ $CaCl_2$.

## II.  Calcium-Specific Lysis of Secretion Granules

Further incubation of granule suspensions in the presence of low concentrations of calcium chloride[2] caused a rapid and extensive lysis (visible clearing and measurable loss of intactness), as shown in Fig. 1. As can be seen, the rate of amylase release at 37°C is dependent on the concentration of $CaCl_2$. Consistently we have observed lysis, using concentrations of calcium as low as 10 $\mu M$. The rate of amylase release is also temperature-dependent (Table I) and ion-specific insofar as $MgCl_2$ tested at concentrations of 2 and 10 m$M$ does not affect granule intactness; results duplicate the line marked Control in Fig. 1. The rate of granule lysis was not dependent on granule concentration over the range $1 \times 10^8 - 3 \times 10^9$ granules/ml (granule concentrations being determined by phase microscopy using a Petroff–Hausser bacteria counter).

[2]The stated $CaCl_2$ concentrations are anticipated to represent upper limits to the actual $Ca^{2+}$ activities, especially since the granules are suspended in phosphate-containing media.

TABLE I

SECRETION GRANULE LYSIS AS A FUNCTION OF TEMPERATURE[a]

| A. Sample | Temperature | Time of incubation | Amylase released (%) |
|---|---|---|---|
| Control | 37°C | 30 sec | 12.4 |
|  |  | 2 min | 11.7 |
|  |  | 5 min | 10.6 |
| +Ca | 20°C | 30 sec | 14.7 |
|  |  | 2 min | 15.3 |
|  |  | 5 min | 16.3 |
|  |  | 30 min | 96.7 |
| +Ca | 37°C | 30 sec | 18.3 |
|  |  | 2 min | 73.8 |
|  |  | 5 min | 78.6 |

| B. Sample | Temperature | Time of incubation | Amylase released (%) |
|---|---|---|---|
| Control | 37°C | 5 min | 15.3 |
| +Ca | 0°C | 10 min | 13.0 |
| +Ca | 20°C | 5 min | 14.9 |
| +Ca | 30°C | 5 min | 51.8 |
| +Ca | 32°C | 5 min | 85.7 |
| +Ca | 37°C | 5 min | 99.7 |

[a] The same experimental protocol was used as described in the legend of Fig. 1. In all $Ca^{2+}$-containing samples the final concentration of $CaCl_2$ was 2 m$M$.

## III.    Phospholipase $A_2$ Activity Associated with Granules

In an attempt to identify processes that could contribute to the observed $Ca^{2+}$-generated granule lysis, we tested for evidence of phospholipase activity. Granule suspensions were incubated for various times in the presence of 2 m$M$ $CaCl_2$, frozen in liquid nitrogen, and subjected to lipid extraction according to the procedure of Folch *et al.* (1957) after EDTA was added in excess of $Ca^{2+}$ to the frozen granule samples. The chloroform-rich phases were evaporated to dryness under nitrogen, redissolved in chloroform, spotted on silica gel G thin-layer plates, and chromatographed in solvent systems designed to resolve either phospholipids or neutral lipids. The developed chromatograms were briefly dried, sprayed with 50% $H_2SO_4$, and charred for 30 minutes at 180°C according to Blank *et al.* (1965). The results presented in Fig. 2 show that $Ca^{2+}$ causes a progressive, rapid degradation of granule membrane phospholipids (PC, PE, and PS) to produce the respective lysophosphatides plus free fatty acid. The activity is that of an A-type phospholipase, and substantial phospholipid degradation is

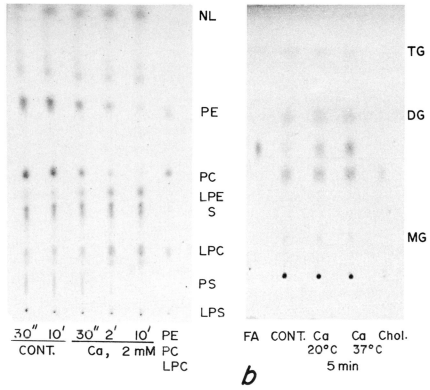

FIG. 2. (a) Thin-layer chromatogram resolving membrane phospholipids of secretion granule fractions; phospholipid composition as a function of time in the absence or presence of 2 m$M$ CaCl$_2$ at 37°C. Granule-containing samples were incubated and processed as described in the text. The samples were applied to silica gel G thin-layer plates (Woelm silica, Analtech Inc., Newark, DE) and developed first in petroleum ether–acetone [3:1 (v/v)] to clear the origin of neutral lipids and to ensure their movement to the top of the plate, and second in chloroform–methanol–acetic acid–water [50:20:3.5:3.5 (v/v)] to resolve the polar lipids (Skipski and Barclay, 1969). The identities of various species were determined according to the appropriate standards. Note the progressive disappearance of PS, PC, and PE and the appearance of LPS, LPC, and LPE during incubation in the presence of Ca$^{2+}$. (b) Thin-layer chromatogram resolving neutral lipids associated with the membranes of secretion granule fractions; composition in the presence and absence of 2 m$M$ CaCl$_2$ as a function of temperature. Lipid extracts were applied in chloroform to silica gel G thin-layer plates (Absorbosil V Prekotes, Applied Sciences, State College, PA) and developed sequentially in ethyl ether–benzene–ethanol–acetic acid [40:50:2:0.2 (v/v)] and hexane–ethyl ether [94:6 (v/v)], according to Skipski and Barclay (1969). Phospholipids remain at the origin, and the identities of neutral lipids determined with standards are indicated in the figure. Note the increased quantity of fatty acid at either 20°C or 37°C in the presence of 2 m$M$ CaCl$_2$.

evident when stimulated granule lysis is barely, if at all, detectable (compare 30 seconds at 37°C, Figs. 1 and 2a, and 5 minutes at 20°C, Table I and Fig. 2b).

In studies intended to identify the source of the phospholipase we found that most of the activity was present in the soluble content extract and not in the membrane subfraction prepared from hypotonic lysates of secretion granules. Furthermore, enzyme activity was found in the incubation medium (as a discharged secretory product) when lobules of parotid tissue were incubated *in vitro* in the presence of the $\beta$-adrenergic agonist, isoproterenol. We therefore feel that the evidence supports the notion that the bulk of the activity is located in the granule content.

We have used the isoproterenol-induced secretion to study some of the properties of the enzyme when incubated in the presence of exogenous phospholipid (sonicated PC vesicles). Enzyme activity requires $Ca^{2+}$; EDTA added in excess of $Ca^{2+}$ immediately and completely prevents further degradation of PC. The cleavage specificity is for fatty acid esterified at the glycerol $\beta$-position as demonstrated by the generation of labeled fatty acid and unlabeled $\alpha$-lysophosphatide using as substrate PC specifically labeled in the $\beta$-position fatty acid [synthesized by one of us—W. L. H. (Hubbell and McConnell, 1971)]. Thus, the activity is characteristic of a phospholipase $A_2$.

## IV.   Activation of Endogenous Phospholipid Hydrolysis

As was already indicated in the comparative examination of Figs. 1 and 2, the $Ca^{2+}$-stimulated hydrolysis of endogenous, granule membrane phospholipids considerably precedes granule lysis. This observation raises a number of fundamental questions: Is the secretory phospholipase $A_2$ the enzyme that is responsible for the observed phospholipid degradation? If so, where is the enzyme located that is being activated upon exposure of isolated granules to exogenous $Ca^{2+}$? Is it still within the secretion granule, or is it among the 10% external secretory protein (by virtue of granule breakage) of a typically 90% intact granule preparation? What is the mechanism by which exogenous $Ca^{2+}$ is activating this $Ca^{2+}$-requiring enzyme? These questions remain unanswered; however, the experiments discussed subsequently have served to identify some interesting leads for further consideration as well as to reveal some major limitations in attempting to study isolated granules that are not completely intact.

In considering the possibility that enzyme that is external to the intact granules in the suspension is responsible for the observed degradation, we have supplemented the incubation medium with excess discharged enzyme to test whether the rate of degradation of membrane phospholipids is increased. The added enzyme was without effect (data not shown), suggesting that either the extragranular phospholipase $A_2$ is not active toward the granule membranes or that extragranular phospholipase $A_2$ is mostly responsible for the degradation of membrane phospholipids and the rate of hydrolysis is already maximal before supplementation with external enzyme.

In a second experiment granules were initially incubated in the presence of $Ca^{2+}$, and at varying times (10 seconds to 5 minutes) after the onset of incubation. $Ca^{2+}$ was "removed" by adding excess EDTA to test whether the continued presence of $Ca^{2+}$ is required to support phospholipid hydrolysis. In all cases addition of EDTA caused an immediate cessation of hydrolysis. The most straightforward interpretation is that degradation of endogenous phospholipids is catalyzed by external enzyme (either the secretory phospholipase liberated from broken granules, or a different membrane-associated enzyme) and that the effect of EDTA is to chelate the $Ca^{2+}$ required directly for enzymatic activity. The experiment does not rule out a more elaborate mechanism in which extragranular $Ca^{2+}$, through some specific interaction with the cytoplasmic aspect of the granule membrane, indirectly activates intragranular phospholipase $A_2$, the $Ca^{2+}$ requirement of the latter being satisfied by the $Ca^{2+}$ known to be located within the granules (Flashner and Schramm, 1977; J. D. Castle, unpublished results).

We have recently purified the secretory phospholipase $A_2$ from both the incubation medium of isoproterenol-stimulated lobules and hypotonic lysates of secretion granules. Clearly it will be of interest to determine the amount of enzyme present within each secretion granule and thereby to estimate the amount of enzyme anticipated to be external in a preparation that is only $\sim$90% intact. Once we know this amount relative to the amount of endogenous phospholipid and have tested the capability of the enzyme to degrade other cellular membranes, we can realistically assess the magnitude to which extragranular secretory phospholipase $A_2$ can potentially participate in the degradation of the membranes of intact granules in the presence of $Ca^{2+}$ in vitro. Further, we should be able to evaluate whether it is worthwhile to use an antibody directed against this phospholipase as a specific and quantitative inhibitor of extragranular enzyme.

In the final series of experiments discussed, we have examined other divalent cations (and also $La^{3+}$) for their ability to serve as substitutes for $Ca^{2+}$ in activating the enzyme present in discharged secretion to hydrolyze exogenous PC and in stimulating degradation of the endogenous phospholipids of secretion granule fractions. Secretion of isoproterenol-stimulated lobules was subjected to gel filtration on Biogel P-6 in the presence of 40 m$M$ Tris buffer + 1 m$M$ EDTA, pH 7.2. The column void volume containing secretory proteins (including phospholipase $A_2$) free of detectable divalent cations was incubated with PC dispersions in the presence of diethylether (Wells and Hanahan, 1969) and either 2 m$M$ $Mg^{2+}$, $Ca^{2+}$, $Sr^{2+}$, $Cd^{2+}$, or $La^{3+}$. Figure 3a is a thin-layer chromatogram of lipids extracted from these preparations after 12 hours of incubation. Evidently, detectable hydrolysis is observed only for $Sr^{2+}$ and $Cd^{2+}$ as $Ca^{2+}$ substitutes. We

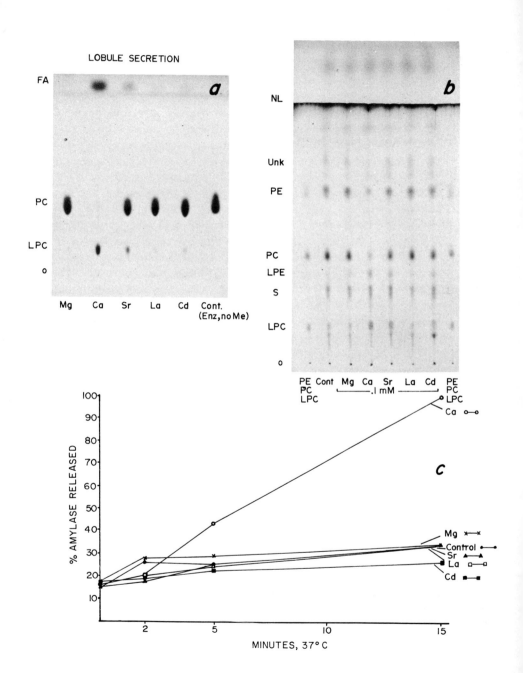

LOBULE SECRETION

emphasize, however, that both of these ions are poor substitutes for $Ca^{2+}$, since $Ca^{2+}$ facilitates the same thorough hydrolysis of PC when the incubation time is reduced to 2 hours. Therefore, activation of secretory phospholipase $A_2$ is highly $Ca^{2+}$-specific. As indicated in Fig. 3c, the lysis of secretion granules shows the same high ion selectivity; at divalent cation concentrations of either 0.1 m$M$ or 2 m$M$, it is catalyzed to a level detectable over control only by $Ca^{2+}$.

Figure 3b shows the unexpected result that 0.1 m$M$ $Cd^{2+}$ or $Sr^{2+}$ is able to stimulate phospholipid hydrolysis, promoting the formation of significant quantities of both LPC and LPE during incubation with secretion granules at 37°C. It is unlikely that extragranular secretory phospholipase $A_2$ is responsible for this degradation, since in the absence of exogenous $Ca^{2+}$ this population of enzyme molecules is expected to be inactive. Furthermore, in the case of $Sr^{2+}$ one can apparently rule out a contribution of trace $Ca^{2+}$ contamination in the divalent cation stock or other solutions, since the EDTA present in the incubation mixture would preferentially (by two orders of magnitude) chelate $Ca^{2+}$ (Schwarzenbach and Flaschka, 1969). Therefore, these particular experiments appear to raise the intriguing possibilities either that there exists another granule-associated phospholipase distinct from the secretory species, which can be activated by other divalent cations, or that intragranular secretory phospholipase $A_2$ using intragranular $Ca^{2+}$ has been activated by exogenously applied $Sr^{2+}$ and $Cd^{2+}$. Clearly this type of approach must be investigated in further detail to determine whether it is legitimate to rule out the participation of secretory phospholipase that has leaked from granule preparations that are less than completely intact. Further, we intend to study other secretory tissues that do not appear to package phospholipase $A_2$ for export to see if the same enhanced phospholipid degradation can be demonstrated.

These experiments emphasize the difficulties that can be encountered when one is attempting to carry out functional studies using isolated cellular organelles under nonphysiological conditions. In the present case involving secretion granules 1 $\mu$m in diameter, even 1% loss of intactness means the externalization of secretory protein probably equivalent in quantity to the amount of protein

---

FIG. 3. (a) Thin-layer chromatogram of extracts of PC incubated as described in the text in the presence of isoproterenol-induced parotid secretion (initially freed of $Ca^{2+}$) and the specified metal ion. Trace quantities of LPC are observed with $La^{3+}$ or $Cd^{2+}$; hydrolysis is more evident with $Sr^{2+}$ (see both LPC and FA); degradation is practically complete for $Ca^{2+}$. (b) Thin-layer chromatogram of polar lipids extracted from secretion granule fractions after their incubation for 5 minutes at 37° in the presence of the specified metal ion at a concentration of 0.1 m$M$. Note the enhanced intensity of the LPC and LPE spots for samples incubated with $Ca^{2+}$, $Sr^{2+}$, and $Cd^{2+}$. The spot of lower mobility than LPC, especially prominent in the sample incubated with $Cd^{2+}$, is not identified and is not present consistently from experiment to experiment. Both chromatograms 3a and 3b were developed as in Fig. 2a. (c) Secretion granule lysis as a function of time of incubation at 37°C in the presence of 0.1 m$M$ of the specified metal ion. Only $Ca^{2+}$ causes release significantly above that of control.

normally found on the outer aspect of the granule membrane. Clearly the utility of *in vitro* studies as functional analogs of *in vivo* events depends on improving the ability to maintain organelle intactness.

Despite the limitations realized with our present preparations, we feel that experiments carried out with $Sr^{2+}$ and $Cd^{2+}$ suggest that it may be worthwhile to pursue further the notion that $Ca^{2+}$-activated phospholipid hydrolysis might have some role in the events that couple stimulus to secretion in exocrine secretory systems.

### ACKNOWLEDGMENTS

These studies were begun when J. D. C. was a postdoctoral fellow of the Cystic Fibrosis Foundation. Subsequently this project was supported in part by: a grant from the Research Corporation; BRSG Grant RR5358 awarded by the Biomedical Research Support Grant Program, Division of Research Resources, National Institutes of Health; American Cancer Society Grant ACS-IN-31-S-2 funded through the Yale Comprehensive Cancer Center; and Grant GM 26524 from the National Institutes of Health.

### REFERENCES

Bedi, K. S., Cope, G. H., and Williams, M. A. (1974). *Arch. Oral Biol.* **19**, 1127–1133.

Blank, M. L., Schmitt, J. A., and Privett, O. S. (1965). *J. Am. Oil Chem. Soc.* **41**, 371–376.

Blaschko, H., Firemark, H., Smith, A. D., and Winkler, H. (1967). *Biochem. J.* **104**, 545–549.

Castle, J. D., Jamieson, J. D., and Palade, G. E. (1972). *J. Cell Biol.* **53**, 290–311.

Castle, J. E., Jamieson, J. D., and Palade, G. E. (1975). *J. Cell Biol.* **64**, 182–210.

Flashner, Y., and Schramm, M. (1977). *J. Cell Biol.* **74**, 789–793.

Folch, J., Lees, M., and Sloane-Stanley, G. H. (1957). *J. Biol. Chem.* **226**, 497–509.

Hubbell, W. L., and McConnell, H. M. (1971). *J. Am. Chem. Soc.* **93**, 314–326.

Meldolesi, J., DeCamilli, P., and Peluchetti, D. (1975). *In* "Secretory Mechanisms of Exocrine Glands" (N. A. Thorn and O. H. Peterson, eds.), pp. 137–148. Academic Press, New York.

Rutten, W. J., DePont, J. H. H. M., Bonting, S. L., and Daemen, F. J. M. (1975). *Eur. J. Biochem.* **54**, 259–265.

Schwarzenbach, G., and Flaschka, H. (1969). "Compleximetric Titrations." Methuen, London.

Skipski, V. P., and Barclay, M. (1969). *In* "Methods in Enzymology" (S. P. Colowick and N. O. Kaplan, eds.), Vol. 14, pp. 530–599. Academic Press, New York.

Wells, M. A., and Hanahan, D. J. (1969). *In* "Methods in Enzymology" (S. P. Colowick and N. O. Kaplan, eds.), Vol. 14, pp. 178–184. Academic Press, New York.

# Chapter 22

# Analysis of the Secretory Process in the Exocrine Pancreas by Two-Dimensional Isoelectric Focusing/Sodium Dodecyl Sulfate Gel Electrophoresis

## GEORGE A. SCHEELE

*Cell Biology Department,*
*The Rockefeller University,*
*New York, New York*

## I.  Introduction

Two-dimensional separation of proteins in slab gels using isoelectric focusing (IEF) in the first dimension and sodium dodecyl sulfate (SDS) gel electrophoresis in the second dimension was introduced by Scheele (1975) for the analysis of complex mixtures of soluble proteins, such as those secreted by the exocrine pancreas (Scheele, 1976). Optimal separation of proteins is achieved with this procedure, since proteins are separated in the first dimension according to their isoelectric point and in the second dimension according to their molecular size. In association with cell fractionation techniques, the two-dimensional gel procedure allows a thorough analysis of the movement of individual secretory proteins through the cell contents and their eventual discharge into the extracellular medium. It allows study of mRNA-directed protein synthesis and posttranslational processing of proteins along the secretory pathway. In association with two-dimensional gel scanning, it allows quantitation of the discharge of individual secretory proteins into the extracellular medium and provides important information regarding the character of the discharge event, whether parallel or nonparallel. This article will emphasize the methodologies used in the study of these events and will summarize, in several instances, the information recently obtained from such studies.

## II. Secretory Product

Figure 1 shows the Coomassie Blue staining pattern of exocrine pancreatic proteins secreted *in vitro* by guinea pig pancreatic lobules. Twenty distinct two-dimensional spots are visualized. Nineteen represent authentic secretory proteins, and one, spot number 19, represents BSA added to the incubation medium. As can be readily seen, proteins are separated in two dimensions to a greater extent than in either one dimension. With this technique, each of the pancreatic proteins has been characterized by isoelectric point, molecular weight, and mass proportion using a mixture of fifteen $^{14}$C-amino acids (Table I). Identifications of proteins in the second-dimension gel were made according to the positions of actual and potential enzyme activities in the first-dimension gel and considerations of molecular weight based on commercially available pancreatic proteins, in the second-dimension gel (Scheele, 1975). Not only are all the major pancreatic hydrolytic enzymes represented, but many are represented in multiple forms, as is the case for chymotrypsinogen, proelastase, procarboxypeptidase A, and lipase.

Similar studies have now been carried out on exocrine pancreatic proteins obtained from other species including the rat, rabbit, and dog (unpublished observations) and man (Scheele *et al.,* 1980). Similar findings of multiple forms of enzymes and zymogens have been encountered. The stability of pancreatic proteins during two-dimensional gel analysis, however, has varied, owing to the potential of these proteins for autoactivation. Analysis of secretory proteins from the rat, rabbit, and dog required the inclusion of soybean trypsin inhibitor (2 $\mu$g/ml) and Trasylol (10 KIU/ml) into the isoelectric focusing gel to achieve optimal results. Exocrine proteins from the human pancreas showed an even greater tendency to autoactivate during the two-dimensional gel procedure, and addition of trypsin inhibitors into the isoelectric focusing gel was inadequate to maintain the integrity of these proteins. The addition of 4–6 $M$ urea to the protein sample and 8 $M$ urea to the isoelectric focusing gel, in addition to the trypsin inhibitors, inhibited this autoactivation process and allowed the analysis of

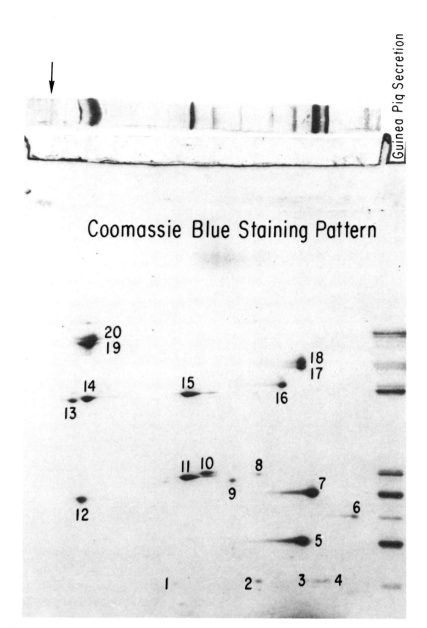

FIG. 1. Separation of guinea pig exocrine pancreatic proteins by two-dimensional IEF/SDS gel electrophoresis. One-dimensional separation of proteins by isoelectric focusing (horizontal strip, above) or gradient polyacrylamide gel electrophoresis in sodium dodecyl sulfate (vertical strip, right) are shown for comparison. Table I gives the characteristics of each numbered spot according to enzyme or potential enzyme activity, isoelectric point, molecular weight, and mass proportion. Stain in lower right corner of two-dimensional gel represents basic ampholytes that were not cleared from the gel during electrophoresis. The position of sample application to the isoelectric focusing gel is given by the arrow at the top. Taken from Scheele (1975).

TABLE I

CHARACTERISTICS OF GUINEA PIG EXOCRINE PANCREATIC PROTEINS

| Spot number | Protein | Isoelectric point | Molecular weight | Mass proportion (%) |
|---|---|---|---|---|
| 1 | — | 6.4 | 13,000 | 0.2 |
| 2 | — | 7.8 | 13,000 | 0.3 |
| 3 | Ribonuclease (RNase) [EC 2.7.7.16] | 8.7 | 13,000 | 0.7 |
| 4 | | 8.8 | 13,600 | 0.4 |
| 5 | Trypsin(ogen) [EC 3.4.4.4] | 8.7 | 24,400 | 33.0 |
| 6 | — | >9.3 | 24,900 | 2.0 |
| 7 | Chymotrypsin(ogen) 2 [EC 3.4.4.5] | 8.7 | 25,850 | 16.4 |
| 8 | — | 7.8 | — | 0.7 |
| 9 | — | 7.6 | — | — |
| 10 | (Pro)elastase 2 [EC 3.4.4.7] | 7.5 | 28,000 | 8.0 |
| 11 | (Pro)elastase 1 [EC 3.4.4.7] | 6.9 | 28,000 | |
| 12 | Chymotrypsin(ogen) 1 [EC 3.4.4.5] | 4.8 | 25,850 | 1.7 |
| 13 | (Pro)carboxypeptidase A1 [EC 3.4.2.1] | 4.6 | 45,000 | 3.5 |
| 14 | (Pro)carboxypeptidase A2 [EC 3.4.2.1] | 4.8 | 45,000 | 8.8 |
| 15 | (Pro)carboxypeptidase B [EC 3.4.2.2] | 6.6 | 47,700 | 8.8 |
| 16 | Lipase (glycerol ester hydrolase) 2 [EC 3.1.1.3] | 8.1 | 49,700 | 3.4 |
| 17 | $\alpha$-Amylase ]Ec 3.2.1.1] | 8.4 | 52,000 | 3.6 |
| 18 | $\alpha$-Amylase [EC 3.2.1.1] | 8.4 | 53,000 | |
| 20 | Lipase (glycerol ester hydrolase) 1 [EC 3.1.1.3] | 5.0 | 66,000 | 8.5 |

*Note:* Multiple forms of enzymes and zymogens are numbered consecutively from anode to cathode in accordance with the IUPAC-IUB Commission on the biochemical nomenclature of multiple forms of enzymes (1978).

human exocrine pancreatic proteins (Bieger and Scheele, 1980). Despite the addition of urea to the isoelectric focusing gel, we were able to identify the majority of proteins after their elution from isoelectric focusing gel fractions. Reduction of the urea concentration to less than $0.5\ M$ in medium optimal for the renaturation of pancreatic secretory proteins allowed reactivation of all proteins tested.

Studies on human exocrine pancreatic proteins also led us to develop a procedure for the direct identification of proteins according to biological activity after their separation by two-dimensional IEF/SDS gel electrophoresis (Scheele and Bieger, 1980). This procedure involved the separation of proteins by the two-dimensional gel technique and the location of protein spots by 0.03% Coomassie Blue G staining in the presence of 100 m$M$ Tris HCl, pH 9.0, and 5% glycerol. Chemical fixatives including methanol and acetic acid were not used. Proteins in the stained spots were simultaneously eluted from the gel matrix and renatured in the appropriate buffer containing 1% Triton X-100. Direct removal and analysis of proteins contained in stained spots on the second-dimension gel confirmed the identities of all human pancreatic proteins separated by the two-dimensional gel procedure. To date, the most thoroughly described systems analyzed by two-dimensional IEF/SDS gel electrophoresis are those comprising guinea pig and human exocrine pancreatic proteins.


## III.   Secretory Pathway


The two-dimensional gel procedure has already been applied to the analysis of secretory proteins contained in tissue homogenates and cell fractions as part of a study to determine the intracellular transport of pulse-labeled proteins from one intracellular compartment to another along the secretory pathway (Scheele *et al.*, 1978). Previous pulse–chase studies in which the intracellular transport of [3H]leucine-labeled proteins was monitored by cell fractionation techniques (Jamieson and Palade, 1967a) had confirmed the findings of tissue radioautography (Jamieson and Palade, 1967b) in the sense that pulse-labeled proteins were first associated with rough microsomes (zero chase), shortly thereafter (about 12 minutes of chase) with the small peripheral vesicles of the Golgi complex, and later (37 and 57 minutes of chase) with the condensing vacuoles and zymogen granules, which cosediment into a single fraction. However, the kinetics of movement of pulse-labeled proteins from one membrane-enclosed compartment to another were neither sharp nor unambiguous in these studies and were additionally complicated by the appearance of significant quantities (about 15–20%) of pulse-labeled protein in the postmicrosomal supernatant fraction at all times. Although others had interpreted these findings as indicating that secretory proteins pass through the cytoplasmic space during their transport through the cell contents under physiological conditions (see Rothman, 1975), the studies of Scheele and associates (1978) have indicated that the ambiguities seen in these earlier studies were due to leakage and adsorption artifacts associated with tissue homogenization and cell fractionation. Their investigation determined the extent to which 3H-labeled secretory proteins, when added to the sucrose homogenizing medium, were absorbed to membrane surfaces of intracellular organelles, and, by employing guinea pig pancreatic lobules pulse-labeled with 14C-amino acids, they were able to calculate (1) the extent of leakage of secretory proteins from individual membrane-enclosed compartments along the secretory pathway and (2) the extent of redistribution of leaked molecules to membrane surfaces by nonspecific adsorption. Analysis of 3H and 14C-labeled secretory proteins in each of the cell fractions by two-dimensional IEF/SDS gel electrophoresis indicated that the appearance of exportable proteins in the postmicrosomal supernatant could be accounted for by the uniform leakage of proteins during tissue homogenization and the preferential removal of positively charged molecules by adsorption to negatively charged surfaces of membrane-bound organelles. Correction of the data for leakage and adsorption artifacts resulted in intracellular transport kinetics that were sharply defined and unambiguous, indicating, for the first time in a precise manner by cell fractionation studies, that secretory proteins are transported successively from the RER to the Golgi complex and from there to zymogen granules.

## IV.  Posttranslational Processing and Identification of Gene-Specific Products

Since multiple forms of enzymes and zymogens are present among guinea pig pancreatic proteins (Table I), it was important to determine which forms represent separate gene products and which represent modified forms of single gene products. Pancreatic lobules received a 10-minute pulse with [$^{35}$S]methionine followed by a varying chase period with [$^{32}$S]methionine. Chase intervals were terminated by rapid freezing in liquid nitrogen, lobules were homogenized in 1% Triton X-100 and 25 m$M$ Tris HCl, pH 9.0, and radioactive proteins contained in tissue homogenates were analyzed by fluorography (Bonner and Laskey, 1974) after their separation by two-dimensional IEF/SDS gel electrophoresis. Figure 2 shows the fluorograms obtained from one such study, which employed a 2.5-

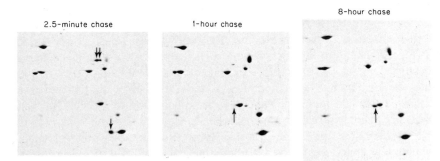

FIG. 2.    Separation of [$^{35}$S]methionine pulse-labeled proteins after varying intervals of chase. Guinea pig pancreatic lobules received a 10-minute pulse with [$^{35}$S]methionine followed by chase in [$^{32}$S]methionine for the times indicated in the figure. After termination of the chase interval by rapid freezing in liquid N$_2$, lobules were homogenized in 1% Triton X-100 and 25 m$M$ Tris HCl, pH 9.0. Proteins were then separated by the two-dimensional gel procedure, and radioactivity was analyzed by fluorography. Explanations for the downward- and upward-pointing arrows are given in the text. Two-dimensional gel patterns are oriented in a manner identical to that in Fig. 1.

minute, a 1-hour, and an 8-hour chase. Three additional radioactive proteins are observed after the 2.5-minute chase interval that are not seen after the longer chase intervals. Each of these precursor forms is marked by downward-pointing arrows. Two represent precursor forms for amylase, and one represents a precursor form for trypsinogen. None of these precursor forms were stained by Coomassie Blue. A study of intermediate chase intervals indicated the disappearance of each of these precursor forms by 10 minutes of chase. Loss of radioactivity associated with the precursor forms were recovered in the corresponding product. Precursor forms for both amylase and trypsinogen are relatively more acidic than the authentic secretory protein or modified product. At present the nature of these processing events is unclear. Should the precursor forms represent secretory proteins with uncleaved transport peptides, it may indicate that transport peptides containing acidic amino acid residues are cleaved with considerably lower efficiency than are those containing neutral or basic residues. Not only are charged residues poorly represented in transport peptides of presecretory proteins, but in such cases the charged residue has generally been a positively charged lysine or arginine residue (Scheele, 1979, 1980). Alternatively, these precursor forms may represent prosecretory forms of amylase and trypsinogen in the guinea pig pancreas. Detailed sequencing studies will be necessary to distinguish among the two possibilities.

Radioactive proteins analyzed after 1-hour and 8-hour chase intervals show the progressive appearance of a modified form of proelastase indicated by the upward-pointing arrows in Fig. 2. In this case the conversion is relatively slow, and both precursor and product forms are observed among secretory proteins using Coomassie Blue stain (Fig. 1). Proelastase 2 is the precursor form, and proelastase 1 is the modified form. These proelastase molecules, then, represent two forms of a single gene product. The nature of the processing event is unclear at present, although several possibilities should be considered. They are the removal of a positively charged peptide or the addition of a negatively charged moiety, possibly a sialic acid residue or a phosphate group. Figure 3 shows the results of a kinetic study of the conversion of proelastase 2 to proelastase 1. Fifty percent conversion occurs near the 12-hour time point. In the unstimulated gland, conversion proceeds slowly and appears to be linear with time. During carbamylcholine stimulation the conversion process is markedly accelerated. Fifteen minutes after the pulse, the modified form represented 80% of the sum of the two. Conversion of this zymgoen is more complete at 15 minutes in the stimulated gland than at 21 hours in the unstimulated gland. The effect of carbamylcholine in the processing of this secretory protein is therefore dramatic. No other precursor–product relationships were observed in these studies, indicating that the two forms each for procarboxypeptidase A, chymotrypsinogen, and lipase represent separate gene products.

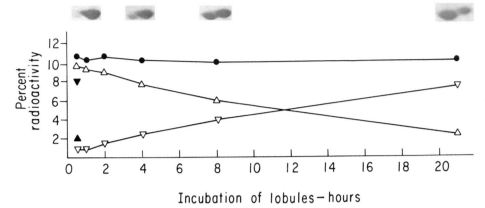

FIG. 3. Posttranslational processing of proelastase 2 to proelastase 1. Guinea pig pancreatic lobules received a 10-minute pulse with [$^{35}$S]methionine followed by a chase in [$^{32}$S]methionine for the times indicated in the figure. Proteins were separated as described in the legend to Fig. 2, and radioactivity contained in the proelastase 1 (-▽-) and proelastase 2 (-△-) spots was quantitated by liquid scintillation spectrometry as previously described (Scheele, 1975). Radioactivity contained in each form of proelastase is given as percentage of radioactivity contained in all secretory proteins. Radioactivity contained in both spots is represented by (-●-). At the top of the figure, insets of two-dimensional fluorograms showing the area of separation of the two proelastase forms indicate, in a visual manner, the extent of processing at the indicated times. After 15 minutes of carbamylcholine stimulation, [$^{35}$S]methionine incorporation into the two proelastase forms is given by (▼) for proelastase 1 and (-▲-) for proelastase 2. A fluorograph inset is not shown for carbamylcholine stimulation.

## V.  Messenger RNA-Directed Protein Synthesis: Effect of Carbamylcholine Stimulation

The complete separation of exocrine pancreatic proteins achieved by two-dimensional IEF/SDS gel electrophoresis allows the precise measurement of biosynthesis of individual secretory proteins. Ninety-five percent of the radioactivity that is incorporated into soluble pancreatic proteins appears in the two-dimensional spots characterized as secretory proteins (Scheele *et al.*, 1978). Identification of separate gene products among secretory proteins as discussed in the preceding section provides the information necessary to determine specific RNA-directed protein synthesis in intact tissue. We have made such measurements in the following manner. Guinea pig pancreatic lobules (Scheele and Palade, 1975) were incubated with [$^{35}$S]methionine for 15 minutes, and the incorporation of radioactive methionine was terminated by rapid freezing in liquid nitrogen. The tissue was then homogenized in 1% Triton X-100 and 25 m$M$ Tris, pH 9.0, and proteins contained in tissue homogenates were separated by two-dimensional IEF/SDS gel electrophoresis. Coomassie Blue stained spots characterized as secretory proteins or unstained regions of the gel representing short-lived precursor forms were then removed from the second-dimension gel and dissolved in hydrogen peroxide (Scheele, 1975), and radioactivity associated with their spots was quantitated by liquid scintillation spectrometry. Table II shows the effect of carbamylcholine stimulation on the mRNA-directed synthesis of selected guinea pig pancreatic secretory proteins. Data are given as percentage incorporation of [$^{35}$S]methionine among all secretory proteins, and values given under carbamylcholine stimulation represent the level of synthesis at the point of maximal change. Trypsinogen and lipase 2 show no change with carbamylcholine stimulation. The synthesis of the two forms of chymotrypsinogen (1 and 2), procarboxypeptidase A1, and lipase 1 increased with carbamylcholine stimulation. Procarboxypeptidase B, procarboxypeptidase A2, and amylase were decreased with stimulation. Except in the case of amylase, the changes observed were maximal within 15 minutes of carbamylcholine stimulation and remained stable for 2 hours of stimulation. The synthesis of amylase progressively decreased with time of stimulation—8.1% at 15 minutes, 7.1% at 30 minutes, and 5.4% at 1 and 2 hours. Other secretory proteins not listed showed no change in relative biosynthetic rates. The ratio of chymotrypsinogen 2 to amylase biosynthesis changed from a value of 1.3 in the control period to

TABLE II

BIOSYNTHESIS OF EXOCRINE PANCREATIC PROTEINS:
EFFECT OF CARBAMYLCHOLINE STIMULATION

| Protein | Control[a] (%) | Carbamylcholine [b] $10^{-5}$ M (%) | Change (%) |
|---|---|---|---|
| Trypsinogen | 23.7 ± 1.2 | 23.2 ± 0.8 | −2.2 |
| Lipase 2 | 8.1 ± 0.3 | 8.1 ± 0.4 | 0 |
| Chymotrypsinogen 1 | 0.76 ± 0.12 | 1.23 ± 0.29 | +62 |
| Chymotrypsinogen 2 | 14.9 ± 1.0 | 19.8 ± 0.5 | +33 |
| Procarboxypeptidase B | 8.9 ± 0.4 | 6.6 ± 0.2 | −26 |
| Procarboxypeptidase A1 | 7.9 ± 0.4 | 8.9 ± 0.3 | +13 |
| Procarboxypeptidase A2 | 7.0 ± 0.8 | 5.9 ± 0.1 | −16 |
| Lipase 1 | 9.2 ± 0.5 | 10.6 ± 0.2 | +15 |
| Amylase | 9.8 ± 0.5 | 5.4 ± 0.2[c] | −45 |

[a] [$^{35}$S]Methionine incorporation into individual proteins over 15 minutes, expressed as percentage of incorporation into all secretory proteins (mean ± S.D.).

[b] [$^{35}$S]Methionine incorporation, as above, after 15 minutes of stimulation with carbamylcholine.

[c] After 1 hour of stimulation with carbamylcholine.

3.8 in the stimulated period. This represents a 2.9-fold increase during carbamyl-choline stimulation. The ratio of chymotrypsinogen 1 to amylase biosynthesis changed similarly (2.9-fold increase) with carbamylcholine stimulation. Similar changes in the ratio of chymotrypsinogen to amylase synthesis with cholecystokinin-pancreozymin stimulation have been observed in studies of the rat pancreas by Dagorn and Mongeau (1977). Changes described here in the synthetic rates of individual secretory proteins are of considerable importance. First, they may explain the changes observed in the discharge of individual secretory proteins with time of stimulation described in the section below. In addition, they indicate that mRNA-directed synthesis of secretory proteins is modulated by carbamylcholine stimulation. Since changes in biosynthetic rates occur rapidly (within 15 minutes) after cholinergic stimulation, these changes are judged to be mediated at the level of mRNA translation.


## VI.   Discharge of Exocrine Pancreatic Proteins: Parallel vs Nonparallel


A number of studies have been conducted to determine if the distribution of secretory proteins can be changed by secretagogue stimulation. In the majority of these studies actual and potential enzyme activities were measured after *in vitro* discharge of proteins into the incubation medium (Scheele and Palade, 1975) or after *in vitro* (Rothman, 1975; Rothman and Wilking, 1978; Steer and Glazer, 1976; Steer and Manabe, 1979) or *in vivo* (Dagorn, 1978) discharge in ductal fluid. With the exception of reports from a single laboratory, secretagogues that show considerable variation in their chemical composition (pancreozymin/caerulein and cholinergic agents) were found to discharge the same mixture of enzymes and zymogens. Comparison of the distribution of secretory proteins collected during basal discharge and pancreozymin-induced discharge showed in some studies no change (Scheele and Palade, 1975; Steer and Manabe, 1979) and in other studies significant changes (Dagorn *et al.*, 1977; Rinderknecht *et al.*, 1978), particularly in the ratio of chymotrypsinogen to amylase discharge. Studies of this kind depend on a careful study of the kinetics of activation of each zymogen tested and a definition of the optimal activation conditions for each (Scheele and Palade, 1975; Glazer and Steer, 1977).

Because of the inherent limitations of the measure of zymogen levels in a mixture of potential proteases, we have sought alternative methods to quantitate the discharge of individual secretory proteins. Measurement of radioactive proteins discharged into the incubation medium and separated by either one-dimensional or two-dimensional SDS gel electrophoresis has proved to be unreliable, since the appearance of radioactive proteins in the incubation medium depends not only on the discharge process, but also on biosynthesis and intracellular transport of these molecules. We have recently developed a method for quantitating secretory proteins separated by two-dimensional IEF/SDS gel electrophoresis by using two-dimensional scanning of Coomassie Blue stained gels and computer analysis of the scanning data. Scanning was performed on photographic reproductions of stained gels with an Optronics two-dimensional gel scanner, and optical density measurements were taken at 100-$\mu$m intervals. Measurements recorded on magnetic tape were analyzed on a Digital Equipment Corp. PDP-11/70 computer. Programs for the analysis, designed by Alan Ezer and Banvir Chaudhary of Rockefeller University Computer Services, included (1) criteria for the selection of two-dimensional gel spots, (2) subtraction of background optical density values, and (3) determination of cumulative densities within and fractional densities among the selected two-dimensional spots. When background levels were low, all spots visualized by Coomassie Blue staining were included in the computer analysis of two-dimensional gel scanning data.

Table III shows the percentage distribution of secretory proteins discharged from guinea pig pancreatic lobules after stimulation with three secretagogues that differ widely in chemical composition. Proteins were separated by two-dimensional IEF/SDS gel electrophoresis and quantitated by two-dimensional gel scanning and computer analysis as described above. During the 3-hour incubation study, amylase release from the tissue was 54.7% for carbamycholine, 60.2% for caerulein, and 37.0% for KCl, compared with a basal release of 12%.

TABLE III

PERCENTAGE DISTRIBUTION OF SECRETORY PROTEINS DISCHARGED BY THREE SECRETAGOGUES;
QUANTITATION BY TWO-DIMENSIONAL GEL SCANNING OF COOMASSIE BLUE STAINED SPOTS

| | Secretagogue | | |
|---|---|---|---|
| Protein | Carbamylcholine, $10^{-5}$ $M$ | Caerulein, $10^{-9}$ $M$ | KCl, 75 m$M$ |
| Trypsinogen | 17.4% | 17.0% | 18.4% |
| Chymotrypsinogen 1 | 3.3% | 3.1% | 2.2% |
| Chymotrypsinogen 2 | 14.6% | 13.6% | 14.3% |
| Proelastase 1 + 2 | 11.2% | 10.8% | 13.1% |
| Procarboxypeptidase A1 + A2 | 14.7% | 13.9% | 13.9% |
| Procarboxypeptidase B | 14.9% | 14.9% | 13.5% |
| Lipase 1 | 9.7% | 11.0% | 10.7% |
| Lipase 2 | 3.0% | 2.8% | 2.6% |
| Amylase | 12.5% | 13.0% | 11.9% |
| Amylase release from tissue at 3 hours (basal, 12%) | 54.7% | 60.2% | 37.0% |

Distributions of individual proteins discharged over this period of time were
remarkably uniform for the three secretagogues studied.

When the discharge process was analyzed in finer detail, two instances of
nonparallel discharge emerged. The percentage of density of both forms of
proelastase changed dramatically with time of carbamylcholine stimulation.
Proelastase 2 showed a decrease and proelastase 1 showed an increase in density
with time (Fig. 4a). This dramatic instance of a change in the distribution of
secreted proteins, or nonparallel discharge, is observed between two forms of
protein derived from a single gene product. As demonstrated above (Section IV),
processing of proelastase 2 to proelastase 1 is markedly accelerated by carbamylcholine stimulation. Consequently, the nonparallel discharge of these two secretory proteins represents the modification of proelastase 2 to proelastase 1
during the process of stimulated discharge. The sum of the two forms of proelastase showed little change over this period of time.

Figure 4b shows the ratio of chymotrypsinogen 2 to amylase discharge as a
function of time by carbamylcholine stimulation in two separate experiments.
With pancreatic lobules obtained from a guinea pig that had been fasted overnight (open symbols), this ratio remained constant with time of *in vitro* stimulation with carbamylcholine. However, pancreatic lobules obtained from a guinea

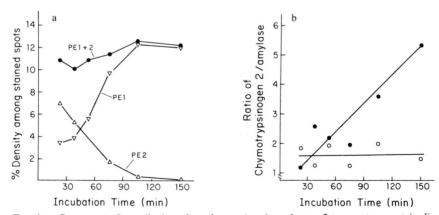

FIG. 4a.   Percentage of contribution of proelastase 1 and proelastase 2 to secretory protein discharged from pancreatic lobules with time after stimulation with $10^{-5}$ $M$ carbamylcholine. Incubation
medium was sampled at the indicated times, and secreted proteins were separated by the two-
dimensional gel procedure. Coomassie Blue R stained spots were quantitated by two-dimensional gel
scanning and computer analysis as described in the text. (b) Ratio of chymotrypsinogen 2 to amylase
in the incubation medium with time after stimulation with $10^{-5}$ $M$ carbamylcholine. Individual proteins were quantitated as given in the legend to Fig. 4a. The open symbols represent ratios derived
in an experiment in which pancreatic lobules were taken from a guinea pig fasted 24 hours. Closed
symbols represent ratios obtained when lobules were taken from a guinea pig fed ad libitum.

pig fed ad libitum (closed symbols) showed a progressive increase in the discharge of chymotrypsinogen 2 to amylase with time of stimulation. In this instance amylase and chymotrypsinogen 2 represent separate gene products.

The nonparallel discharge findings shown here and described by others (Dagorn, 1978; Dagorn et al., 1977; Rinderknecht et al., 1978) are of considerable importance because they indicate that the contents of individual zymogen granules may show quantitative variations in the distribution of secretory proteins. Further, they indicate that functional subpopulations of zymogen granules exist, with respect to their potential for discharge during the varying functional states of the gland (resting vs. stimulation). As described in Section V above, the proportion of secretory proteins synthesized in the exocrine pancreas changes with carbamylcholine stimulation. Similar changes have been observed with pancreozymin stimulation (Dagorn and Mongeau, 1977). It is likely that, under in vivo conditions, the proportions of secretory proteins synthesized in the pancreas change with time during the transition from feeding (hormone stimulation) to fasting (absence of hormone stimulation). The pool of zymogen granules from which basal discharge is derived may not be identical to that from which secretagogue-stimulated discharge is derived. Should these two pools of zymogen granules be further distinguished on the basis of granule maturity, and contain secretory proteins synthesized at different points in time (under different conditions of hormone stimulation as described above), then basal discharge would be expected to contain a different mixture of enzymes and proenzymes from those elicited during secretagogue stimulation. Subpopulations of zymogen granules may also exist in the pancreas, not in relation to time, but in relation to space. The proportional mix of digestive enzymes synthesized and packaged into secretory granules may vary from one acinar cell to another, related to the vascular supply of the tissue and therefore the availability of secretagogues and substrate molecules, or from one region of the pancreas to another, related to the proximity of islet tissue and therefore exposure to islet hormones (Malaisse-Lagae et al., 1975). One population of acinar cells might be responsible primarily for basal discharge and another for stimulated discharge. Further studies are necessary to confirm or reject these hypotheses. At the moment, most of the lines of evidence suggest a single secretory pathway for the exocrine pancreas and synchronous discharge of enzymes and proenzymes at the level of individual zymogen granules. Under conditions of steady state, the contents of zymogen granules are relatively uniform and secretory proteins are discharged in constant proportions (parallel secretion). During conditions of change in the biosynthesis and intracellular processing of secretory proteins, the contents of individual zymogen granules are varied, and, since mixing of granules is apparently incomplete, proportions of secretory proteins are changed during the discharge process (nonparallel secretion).

## Acknowledgments

I wish to acknowledge the expert technical assistance of Jessica Pash for both the biochemical experiments and the computer analyses, and to thank Dr. Thomas Carne for his careful reading of the manuscript. The research was supported in part by Grant AM 18532 from the National Institutes of Health.

## References

Bieger, W., and Scheele, G. A. (1980). *Anal. Biochem.* **109,** 222–230.
Bonner, W. M., and Laskey, R. A. (1974). *Eur. J. Biochem.* **46,** 83–88.
Dagorn, J.-C. (1978). *J. Physiol.* (*London*) **280,** 435–448.
Dagorn, J.-C., and Mongeau, R. (1977). *Biochim. Biophys. Acta* **498,** 76–82.
Dagorn, J.-C., Sahel, J., and Sarles, H. (1977). *Gastroenterology* **73,** 42–45.
Glazer, G., and Steer, M. L. (1977). *Anal. Biochem.* **77,** 130–140.
IUPAC-IUB Commission (1978). *Eur. J. Biochem.* **82,** 1–30.
Jamieson, J. D., and Palade, G. E. (1967a). *J. Cell Biol.* **34,** 577–596.
Jamieson, J. D., and Palade, G. E. (1967b). *J. Cell Biol.* **34,** 597–615.
Malaisse-Lagae, F., Ravazzola, M., Robberecht, P., Vandermeers, A., Malaisse, W. J., and Orci, L. (1975). *Science* **190,** 795–797.
Rinderknecht, H., Renner, I. G., Douglas, A. P., and Adham, N. F. (1978). *Gastroenterology* **75,** 1083–1089.
Rothman, S. S. (1975). *Science* **190,** 747–753.
Rothman, S. S., and Wilking, H. (1978). *J. Biol. Chem.* **253,** 3543–3549.
Scheele, G. A. (1975). *J. Biol. Chem.* **250,** 5375–5385.
Scheele, G. A. (1976). *In* "Cell Biology" (P. L. Altman and D. D. Katz, eds.), pp. 334–336. Fed. Am. Soc. Exp. Biol., Bethesda, Maryland.
Scheele, G. A. (1979). *Mayo Clin. Proc.* **54,** 420–427.
Scheele, G. A. (1980). *Am. J. Physiol.* (in press).
Scheele, G. A., Pash, J., and Bieger, W. (1981). *Anal. Biochem.* **112,** 304–313.
Scheele, G. A., and Palade, G. E. (1975). *J. Biol. Chem.* **250,** 2660–2670.
Scheele, G. A., Palade, G., and Tartakoff, A. (1978). *J. Cell Biol.* **78,** 110–130.
Scheele, G. A., Bartelt, D., and Bieger, W. (1981). *Gastroenterology,* **80,** 461–473.
Steer, M. L., and Glazer, G. (1976). *Am. J. Physiol.* **231,** 1860–1865.
Steer, M. L., and Manabe, T. (1979). *J. Biol. Chem.* **254,** 7228–7229.

# Chapter 23

# Inhibition of Enzyme Secretion and Autophagy of Secretory Granules Caused by Action of High Concentration of Secretory Hormones on Rat Pancreatic Slices

## NAPHTALI SAVION[1] AND ZVI SELINGER

*The Department of Biological Chemistry,*
*The Hebrew University of Jerusalem,*
*Jerusalem, Israel*

## I. Introduction

The exocrine pancreas secretes into the duodenum digestive enzymes together with water and electrolytes. Two receptors mediate enzyme secretion—one a

[1]Present address: The Lautenberg Center for General and Tumor Immunology, Hadassah Medical School, The Hebrew University of Jerusalem, Jerusalem, Israel.

muscarinic cholinergic (Hokin, 1951), the other a receptor for the peptide hormone cholecystokinin-pancreozymin (CCK-PZ) (Harper and Raper, 1943). On the other hand, secretion of water and electrolytes is controlled by secretin (Bayliss and Starling, 1902; Brown *et al.*, 1967). It has been shown that during stimulation of enzyme secretion the secretory granule fuses with the cell membrane facing the lumen. The content of the secretory granule is then emptied to the cell exterior, while the granule membrane becomes part of the cell membrane, resulting in enlargement of the lumen perimeter (Palade, 1959, 1975).

Secretagogues have a complex effect on enzyme secretion in the pancreas. Both cholinergic agonists and CCK-PZ and its peptide analogs have an optimal concentration that produces a maximal rate of enzyme secretion. At higher concentration of secretagogues, however, the rate of enzyme secretion declines, resulting in a biphasic dose-response curve. This pattern of pancreatic enzyme secretion has been demonstrated *in vivo* for both CCK-PZ (Debray *et al.*, 1963; Folsch and Wormsley, 1973; Leroi *et al.*, 1971) and acetylcholine (Hokin, 1968). By using pancreatic fragments, a biphasic dose-response curve has been found for acetylcholine (Debray *et al.*, 1963), carbamylcholine (Kulka and Sternlicht, 1968; Robberecht and Cristophe, 1971; Scheele and Palade, 1975), pilocarpine (Nevalainen and Janigan, 1974), CCK-PZ (Scheele and Palade, 1975), and its analog caerulein (Scheele and Palade, 1975; Vincent and Bauduin, 1972).

The basic assumption in the present study was that by analyzing a disturbance in the normal process of enzyme secretion we would be able to learn something about the normal process of enzyme secretion. We therefore defined the morphological and biochemical characteristics of the inhibition of enzyme secretion at high concentrations of secretagogues. It is shown that the inhibition occurs at the cellular rather than at the receptor level and that it is associated with morphological changes at the luminal membrane. Concomitantly, the high concentration of secretagogues induces the activation of the lysosomal system, which takes up secretory granules to form autophagic vacuoles and subsequently releases a myocardial depressant factor (MDF) to the incubation medium by the pancreatic slices.

## II. Does the Inhibition of Enzyme Secretion Induced by High Concentration of Secretagogues Occur at the Receptor Level?

Inhibition of pancreatic enzyme secretion at high concentrations of secretagogues in the pancreatic slice system is shown in Fig. 1. This inhibition was thought to be one aspect of desensitization of the receptor, as described by Foreman and Garland (1974) for the process of histamine secretion by mast cells.

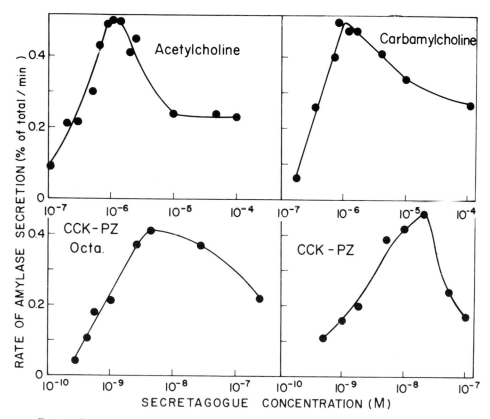

FIG. 1. Dose-response curve of amylase secretion stimulated by various secretagogues. Rat pancreatic slices were prepared by cutting the pancreas into small pieces (1-2 mm³). The slices were divided into equal portions, each equivalent to about one-third of a pancreas. Each portion of slices was placed in a 25-ml Erlenmeyer flask containing 3 ml of KRB medium supplemented with β-hydroxybutyrate (5 mM). The vessel was gassed with a mixture of 95% $O_2$ and 5% $CO_2$ and incubated at 37°C with shaking at 80 rpm (Savion and Selinger, 1978). Secretagogues were added at zero time. The systems stimulated by acetylcholine received eserine (physostigmine, 0.1 mM) 10 minutes before addition of acetylcholine. At 20, 40, and 60 minutes, 50-μl aliquots of the medium were removed for amylase determination. At the end of the experiment, the tissue was ground with a Polytron homogenizer, and amylase activity was determined in aliquots of the homogenate. Amylase was assayed according to Bernfeld (1955). The rate of amylase secretion into the medium during any time interval is expressed as a percentage of the total per minute.

In contrast, in the exocrine pancrease several lines of evidence indicate that the inhibition of pancreatic enzyme secretion does not occur at the receptor level. A high concentration of cholinergic agonist inhibited stimulation of enzyme secretion by an optimal concentration of CCK-PZ analog, known to act on a different receptor (Fig. 2). Furthermore, as demonstrated in Fig. 3, when inhibitory con-

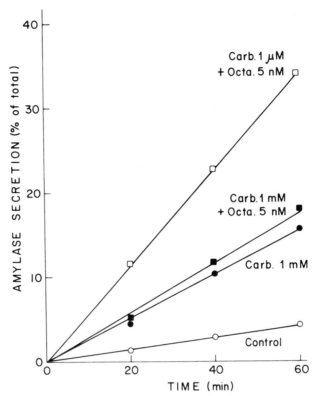

FIG. 2.   Effect of a high concentration of carbamylcholine on induction of secretion of an optimal concentration of CCK-PZ octapeptide. Pancreatic slices were prepared and incubated as described in Fig. 1. The following secretagogues were added at zero time: carbamylcholine 1 $\mu M$ + CCK-PZ octapeptide 5 n$M$ (□); carbamylcholine 1 m$M$ + CCK-PZ octapeptide 5 n$M$ (■); carbamylcholine 1 m$M$ (●); and a control system with no addition (○).

centrations of either carbamylcholine or CCK-PZ octapeptide were used, it was not possible to increase the rate of enzyme secretion by the calcium ionophore A23187, which, in the presence of calcium, bypasses the receptor and elicits efficient enzyme secretion (Eimerl et al., 1974). These experiments led to the conclusion that the inhibition of enzyme secretion occurs at a stage later than activation of the receptor or introduction of calcium into the cell.

The extremely high concentrations of secretagogues required to cause the inhibitory effect rasies the question of whether the secretagogues enter the cell and inhibit enzyme secretion through action on receptors inside the cells. In this regard, the work of Cheng and Farquhar (1976) is relevant. These authors demonstrated the presence of adenylate cyclase in intracellular membranes of rat liver, which suggests that components usually found on the cell membrane could also be present inside the cell. The following experiment suggests, however, that the inhibition of enzyme secretion is initiated by an action of the secretagogues on receptors at the cell membrane. Atropine, a competitive inhibitor that specifically blocks the muscarinic cholinergic receptor, displaced the dose-response curve for carbamylcholine to the right but did not change the biphasic characteris-

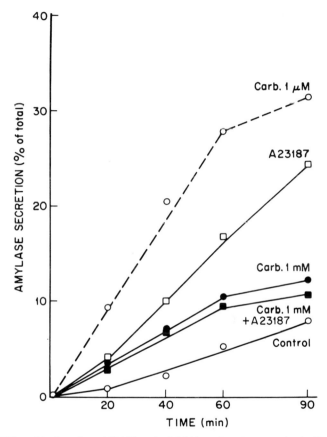

FIG. 3.    Effect of the ionophore A23187 on the inhibition of amylase secretion by high concentration of carbamylcholine. Pancreatic slices were incubated for 15 minutes in KRB medium not containing calcium, but containing 2 mM ethylene glycol-bis(aminoethyl ether) $N,N'$-tetraacetic acid (EGTA), followed by incubation for 5 minutes in fresh KRB medium containing neither calcium nor EGTA. The slices were then divided into equal portions, placed into 25-ml Erlenmeyer flasks with 3 ml of KRB medium from which calcium was omitted but which contained 0.1 mM EGTA, and incubated as described in Fig. 1. The concentrations of secretagogues were: carbamylcholine 1 $\mu M$ (O---O) or 1 mM (●——●); the ionophore A23187, 5 $\mu g/ml$ (□——□), or A23187, 5 $\mu g/ml$, plus carbamylcholine 1 mM (■——■); and a control system without any addition (O——O). After 20 minutes of preincubation, 2.5 mM calcium was added to all systems at zero time of the figure. The amount of amylase present at zero time in each system was subtracted.

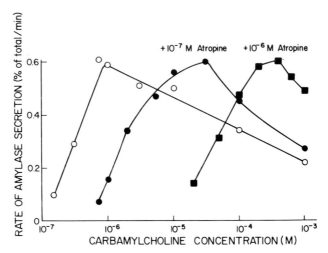

FIG. 4. Effect of atropine on stimulation of amylase secretion by various concentrations of carbamylcholine. Pancreatic slices were prepared and incubated as described in Fig. 1. Pancreatic slices were preincubated for 10 minutes in the absence (O———O) or in the presence of atropine 0.1 $\mu M$ (●———●) and 1 $\mu M$ (■———■). At zero time, carbamylcholine was added to give the indicated concentrations. Rates of amylase secretion were measured as in Fig. 1.

tics of the response (Fig. 4). Taken together, the experiments suggest that, although the inhibition of enzyme secretion apparently occurs inside the cell, it is initiated by interaction of the secretagogues with their receptors on the cell membrane. The experiments also indicate that, in order to produce maximal enzyme secretion, only a fraction of the pancreatic cell receptors have to be occupied. In support of this suggestion were the observations that depolarization, $^{45}$Ca efflux (Matthews et al., 1973), and the enhanced incorporation of $^{32}$P into phosphatidylinositol (Hokin, 1968) required a higher concentration of secretagogues than that required to produce a maximal stimulation of enzyme secretion.

A possible explanation based on a reduction in the intracellular level of ATP induced by high concentration of secretagogues was excluded, since the level of ATP was at the range of 8–10 nmole per milligram of protein when either an inhibitory or an optimal concentration of secretagogue was used (Savion and Selinger, 1978).

## III.  Ultrastructural Changes in Pancreatic Slices Incubated in the Presence of High Concentrations of Secretagogues

Morphological studies of the exocrine pancreas were of great importance in elucidating the various intracellular steps in the secretory pathway (Palade,

1975). Since the biochemical studies indicate that inhibition of enzyme secretion occurs at the cellular rather than at the receptor level, it was hoped that in this case, too, the inhibition would be characterized by distinct morphological changes. It was found that slices incubated for 1 hour in the presence of an inhibitory concentration of carbamylcholine (1 m$M$) showed severe reduction in the size of the lumen (Fig. 5), in contrast to the enlargement of the lumen under optimal concentrations of carbamylcholine (1 $\mu M$) (Fig. 6), which has been shown to be due to fusion of secretory granules with the cell membrane lining the lumen (Palade, 1959). At present, it is not known what precise molecular mechanism causes the collapse of the acinar lumen, aside from the fact that the morphological changes occur only at the apical part of the cell membrane.

A clue to the mechanism comes from experiments on the early morphological

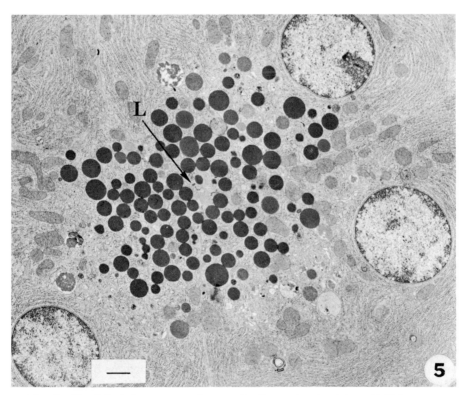

FIG. 5. Morphological appearance of pancreatic acinus under stimulation with high concentration of carbamylcholine. Pancreatic slices were prepared and incubated in the presence of carbamylcholine (1 m$M$) for 1 hour as described in Fig. 1. The slices were fixed for electron microscopy in formaldehyde-glutaraldehyde fixative followed by postfixation in 1% osmium tetroxide, dehydrated in graded ethanol solutions, and embedded in Epon (Karnovsky, 1965). The lumen (L) is small, is surrounded by numerous secretory granules, and contains dense secretory material. Microvilli cannot be seen. Bar, 1 $\mu$m.

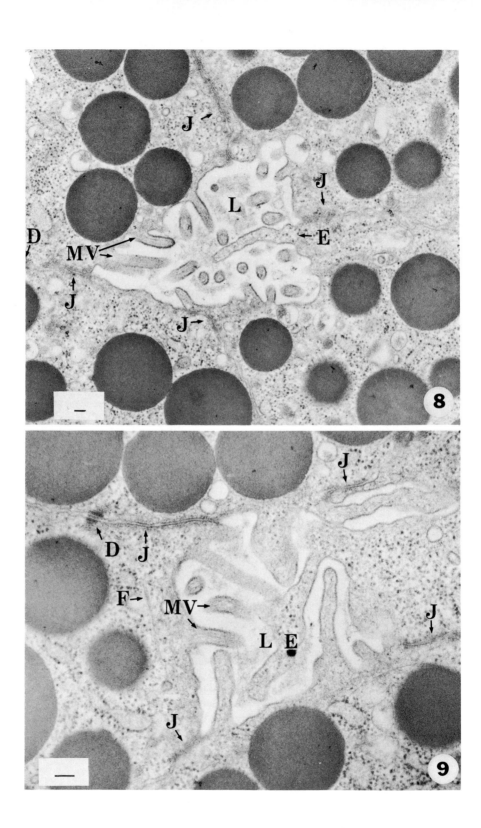

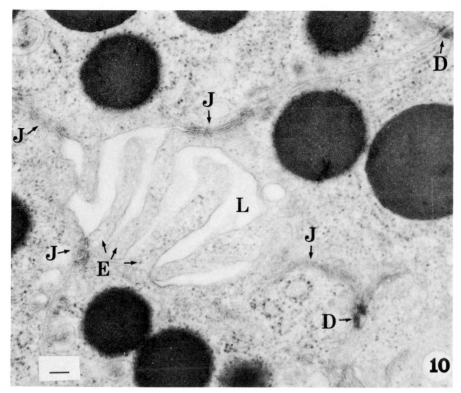

FIG. 10. Acinar lumen of pancreatic slices incubated for 15 minutes in the presence of an inhibitory concentration of carbamylcholine. Pancreatic slices were prepared, incubated for 15 minutes in the presence of high concentration of carbamylcholine (1 m$M$), and processed for electron microscopy as described in Fig. 5. About half of the lumen area (L) is occupied by multiple, distended evaginations (E) filled with cellular material. No microvilli can be seen. Junctional complexes, J. Desmosomes, D. Bar, 0.1 $\mu$m.

FIGS. 8 AND 9. Acinar lumina of pancreatic slices incubated for 5 minutes in the presence of an inhibitory concentration of carbamylcholine. Pancreatic slices were prepared, incubated for 5 minutes in the presence of high concentration of carbamylcholine (1 m$M$), and processed for electron microscopy as described in Fig. 5. The lumen (L) is identified by the junctional complexes (J) all around it. Bar, 0.1 $\mu$m. In Fig. 8, the lumen contains only one evagination (E), which can be recognized by its irregular shape. The evagination is filled with cellular material including ribosomes and is devoid of filamentous structure. Many microvilli (MV) containing filaments are still preserved. In Fig. 9, the microvilli and the filamentous structure (F) are preserved in only one cell (at the left side of the lumen). In the other acinar cells, the microvilli are replaced by irregular evaginations containing ribosomes. Desmosomes, D.

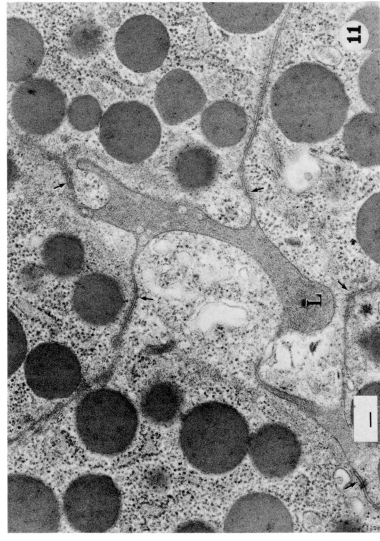

Fig. 11. Acinar lumen of pancreatic slices incubated for 1 hour in the presence of an inhibitory concentration of carbamyl-choline. Pancreatic slices were prepared, incubated for 1 hour in the presence of 1 m$M$ carbamylcholine, and processed for electron microscopy as described in Fig. 5. The lumen (L), surrounded by junctional complexes (arrows), is filled with dense secretory material. There is severe reduction in the size of the lumen, apparently due to pushing of the cellular content on the lumen membrane. Many ribosomes are located just adjacent to the luminal membrane. Bar, 0.1 $\mu$m.

tion of the cytoskeleton, which supports the luminal membrane. This leads to weakening and protrusion of the cell membrane to form evaginations, which eventually severely reduce the lumen diameter. In view of these results, we agree with the hypothesis of Orci *et al.* (1972) that the cell web has a role in the secretory process. In the absence of a secretagogue, the microfilamentous system acts as a barrier preventing the fusion of secretory granules with the luminal membrane. Upon stimulation of secretion at optimal concentration of secretagogues, a partial disaggregation of the filamentous system takes place, thus allowing the movement of a secretory granule toward the luminal membrane, which leads to a fusion of the secretory granule with the luminal membrane. This putative mechanism also demands that disaggregation of the microfilamentous structure during optimal secretion would be counteracted by a repair process, which maintains the acinar lumen morphology. At high secretagogue concentration the disaggregation process is greatly increased, thus leading to the disruption of the filamentous system. This can be due to high levels of cellular calcium introduced into the cell through an excessive stimulation by high concentrations of secretagogues.

## IV.  Secretagogue-Induced Lysosomal Activation and Production of a Myocardial Depressant Factor in Rat Pancreatic Slices

It was observed by electron microscopic studies that, at high concentrations of secretagogues, the lysosomal system was activated. Lefer and Glenn (1972) demonstrated that MDF is produced in the pancreas of various species under conditions of splanchnic ischemia by lysosomal proteolytic activity. The MDF is a small polypeptide released into the circulatory system and has a negative ionotropic effect on the heart. It is identified by bioassay using a heart papillary muscle (Lefer, 1966) and by paper chromatography (Barenholz *et al.*, 1973). We found that high concentrations of secretagogues also induced the production of MDF in pancreatic slices and its release into the medium. Our observation, taken together with those of Lefer and Glenn (1972), about the MDF production leads to the conclusion that the MDF can serve as a marker for the lysosomal activity in the pancreatic system. Both the production of MDF and the process of inhibition of enzyme secretion showed similar requirements for high concentrations of secretagogues (Fig. 12). A concentration of 1 m$M$ carbamylcholine or 1 $\mu M$ CCK-PZ octapeptide causes production of MDF and its release into the medium. The release of MDF is not a part of the process of enzyme secretion, since there was no correlation between release of MDF and amylase secretion. The mus-

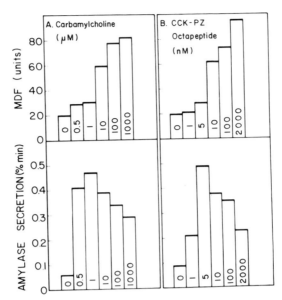

FIG. 12.    MDF release and amylase secretion stimulated by secretagogues in rat pancreatic slices. Pancreatic slices were prepared and incubated for 2 hours, and the rate of amylase secretion was determined as described in Fig. 1. The amount of MDF released into the incubation medium was determined by the paper chromatographic method described by Barenholz *et al.* (1973). The effect of different concentrations of carbamylcholine (left side of the figure) and CCK-PZ octapeptide (right side of the figure) on either MDF release (upper part of the figure) or amylase secretion (lower part of the figure). The carbamylcholine and CCK-PZ octapeptide concentrations are written inside the bars.

carinic cholinergic blocker, atropine, blocks production of MDF induced by supraoptimal concentrations of carbamylcholine, while the induction by anoxic conditions, namely nitrogen atmosphere, is not blocked by atropine (Fig. 13). The induction of MDF production by supraoptimal concentrations of carbamylcholine is mediated by the cholinergic receptor, in contrast to the induction by anoxic conditions, which is not mediated through the cholinergic receptor.

Pancreatic slices incubated in calcium-free medium do not release MDF upon exposure to supraoptimal concentration of carbamylcholine (Fig. 14). It appears that the induction of MDF production by high concentration of secretagogue is dependent on extracellular calcium. Moreover, treatment of pancreatic slices with the ionophore A23187, which is known to introduce calcium into the cell (Eimerl *et al.*, 1974), induces at concentrations as high as 50 $\mu$g/ml the production and release into the medium of MDF (Fig. 15). These results indicate the role of calcium as second messenger mediating the process of MDF production induced by carbamylcholine at supraoptimal concentrations.

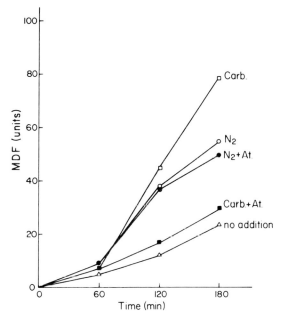

FIG. 13. The effect of atropine on MDF release induced by carbamylcholine or nitrogen atmosphere. Pancreatic slices were prepared and incubated as described in Fig. 1. At zero time, atropine (10 $\mu M$) was added to a few systems (■, ●). At 60 minutes the atmosphere in part of the systems was changed to N$_2$ atmosphere (○, ●), and to others carbamylcholine (1 m$M$) was added (□, ■). A control system without any addition was also included (△). At the times indicated in the figure, aliquots were taken for MDF determination as described by Barenholz *et al.* (1973).

FIG. 14. The effect of extracellular calcium on the release of MDF induced by carbamylcholine or N$_2$ atmosphere. Pancreatic slices were prepared and incubated as described in Fig. 1. At the indicated time, aliquots of medium were taken for MDF determination by paper chromatographic technique as previously described (Barenholz *et al.*, 1973). The systems are: nitrogen atmosphere (□, ■), carbamylcholine (1 m$M$; ○, ●), and control systems without any addition (△, ▲). The systems marked with open symbols contained EGTA (0.5 m$M$) in the incubation medium; those with closed symbols contained calcium (2.5 m$M$) and EGTA.

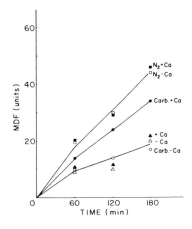

## ACKNOWLEDGMENTS

This work was supported by a grant from the National Institutes of Health (Grant AM-10451) and from the Chief Scientist Office, Ministry of Health, Israel. We wish to thank Mr. Harvey Scodel for his invaluable help in the preparation of this manuscript.

## REFERENCES

Barenholz, Y., Leffler, J. N., and Lefer, A. M. (1973). *Isr. J. Med. Sci.* **9**, 640–647.
Bayliss, W. M., and Starling, E. (1902). *J. Physiol. (London)* **28**, 325–353.
Bernfeld, P. (1955). *In* "Methods in Enzymology" (S. P. Colowick and N. O. Kaplan, eds.), Vol. 1, pp. 149–158. Academic Press, New York.
Brown, J. C., Harper, A. A., and Scratcherd, T. (1967). *J. Physiol. (London)* **190**, 519–530.
Cheng, H., and Farquhar, M. G. (1976). *J. Cell Biol.* **70**, 660–670.
Debray, C., Vaille, C., de la Tour, J., Roze, C., and Souchard, M. (1963). *Rev. Int. Hepatol.* **13**, 473–499.
Eimerl, S., Savion, N., Heichal, O., and Selinger, Z. (1974). *J. Biol. Chem.* **249**, 3991–3993.
Folsch, U. R., and Wormsley, K. G. (1973). *J. Physiol. (London)* **234**, 79–94.
Foreman, J. C., and Garland, L. G. (1974). *J. Physiol. (London)* **239**, 381–391.
Glenn, T. M., and Lefer, A. M. (1971). *Circ. Res.* **29**, 338–349.
Harper, A. A., and Raper, H. S. (1943). *J. Physiol. (London)* **102**, 115–125.
Hokin, L. E. (1951). *Biochem. J.* **48**, 320–326.
Hokin, M. R. (1968). *Arch. Biochem. Biophys.* **124**, 280–284.
Karnovsky, M. J. (1965). *J. Cell Biol.* **27**, 137–138a.
Kulka, R. G., and Sternlicht, E. (1968). *Proc. Natl. Acad. Sci. U.S.A.* **61**, 1123–1128.
Lefer, A. M. (1966). *J. Pharmacol. Exp. Ther.* **151**, 294.
Lefer, A. M., and Glenn, T. M. (1972). *In* "Shock in Low and High Flow States" (B. K. Forscher, R. C. Lillehei, and S. S. Stubbs, eds.), pp. 88–105. Excerpta Medica, Amsterdam.
Leroi, J., Morisset, J. A., and Webster, P. D. (1971). *J. Lab. Clin. Med.* **78**, 149–157.
Matthews, E. K., Peterson, O. H., and Williams, J. A. (1973). *J. Physiol. (London)* **234**, 689–701.
Miller, R., and Palade, G. E. (1964). *J. Cell Biol.* **23**, 519–552.
Nevalainen, T. J., and Janigan, D. T. (1974). *Rex. Exp. Med.* **162**, 161–167.
Orci, L., Gabbay, K. H., and Malaisse, W. J. (1972). *Science* **175**, 1128–1130.
Palade, G. E. (1959). *In* "Subcellular Particles" (T. Hayashi, ed.), pp. 64–83. Ronald Press, New York.
Palade, G. E. (1975). *Science* **189**, 347–358.
Robberecht, P., and Cristophe, J. (1971). *Am. J. Physiol.* **220**, 911–917.
Savion, N., and Selinger, Z. (1978). *J. Cell Biol.* **76**, 467–482.
Scheele, G. A., and Palade, G. E. (1975). *J. Biol. Chem.* **250**, 2660–2670.
Smith, R. E., and Farquhar, M. G. (1966). *J. Cell Biol.* **31**, 319–347.
Vincent, D., and Bauduin, H. (1972). *Biol. Gastroenterol.* **5**, 85–90.

# Part V.   Membrane Dynamics

# Chapter 24

## *Membrane Circulation: An Overview*

### ERIC HOLTZMAN

*Department of Biological Sciences,*
*Columbia University, New York, New York*

## I.   Introduction

In the past few years there have been encouraging advances in the analysis of the genesis and circulation in the cell of the membranes involved in secretory processes. But the gaps remaining are still at least as broad as the areas of understanding. The articles in this segment of the book concern several specific aspects of membrane formation and behavior. To provide a background for these discussions, I shall here briefly survey major routes of membrane circulation thought to be important in secretory cells and elsewhere. Then I shall outline a number of the significant areas of uncertainty about these routes. More detailed

379

discussions and a more extensive bibliography are presented in Holtzman *et al.* (1979) and Holtzman and Mercurio (1980).

## II.   Membrane Routes and Cycles

The concept that membranes move from one compartment of the cell to another derives primarily from microscopic studies. Present views have evolved from earlier formulations about membrane "flow" or "shuttling" (see, e.g., Bennett, 1969; Palade, 1959). For our purposes, special attention must be paid to the following routes and cycles of presumed membrane movement (Fig. 1).

## A.   Movement between the Endoplasmic Reticulum and the Golgi Apparatus

As discussed earlier in this book, transport of secretory proteins from the endoplasmic reticulum (ER) to the sacs or condensing vacuoles of the Golgi apparatus evidently can occur by way of vesicles that bud from the ER and fuse with the Golgi-associated structures. It is widely assumed that there is a corresponding shuttling of vesicles back to the ER for reutilization in transport. This

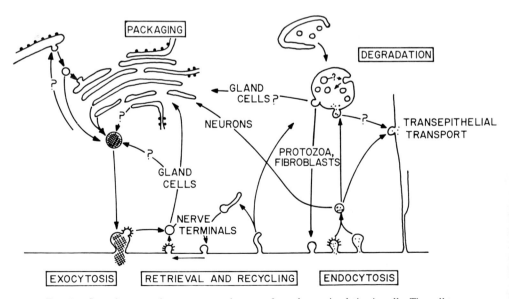

Fig. 1.   Some known and some suspected routes of membrane circulation in cells. The cell types named are examples of those in which the particular routes indicated have been studied.

shuttling has not been definitively demonstrated but does seem likely from several considerations. For example, the contents of the secretory packages leaving the Golgi apparatus are a good deal more concentrated than are the contents of the ER. The process of concentration occurs in Golgi sacs or in forming secretory structures (Palade, 1975; Whaley, 1975). Thus, the transport vesicles would seem to deliver a considerable "excess" of membrane to the Golgi region by comparison with the amount of membrane that leaves the region in the form of the membrane delimiting secretory bodies. Vesicle shuttling and reuse would deal with this "excess" in "economical" fashion.

In addition to transport in vesicles, other types of ER–Golgi apparatus relations have been thought to provide routes for movement of material synthesized in the ER to Golgi-associated structures (e.g., direct continuities, or transformation of ER cisternae into Golgi sacs: Fig. 1; Holtzman et al., 1977; Morré, 1977; Novikoff, 1976; Whaley, 1975). These routes could also serve for movement of membrane. Circumstantial evidence hinting that they do, comes, for example, from studies on the formation and growth of the Golgi apparatus (see, e.g., Flickinger, 1978).

## B. Endocytosis Coupled to Exocytosis

The exocytic release of secretions results in the incorporation, into the cell surface, of the membrane participating in the transport of the secretions. Morphological studies and work with endocytosis tracers such as horseradish peroxidase (HRP) on gland cells and neurons have made it clear that there is a compensatory endocytosis-like retrieval of membrane from the cell surface (reviewed in Holtzman, 1977). This retrieval eventually returns the surface to its original area. It is presumed, on still fragmentary evidence, that the membrane retrieved endocytically is the same as, or is equivalent in composition to, that added by exocytosis (see Section III,C below and Meldolesi et al., 1978). Other mechanisms of removal of membrane from the cell surface after exocytosis have been discussed, such as shedding to the extracellular space (Lawson et al., 1977; for a dramatic case see the preliminary report by Specian and Neutra, 1979, on goblet cells) or withdrawal of membrane molecules rather than of intact membrane (Meldolesi et al., 1978). But for the most part, such proposed mechanisms have yet to be shown to play quantitatively significant or widespread roles. (As Lawson et al. point out, there are also some problems of potential artifact that may affect the evidence for shedding.)

## C. Reutilization of Retrieved Membrane

Experimental support for the idea that membrane retrieved from the cell surface can be reused for the packaging of secretions has grown increasingly strong.

Biochemical investigations on several glands (Meldolesi *et al.*, 1978; Winkler, 1977) showed that the membranes delimiting secretory granules turn over at a considerably slower rate than do the granule contents. This by itself would be an ambiguous finding, since the contents and the membranes probably draw their components from different intracellular pools of macromolecules (Holtzman, 1976; Wallach *et al.*, 1975). However, the reservoirs of relevant membrane proteins in the cell are not sufficiently large to account for the slow turnover of secretory granule membranes, were the membranes to be used only once and then degraded (Meldolesi *et al.*, 1978; Winkler, 1977). Moreover, the notion of reuse of retrieved membrane has received strong support from microscopic studies using several endocytosis tracers (dextrans, cationic ferritin, HRP). Of central weight were the findings by Heuser and Reese for neurons, that HRP endocytized into synaptic vesicles as a result of membrane retrieval linked to neurotransmission, could be depleted from the vesicles in subsequent rounds of transmission (Heuser and Reese, 1977; Zimmerman, 1979). For gland cells, studies by Herzog and Farquhar (1977) and by Farquhar (1978; see also Pelletier, 1973) have provided evidence that endocytically retrieved membrane can return to the Golgi region for reuse in packaging secretions (see also Chapter 25 by M. G. Farquhar).

## D. Lysosomes

Early studies by our group and others demonstrated that tracers such as horseradish peroxidase, taken into the cell through endocytosis coupled to exocytosis, accumulate primarily in lysosomes (reviewed in Holtzman, 1976; Holtzman *et al.*, 1977). This led to the suggestion that the lysosomes function as a degradative terminal depot for the membranes circulating during the secretory cycle. The involvement of multivesicular bodies (MVBs) was emphasized, since these bodies are major sites of accumulation of the tracers and since they seem to form through mechanisms by which the circulating membranes could gain access to the lysosome interior. More recent work, however, has turned up major complexities. For example, when cationic ferritin or dextrans are used to study membrane retrieval, the prominence of lysosomes in tracer accumulation can be less marked than with HRP or native, anionic ferritin (Herzog and Farquhar, 1977; Herzog and Miller, 1979; Farquhar, 1978). In some gland cells, with dextrans or cationic ferritin, the Golgi apparatus and secretion granules are found to be labeled rapidly and extensively, which is not generally true in cells studied with HRP or native ferritin. The bases of these differences in tracer distribution are not yet thoroughly understood (see Section III,A below). However, since cationic ferritin adheres much more tightly to membranes than does HRP, they probably are associated, in part, with differences between the fate of the *contents* and the fate of the *membranes* of endocytic structures.

Related to this last point are findings on cells engaged in pinocytosis and phagocytosis, which have suggested that the encounters between lysosomes and membranes transporting materials from the cell surface need not be fatal to the membranes (Allen, 1974; Dean, 1977; W. A. Muller *et al.*, 1980; Schneider *et al.*, 1979). The traffic between the lysosomes and the plasma membrane is still to be described (Fig. 2), but in some cases, at least, it is very likely to involve a two-way movement of membrane (W. A. Muller *et al.*, 1980). This, in turn, raises the possibility that membrane retrieved from the cell surface after secretion could fuse with a lysosome, depositing the contents acquired at the cell surface, and then bud off and move to the Golgi apparatus for use in packaging of secretions (Farquhar, 1978; Herzog and Miller, 1979). Such a sequence of events might, for example, serve to permit reuse of membrane for packaging of secretions while minimizing entry of extracellular materials into secretory structures. The preliminary report by Abrahamson and Rodewald (1979) on the differential fate of membrane-bound vs. non-membrane-bound tracers incorporated in the same endocytic vesicles lends plausibility to such speculations.

## E.   Uses of Membrane Circulation

The circulation of membranes in cells has provided the basis for some interesting techniques. Particularly important for neurobiology has been the development of "retrograde transport" mapping methods (reviewed in Cowan and Cuénod, 1975). Exogenous, microscopically detectable tracers endocytized at nerve terminals are transported up the axon to the perikarya. Thus, the location in

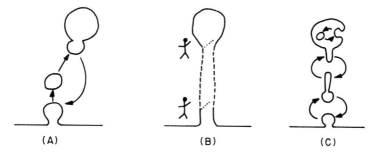

(A)                         (B)                        (C)

Fig. 2.   Macromolecules such as tracers moving from extracellular spaces to intracellular compartments might conceivably move by several mechanisms (see text; cf. Schneider *et al.*, 1979; Jaken and Thinès-Sempoux, 1978). (A) Direct transport in vesicles. Recycling of membrane to the cell surface by a similar mechanism is also illustrated. (B) Passage through connecting channels. This might not require actual movement of membrane, but would involve channel systems, gates, and gatekeeping mechanisms ("gatekeepers") that have yet to be identified. (C) Transfer via intermediate structures, such that a vesicle budding from the cell surface would transfer its contents to an intermediate carrier. The carrier would then move on to interact with other compartments while the vesicle returned to the cell surface. [Wehland *et al.* (1981) also speculate that the coats of coated pits may remain at the cell surface throughout endocytosis.]

the nervous system of the perikarya corresponding to a particular set of terminals can be determined.

In our laboratory, we have been exploring the use of the uptake of endocytosis tracers during membrane retrieval, as a means of evaluating, semiquantitatively, the intensity of synaptic activity under varying physiological conditions (Schacher *et al.*, 1976; Evans *et al.*, 1978). We have found, for example, that the uptake of HRP by terminals of retinal photoreceptors varies with different illumination conditions and in different ionic environments. These changes in tracer uptake provide information that usefully complements more conventional physiological data.

# III.  Some Unresolved Issues

Other contributors to this book have already considered such much-debated matters as the mechanisms of membrane movement, fusion, and fission, and the maintenance of differences among the membranes of the several cellular compartments that interact in the secretory cycle. Here we shall try to provide some additional perspectives on a number of these matters, and also to raise a few other issues.

## A.  Technical Problems

The analysis of membrane circulation has depended heavily upon the extrapolation to dynamic events from static electron micrographs, and upon consideration of the behavior of materials present in the interior of membrane-delimited compartments—endogenous materials in the case of secretory granules, and macromolecular tracers in the case of endocytic compartments. Biochemical and autoradiographic studies have also contributed importantly, as have newer microscopic methods, such as the evolving freeze-fracture approaches. However, in general, we still lack markers or methods adequate to follow specific membranes in detail throughout their life history. For example, in considering endocytosis, we most often think in terms of direct transport of materials in vesicles or vacuoles (Fig. 2A). This bias seems reasonable for many cases, and especially for such thoroughly studied endocytic phenomena as phagocytosis and pinocytosis by polymorphs or macrophages (Silverstein *et al.*, 1977). But for other cases, with present methods it is difficult to distinguish definitively among several possible endocytic routes (Fig. 2; cf. Jaken and Thinès-Sempoux, 1978; Schneider *et al.*, 1979).

Some of the pitfalls in the use of particular tracers have been identified or are being studied. For example, different peroxidases seem to be handled differently by neurons, perhaps hinting at effects of peroxidases on membrane behavior (Bunt and Haschke, 1978). The employment of bulk-phase endocytosis tracers such as HRP permitted analysis of the coupling of exocytosis and endocytosis. But, as we have already mentioned, other tracers were needed—especially those that bind to membranes—to reveal additional phases of membrane circulation. Although clearly moving the analysis forward in major fashion, these latter tracers have their own potential problems that require more adequate evaluation. Dextrans induce swelling of elements of the Golgi apparatus, for reasons that are not entirely clear. In our laboratory, efforts to employ them to study photoreceptors have been hampered by what seems to be tracer-induced severe damage to rod cells (L. Liscum and E. Holtzman, unpublished). When lectins, cationic ferritin, or antibodies bind to membranes, they can produce alterations that might be germane to subsequent membrane behavior. These range from induction of endocytosis, redistribution of membrane macromolecules, or changes in cyclic nucleotide levels, to toxic effects and apparent alterations in patterns of fusions among intracellular membranes, including apparent inhibition of fusions of endocytic structures with lysosomes (see, e.g., Edelson, 1974; Goldman, 1974). Some of these problems can be minimized by use of low concentrations of tracer (see, e.g., Farquhar, 1978) or by suitable control procedures, such as the use of cholera toxin subunits that bind to membranes but have much less dramatic effects on the cell's biochemistry than does the intact toxin (Joseph *et al.*, 1979). However, circumspection is still needed in interpreting results.

More information is also required on the effects of the intensity of secretory activity on membrane circulation. For example, in adrenal medulla cells stimulated to secrete much of their epinephrine store rapidly, lysosomes seem to play a more prominent role in disposing of retrieved membrane (Abrahams and Holtzman, 1973) than may be the case in less active cells (Winkler, 1977). Does this reflect a response of the cell to the presence of a membrane "excess"? In nerve terminals the prominence of large "cisternal" compartments as intermediates in membrane retrieval varies considerably with intensity of stimulation, and also with temperature (see, e.g., Llinas and Heuser, 1977). Several of the papers presented in this volume report other effects of extensive or "excessive" stimulation that merit consideration in the design of experiments on membrane cycling.

Finally, the use of cultured or dissociated eells for analysis of membrane circulation during secretion poses some potential problems. Notably, the restriction of exocytosis or of access of endocytosis tracers to particular cell poles, as is true *in vivo,* may be relaxed *in vitro.* Direct comparisons of endocytic membrane retrieval in *in vivo* and *in vitro* preparations have revealed differences (Herzog and Farquhar, 1977) that warn against too rapid extrapolation from the one situation to the other.

## B.  Molecular Events

Thus far we have stressed the movement of membrane as such from one cell compartment to another. Especially for microscopists, like me, it is seductive to think in such terms. Still to be determined, however, are the details of coexistence between bulk membrane movement and molecular events involved in the assembly, maintenance, or alteration of membranes (for review and references, see Holtzman et al., 1979). For example, several contributions to this book have emphasized that both free and bound ribosomes can contribute proteins to preexisting membranes. Presumably, peripheral membrane proteins, and perhaps some integral ones, can exit from membranes. Lipid molecules can exchange among membranes (Wirtz and van Deenen, 1977), and portions of lipid molecules within membranes can be modified or turned over differentially. In addition, lateral mobility and reassortment phenomena can markedly alter the composition of local membrane regions, producing perhaps, a kind of dynamic mosaicism superimposed upon the longer-lasting heterogeneity of the different membrane regions of a cell. To what extent do these various phenomena and mechanisms engender modifications of membranes during their circulation in the cell? For example, do the vesicles that bud from one cellular compartment and fuse with another carry a selected or a random sample of the "parent" membrane's molecules (evidence from the study of endocytosis suggests that both situations may occur: Holtzman et al., 1979; Silverstein et al., 1977)? When different compartments are continuous with one another, do the continuities permit actual movement of membrane per se, or is there a sort of flow of molecules within the plane of the membrane, perhaps permitting differential movement of different components (cf. Fujii-Kuriyama et al., 1979)?

In considering the biogenesis of secretory structures, one might think of ER membrane moving to Golgi compartments and, after modification, giving rise to the membranes delimiting secretory granules. This would fit nicely with known or suspected morphological relations and biosynthetic capabilities (Morré, 1977; Whaley, 1975; see also Howell et al., 1978, for their discussion of possible biochemical similarities between ER and Golgi membranes). But the differences in composition among the compartments involved, and the absence of unambiguous evidence as to how membranes transform, warn against too ready an acceptance of the most straightforward biogenetic schemes. This is accentuated by the occurrence of membrane cycling from the cell surface to the Golgi apparatus. How much "replenishment" from the ER does the Golgi apparatus actually require under varying circumstances, and could this be provided on a molecule-by-molecule basis rather than by membrane movement? To what extent do the membranes coming back from the surface mix with those reaching the Golgi region from the ER? Is "older" and "newer" membrane kept separate, or does intermingling of molecules take place so that a given membrane really has no well-defined age? Do the regional differences among and within Golgi sacs (Farquhar, 1978; Whaley, 1975) reflect segregation of membrane regions involved in enzymatic processing of secretions, from membrane destined to delimit secretory bodies? More generally, does the cell use the same basic biosynthetic apparatus to generate different categories of membranes, and, if so, how is this achieved?

One set of questions needing further study concerns the sites and mechanisms by which membrane lipids are brought together with membrane proteins (Holtzman and Mercurio, 1980). At what stage in its life history does a newly synthesized membrane protein first come to be associated with the lipids that will accompany it at its destination or site of function in the cell? Is the rough ER the major source of membrane lipids, and is this true even in cells with abundant smooth ER? If the smooth ER can synthesize membrane lipids, do these simply intermix with the lipids made in the rough ER via the continuities between the compartments, or is their transport and utilization more complex?

One system that is proving useful for exploring issues of membrane assembly, differentiation, and turnover, and the relations of bulk movement and molecular events, is the retinal rod cell. This cell type maintains distinctly different populations of membranes at its two ends, the photoreceptive outer segment and the presynaptic terminal (for review and references, see Young, 1976; Holtzman and Mercurio, 1980). The outer-segment discs of rods arise through a series of incompletely understood processes thought to include exocytosis-like addition of membrane to the cell surface, and to involve the origin of the discs from infoldings (or outpouchings: Steinberg et al., 1980) of the plasma membrane, through events that resemble endocytosis at least superficially. The discs, once they have formed, exhibit membrane movement in bulk. They migrate up the outer segment, eventually to be shed and phagocytosed by the pigment epithelium, which degrades them in lysosomes. They also show differential molecular exchanges (lipid molecules seem to move in and out of a disc as it migrates, while much of the protein remains fixed in the disc: cf. Young, 1976). For frog rod cells, we have identified morphologically and cytochemically distinctive regions of rough and smooth ER located in different areas of the cell (Holtzman and Mercurio, 1980). Through EM autoradiographic studies (Holtzman and Mercurio, 1980; Mercurio and Holtzman, unpublished) we have found that both the rough ER and the smooth ER seem to participate in synthesis of lipids or in the accumulation of recently made lipids; the rough ER plays the major role in quantitative terms. Interestingly, there appear to be differences in the relative rates and patterns of transport of ER-synthesized lipids as contrasted to proteins, from the ER through compartments such as the Golgi apparatus to other cellular membrane systems. We have also found that both rough ER and smooth ER show calcium-containing deposits in photoreceptors incubated with cytochemical procedures thought to demonstrate sites of calcium binding or accumulation in cells. Presumably, we are detecting roles of the ER either in transport of calcium binding components or in control of cytosol calcium levels (Ungar and Holtzman, 1981). Eventually such studies should help us determine whether different zones of the rough ER or of the smooth ER are functionally distinctive in terms that might relate to the polarity of the cell.

## C.   Details of Retrieval

A key question about the endocytic retrieval of membrane following exocytosis concerns the degree to which components of the plasma membrane and those of the membranes of secretory structures intermingle during the time the membranes are fused. Biochemical studies suggest that the plasma membranes of nerve terminals or secretory cells maintain compositions different from that of the synaptic vesicles or secretory granule membranes (for discussion, see Holtzman *et al.*, 1977; Llinas and Heuser, 1977; Wagner *et al.*, 1978). This might mean that the components of the interacting membranes remain segregated during exocytosis and endocytosis. But the evidence needs firming, especially since we know little about the composition of the specific local plasma membrane regions where release of secretions and retrieval actually occur; these regions might differ from the remainder of the plasma membrane. Observations of coated vesicles budding from exocytic figures (Grynzspan-Winograd, 1971; Douglas *et al.*, 1971) and freeze-fracture studies (Meldolesi *et al.*, 1978) do suggest that membrane retrieval can involve the same membrane that has entered the cell surface during exocytosis. But these observations are difficult to interpret in quantitative terms, and they cannot rule out some intermingling of molecules prior to retrieval.

How synaptic vesicles fill and refill with transmitters is still in dispute (for references, see Holtzman, 1977; Holtzman *et al.*, 1977). But many current hypotheses rely upon presumed or known properties of the vesicle membrane: its permeability, or the presence of transmitter-synthesizing enzymes, of carriers, or of pumps that establish ionic and potential gradients controlling the distribution of transmitters. Retrieval processes at nerve terminals thus would have to provide recycled synaptic vesicles with specific membrane components needed for proper function. [Other hypotheses about transmitter storage center on participation of vesicle contents, such as the nucleotides, macromolecules, and divalent cations thought to form storage complexes with the transmitters. But how "used" vesicles refill with these is not known (see, e.g., Zimmerman, 1979).]

Heuser and Reese (1977; also Heuser, 1977; Llinas and Heuser, 1977) have argued that retrieval of synaptic vesicle membrane occurs via coated vesicles that form at some distance from sites of exocytosis. From this, and from studies on the distribution of intramembranous particles (IMPs) in the plasma membranes of active terminals, they suggest that, prior to retrieval, synaptic vesicle components diffuse for appreciable distances within the plasma membrane. That the IMPs are reliable tracers is still to be determined (e.g., changes in the local abundance of large particles could sometimes involve assembly from smaller ones, rather than movement: Tokunaga et al., 1979; Holtzman et al., 1979). But if Heuser and Reese are correct, their conclusions raise questions about the reconstitution of functional synaptic vesicles. Does the accumulation of large IMPs in the membrane of forming coated vesicles noted by Orci (see Chapter 18 by L. Orci et al.) and by others (e.g., Heuser and Reese, 1977) reflect reassortment processes important for selective retrieval? Is the membrane retrieved by coated vesicles at presynaptic terminals made of the same components as were added to the surface from synaptic vesicles during the most recent rounds of exocytosis? Or can it include molecules from the plasma membrane or ones persisting in the cell surface from earlier bouts of exocytosis? Do processes occurring inside the cell reorganize retrieved membrane, perhaps selectively recruiting the components needed for functional synaptic vesicles, or replenishing materials lost in exocytosis, from intracellular stores (see the comments below on cisternae and, in the next section, on agranular reticulum)?

The participation of coated vesicles in postexocytic membrane retrieval has been demonstrated in several systems (see, e.g., Douglas et al., 1971; Heuser and Reese, 1977). But the significance of the vesicle coating is not established, and it has yet to be shown that the participation of coated vesicles is obligatory. For nerve terminals there still is active exploration of the possibility that under some physiological conditions synaptic vesicles may fuse only very briefly with the surface and then bud more or less directly back into the axoplasm, perhaps with no coat being involved (see, e.g., Ceccarelli et al., 1979). A related matter is the participation of large vacuolar and cisternal elements in retrieval, seen, for example, under conditions of intense stimulation (cf. Section III,B). Both in neurons and in gland cells some of these structures probably can arise directly from the cell surface (Abrahams and Holtzman, 1973; Llinas and Heuser, 1977). How is their formation and composition controlled? When they give rise to "functional" membrane such as synaptic vesicles, is there a sorting-out process, perhaps reflected in the involvement of vesicle coating, by which membrane components "improperly" internalized by the cell, or present in excess, are excluded? [Heuser (1977) has suggested that IMP-poor portions of cisternae in neuromuscular junctions may be slated for degradation.] For terminals of frog retinal photoreceptors, we (Liscum et al., 1980) have found that changes in the cells ionic environment (presence of $Ba^{2+}$ ions; blockage of $K^+$ channels) markedly affect the abundance of cisternae, which hints at the possible existence of ion-related control mechanisms. Also of interest in these connections are observations on the terminals of neurosecretory cells. These terminals seem unlikely to reuse retrieved membrane locally, since there are no obvious routes in the terminals for repackaging of neurosecretory materials. Recent studies suggest that much of membrane retrieval in neurosecretory terminals leads to the rapid formation of large vacuoles, rather than formation of synaptic vesicle-like structures (reviewed in Morris *et al.*, 1978). The vacuoles apparently are destined for transport up the axon and subsequent degradation (or perhaps reuse) in the perikaryon. (See also Fig. 2 legend, p. 383.)

## D.   Membrane Fate

What are the extent and mechanisms of membrane degradation pertinent to the circulation processes being discussed? Secretory granule membranes and other relevant cellular membranes do turn over, but relatively slowly (Melodolesi *et al.*, 1978; Winkler, 1977). Winkler, for example, has estimated that the adrenal medulla cell reuses about 80% and degrades about 20% of chromaffin granule membranes cycling in adrenal glands that have not been extensively stimulated. Our own investigations on nerve terminals suggest that removal of synaptic vesicle membrane from the terminals via retrograde transport is a relatively leisurely affair, requiring many hours to clear the terminals of vesicles that have cycled through the surface (Teichberg *et al.*, 1975), which would permit multiple reuse of the vesicles before their destruction.

With few exceptions, such as the retinal photoreceptor outer-segment discs mentioned in Section III,B, the mechanisms by which membranes or their components are broken down remain to be determined. For example, the relative importance of the entry or loss of individual membrane molecules vs. the destruction of membrane in bulk is quite unclear (Meldolesi *et al.*, 1978; Holtzman, 1976). Lysosomes, however, remain prime suspects as central participants in intracellular membrane degradation. There is, for instance, evidence that membrane circulating through the cell surface can become internalized within lysosomes. The characteristic thick luminal plasma membrane of the mammalian bladder can be recognized within lysosomes of bladder cells (Hicks, 1966; Porter *et al.*, 1967). In crayfish eyes, where multivesicular bodies are thought to participate in degrading membrane from the microvilli of the light-sensitive rhabdomere, the membranes delimiting the internal vesicles of the MVBs closely resemble those of the microvilli, in freeze-fracture appearance (Eguchi and Waterman, 1976; see also Section III,E). Studies on the circulation of epithelial growth factor receptors (McKanna *et al.*, 1979) also demonstrate what seems to be the entry of receptor-rich cell surface membrane into the interior of MVBs. More circumstantial, but still suggestive, are observations of the accumulation of large MVBs with much internal membrane in cells of the adrenal medulla when the cells are stimulated to rapid, extensive secretion (Abrahams and Holtzman, 1973).

In Section II,A we mentioned the "surplus" of membrane that might pass through the Golgi region during ER–Golgi transport. Given the abundance of lysosomes in the Golgi region and the evidence that autophagic incorporation of membrane occurs there (see, e.g., Novikoff, 1976), it would be worth exploring the possibility that some of this "surplus" membrane is degraded via lysosomes.

Related to matters of degradation are questions of selectivity. Our observations on the distribution of HRP-labeled vesicles in photoreceptor terminals (Schacher et al., 1976) suggest that vesicles are reused at random with respect to their recent history (i.e., whether they recently originated from endocytized membrane). More extensive and detailed such analyses are needed, and they must be extended to membrane breakdown. If membrane recycling results in constant intermingling of older and newer molecules, does this help explain the apparent lack of age-dependency in turnover of cellular structures (cf. DeDuve, 1973)? Can the cell identify and segregate special membrane slated for degradation, such as the particle-poor regions of nerve-terminal cisternae mentioned in the last section? Or is degradation also largely a random process? There has been one study suggesting that the endogenous dopamine $\beta$-hydroxylase transported in retrograde direction by neurons may be enzymatically inactive (Nagatsu et al., 1976). Does inactivity somehow trigger transport and subsequent degradation, or is the enzyme inactivated as a result of its entry into the retrograde transport route?

The nature of some of the intracellular compartments that can participate in membrane transport needs further study from the viewpoints of concern to us in this section. For example, in neurons, Gonatas and co-workers (see Joseph et al., 1979) have demonstrated, and we have confirmed (Fig. 3), that certain tracers that bind to the plasma membrane and are endocytically internalized accumulate in Golgi-associated membrane systems. These systems show both the morphology and the acid phosphatase activity of GERL (Novikoff, 1976). Do these observations point toward a pathway for membrane cycling back and forth between such Golgi-associated structures and the cell surface? Or is a degradative pathway involved? How do the observations relate to the possibility that some hydrolases may also enter GERL via endocytosis (Rome et al., 1979) or that the hydrolases of GERL might help modify secretory molecules or other materials (Novikoff, 1976). Is GERL best thought of as part of the ER, as a network of lysosomal compartments, or as a product of the Golgi apparatus (Holtzman and Mercurio, 1980).

Questions of a similar sort arise with respect to axonal compartments. Tracers endocytized by axons or terminals are found to enter axonal sacs and tubules. These latter compartments resemble the structures of the axonal agranular reticulum that are thought to participate in transport of macromolecules from the perikaryon to the terminals (reviewed in Holtzman et al., 1977). Does this reflect a recycling route whereby endocytized membrane is reused to package newly transported material? Such a possibility remains open, but at present the evidence is somewhat stronger that the sacs and tubules in question are lysosome-related and/or are participants in retrograde transport (reviewed in Broadwell and

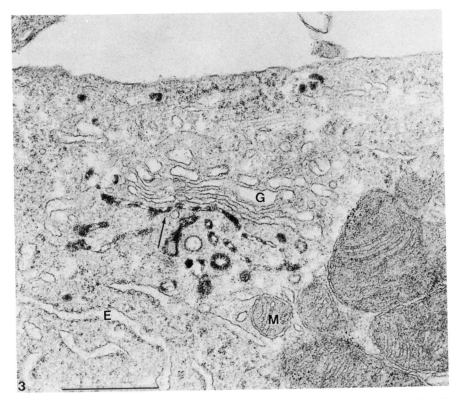

Fig. 3.   Portion of the myoid region of a frog rod photoreceptor from an isolated retina fixed 60 minutes after exposure to cholera toxin conjugated to horseradish peroxidase (Joseph *et al.*, 1979; we thank Dr. N. K. Gonatas for providing the conjugate) and incubated to demonstrate peroxidase activity. The experiment is from an unpublished series by M. L. Matheke and E. Holtzman. E indicates endoplasmic reticulum; M, mitochondria; and G, Golgi apparatus. Reaction product is seen at one face of the Golgi apparatus, in an elongate sac (arrow) and in vesicles and other membrane-delimited elements. Bar, 0.5 µm.

Brightman, 1979, Holtzman *et al.*, 1977; Holtzman and Mercurio, 1980) and as such are at least partially distinct from the agranular reticulum.

## E.   Membrane Circulation and the Cell Surface

Exocytosis and endocytosis obviously affect the cell surface during the secretory cycle. Moreover, these processes, or ones closely akin to them, are widely held to be important for the growth, maintenance, and turnover of the plasma membrane and cell coats (Whaley, 1975; Morré, 1977).

The extensive movement of membrane through the cell surface during secretion could produce large-scale transient changes in surface permeability or other properties. For gland cells it is suggested that secretory granule membranes are so constituted as, for example, to avoid making the cell surface more permeable to ions or water (Palade, 1975). For nerve terminals the situation is less clear. Although it is not certain how synaptic vesicles fill with transmitter, most of the proposals referred to in Section III,C suggest that transmitters or immediate precursors can cross the vesicle membrane. Thus, in the absence of controls or other contravening factors, incorporation of vesicle membrane into the cell surface might open channels for passage of transmitter between axoplasm and external space. Does this bear on speculations that neurotransmission actually depends on release of transmitters from the axoplasm rather than from synaptic vesicles (see, e.g., Marchbanks, 1979). Or does it help explain the recently documented "leakage" of acetylcholine from the axoplasm of the neuromuscular junction (Katz and Miledi, 1977)?

Cells engaged in extensive receptor-mediated absorptive endocytosis might be expected to exhibit depletion of cell surface receptors, as these receptors are internalized through endocytosis. In some cases this does occur and replacement takes some time, as if new receptor molecules have to be synthesized (Haigler and Cohen, 1979). But in other cases, such as the uptake of low-density lipoproteins (LDL) by fibroblasts, the surface population of receptors remains abundant over prolonged periods despite high rates of endocytosis and in the face of conditions (e.g., the presence of inhibitors of protein synthesis) that seem to preclude replacement by newly made receptor (Brown and Goldstein, 1979). The simplest explanation for such cell surface "constancy" probably is the recycling of receptor-rich membrane to the surface, though this has yet to be demonstrated directly. But if such return is from the lysosomes, where the LDL is deposited, how are the receptors protected from damage during their passage through the lysosome surface? Is it possible that the receptors never get that far, either somehow escaping internalization from the cell surface, or transferring their load to an intermediate carrier before returning to the plasma membrane (cf. Fig. 2C and Brown and Goldstein, 1979)?

There are several situations in which cyclical changes in the amount of membrane present at the cell surface are accomplished, at least in part, through bulk membrane movement. One example is the rhabdomere—the microvillar system of the photoreceptor cells of a variety of invertebrates. The extent of the rhabdomere varies considerably with illumination conditions (e.g., during the diurnal cycle: for references, see Eguchi and Waterman, 1979; Holtzman and Mercurio, 1980). How membrane is added to the microvilli is still unclear; the ER and Golgi apparatus are thought to be involved. In some species, loss of membrane may occur through shedding to the extracellular space, somewhat analogous to the disc-shedding of the vertebrate eye. However, for a number of species, the endocytic withdrawal of membrane from the microvilli into the photoreceptor cytoplasm is reported. As Section III,D indicated, MVBs seem to represent one of the sites to which the endocytized membrane is delivered (but see also Hafner and Bok, 1977, for data suggesting that the origin of the MVB contents may be complex). Is the cycle of membrane in these cells simply insertion, endocytic withdrawal, then destruction? Or is some retrieved membrane stored for later reinsertion in the surface? A similar question arises with respect to the mammalian bladder, where membrane is inserted and removed from the luminal surface as the bladder changes its geometry during filling and emptying (Hicks, 1966; Minsky and Chlapowski, 1978; Porter et al., 1967).

Can qualitative changes in the cell surface, important for the cell's physiological properties, be mediated by bulk membrane movements? The toad bladder's permeability to water increases in response to antidiuretic hormone. We have shown that this is accompanied by movement of membrane between the luminal surface of the cells involved in this response, and intracellular compartments: there is increased endocytosis of HRP, as well as exocytosis of carbohydrate-containing granules (Masur et al., 1972; Gronowicz et al., 1980; we find exocytic figures within 5 minutes of hormone addition). Interesting subsequent findings by others (J. Muller et al., 1980; Wade, 1979) suggest that exocytosis-like addition of membrane bearing characteristic arrays of intramembranous particles also takes place. How these morphological changes relate to the physiological ones is not known. However, it is noteworthy that in several situations—the toad bladder, ova undergoing fertilization (see, e.g., Chandler and Heuser, 1979), and perhaps the oxyntic cell (Carlisle et al., 1978)—changes in the permeability or transport properties of the cell surface are accompanied by circulation of membrane through the surface. It may also be significant that some regions of membrane growth, such as neuronal growth cones (references in Holtzman and Mercurio, 1980), show relatively high levels of endocytosis, raising the possibility that membrane retrieval here might contribute to alternation or "maturation" of the growing membrane.

## IV.   Final Comments

My intention in this article is to provide a framework for those that follow. Thus, perhaps the most important conclusion to be drawn from the material reviewed is that the dynamics of membranes remain elusive and difficult to study. Despite this, progress seems increasingly rapid. It was not very long ago that the mechanisms of biogenesis and insertion of membrane macromolecules were almost entirely matters of speculation. Similarly, the analysis of the mobility of membrane molecules is a quite recent development. And clear demonstrations of membrane recycling date back only a few years. The accomplishments described in the chapters that follow and those presented earlier sample highlights of recent progress. They also give promise of much more to come.

## ACKNOWLEDGMENTS

Work in my laboratory is currently supported by National Institutes of Health Grant EY 03168 from the National Eye Institute. Prior support was from the National Institute for Neurological and Communicative Disorders and Stroke. Technical assistance was provided ably by Fe Reyes, and photographic assistance by Tana Ross. Associates and students currently in the laboratory who have contributed material underlying this paper include S. K. Masur, A. Mercurio, L. Liscum, F. Ungar, and M. L. Matheke.

## REFERENCES

Abrahams, S., and Holtzman, E. (1973). *J. Cell Biol.* **56,** 540–558.

Abrahamson, D. R., and Rodewald, R. (1979). *J. Cell Biol.* **83,** 284A.

Allen, R. D. (1974). *J. Cell Biol.* **63,** 904–922.

Bennett, H. S. (1969). *In* "Handbook of Molecular Cytology" (A. Lima-da-Faria, ed.), pp. 1294–1319. Elsevier/North-Holland Publ., Amsterdam and New York.

Bunt, A. H., and Haschke, R. H. (1978). *J. Neurocytol.* **7,** 665–678.

Broadwell, R. D., and Brightman, M. W. (1979). *J. Comp. Neurol.* **185,** 31–74.

Brown, M. S., and Goldstein, J. L. (1979). *Proc. Natl. Acad. Sci. U.S.A.* **76,** 3330–3337.

Carlisle, K. S., Chew, C. S., and Hersey, S. J. (1978). *J. Cell Biol.* **76,** 31–42.

Ceccarelli, B., Grohovaz, F., and Hurlburt, W. P. (1979). *J. Cell Biol.* **81,** 178–192.

Chandler, D. E., and Heuser, J. S. (1979). *J. Cell Biol.* **82,** 91–108.

Cowan, W. M. and Cuénod, M., eds. (1975). "The Use of Axonal Transport in the Study of Neuronal Connectivity." Elsevier, Amsterdam.

Dean, R. T. (1977). *Biochem. J.* **168,** 603–605.

DeDuve, C. (1973). *J. Histochem. Cytochem.* **21,** 941–948.

Douglas, W. W., Nagasawa, J., and Schulz, R. (1971). *Mem. Soc. Endocrinol.* **19,** 353–378.

Edelson, P. J. (1974). *J. Exp. Med.* **140,** 1364–1386, 1387–1398.

Eguchi, E., and Waterman, T. H. (1976). *Cell Tissue Res.* **169,** 419–434.

Eguchi, E., and Waterman, T. H. (1979). *J. Comp. Physiol.* **131,** 191–203.

Evans, J. A., Hood, D. C., and Holtzman, E. (1978). *Vision Res.* **18,** 145–151.

Farquhar, M. G. (1978). *J. Cell Biol.* **77,** R35–R42.

Flickinger, C. J. (1978). *J. Cell Sci.* **34,** 53–63.

Fujii-Kuriyama, Y., Negishi, M., Mikawa, R., and Tashiro, Y. (1979). *J. Cell Biol.* **81,** 510–519.

Goldman, R. (1974). *FEBS Lett.* **46,** 203–208.

Gronowicz, G., Masur, S. K., and Holtzman, E. (1980). *J. Membr. Biol.* **52,** 221–235.

Grynszpan-Winograd, U. (1971). *Philos. Trans. R. Soc. London, Ser. B* **261,** 291–298.

Hafner, G. S., and Bok, D. (1977). *J. Comp. Neurol.* **174,** 397–416.

Haigler, H. T., and Cohen, S. (1979). *Trends Biochem. Sci.* **4,** 132–134.

Herzog, V., and Farquhar, M. G. (1977). *Proc. Natl. Acad. Sci. U.S.A.* **74,** 5073–5077.

Herzog, V., and Miller, F. (1979). *Eur. J. Cell Biol.* **19,** 203–215.

Heuser, J. E. (1977). *In* "Motor Innervation of Muscle" (S. Thesleff, ed.), pp. 51–116. Academic Press, New York.

Heuser, J. E., and Reese, T. S. (1977). *In* "Handbook of Physiology" (E. Kandel, ed.), Rev. ed. Sect. 1, pp. 261–294. Waverly Press, Baltimore, Maryland.

Hicks, R. M. (1966). *J. Cell Biol.* **30,** 623–643.

Holtzman, E. (1976). "Lysosomes; A Survey," Cell Biol. Monogr., Vol. 3, Springer-Verlag, Berlin and New York.

Holtzman, E. (1977). *Neuroscience* **2,** 337–355.

Holtzman, E., and Mercurio, A. (1980). *Int. Rev. Cytol.* **67,** 1–67.

Holtzman, E., Schacher, S., Evans, J., and Teichberg, S. (1977). *Cell Surf. Rev.* **4,** 165–246.

Holtzman, E., Gronowicz, G., Mercurio, A., and Masur, S. K. (1979). *Biomembranes* **10,** 77–139.

Howell, K. E., Ito, A., and Palade, G. E. (1978). *J. Cell Biol.* **79,** 581–589.

Jaken, L., and Thinès-Sempoux, D. (1978). *Cell Biol. Int. Rep.* **2,** 515–524.

Joseph, K. C., Stieber, A., and Gonatas, N. K. (1979). *J. Cell Biol.* **81,** 543–554.

Katz, B., and Miledi, R. (1977). *Proc. R. Soc. London, Ser. B* **196,** 59–72.

Lawson, D., Raff, M. C., Gomperts, B., Fewtrell, C., and Gilula, N. B. (1977). *J. Cell Biol.* **72,** 242–259.

Liscum, L., Holtzman, E., and Hood, D. C. (1980). *Invest. Ophthalmol.,* Suppl., April, 1980, pp. 131–132 (ARVO abst.).

Llinas, R. R., and Heuser, J. E., eds. (1977). "Depolarization-Release Coupling Systems in Neurons," Neurosci. Res. Program Bull. 15, No. 4. MIT Press, Cambridge, Massachusetts.

McKanna, J. A., Haigler, H. T., and Cohen, S. (1979). *Proc. Natl. Acad. Sci. U.S.A.* **76,** 5689–5693.

Marchbanks, R. M. (1979). *Trends Neurosci.* **2,** 56.

Masur, S. K., Holtzman, E., and Walter, R. (1972). *J. Cell Biol.* **52,** 211–219.

Meldolesi, J., Borgese, N., DeCamilli, P., and Ceccarelli, B. (1978). *Cell Surf. Rev.* **5,** 510–588.

Minsky, B. D., and Chlapowski, F. J. (1978). *J. Cell Biol.* **77,** 685–697.

Morré, D. J. (1977). *Cell Surf. Rev.* **4,** 1–83.

Morris, J. F., Nordmann, J. J., and Dyhall, R. E. J. (1978). *Int. J. Exp. Pathol.* **18,** 1–28.

Muller, J., Kachadorian, W. A., and Scala, V. A. (1980). *J. Cell Biol.* **85,** 83–95.

Muller, W. A., Steinman, R. M., and Cohn, Z. A. (1980). *In* "Mononuclear Phagocytes: Functional Aspects." Nijhoff, The Hague (in press).

Nagatsu, I., Kondo, Y., Kato, T., and Nagatsu, T. (1976). *Brain Res.* **116,** 277–285.

Novikoff, A. B. (1976). *Proc. Natl. Acad. Sci. U.S.A.* **73,** 2781–2787.

Palade, G. E. (1959). *In* "Subcellular Particles" (T. Hayashi, ed.), pp. 64–83. Ronald Press, New York.

Palade, G. E. (1975). *Science* **189,** 347–357.

Pelletier, G. (1973). *J. Ultrastruct. Res.* **43,** 445–459.

Porter, K. R., Kenyon, K., and Badenhausen, S. (1967). *Protoplasma* **63,** 262–274.

Rome, L. H., Garvin, A. H., Allietta, M. M., and Neufeld, E. (1979). *Cell* **17,** 143–155.

Schacher, S., Holtzman, E., and Hood, D. C. (1976). *J. Cell Biol.* **70,** 178–192.

Schneider, Y.-J., Tulkens, P., DeDuve, C., and Trouet, A. (1979). *J. Cell Biol.* **82,** 449–465, 466–474.

Silverstein, S. C., Steinman, R. M., and Cohn, Z. A. (1977). *Annu. Rev. Biochem.* **46,** 669–722.

Specian, R. D., and Neutra, M. R. (1979). *J. Cell Biol.* **83,** 429A.

Steinberg, R. H., Fisher, S. K., and Anderson, D. H. (1980). *J. Comp. Neurol.* **190,** 501–518.

Teichberg, S., Holtzman, E., Crain, S. M., and Peterson, E. R. (1975). *J. Cell Biol.* **67,** 215–230.

Tokunaga, A., Sandri, C., and Akert, K. (1979). *Brain Res.* **174,** 207–219.

Ungar, F., Piscopo, I., and Holtzman, E. (1980). *Brain Res.* **205,** 200–206.

Wade, J. B. (1979). *Curr. Top. Membr. Transp.* (in press).

Wagner, J. A., Carlson, S. S., and Kelly, R. B. (1978). *Biochemistry* **17,** 1199–1206.

Wallach, D., Kirchner, N., and Schramm, M. (1975). *Biochim. Biophys. Acta* **375,** 87–105.

Wehland, J., Willingham, M. C., Dickson, R., and Pastan, I. (1981). *Cell* **25**, 105–120.
Whaley, W. G. (1975). "The Golgi Apparatus," Cell Biol. Monogr., Vol. 2. Springer-Verlag, Berlin and New York.
Winkler, H. (1977). *Neuroscience* **2**, 657–683.
Wirtz, K. W. A., and van Deenen, L. L. M. (1977). *Trends Biochem. Sci.* **2**, 49–51.
Young, R. W. (1976). *Invest. Ophthalmol.* **15**, 700–725.
Zimmerman, H. (1979). *Prog. Brain Res.* **49**, 141–151.

# Chapter 25

# Membrane Recycling in Secretory Cells: Implications for Traffic of Products and Specialized Membranes within the Golgi Complex

## MARILYN GIST FARQUHAR

*Section of Cell Biology,*
*Yale University School of Medicine,*
*New Haven, Connecticut*

# I.   Introduction

Release of secretory products by exocytosis involves the continual insertion into the plasmalemma of a considerable amount of membrane derived from secretory granules or vesicles. At the time that exocytosis was first discovered (Palade, 1959), it was recognized that membrane must be continually removed from the cell surface to compensate for that added during exocytosis in order to maintain cell size constant. To explain the mechanism involved, several hypotheses have been proposed: (1) recycling, or recovery of the same membrane and recirculation back to the Golgi (Palade, 1959, 1975); (2) dismantling to macromolecular components, which are subsequently reassembled (Fawcett, 1962); or (3) complete degradation (Amsterdam *et al.*, 1971). The available evidence that bears on this problem has been contradictory and conflicting (see below).

In this article a brief summary of the work on this problem done by our group in the past few years is presented, which points to the existence of considerable membrane recycling involving Golgi and lysosomal elements in secretory cells. The results summarized also provide some new information on the traffic of membranes and products through the Golgi complex.

# II.   General Background

Evidence that has contributed to our understanding of the fate of granule membrane has come from two main sources: (1) biochemical data in which the turnover of granule membrane components has been studied and compared with that of their content, and (2) morphological data obtained with the electron microscope in which various electron-opaque tracers were used to trace the fate of the recovered membrane.

## A.   Results Obtained by Studying Rates of Turnover of Granule Membranes vs  Their Contents

To obtain biochemical data on the turnover of granule membranes, the following type of experiment has been done: a pulse of a radioactive amino acid is administered (*in vivo* or *in vitro*), secretory granules are isolated after 1–2 hours of chase when the label is known to be concentrated in the granules, the granules are lysed, and the amount of labeling of membrane proteins vs. that of the content proteins is determined. The early data obtained in this type of experiment were misleading because the results indicated that the granule membranes and the content were both labeled to the same extent, suggesting that they turn over at the same rate (Amsterdam et al., 1971). It is now clear that the granule membranes used in these studies were heavily contaminated with sticky content proteins. More recently, data have been obtained in several laboratories (Meldolesi, 1974; Castle et al., 1975; Wallach et al., 1975) on granule membrane fractions (in which great care was taken to assure removal of contaminating content proteins) that show a distinct difference in the labeling pattern. In all these cases the membrane proteins were much less heavily labeled than the content proteins, indicating that the membrane proteins turn over at a much slower rate. Thus, the recent turnover data imply that the membrane or its constituents are reutilized rather than degraded with each round of exocytosis; however, these data do not provide any insight as to the mechanism involved—i.e., whether the membranes are recovered intact, or whether they are partially or completely dismantled and subsequently reassembled before being reutilized. Moreover, they do not indicate the fate of the recovered membrane or membrane constituents within the cell.

## B.   Results Obtained with Electron-Dense Tracers

Direct information on these points has been provided by tracer studies carried out by electron microscopy. In this type of experiment, granule discharge is stimulated in a secretory cell with an appropriate secretagogue, and an electron-dense marker is introduced to trace the fate of the recovered membrane.

In the 1970s, experiments of this type were carried out by a number of investigators, using, for the most part, native ferritin and horseradish peroxidase (HRP) as tracers. In all cases, the tracers were found in invaginations of the cell membrane or in small vesicles located near the cell surface at early time points (2–15 minutes) and in lysosomes at later time points (1–2 hours). Results of this type were obtained on a number of secretory cells, including neurosecretory neurons of the posterior pituitary (Nagasawa et al., 1971; Douglas, 1973; Theodosis et al., 1976), and cells of the adrenal medulla (Nagasawa and Douglas, 1972; Abrahams and Holtzman, 1973), anterior pituitary (Pelletier, 1973; Farquhar et al., 1975), endocrine pancreas (Orci et al., 1973), exocrine pancreas (Geuze and Kramer, 1974), seminal vesicle (Mata, 1976), and parotid gland (Kalina and Rabinovitz, 1975; Oliver and Hand, 1978). These tracer experiments clearly established that after exocytosis surface membrane is recovered intact, in the form of small vesicles—that is, exocytosis is coupled to endocytosis.

The fact that the tracers taken up by endocytosis eventually became concentrated in lysosomes led the majority of these workers to conclude that after recovery the surface membrane is degraded in lysosomes.[1] In retrospect, it is clear that, since native ferritin and HRP are found in the fluid contents of endocytic vesicles and serve as content rather than membrane markers, the information provided by these tracers is limited to the identification of compartments with which the incoming vesicles fuse. Interpretations of the data should therefore be tempered by this consideration. However, the tracer data appeared to fit nicely with the erroneous early turnover data (discussed above). Both supported the conclusion that granule membranes are degraded rather than reutilized after exocytosis; as a result, this idea gained wide acceptance. With the publication of more valid turnover data based on improved membrane preparations, a discrepancy emerged between implications of the turnover data (which were compatible with extensive reutilization of membrane constituents) and most of the tracer data, which pinpointed the lysosomes as a dead end and implied membrane degradation.

To summarize, the question of the fate of granule membranes after exocytosis has been a controversial and debated topic, with most of the available experimental data apparently supporting the notion that they are degraded or partially dismantled. As a consequence, the idea that granule membranes are recovered intact and degraded in lysosomes has prevailed, and the existence of recycling has been contested up until quite recently (see Chapter 24 by E. Holtzman; for recent reviews see Holtzman *et al.,* 1978); Oliver and Hand, 1978; Meldolesi *et al.,* 1978.

## C.   Initial Results on Anterior Pituitary Cells

Several years ago we obtained results (referred to in footnote 1) on anterior pituitary cells using HRP (Farquhar *et al.,* 1975) that suggested that, in addition to lysosomes, other cell compartments, primarily the Golgi complex, might be involved in recovery of surface membrane. As a result, we began to systematically study membrane recovery in secretory cells, using tracers other than HRP or native ferritin.

The first tracers we worked with were dextrans, which we selected because

---

[1]Among these studies the only well-characterized glandular cells in which an extracellular tracer appeared in a cell compartment other than endocytic vesicles and lysosomes were those of the anterior pituitary (cells), where HRP was found in a single Golgi cisterna along the trans side of the Golgi stack and in forming granules (Pelletier, 1973; Farquhar *et al.,* 1975). In addition, Gonatas and his co-workers, using labeled lectins (Gonatas *et al.,* 1977) or cholera toxin (Joseph *et al.,* 1979), have traced surface membranes to a single cisterna along the trans side of the Golgi stack (referred to as GERL) in cultured neurons, and Mata (1976) has found uptake of HRP into Golgi cisternae in the secretory epithelium of the hamster seminal vesicle. However, it is not clear whether this traffic in cultured neurons or in the seminal vesicle is related to secretion or to pinocytosis.

they are uncharged, relatively inert, and available in a wide range of sizes. The systems we studied were cells of the parotid and lacrimal glands, which we chose because these cells are very active exocrine secretory elements, they can be maintained satisfactorily *in vitro* (as isolated acini, lobules, or dissociated cells), and exocytosis of their granules can be readily stimulated with appropriate secretagogues.

# III.   Studies on Exocrine Cells of the Parotid and Lacrimal Glands Using Dextrans

The first informative results we obtained on pathways of membrane retrieval were on cells from the parotid and lacrimal glands using dextrans (Herzog and Farquhar, 1977).

## A.   Experiments Carried Out *in Vitro*

We prepared isolated acini (by collagenase digestion) and incubated the acini *in vitro* in the presence of dextrans, adding either isoproterenol (parotid) or carbamylcholine (lacrimal) to induce granule discharge (Amsterdam *et al.*, 1969). Electron microscopy revealed that, as in the case of previous work with HRP and ferritin, exocytosis was followed by endocytosis. At early time points (30 minutes) dextran was found in coated pits or in endocytic vesicles located in the apical cytoplasm (between the lumen and the Golgi complex) (see Figs. 1–3). However, the destination of the incoming endocytic vesicles differed from that traced in the previous studies: after 1 hour, dextran particles were frequently seen in multiple Golgi cisternae (see Fig. 3) and in condensing vacuoles on the trans side of the Golgi stack as well as in lysosomes. The main new finding was that the incoming vesicles carrying the dextran fused with multiple Golgi cisternae and condensing vacuoles as well as with lysosomes. The fact that the trans cisternae were usually much more heavily labeled than those on the cis side, and that the tracer was concentrated in their dilated rims (refer to Fig. 3), suggested that the incoming vesicles fuse preferentially with the dilated rims of the transmost cisternae. It should be noted that in many tissues it is the transmost cisternae that has been identified by Novikoff (1976) (on the basis of cytochemical reactivity for acid phosphatase) as the principal component of the GERL system, which he assumes connects the Golgi with the ER and lysosomes.

## B.   Findings Obtained *in Vivo*

A limitation of the *in vitro* incubation system was that the dextran present in the medium had access to all fronts of the cell. Therefore, the dextran found in

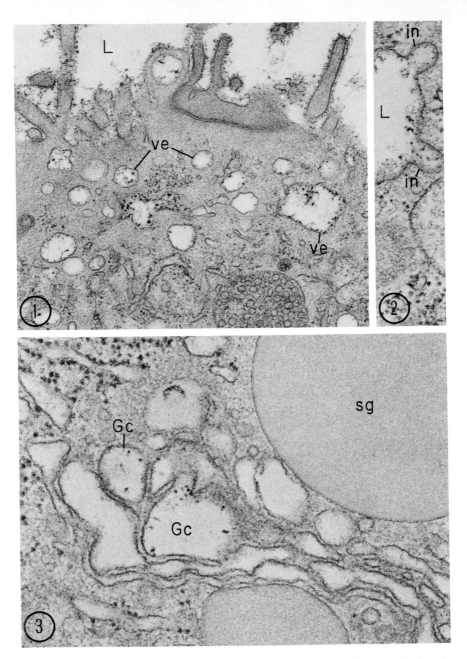

FIGS. 1–3.   Results obtained using dextrans to trace the fate of surface membrane in cells from the lacrimal (Fig. 1) and parotid (Figs. 2 and 3) glands. These figures show that, after exocytosis, membrane is recovered by endocytosis and the *resultant* endocytic vesicles subsequently fuse with Golgi cisternae. Figure 1 is from an acinus isolated from the rat lacrimal gland and incubated *in vitro* with dextran (T-10) for 10 minutes. Multiple endocytic vesicles (ve) containing dextran are seen in

the Golgi and condensing vacuoles could have been derived from the basal or lateral cell surfaces, rather than the luminal front, and could represent traffic other than that related to recovery of membrane from the luminal aspect of the cell. To eliminate this possibility we exposed parotid cells to dextran exclusively along their luminal surface by infusing rats with dextran retrograde up the parotid duct. The results were very similar to those obtained *in vitro* except that (1) uptake into Golgi cisternae occurred much faster and (2) much less tracer was seen in lysosomes. The surprising finding was that, as early as 5 minutes after an *in vivo* infusion, dextran was found in the stacked Golgi cisternae (Fig. 3).

To summarize the results of both the *in vivo* and the *in vitro* experiments, the two novel findings were, first, the demonstration that the tracer can reach—directly or indirectly—most of the cisternae in a given Golgi stack, and, second, the demonstration of the rapidity with which movement of the vesicles and their fusion with the Golgi cisternae takes place—i.e., within 5 minutes. An additional finding of interest was that *in vivo* this rapid traffic to Golgi cisternae and to condensing vacuoles clearly predominated over that to lysosomes. The limitation in this work is the fact that dextran (like HRP and native ferritin), as far as is known, is a content marker. This means that we could identify the compartments with which the incoming vesicles fuse, but we could not trace the fate of specific membrane patches after fusion with the receiving compartment.

The results with dextrans differed from those obtained by others on the same cell type (parotid gland cell) with other tracers, primarily native ferritin (Kalina and Rabinovitz, 1975) and HRP (Oliver and Hand, 1978). The reasons for these differences are still not clear. It is worth mentioning, however, that dextrans are neutral (uncharged) particles, whereas native ferritin is anionic and HRP is a mixture of isozymes with different isoelectric points $(pI = 4.2–8.6)$ (Rennke *et al.*, 1978; Ottosen *et al.*, 1980).

## IV. Studies on Anterior Pituitary Cells Using Anionic and Cationic Ferritins

The next series of observations pertaining to the problem of membrane recovery in secretory cells was made using ferritins, especially cationic ferritin (CF), to study the mammotroph or prolactin-secreting cell of the rat anterior pituitary

---

the apical cytoplasm. Figures 2 and 3 are fields from parotid acinar cells of rats given isoproterenol (to stimulate exocytosis) followed by an infusion of dextran T-40 into the duct, and fixed 5 minutes after the infusion. Dextran is seen within coated invaginations (in) of the luminal (L) cell membrane and within several of the stacked Golgi cisternae (Gc). sg = secretion granule. Fig. 1, ×17,000; Fig. 2, ×48,000; Fig. 3, ×57,000. From Herzog and Farquhar (1977).

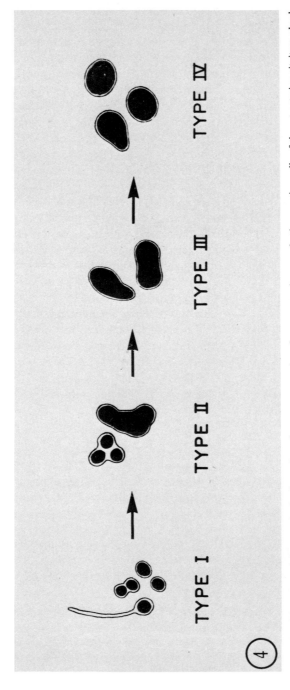

TYPE I          TYPE II          TYPE III          TYPE IV

Fig. 4. Diagram to show steps in the formation of secretory granules of mammotrophs or prolactin-secreting cells of the rat anterior pituitary gland. Prolactin is initially concentrated into small (type I) granules within the transmost Golgi cisternae. These pinch off and several aggregate into polymorphous forms (type II), which round up and take on simpler contours (type III) before achieving their mature rounded or ovoid shape (type IV). From Farquhar *et al.* (1978).

gland (Farquhar, 1978a). Cationic ferritin was chosen as a tracer because, owing to it net positive charge, it binds to negatively charged groups on the cell membrane (Danon *et al.*, 1972); therefore, it might be expected to act as a membrane marker under appropriate conditions. We chose the mammotroph as the cell to study because for some time we have been using it as a model for investigating mechanisms and pathways of secretion in endocrine cells (Farquhar, 1971, 1977). Based on a combination of morphological, cytochemical, cell fractionation, and autoradiographic studies carried out over the last 15 years, a good deal of information has been accumulated on the secretory process in this cell type. In particular, information is available concerning the steps in the formation of mammotroph secretory granules and on the composition of the granule membranes and contents. The granules originate in the dilated rims of the transmost Golgi cisternae and are assembled in a stepwise fashion over a period of 1–3 hours, as depicted in Fig. 4 (Farquhar, 1971; Farquhar *et al.*, 1978). They contain prolactin as the main component (>80%) plus sulfated glycopeptides and glycosaminoglycans as minority components (Zanini and Giannattasio, 1974; Giannattasio and Zanini, 1976; Slaby and Farquhar, 1980, Zanini *et al.*, 1980).

For our recycling studies we used a suspension of dissociated cells maintained in culture (Farquhar *et al.*, 1975). A cell suspension was prepared (by enzyme digestion) from pituitaries of animals pretreated with estrogen *in vivo* (to stimulate prolactin secretion), tracers were then added to a suspension of cultured cells, and the fate of surface membrane was followed. A number of tracers were tried, but the results with CF were the most informative.

## A.   Results Obtained with Cationized Ferritin (p*I* > 7.8)

When CF (0.05 mg/ml) was added to a pituitary cell suspension at 0°C or 25°C, it immediately bound to the cell surfaces (where it was found in a continuous layer one to two molecules deep) (Fig. 5). If the cells were subsequently incubated at 37°C, the findings were quite similar to those obtained previously with dextrans in the parotid and lacrimal glands: CF was taken up by endocytosis (Fig. 5), and, after 10–60 minutes of incubation, it was found in increasing amounts within multiple stacked Golgi cisternae, in forming granules, and in lysosomes, indicating that the incoming vesicles containing CF had fused with all these compartments (Figs. 6–8). Some of the CF within vesicles and Golgi cisternae was located in close proximity to the membrane, apparently adhering to it (Fig. 8); however, in other cases the CF formed part of the vesicle content (Figs. 5 and 7).

Additional information on the route of the incoming vesicle traffic was obtained by studying the kinetics of CF uptake into forming granules. At 15–30 minutes, CF was limited to the earliest (type I) granules forming within the dilated rims of the transmost Golgi cisternae (Fig. 7) or located in close prox-

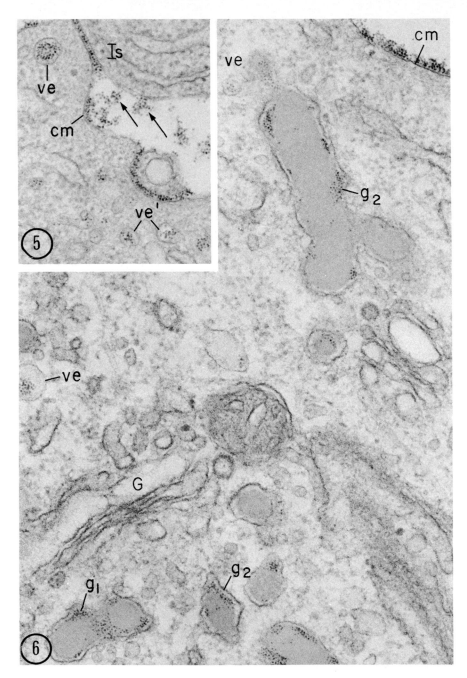

FIGS. 5 AND 6.   Mammotrophs from estrogen-treated female rats incubated with CF (0.1 mg/ml) to trace the surface membrane. Figure 5 shows that initially (after 15 minutes of incubation at 37°C)

imity to the latter (Fig. 8); and only later was it regularly found in more mature granule forms (Fig. 6) (in type II and type III granules after 1 hour, and in fully mature type IV granules after 2–3 hours). Within the granules CF was typically located between the membrane and the dense content, adhering to the periphery of the dense content (Figs. 6–8). Similar findings were obtained in other pituitary cell types (Fig. 9). Figure 9 illustrates a somatotroph with heavy accumulations of CF in multiple Golgi cisternae and granules.

These findings indicated that CF is added to the granules at the earliest steps in granule formation and appears in more mature forms only after maturation and assembly of these early labeled forms. Based on these results the following conclusions were reached concerning the route of the incoming CF-labeled vesicles: (1) They preferentially fuse with Golgi cisternae and closely associated small, immature granules; (2) although they can fuse with all the Golgi cisternae, fusion with the transmost cisternae from which granules arise is more frequent; (3) fusion with more mature granules is infrequent; and (4) fusion with cell compartments other than Golgi, lysosomes, or forming granules (e.g., ER, mitochondria, nuclear envelope) is rare or nonexistent.

## B. Results Obtained with Native (Anionic) Ferritin (p$I$ ~4.8)

When native (anionic) ferritin was used as a tracer under similar conditions, the results were quite different. In contrast to CF, native ferritin did not bind to the cell membrane; it was taken up by endocytosis in much smaller amounts (as part of the vesicle content) and was not seen in Golgi cisternae or condensing granules. Even when added to the incubation medium at 100× the concentration of CF (5 mg/ml, as opposed to 0.05 mg/ml), it remained confined to the lysosomal system (i.e., endocytic vesicles, lysosomes, or multivesicular bodies) and was never seen in elements along the secretory pathway (Golgi cisternae or secretion granules).

---

CF binds to the cell membrane (cm) and is taken up by endocytosis. Numerous endocytic vesicles (ve) containing CF are seen in the cytoplasm near the plasmalemma. The CF is aggregated on the free cell surface (arrows) but forms a regular layer one to two molecules deep in the intercellular spaces (Is). Inside the vesicles (ve') the CF is also aggregated; as a consequence some CF molecules are closely associated with the vesicle membrane, whereas others are not. Figure 6 shows that after longer periods of incubation (1 hour at 37°C) CF is seen within immature (types I and II) secretory granules ($g_1$ and $g_2$) as well as in endocytic vesicles (ve). The incoming CF-labeled vesicles apparently fuse preferentially with the dilated rims of the transmost Golgi cisternae (see Fig. 7), and the CF becomes trapped within the forming granules, where it is typically located at the periphery of the dense content. cm = Cell membrane; G = Golgi cisternae. Fig. 5, ×70,000; Fig. 6, ×76,000. From Farquhar (1978a).

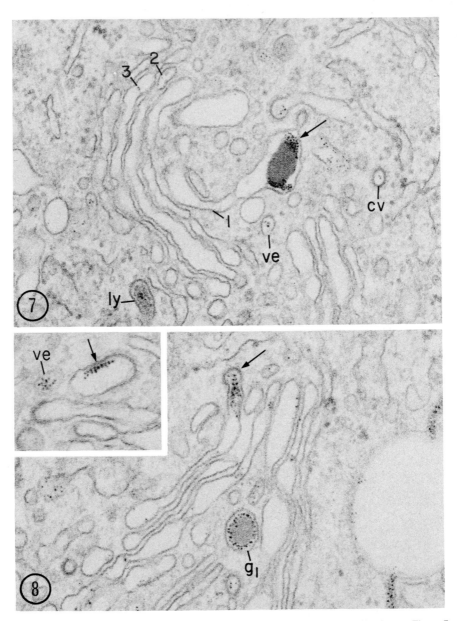

FIGS. 7 AND 8. Fields from pituitary mammotrophs incubated with CF for 60 minutes. Figure 7 shows CF within three Golgi cisternae (1–3). The tracer is particularly concentrated around a secretory granule forming within the transmost Golgi cisterna (arrow), suggesting that the incoming vesicles carrying CF fuse preferentially with these Golgi elements. Note that the CF is concentrated at the periphery fo the forming granule adhering to its dense contents. CF is also seen within

## C.   Experiments Using Cationic Ferritins with Different Isoelectric Points (p$I$ = 6.0–8.8)

Further experiments were carried out with CF fractions with isoelectric points intermediate between that of native ferritin (p$I$ ~4.8) and commercially available CF (p$I$ > 8.0), prepared as described previously (Kanwar and Farquhar, 1979). These indicated that the key factor in determining how the tracer is processed intracellularly is its ability to interact with, and bind to, cell membranes. This conclusion is based on the finding that narrow-range fractions with isoelectric points of 4.8–7.3 do not bind to the membranes of pituitary cells under the conditions tested, act as fluid-phase or content markers, and appear only in lysosomes, whereas CF fractions with a p$I$ > 7.5 (which includes the CF preparation commercially available from Miles) binds to the cell membrane, acts at least initially as a membrane marker,[2] and appears in Golgi cisternae and forming secretory granules as well as in lysosomes.

Although CF binds to membranes, it is not an ideal membrane marker, since its binding depends primarily on charge interaction, and thus the tracer is susceptible to detachment if the vesicle fuses with a compartment in which there is competition for binding with acidic groups of higher charge density than those of the vesicle membrane. For example, one can speculate that the location of CF in association with the content (rather than the membrane) of forming granules is due to the fact that, when the incoming vesicles containing CF fuse with immature granules, the CF interacts with and forms complexes with the highly negatively charged sulfated macromolecules (glycosaminoglycans and glycopeptides) present in the mammotrophs' granules, thereby displacing the tracer from the vesicle membrane and trapping the resultant aggregates within the granules.

## D.   Conclusions

The findings with CF in pituitary cells were reassuring because they provided another example obtained with a different tracer on a very different (endocrine)

---

[2]It should be pointed out, however, that, since the CF tends to form aggregates at the cell surface (see Fig. 5), only a fraction of the CF present (presumably unaggregated CF or that located at the periphery of the aggregates) binds to the cell membrane and remains associated with the membrane of the endocytic vesicle after internalization.

---

several vesicles (ve), one of which is coated (cv) in the Golgi region and within a lysosome (ly). Figure 8 shows CF within multiple stacked Golgi cisternae and within a forming granule ($g_1$). One cisterna loaded with CF has a coated region at its tip (arrow), suggesting that a CF-loaded vesicle has just fused with the Golgi cisterna. The inset depicts another Golgi cisterna with a row of CF molecules attached to its membrane, which appears coated on part of its surface (arrow). Fig. 7, ×70,000; Fig. 8, ×87,000; inset, ×100,000. From Farquhar (1978a).

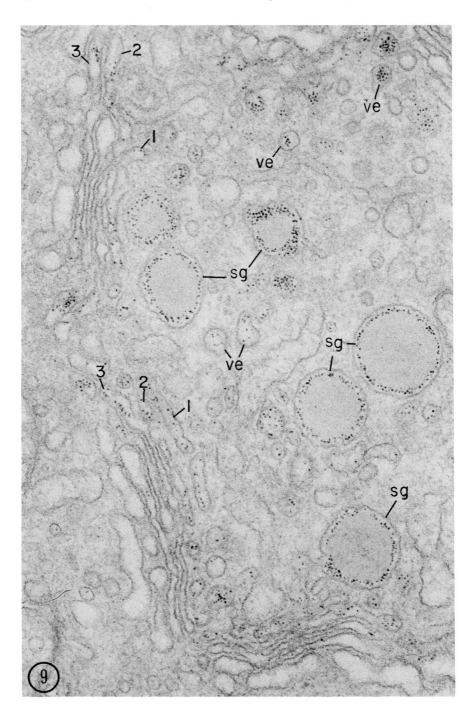

type of a secretory cell in which recovered surface membrane reaches multiple stacked Golgi cisternae and condensing granules. The fact that the traffic of incoming vesicles was heaviest to Golgi cisternae and the earliest (type I) condensing granules strongly suggests that this traffic is connected with recycling of granule membranes. In addition, the different results obtained with the anionic and cationic forms of the same molecule differing only in net charge were very instructive. They showed that, depending on the charge of the ferritin and its ability to bind to the cell membrane, the vesicle can be traced to a different destination.

## E. Are There Two Recovery Routes, or One Route with Two Stops?

To explain the different results obtained with differently charged ferritins, the simplest assumption is that two different recovery routes (plasmalemma → lysosomes, and plasmalemma → Golgi) exist for native and cationic ferritin, respectively. However, another intriguing possibility that was considered earlier (Farquhar, 1978a) is that there is a single recovery route with two stations—lysosomes representing the first station and Golgi cisternae the second (see Fig. 16). This would require only that the incoming vesicle lose its content at the first station (lysosome) and carry the membrane marker on to the next station (Golgi and/or forming granules), where the interaction of the marker with the local content (e.g., glycosaminoglycans) would be stronger than with the membrane. Specifically, any free or loosely associated ferritin included in the vesicle contents would be left behind at the time of fusion with lysosomes, whereas the CF bound to the membrane would be carried on to the Golgi cisternae, where it presumably complexes with sulfated macromolecules. Our results did not allow us to choose between the possibility of one vs. two recovery routes. To answer these questions more precise kinetic data were needed, and information on the types and quantities of membrane involved. Such questions are not so easily investigated when dissociated pituitary cells (or isolated parotid acini) are used because the amount of tissue available for analysis is limited. Hence, we searched for a system of secretory cells that would not present such limitations and that would be suitable for investigation of these problems. We turned to myeloma cells in culture for this purpose.

FIG. 9. Somatotroph or growth hormone-secreting cell from a male rat incubated for 60 minutes in CF (0.05 mg/ml), showing uptake of CF into multiple Golgi cisternae and secretion granules (sg). CF molecules are present within several stacked Golgi cisternae 1–3), multiple smooth vesicles (ve) in the Golgi region, and numerous secretion granules (sg) of varying size. Note that the CF molecules are located exclusively at the periphery of the dense granule contents. ×85,000. From Farquhar (1978a).

## V.  Pathways Followed by Surface Membrane in Plasma Cells and Myeloma Cells

The production of immunoglobulins by plasma cells involves the same operations and pathways as the production of secretory proteins by cells of exocrine and endocrine glands except that these cells do not concentrate their secretory product into morphologically recognizable secretory granules (Palade, 1975; Tartakoff and Vassalli, 1977). Instead, immunoglobulins are packaged in relatively dilute solution in small, membrane-limited vesicles that continually release their contents by exocytosis. Since these cells do not concentrate their product and since exocytosis is a continual, ongoing process, it seemed likely that membrane traffic (from Golgi to cell surface and back) might be relatively extensive. If such were the case, the immunoglobulin-secreting cells would represent a favorable system for investigating in further detail mechanisms and pathways of membrane recycling in secretory cells, especially because numerous established cell lines are available and they can easily be maintained in culture.

With this in mind we undertook studies on both mature plasma cells (harvested from lymph nodes of animals immunized with HRP) and myeloma cell lines. We used immunocytochemistry to identify the secretory compartments and several tracers (HRP and anionic and cationic ferritin) to trace the pathway of membrane removed from the cell surface (Ottosen *et al.*, 1980). A number of myeloma cell lines were screened, and, because of the presence of well-organized Golgi complexes and their ability to take up CF, two cell lines—RPC 5.4 and X63 Ag8 (both of which secrete IgG)—were selected for study.

### A.  Results Obtained with Cationized Ferritin

When plasma cells harvested from lymph nodes (Figs. 10 and 11) or cultured myeloma cells (Figs. 12 and 13) were incubated with CF (0.05 mg/ml) under the same conditions as anterior pituitary cells, the CF bound to the cell membranes, and upon subsequent incubation at 37°C the uptake and fate of the tracer was the same as that demonstrated previously in pituitary cells. It bound to the cell membrane and was internalized by endocytosis, often within coated vesicles (Fig. 12), and the endocytic vesicles containing the tracer fused with lysosomes, with multiple stacked Golgi cisternae, and with secretory vacuoles present in the Golgi region. After 40–60 minutes of incubation, CF was typically located in multiple stacked Golgi cisternae (Fig. 13). The presence of CF in the medium did not appear to perturb the cells, because cell viability [assessed by trypan blue exclusion and by release of lactate dehydrogenase (LDH) into the medium] and the rate of release of radiolabeled secretory products were the same in cells incubated with and without CF. In specimens in which the secretory compart-

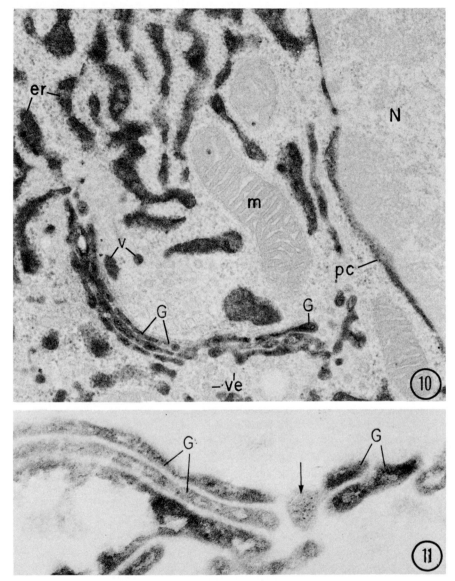

Figs. 10 and 11. Portions of plasma cells from the lymph nodes of rats immunized with HRP. The distribution of anti-HRP immunoglobulin, the cell's secretory product, is demonstrated by immunostaining. Figure 10 illustrates the presence of immunoglobulin throughout the rough ER (er) including the perinuclear cisternae (pc), in the stacked Golgi cisternae (G), and in small vesicles or vacuoles (v) in the Golgi region. Other small vesicles (ve) are not stained. Figure 11 is an enlargement of the Golgi region from the plasma cell from the same experiment as in Fig. 10, except that the cell had been incubated with CF for 60 minutes prior to fixation and immunostaining. Immunoglobulin is present throughout the stacked Golgi cisternae (G). Note the presence of CF molecules in one cisternae, which also contains immunoglobulins (arrow). Specimens fixed in glutaraldehyde, sectioned on a cryostat, and incubated with HRP, followed by incubation in diaminobenzidine. N = Nucleus; m = mitochondrion. Fig. 10, ×22,000; Fig. 11, ×75,000. From Ottosen et al. (1980).

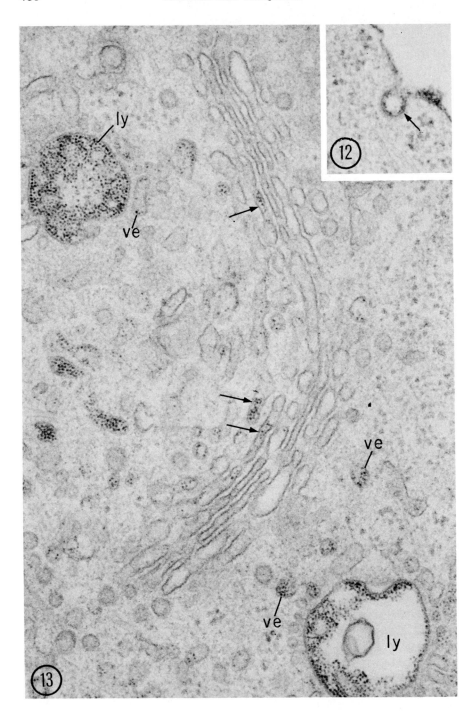

ments were identified by immunocytochemistry, the secretory product and CF could be identified within the same elements (Fig. 11).

## B.   Results Obtained with Horseradish Peroxidase and Native Ferritin

When myeloma cells were incubated with either HRP (1 mg/ml) or native ferritin (up to 5 mg/ml), these tracers, which act as content markers, were taken up by endocytosis (in the fluid content of the endocytic vesicles) and, upon prolonged incubation (1 hour), appeared only in lysosomes (Fig. 14). Neither tracer was ever seen in the stacked Golgi cisternae in these cells. The most surprising results were obtained when cells were incubated in series with both a membrane (CF) and a content (HRP) marker as follows: cells were first incubated in CF at 0°C (to allow binding of CF to the cell surface) and then washed and incubated at 37°C with HRP (up to 1 hour). The CF bound to the cell surfaces, and both the HRP and CF were taken up usually in the same vesicles; i.e., most incoming vesicles contained both HRP and CF, although a few contained CF alone. After incubation for 1 hour at 37°C, there was a distinct difference in the distribution of the two tracers: both were present in lysosomes, but only CF was seen in the stacked Golgi cisternae (Fig. 15). Thus, the eventual destination of each tracer was the same in the double tracer experiments as when the cells were incubated in each tracer alone, indicating that the cell can somehow sort the two tracers in transit and direct them to different compartments. Such results are easier to explain if the incoming endocytic vesicles first fuse with lysosomes, where they lose their contents (containing HRP and any loosely bound or free CF), and the membrane with the bound CF moves on to the Golgi.

## C.   Conclusions

The results have demonstrated the existence in plasma cells and myeloma cells of considerable membrane traffic from the cell surface to Golgi cisternae and secretory vacuoles (as well as to lysosomes), which is quite comparable to that detected previously in exocrine and endocrine glandular cells. This raises the hope that appropriately selected myeloma cell lines represent suitable and promising systems in which to investigate membrane recycling in secretory cells in further detail. In particular, the availability of a homogeneous cell population that

---

Figs. 12 AND 13.   Portion of mouse myeloma cells (RPC 5.4 cell line), fixed after incubation in CF (0.05 mg/ml) for 10 minutes (Fig. 12) or 60 minutes (Fig. 13). Figure 12 shows that CF binds to the cell surface and is taken up by endocytosis, often in coated invaginations as seen here (arrow). Figure 13 shows CF in several stacked Golgi cisternae (arrows) and in small vesicles (ve) located near the Golgi cisternae or near lysosomes (ly). The two large lysosomes in the field also contain considerable CF. Fig. 12, ×76,000; Fig. 13 ×57,000. From Ottosen et al. (1980).

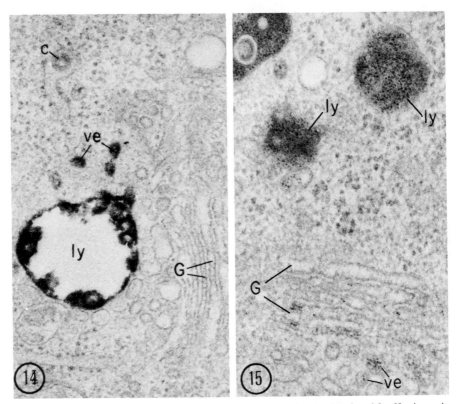

Figs. 14 and 15.    Figure 14 shows a mouse myeloma cell (RPC 5.4) incubated for 60 minutes in HRP at 37°C. Peroxidase is seen in a large lysosome (ly) located near the Golgi cisternae (G) and in a few small vesicles (ve) located nearby. Note that no peroxidase is present in the Golgi cisternae or in vesicles associated with them. Figure 15 shows a similar cell incubated in CF followed by HRP. Both CF and HRP are present in two large lysosomes (ly); CF is also seen within several Golgi cisternae (G) and associated vesicles (ve), but these elements do not contain HRP. c = virus particle. Fig. 14, ×65,000; Fig. 15, ×75,000. From Ottosen *et al.* (1980).

can be propagated in culture in large quantities makes this a more feasible system in which to carry out cell fractionation and subsequent biochemical studies than the acini and dissociated cell systems used previously.

## VI.    General Conclusions and Implications of Findings

### A.    Widespread Nature of Membrane Recycling

We have now obtained results (reviewed above) on five different cell types using two different tracers, which indicate that, following exocytosis, surface membrane is recovered and returned to multiple Golgi cisternae; moreover, we have shown that this membrane traffic is especially heavy to those portions of the Golgi (condensing vacuoles or dilated rims of transmost cisternae) in which secretory products are normally concentrated. Similar data have also been obtained by others on three additional cell types—$\beta$ cells of the endocrine pancreas (Orci *et al.*, 1978), thyroid epithelial cells (Herzog and Miller, 1979), and exocrine pancreatic cells (Herzog and Reggio, 1980). The most likely explanation of these findings is that granule membrane is recovered and reutilized in the packaging of newly synthesized secretory products; i.e., it is recycled. In this respect the tracer data and the turnover data are in accord, and, although the evidence is still indirect, there seems to be little reason at present to doubt that the membranes of discharged secretory granules are recovered and reutilized in the packaging of successive generations of secretion granules.

In addition to granule membranes, there is also evidence for recycling of synaptic vesicle membranes (reviewed by Heuser, 1978; Meldolesi *et al.*, 1978; Holtzman *et al.*, 1978) and membranes of pinocytic vesicles in macrophages and fibroblasts (Steinman *et al.*, 1976; Schneider *et al.*, 1979). All these represent particularly favorable situations for following the fate of the recovered and recycled surface membrane, owing to the large quantities of membrane involved and to the fact that, at least in the first two cases (secretory granule and synaptic vesicle membranes), relocation of the membrane can be controlled experimentally. In many other cases in which membrane is relocated intracellularly (e.g., vesicles that transport secretory products from ER to Golgi or lysosomal enzymes from Golgi to lysosomes), the situation is much less favorable for study, owing either to difficulty in tracer access or to the smaller amounts of membrane involved, but a similar recycling mechanism seems likely. Hence, at present it can be assumed that membrane reutilization is a widespread phenomenon, and in all likelihood it is even more widespread than can be documented at present. This being the case, it is appropriate to consider some of the implications of the recycling data in secretory cells.

## B. Multiplicity of Routes Followed by Recovered Membrane

Note has already been made of the fact that in the work on anterior pituitary cells the variations in traffic detected with anionic and cationic ferritins could be explained by the existence of two possible recovery routes (plasmalemma → Golgi *and* plasmalemma → lysosomes → Golgi) (diagrammed in Fig. 16). The data did not allow us to distinguish between the two. In the meantime, Herzog and Miller (1979) have obtained evidence for the utilization of such an indirect Golgi route in thyroid epithelial cells; moreover, the results obtained when myeloma cells are incubated in series in both a membrane marker (CF) and a content marker (HRP) are easier to explain by this kind of two-station traffic. However, results obtained on exocrine cells of the parotid (Herzog and Farquhar, 1977) and on exocrine pancreatic cells (Herzog and Reggio, 1980) indicate that

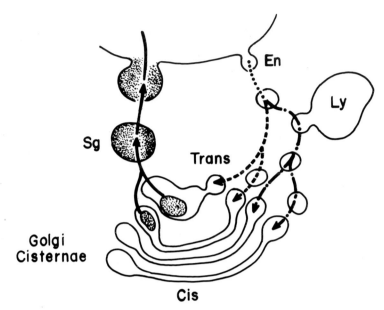

FIG. 16.    Diagram showing two possible routes that could be taken by surface membrane to reach the stacked Golgi cisternae. Following exocytosis of secretory granules (———), patches of surface membrane are recovered by endocytosis and fuse with the dilated rims of multiple stacked Golgi cisternae. Membrane recovered by endocytosis can either fuse directly with the cisternae (– – –) or fuse first with lysosomes and then with the Golgi (— · —). The available evidence suggests that both routes may be used in different cell types.

membrane recovery and its fusion with stacked Golgi cisternae is rapid and extensive, rendering it likely that a direct route (plasmalemma → Golgi) is used in these cells, at least *in vivo*. In addition to the pathways outlined (involving Golgi and lysosomes), it has been reported that in basophilic granulocytes (Dvorak *et al.*, 1972) and in $\beta$ cells of the pancreatic islets (see Chapter 18 by L. Orci *et al.*) surface membrane fuses with mature secretion granules.

Thus, the most likely possibility is that multiple recovery routes are used, and individual cell types direct the traffic according to their activities or functional state. The precise pathways and the types and amounts of membrane involved remain to be worked out in individual cell types.

## C.   Implications of Membrane Recycling Data for Golgi Traffic

Up to now our concepts of pathways taken by membranes and products in traversing the Golgi complex in secretory cells have been dominated by a consideration of biogenetic traffic and have been based on the assumption that granule membranes are degraded after exocytosis rather than being reutilized. According to the current "dogma," both membranes and secretory products are assumed to enter the Golgi on one face (called the cis face), and the cisternae are believed to be progressively displaced one by one, with the membranes being used up on the opposite face (trans) as containers for secretory products. New cisternae are believed to arise on the cis face. This dogma is reflected in the naming of the Golgi complex with one side (the cis side) being referred to as the "entry" or "immature" face and the other (the trans side) as the "mature" or "exit" face. Thus, a continual, unidirectional membrane flow from the ER → Golgi → plasmalemma has been envisaged and frequently depicted in diagrams (Leblond and Bennett, 1977; Morré et al., 1974; Schacter, 1974). We have presented elsewhere (Farquhar, 1978b) our reasons for questioning this flow diagram and have pointed out that there is a lack of experimental evidence to support it. Indeed, the recycling evidence constitutes part of the evidence that has led us to question it, since the main flow of incoming membrane traffic from the cell surface to the Golgi complex is *not* to the cis cisternae, which would be expected if a cis → trans flow of membrane took place. Rather, the incoming vesicle traffic is directed to the dilated rims of multiple Golgi cisternae and is heaviest along the transmost cisternae.

It is worth emphasizing that the dilated rims of the Golgi cisternae represent particularly active zones of membrane traffic and exchange. These particular segments of each Golgi cisterna are specialized in several other respects. First, the morphological (Farquhar, 1971) as well as the autoradiographic data (Farquhar et al., 1978) indicate that they represent the sites where secretory products are concentrated. Second, cytochemical evidence indicates that they have a different composition than the remainder (centers) of the Golgi cisternae, since several enzymes—e.g., 5'-nucleotidase (Farquhar et al., 1974) and adenylate cyclase (Cheng and Farquhar, 1976)—are more concentrated along the dilated peripheries of the cisternae than along their more collapsed centers, at least in hepatocytes (Figs. 17 and 18).

To summarize, the membrane recycling data plus other data mentioned point to the specialized nature of the dilated rims of the Golgi cisternae and do not support a simple cis → trans flow diagram. They are compatible with the following hypotheses:

1. The main flow of both granule membrane (recycled from the cell surface) and content (from the rough ER) is to the dilated rims of multiple Golgi cisternae.

2. In cells that concentrate their secretory products, traffic from the cell surface is heaviest to the transmost Golgi cisternae where concentration takes place.

3. The dilated rims of the Golgi cisternae represent a special subcategory of Golgi membranes that differ in their protein composition from the rest.

4. Each Golgi cisterna is a mosaic in which differentiated domains are maintained in the plane of the membrane.

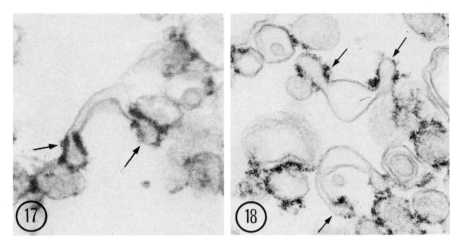

FIGS. 17 AND 18.   Isolated Golgi cisternae from a Golgi fraction (unfixed) incubated for 5'-nucleotidase (Fig. 17) or adenylate cyclase (Fig. 18), showing that the reaction product for these two enzymes is concentrated in the dilated rims of the cisternae (arrows). Fig. 17, ×80,000; Fig. 18, ×85,000. Figure 17 is from Farquhar *et al.* (1974), and Fig. 18 is from Cheng and Farquhar (1976).

5. The individual Golgi cisternae (like other cell components) retain the specificity of their membranes, and transport of both secretory products and membrane components is largely effected by vesicular carriers, which fuse preferentially with the dilated rims.

## D.   Implications of Membrane Recycling to Golgi for Cell Metabolism

Currently there is a great deal of interest and excitement, stimulated by the work of Goldstein and Brown (Goldstein *et al.*, 1979), concerning the role of traffic directed to lysosomes (plasmalemma → lysosomes) in regulation of intracellular metabolic events (through receptor-mediated endocytosis and lysosomal digestion). The validation of the existence of a pathway from the cell surface to the Golgi is equally exciting and has even broader implications because it provides a route whereby molecules from the cell surface can reach a biosynthetic compartment. Thus, a mechanism is provided whereby various informational molecules from the extracellular environment (hormones, enzymes, catacholamines) can reach the Golgi and influence intracellular events. Indeed, several hormones—insulin (Josefsberg *et al.*, 1979) and β-MSH (Moellman *et al.*, 1978)—have already been demonstrated to reach the Golgi complex, although to date the biological consequences of their uptake into Golgi elements is not clear.

The existence of recycling membrane traffic also provides a mechanism whereby surface membrane components (receptors, enzymes, and other membrane proteins) could potentially be modified (e.g., reglycosylated, sulfated, phosphorylated) in passage. To date no examples of this type of phenomenon are available, but there is no apparent reason why they could not take place if the molecules are brought into contact with the right Golgi compartments or subcompartments.

## E. Problems and Questions for the Future

Among the many interesting questions relating to recycling in addition to those already raised, which remain to be established by further work, are the following:

1. *What is the nature of the membrane removed from the cell surface?* That is, is the segment of membrane removed from the cell surface by endocytosis the same membrane as that inserted during exocytosis? Does the membrane inserted into the plasma membrane at the time of exocytosis mix with the rest? Freeze-fracture observations on rat parotid suggest that the former granule membrane may *not* mix with the rest (DeCamilli *et al.*, 1976). However, definitive answers to these questions cannot be given at present. They await the isolation of recycled membrane to enable a comparison between its components and those of the granule membrane. It is clear that in those cases in which data are available (see Meldolesi *et al.*, 1978) the membrane of secretion granules differs from that of other cellular membranes (ER, plasmalemma).

2. *What is the magnitude of the recycling process?* To data there are no data on this point, but in the future an estimate of the amount of membrane involved could be obtained either directly by labeling the surface membrane, or indirectly by determining the rate of granule discharge and correlating this information with morphometric data (on the amount of the granule membrane present) and with the turnover data.

3. *What is the nature of the carrier?* Are coated vesicles involved? We have obtained evidence in both parotid acinar cells and myeloma cells that coated vesicles may be involved in recovery of surface membrane; however, since coated vesicles are involved in a variety of transport functions [transport of secretory proteins from ER → Golgi (Palade and Fletcher, 1977); transport of lysosomal enzymes from Golgi → lysosomes (Friend and Farquhar, 1967)], the presence of clathrin coats is not specific to this population of vesicles.

4. *Need for stable, nonperturbing labels.* What is needed to answer all these questions are (1) means to introduce stable (covalently bound), nonperturbing membrane labels that will not influence the destiny of the tagged components and (2) systems amenable to their use and analysis. In practice this has proved to be difficult, since many of the methods of tagging or labeling (e.g., iodination)

result in partial denaturation or oxidation of proteins (Rutherford and Gennaro, 1979; Opresco *et al.*, 1979), alteration of their charge, or modification of their terminal hexose groups. Any or all of these changes might be expected to have an adverse affect on the recognition of a given membrane component and result in redirecting or modifying membrane traffic. Recently some promising new approaches aimed at solving or circumventing these problems have appeared (Mellman *et al.*, 1980; Muller *et al.*, 1980; Thilo and Vogel, 1980). (See Note Added in Proof.)

## VII.   Summary and Conclusions

Although the concept of reutilization or recycling of secretory granule membrane was proposed over 20 years ago, there has been little or no experimental evidence to support this idea until quite recently. In fact, the available evidence was to the contrary: the early turnover data (based on the similarity in the turnover of the granule membranes were degraded rather than reutilized; and the early tracer data suggested that surface membrane was recovered intact (by endocytosis) and subsequently degraded in lysosomes. As a result of these negative data, the existence of membrane recycling has been denied by most workers until quite recently.

In this chapter we summarize evidence obtained on several glandular cell types using electron-dense tracers, which points to the existence of recycling of granule membranes as containers for secretory products. Work done on five different cell types of varied nature (exocrine, endocrine, and immunoglobulin-secreting cells) indicates that following exocytosis surface membrane is recovered by endocytosis and fuses—directly or indirectly—with multiple stacked Golgi cisternae as well as with lysosomes. Fusion takes place preferentially with the dilated rims of the Golgi cisternae, where concentration of secretory products usually takes place. The tracers used to follow the fate of the surface membrane were dextrans for parotid and lacrimal gland cells and cationized ferritin for anterior pituitary and immunoglobulin-secreting cells. In all these cases, the tracer was found at early intervals (2–15 minutes) in endocytic vesicles near the discharging surface, and at later intervals (1–2 hours) in multiple stacked Golgi cisternae, in forming secretion granules or vacuoles, and in lysosomes, indicating that the incoming vesicles had fused with these compartments. The presence of tracers within forming secretion granules suggests that this traffic is connected with the recycling of the membrane containers for secretory products—i.e., granule or secretory vesicle membranes. Recent data obtained on improved granule membrane preparations (in which care was taken to remove contaminating content proteins) indicate that the turnover of the granule membrane proteins is much slower than that of the content proteins, a finding compatible with the membrane or its constituents being reutilized. Thus, at present the tracer data and turnover data are in accord, and although the evidence is indirect, there is little reason to doubt the existence of recycling of granule membranes (as well as other types of membranes) in secretory cells. The precise routes taken by surface membrane to reach the Golgi complex remain to be established; however, the available evidence renders it likely that both a direct route (plasmalemma → Golgi) and an indirect route involving lysosomes (plasmalemma → lysosomes → Golgi) are used in different cell types.

According to current concepts, to traverse the Golgi complex membranes and products follow a unidirectional course from the rough ER → cis Golgi cisternae → trans Golgi cisternae → plasmalemma. The fact that the main flow of incoming membranes from the cell surface is to the dilated rims of multiple Golgi cisternae (rather than preferentially to cis Golgi cisternae) had led us to question this flow diagram.

The existence of a pathway from the cell surface to the Golgi complex has a number of important implications. It provides a route whereby informational molecules (hormones, catecholamines, and other agents) from the extracellular environment can reach a biosynthetic compartment and thereby influence additional intracellular events. It also provides a mechanism whereby surface membrane components (receptors, enzymes, and other membrane proteins) could be modified (reglycosylated, sulfated, phosphorylated) in passage through the Golgi complex.

The precise pathways and amounts and types of membrane involved in recycling and the physiological implications for cell metabolism remain to be worked out in individual cell types. This work will be facilitated by the availability of appropriate, stable, nonperturbing labels.

### ACKNOWLEDGMENT

The original research reported in this paper was supported by Grant AM 17780 from the National Institutes of Health.

### REFERENCES

Abrahams, S. J., and Holtzman, E. (1973). *J. Cell Biol.* **56,** 540–558.
Amsterdam, A., Ohad, I., and Schramm, M. (1969). *J. Cell Biol.* **41,** 753–773.
Amsterdam, A., Schramm, M., Ohad, I., Solomon, Y., and Selinger, Z. (1971). *J. Cell Biol.* **50,** 187–200.
Castle, J. D., Jamieson, J. D., and Palade, G. E. (1975). *J. Cell Biol.* **64,** 182–210.
Cheng, H., and Farquhar, M. G. (1976). *J. Cell Biol.* **70,** 671–684.
Danon, D., Goldstein, L., Marikovsky, Y., and Skutelsky, E. (1972). *J. Ultrastruct. Res.* **38,** 500–510.

DeCamilli, P., Peluchetti, D., and Meldolesi, J. (1976). *J. Cell Biol.* **70,** 59–74.

Douglas, W. W. (1973). *Prog. Brain Res.* **39,** 21–39.

Dvorak, A. M., Dvorak, H. F., and Karnovsky, M. J. (1972). *Lab. Invest.* **26,** 27–39.

Farquhar, M. G. (1971). *In* "Subcellular Structure and Function in Endocrine Organs" (H. Heller and K. Lederis, eds.), pp. 79–122. Cambridge Univ. Press, London and New York.

Farquhar, M. G. (1977). *Adv. Exp. Med. Biol.* **80,** 37–94.

Farquhar, M. G. (1978a). *J. Cell Biol.* **77,** R35–R42.

Farquhar, M. G. (1978b). *In* "Transport of Macromolecules in Cellular Systems" (S. C. Silverstein, ed.), pp. 341–362. Dahlem Konferenzen, Berlin.

Farquhar, M. G., Bergeron, J. J. M., and Palade, G. E. (1974). *J. Cell Biol.* **60,** 8–25.

Farquhar, M. G., Skutelsky, E., and Hopkins, C. R. (1975). *In* "The Anterior Pituitary Gland" (A. Tixier-Vidal and M. G. Farquhar, eds.), pp. 81–135. Academic Press, New York.

Farquhar, M. G., Reid, J. J., and Daniell, L. W. (1978). *Endocrinology* **102,** 296–311.

Fawcett, D. W. (1962). *Circulation* **26,** 1105–1132.

Friend, D. S., and Farquhar, M. G. (1967). *J. Cell Biol.* **35,** 357–376.

Geuze, J. J., and Kramer, M. F. (1974). *Cell Tissue Res.* **156,** 1–20.

Giannattasio, G., and Zanini, A. (1976). *Biochim. Biophys. Acta* **439,** 349–357.

Goldstein, J. L., Anderson, R. G. W., and Brown, M. S. (1979). *Nature (London)* **279,** 679–685.

Gonatas, N. K., Kim, S. U., Stieber, A., and Avrameas, S. (1977). *J. Cell Biol.* **73,** 1–13.

Herzog, V., and Farquhar, M. G. (1977). *Proc. Natl. Acad. Sci. U.S.A.* **74,** 5073–5077.

Herzog, V., and Miller, F. (1979). *Eur. J. Cell Biol.* **19,** 203–215.

Herzog, V., and Reggio, H. (1980). *Eur. J. Cell Biol.* **21,** 141–150.

Heuser, J. E. (1978). *In* "Transport of Macromolecules in Cellular Systems" (S. C. Silverstein, ed.), pp. 445–464. Dahlem Konferenzen, Berlin.

Holtzman, E., Schacher, S., Evans, J., and Teichbert, S. (1978). *Cell Surf. Rev.* **4,** 165–246.

Josefsberg, Z., Posner, B. I., Patel, B., and Bergeron, J. J. M. (1979). *J. Biol. Chem.* **254,** 209–214.

Joseph, K. C., Stieber, A., and Gonatas, N. K. (1979). *J. Cell Biol.* **81,** 543–554.

Kalina, M., and Rabinowitz, R. (1975). *Cell Tissue Res.* **163,** 373–382.

Kanwar, Y. S., and Farquhar, M. G. (1979). *J. Cell Biol.* **81,** 137–153.

Leblond, C. P., and Bennett, G. (1977). *In* "International Cell Biology 1976–1977" (B. R. Brinkley and K. R. Porter, eds.), pp. 326–336. Rockefeller Univ. Press, New York.

Mata, L. R. (1976). *J. Microsc. Biol. Cell.* **25,** 127–132.

Meldolesi, J. (1974). *J. Cell Biol.* **61,** 1–13.

Meldolesi, J., Borgese, N., De Camilli, P., and Ceccarelli, B. (1978). *In* "Membrane Fusion" (G. Poste and G. L. Nicholson, eds.), pp. 509–627. Elsevier/North-Holland Publ., Amsterdam and New York.

Mellman, I., Steinman, R. M., Unkeless, J., and Cohn, Z. (1980). *J. Cell Biol.* **86,** 712–722.

Moellmann, G. E., Varga, J. M., Godawaka, E. W., Lambert, D. T., and Lerner, A. B. (1978). *J. Cell Biol.* **79,** 196a.

Morré, D. J., Keenan, T. W., and Huang, C. M. (1974). *Adv. Cytopharmacol.* **2,** 107–125.

Muller, W. A., Steinman, R. M., and Cohn, Z. A. (1980). *J. Cell Biol.* **86,** 292–303.

Nagasawa, J., and Douglas, W. W. (1972). *Brain Res.* **37,** 141–145.

Nagasawa, J., Douglas, W. W., and Schulz, R. A. (1971). *Nature (London)* **232,** 341–342.

Novikoff, A. B. (1976). *Proc. Natl. Acad. Sci. U.S.A.* **73,** 2781–2787.

Oliver, C., and Hand, A. R. (1978). *J. Cell Biol.* **76,** 207–220.

Opresco, L., Wiley, H. S., and Wallace, R. A. (1980). *J. Cell Biol.* **86,** 712–722.

Orci, L., Malaisse-Lagae, F., Ravazzola, M., Amherdt, M., and Renold, A. E. (1973). *Science* **181,** 561–562.

Orci, L., Perrelet, A., and Gordon, P. (1978). *Recent Prog. Horm. Res.* **34,** 95–121.

Ottosen, P. D., Courtoy, P. J., and Farquhar, M. G. (1980). *J. Exp. Med.* **152,** 1-19.
Palade, G. E. (1959). *In* "Subcellular Particles" (T. Hayashi, ed.), pp. 64-80. Ronald Press, New York.
Palade, G. E. (1975) *Science* **189,** 347-358.
Palade, G. E., and Fletcher, M. (1977). *J. Cell Biol.* **75,** Part 2, 371a.
Pelletier, G. (1973). *J. Ultrastruct. Res.* **43,** 445-459.
Rennke, H. G., Patel, Y., and Venkatachalam, M. A. (1978). *Kidney Int.* **13,** 278-288.
Rutherford, D. T., and Gennaro, J. F., Jr. (1979). *J. Cell Biol.* **83,** Part 2, 438a.
Schachter, H. (1974). *Biochem. Soc. Symp.* **40,** 57-71.
Schneider, Y. J., Tulkens, P., deDuve, C., and Trouet, A. (1979). *J. Cell Biol.* **82,** 466-474.
Slaby, F., and Farquhar, M. G. (1980). *Mol. Cell. Endocrinol.* **18,** 33-48.
Steinman, R. M., Brodie, S. E., and Cohn, Z. A. (1976). *J. Cell Biol.* **68,** 665-687.
Tartakoff, A., and Vassalli, P. (1977). *J. Exp. Med.* **146,** 1332-1345.
Theodosis, D. T., Dreifuss, J. J., Harris, M. C., and Orci, L. (1976). *J. Cell Biol.* **70,** 294-303.
Thilo, L., and Vogel, G. (1980). *Proc. Natl. Acad. Sci. U.S.A.* **77,** 1015-1019.
Wallach, D., Kirshner, N., and Schramm, M. (1975). *Biochim. Biophys. Acta* **375,** 87-105.
Zanini, A., and Giannattasio, G. (1974). *Adv. Cytopharmacol.* **2,** 329-339.
Zanini, A., Giannattasio, G., and Meldolesi, J. (1980). *In* "Biochemical Endocrinology: Synthesis and Release of Adenohypophyseal Hormones, Cellular and Molecular Mechanisms" (K. McKerns and M. Jutisz, eds.), pp. 105-123. Plenum, New York.

NOTE ADDED IN PROOF. Recently, we have covalently labeled myeloma cells (by lactoperoxidase iodination procedures) and have obtained autoradiographic evidence that iodinated cell surface components are internalized, and a considerable proportion (20-50%) of the surface label reaches Golgi components or forming secretion vacuoles (Wilson *et al.*, Abstracts, American Society for Cell Biology, *J. Cell Biol.*, November, 1981).

# Chapter 26

# Membrane Retrieval in Exocrine Acinar Cells

## CONSTANCE OLIVER AND ARTHUR R. HAND

*Laboratory of Biological Structure,*
*National Institute of Dental Research,*
*National Institutes of Health,*
*Bethesda, Maryland*

## I. Introduction

When exocrine acinar cells secrete, the secretory granule membrane is inserted into the apical plasma membrane and the granule content released. Although this process results in the addition of considerable membrane to the surface, the cells maintain a relatively constant volume. Until recently the manner in which the cells accomplished this was poorly understood. Retrieval of surface membrane had been suggested as the primary mechanism involved. Previous morphological studies had shown the presence of smooth-surfaced vesicles in the apical cytoplasm of exocrine acinar cells (Amsterdam *et al.*, 1969; Hand, 1973; Geuze and Kramer, 1974; Bogart, 1975; Palade, 1975). Since these vesicles increased in number following stimulated secretion, it had been proposed that they represented apical membrane endocytosed by the cells. Further support for the concept of membrane retrieval and possible reutilization came from biochemical studies of intracellular membrane compartments of exocrine acinar cells (Meldolesi and Cova, 1971; Meldolesi, 1974a,b; Wallach *et al.*, 1975). In both pancreatic and parotid acinar cells the rate of *de novo* synthesis of secretory granule membrane was too low to account for all the membrane required for intracellular protein transport and secretory granule production. Thus, there was both morphological and biochemical evidence suggesting that membrane retrieval and reutilization occurred in exocrine acinar cells.

Our initial studies (Oliver and Hand, 1978) were directed toward establishing that membrane was retrieved from the apical cell surface of parotid acinar cells and characterizing this process. Because exocrine acinar cells are so highly polarized, we then examined the uptake of material from those cell surfaces not involved with secretory granule release. More recent studies (Oliver, 1980) have compared the mechanisms of endocytosis at the lateral and basal cell surfaces with those at the apical surface. In all our studies we have employed soluble phase (content) markers, primarily horseradish peroxidase (HRP), to follow the uptake and fate of endocytosed tracer.

## II.   Membrane Retrieval at the Apical Cell Surface

Horseradish peroxidase was delivered to the apical cell surface by retrograde infusion into the parotid glands of adult male and female Sprague–Dawley rats. A cannula was inserted in the main excretory duct, and a solution of HRP (type II; 10 mg/ml sterile saline; Sigma Chemical Company, St. Louis, MO) was allowed to flow into the gland. The infusion pressure was calculated to be 15–16 mm Hg. Some animals received isoproterenol (30 mg/kg body weight, i.p.) either before or after the HRP in order to stimulate secretion. Animals were sacrificed at varying intervals after HRP administration. Glands were fixed by vascular perfusion with a modified Karnovsky (1965) fixative containing 2% glutaraldehyde (Ladd Research Industries, Burlington, VT), 2% formaldehyde (Ladd), and 0.025% $CaCl_2$ in 0.1 $M$ cacodylate buffer (pH 7.4). Following perfusion, glands were removed, cut into 1 × 5-mm strips, and immersed in fresh fixative. Total fixation time was 1 hour. The glands were then rinsed in 0.1 $M$ cacodylate buffer (pH 7.4) containing 7% sucrose (sucrose buffer) and stored overnight at 4°C. The strips were cut into 75-$\mu$m-thick sections with a Smith–Farquhar TC-2 tissue sectioner, and the sections incubated in Fahimi's (1970) diaminobenzidine (Sigma)–hydrogen peroxide medium for 45 minutes at 25°C. The tissue was rinsed in sucrose buffer and postfixed in 1% osmium tetroxide for 1 hour at 25°C, rinsed in sucrose buffer, dehydrated through a graded series of ethanols and embedded in Spurr's (1969) resin. Thin sections were cut with a diamond knife, mounted on bare copper grids, stained lightly with Reynold's lead citrate (1963), and examined in either a JEM 100-B or 100-C electron microscope.

In resting and isoproterenol-stimulated parotid acinar cells, the pattern of uptake and fate of the tracer was essentially the same. The HRP was endocytosed from the apical surface in small smooth-surfaced C- or ring-shaped vesicles (Fig. 1). These vesicles accumulated in the region of the Golgi apparatus, so that at 1–2 hours after HRP administration the majority of the reaction product was localized in small vesicles, dense bodies, and multivesicular bodies adjacent to the Golgi appartus (Fig. 2). At later times, reaction product was found in lysosomes adjacent to the Golgi apparatus (Fig. 3), as well as in isoproterenol-

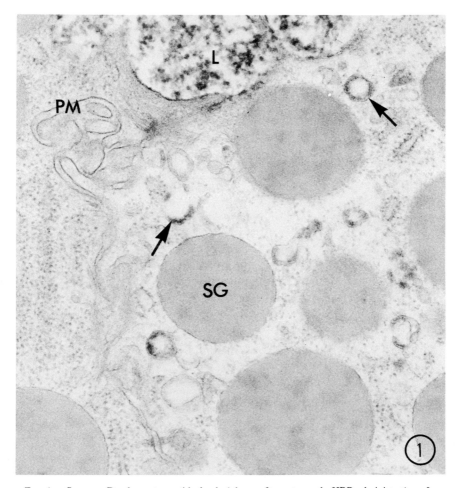

FIG. 1. Sprague–Dawley rat parotid gland, 1 hour after retrograde HRP administration. Junctional complexes at the apical cell surface prevent the passage of the HRP from the lumen (L) to the intercellular space. Intracellulary, reaction product is localized to small ring- or C-shaped vesicles (arrows). PM, plasma membrane; SG, secretory granule. ×42,500.

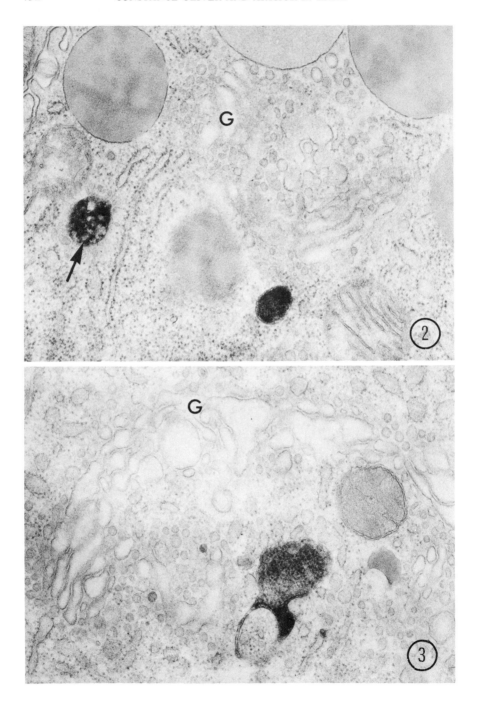

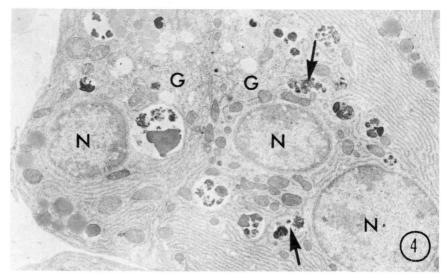

FIG. 4.   Sprague–Dawley rat parotid gland, 8 hours after retrograde HRP administration and 7 hours after isoproterenol injection. Reaction product is present in isoproterenol-induced autophagic vacuoles (arrows). G, Golgi apparatus; N, nucleus. ×5200.

induced autophagic vacuoles (Simson, 1972) (Fig. 4). In addition to inducing autophagic vacuoles, isoproterenol administration altered the permeability of the apical junctional complexes such that, when isoproterenol was given prior to the HRP, the tracer gained access to the lateral and basal extracellular spaces (Fig. 5). This is in contrast to the unstimulated glands, where HRP was confined to the lumen (Fig. 1). Intracellularly in stimulated glands, HRP reaction product was found in lysosomes and vesicles adjacent to the Golgi apparatus (Fig. 6). At no time in either resting or stimulated glands was reaction product found in Golgi saccules, GERL, or secretory granules.

## III.   Membrane Retrieval at the Lateral and Basal Cell Surfaces

Uptake of HRP (Sigma, types II and VI) at the lateral and basal cell surfaces was examined in rat and mouse pancreatic and parotid acinar cells. Tracer (1 mg/gm body weight in sterile saline) was injected intravenously via the saphen-

FIGS. 2 AND 3.   Sprague–Dawley rat parotid gland. Fig. 2: 1 hour after retrograde HRP administration. Reaction product is localized in a dense body and a multivesicular body (arrow) adjacent to the Golgi apparatus (G). × 35,300. Fig. 3: 6 hours after retrograde HRP administration and 5 hours after isoproterenol injection. Reaction product is present in a few small vesicles and a lysosome lying near the Golgi apparatus (G). ×35,400.

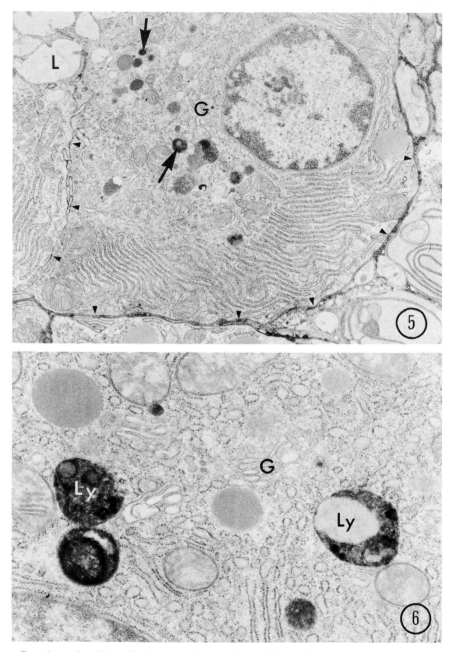

FIGS. 5 AND 6.    Wistar–Furth rat parotid gland. Fig. 5: 2 hours after isoproterenol injection and 1 hour after retrograde HRP administration. When isoproterenol is given prior to the HRP, the peroxidase is not confined to the lumen (L), and reaction product is found in the lateral and basal extracellular spaces (arrowheads). Intracellularly, reaction product is found in vesicles and dense bodies (arrows) adjacent to the Golgi apparatus (G).  ×6200.

ous vein into adult Wistar–Furth rats or NIH Swiss mice. Some animals received isoproterenol (20 mg/kg body weight, i.p.) or pilocarpine (40 mg/kg body weight, i.p.) to stimulate secretion from the parotid gland or the pancreas, respectively. Animals were sacrificed at varying intervals after HRP administration. Tissue was fixed and processed as described in Section II, except that some samples were postfixed in reduced osmium (Karnovsky, 1971) and dehydrated without rinsing.

Unlike uptake from the apical surface, comparatively little tracer was endocytosed from the lateral and basal cell surfaces in resting glands (Fig. 7). Following stimulated secretion, however, considerable tracer could be localized intracellularly (Fig. 8). Despite the difference in the quantity of tracer internalized, the endocytic process and localization of tracer were essentially the same in both resting and stimulated cells. Tracer was endocytosed in vesicles primarily in regions of plasma membrane infoldings (Fig. 9). Especially in stimulated cells, uptake appeared to occur predominantly in coated vesicles (Fig. 10). Once internalized, the tracer was rapidly localized in a system of anatomosing tubules located adjacent to the plasma membrane (Figs. 9 and 11). This system has been identified as a series of pleomorphic tubular lysosomes (Fig. 12) located in the basal portion of exocrine acinar cells. These lysosomes possess trimetaphosphatase activity but not acid phosphatase activity (Oliver, 1979). Additionally, HRP reaction product could be localized in multivesicular bodies both adjacent to the plasma membrane and adjacent to the Golgi apparatus (Fig. 13). Small Golgi-associated vesicles and tubular lysosomes in close proximity to the cis Golgi saccules (Fig. 14) also contained reaction product. At later time intervals, apically located vesicles and tubules became reactive (Fig. 15), and the number of reactive lysosomes near the Golgi apparatus increased. In glands that received HRP intravenously and native ferritin by retrograde infusion, both tracers could be localized in the same lysosome (Fig. 16), indicating that mixing of material brought in from all cell surfaces had occurred. As with retrograde administration of tracer, reaction product was not found in Golgi saccules, GERL, or secretory granules.

# IV. Conclusions

The results of our studies on uptake of HRP from the apical and lateral-basal cell surfaces clearly show that there are two distinct endocytic pathways in

Fig. 6: 5½ hours after isoproterenol injection and 4½ hours after retrograde HRP administration. At later times, intracellular reaction product is found primarily in lysosomes (Ly) near the Golgi apparatus (G). ×20,000

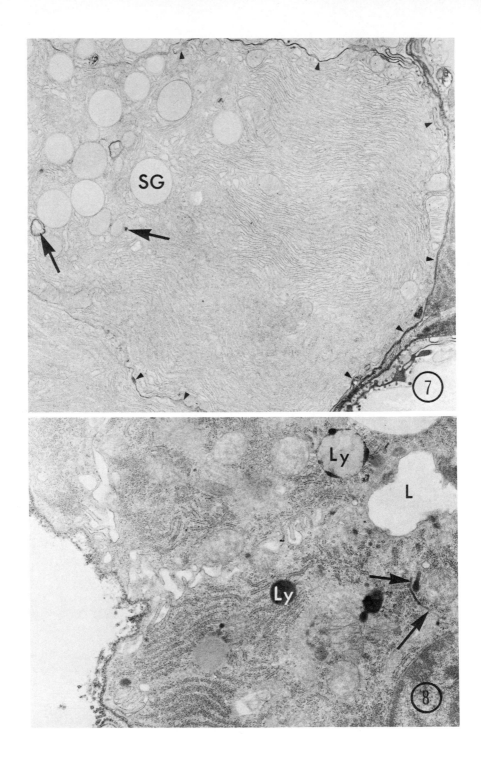

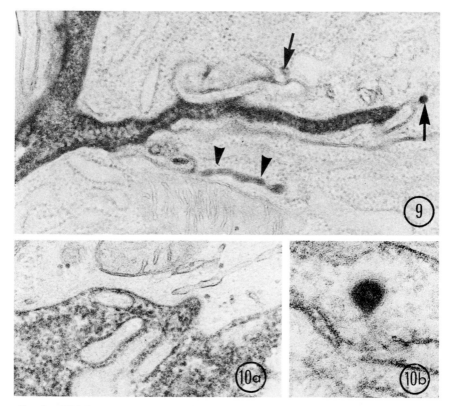

FIG. 9.   NIH Swiss mouse pancreas, 3 hours after pilocarpine and intravenous HRP injection. The HRP appears to be endocytosed in vesicles (arrows), which form predominantly in areas of plasma membrane infolding. Note tubular lysosome adjacent to membrane infolding (arrowheads). ×43,300.

FIG. 10.   Wistar–Furth rat pancreas, 30 minutes after pilocarpine and intravenous HRP injection. (a) Especially in stimulated cells, many of the endocytic vesicles are coated. ×50,000. (b) Once internalized, the vesicles rapidly lose their coats. ×133,000.

FIG. 7.   NIH Swiss mouse pancreas, 3 hours after intravenous HRP injection. In unstimulated glands, although HRP is present at the lateral and basal cell surfaces (arrowheads), relatively little is internalized (arrows). SG, secretory granule. ×9000.

FIG. 8.   Wistar–Furth rat parotid gland, 3 hours after isoproterenol and intravenous HRP injection. More HRP is internalized in stimulated cells. Reaction product is present in lysosomes (Ly), vesicles, and tubules (arrows). L, lumen. ×11,000.

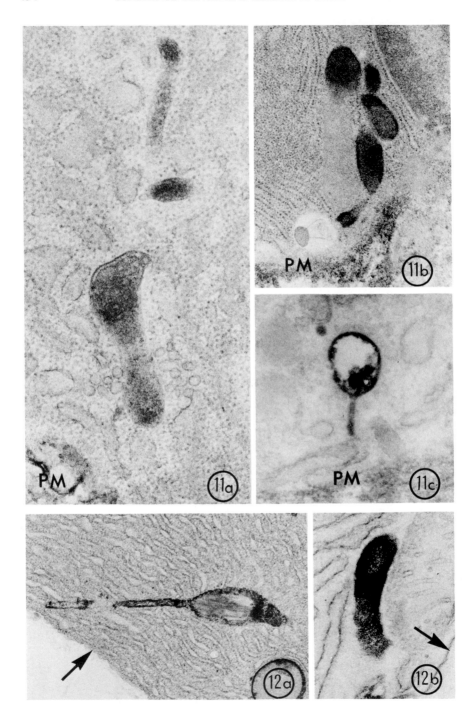

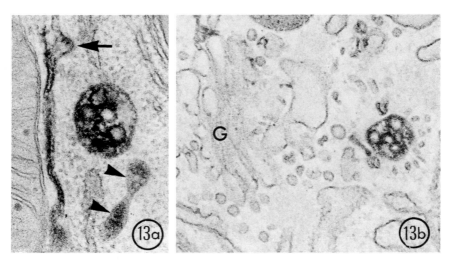

FIG. 13. Multivesicular bodies near (a) the plasma membrane and (b) the Golgi apparatus (G) contain HRP reaction product. Arrow, coated vesicle; arrowheads, basal lysosome. (a) Wistar–Furth rat pancreas, 30 minutes after intravenous HRP injection. ×50,000. (b) NIH Swiss mouse pancreas, 1 hour after intravenous HRP injection. ×56,000.

exocrine acinar cells (Fig. 17). One involves uptake of tracer from the apical surface into smooth-surfaced ring or C-shaped vesicles. These vesicles ultimately deliver their contents to lysosomes lying adjacent or basal to the Golgi apparatus. The other pathway, from the lateral and basal cell surfaces, is much more dependent upon stimulated secretion for uptake of tracer. Endocytosis occurs primarily in coated vesicles, and tracer rapidly enters a system of basal tubular lysosomes. At later times tracer is localized in lysosomes lying adjacent to the Golgi apparatus. Furthermore, the pattern of uptake from the lateral-basal cell surface—i.e., dependence on stimulation and endocytosis in coated vesicles, suggests that this process may be related to receptor-mediated endocytosis. Conversely, endocytosis at the apical cell surface may be more concerned with retrieval of secretory granule membrane.

The sequestration of exogenously administered HRP in lysosomes, but not in Golgi saccules, GERL, or secretory granules, is in agreement with most other

FIG. 11. Wistar–Furth rat. Following intravenous injection, HRP is localized in a pleomorphic system of anastomosing tubules lying adjacent to the lateral and basal plasma membranes (PM). (a) Pancreas, 30 minutes after intravenous HRP injection. ×46,000. (b) Parotid gland, 1 hour after intravenous HRP injection. ×32,000. (c) Pancreas, 4 hours after intravenous HRP injection. ×47,000.

FIG. 12. The basal lysosomes are reactive when incubated for trimetaphosphatase activity (arrows, plasma membrane). (a) Wistar–Furth rat pancreas. ×22,000. (b) NIH Swiss mouse pancreas. ×56,000.

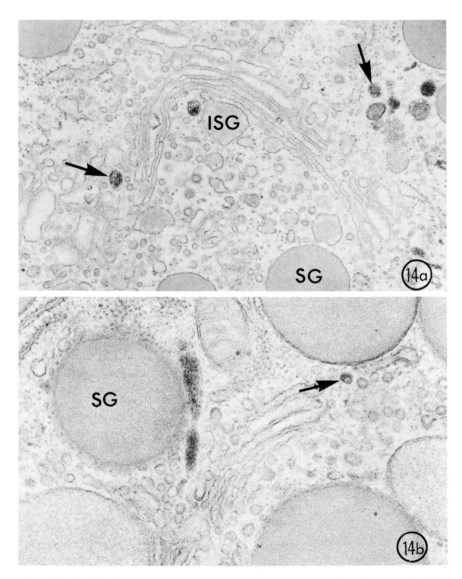

FIG. 14.    Small vesicles (arrows, a and b) and tubular lysosomes (b) adjacent to the Golgi apparatus contain HRP reaction product. ISG, immature secretory granule; SG, secretory granule. (a) Wistar–Furth rat pancreas, 4 hours after intravenous HRP injection. ×40,000. (b) Wistar–Furth rat parotid gland, 30 minutes after intravenous HRP injection. ×33,000.

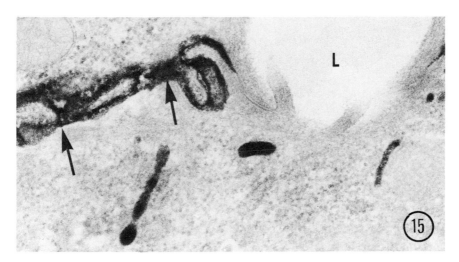

FIG. 15. Wistar–Furth rat parotid gland, 6 hours after isoproterenol and intravenous HRP injection. At these later times, apically located tubules frequently contain HRP reaction product. Even though much HRP remains in the intercellular space (arrows), the junctional complexes have prevented the passage of HRP into the lumen (L). ×41,000.

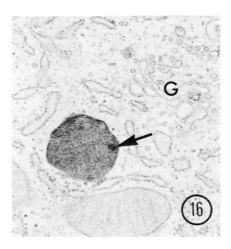

FIG. 16. Wistar–Furth rat pancreas, 4 hours after retrograde administration of native ferritin, intravenous injection of HRP, and pilocarpine stimulation. A secondary lysosome adjacent to the Golgi apparatus (G) contains both HRP reaction product and patches of native ferritin (arrow). ×29,000.

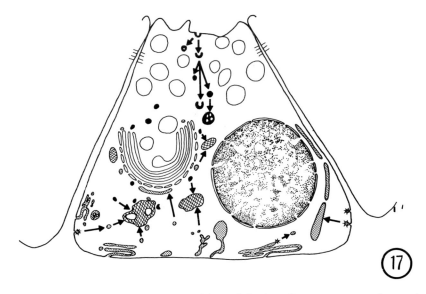

FIG. 17.   Diagram summarizing uptake and intracellular fate of exogenous tracers in exocrine acinar cells. Endocytosis of tracer (black) from the apical cell surface occurs in smooth C- or ring-shaped vesicles. The vesicles then move basally where they may fuse with secondary lysosomes (hatched). Tracer (stippled) is endocytosed from the lateral and basal cell surfaces primarily in coated vesicles. The tracer is rapidly localized in a system of tubular lysosomes. Some of the tracer is also delivered to secondary lysosomes (hatched) located adjacent to the Golgi apparatus. Material brought in from all cell surfaces may mix in these lysosomes.

investigations employing soluble phase markers (Becker *et al.*, 1967; Friend and Farquhar, 1967; Holtzman and Dominitz, 1968; Holtzman, 1971; Steinman and Conn, 1972; Abrahams and Holtzman, 1973; Geuze and Poort, 1973; Turner and Harris, 1974; Kalina and Robinovitch, 1975; Teichberg *et al.*, 1975; Tixier-Vidal *et al.*, 1976; Broadwell and Brightman, 1979; Broadwell *et al.*, 1980; Herzog and Reggio, 1980; see also chapters by E. Holtzman and M. G. Farquhar). These studies suggest that membrane retrieved from the cell surface is sequestered in lysosomes and is not directly reutilized in secretory granule formation. As HRP is a soluble phase marker and does not bind directly to membranes, this evidence does not preclude that membrane or membrane components ultimately may be removed from the lysosomes for reutilization in secretory granule formation (see also chapter by M. G. Farquhar).

In a few instances, such as uptake into cells of the small intestine (Cornell *et al.*, 1971), seminal vesicles (Mata and David-Ferriera 1973; Mata, 1976), and somatotrophs in the anterior pituitary (Pelletier 1973; Broadwell and Oliver, 1980), HRP reaction product has been localized in Golgi saccules or GERL in addition to lysosomes. These findings suggest that direct reutilization of membrane may occur in some cell types. The reasons for the discrepancies in the pattern of uptake are not known. They may be due to basic metabolic differences among the cell types, or the various cells may each utilize unique endocytic pathways.

The fact that cells may have more than one endocytic pathway has been shown not only by our studies but by others employing multiple types of tracer in the same cell (Farquhar, 1978; Ottosen *et al.*, 1980; Herzog and Reggio, 1980; Farquhar, this volume). Uptake through different endocytic pathways may explain the discrepancies observed in the localization of HRP and dextrans in parotid and pancreatic acinar cells (Oliver and Hand, 1978; Herzog and Farquhar, 1977; Herzog and Reggio, 1980), and may play a role in the intracellular fate of native ferritin vs. cationized ferritin in mammotrophs (Farquhar, 1978) and thyroid follicular cells (Herzog and Miller, 1979).

To date, little is known about the precise events controlling or directing endocytic uptake. It is evident that cell type, cell surface, and the nature of the tracer are all important in determining the intracellular localization of a marker. Although the use of tracers can yield substantial information on membrane retrieval and reutilization, much remains to be learned about tracer–cell interactions and the endocytic process.

## REFERENCES

Abrahams, S. J., and Holtzman, E. (1973). *J. Cell Biol.* **56,** 540–558.
Amsterdam, A., Ohad, I., and Schramm, M. (1969). *J. Cell Biol.* **41,** 753–773.
Becker, N. H., Novikoff, A. B., and Zimmerman, H. M. (1967). *J. Histochem. Cytochem.* **15,** 160–165.
Bogart, B. I. (1975). *J. Ultrastruct. Res.* **52,** 139–155.
Broadwell, R. D., and Brightman, M. W. (1979). *J. Comp. Neurol.* **185,** 31–74.
Broadwell, R. D., and Oliver, C. (1980). *J. Cell Biol.* **87,** 206a.
Broadwell, R. D., Oliver, C., and Brightman, M. W. (1980). *J. Comp. Neurol.* **190,** 519–532.
Cornell, R., Walker, W. A., and Isselbacher, K. (1971). *Lab. Invest.* **25,** 42–48.
Fahimi, H. D. (1970). *J. Cell Biol.* **47,** 247–262.
Farquhar, M. G. (1978). *J. Cell Biol.* **77,** R35–R42.
Friend, D. S., and Farquhar, M. G. (1967). *J. Cell Biol.* **35,** 357–376.
Geuze, J. J., and Kramer, M. F. (1974). *Cell Tissue Res.* **156,** 1–20.
Geuze, J. J., and Poort, C. (1973). *J. Cell Biol.* **57,** 159–174.
Hand, A. R. (1973). *In* "Symposium on the Mechanism of Exocrine Secretion" (S. S. Han, L. Sreebny, and R. Suddick, eds.), pp. 129–151. Univ. of Michigan Press, Ann Arbor.
Herzog, V., and Farquhar, M. G. (1977). *Proc. Natl. Acad. Sci. U.S.A.* **74,** 5073–5077.
Herzog, V., and Miller, F. (1979). *Eur. J. Cell Biol.* **19,** 203–215.
Herzog, V., and Reggio, H. (1980). *Eur. J. Cell Biol.* **21,** 141–150.
Holtzman, E. (1971). *Philos. Trans. R. Soc. London, Ser. B* **261,** 407–421.
Holtzman, E., and Dominitz, R. (1968). *J. Histochem. Cytochem.* **16,** 320–336.
Kalina, M., and Robinovitch, M. (1975). *Cell Tissue Res.* **163,** 373–382.
Karnovsky, M. J. (1965). *J. Cell Biol.* **27,** 137a.

Karnovsky, M. J. (1971). *J. Cell Biol.* **51,** 146a.

Mata, L. R. (1976). *J. Microsc. Biol. Cell.* **25,** 127–132.

Mata, L. R., and David-Ferriera, J. F. (1973). *J. Microsc. (Paris)* **17,** 103–106.

Meldolesi, J. (1974a). *Philos. Trans. R. Soc. London, Ser. B* **268,** 38–53.

Meldolesi, J. (1974b). *Adv. Cytopharmacol.* **2,** 71–85.

Meldolesi, J., and Cova, D. (1971). *J. Cell Biol.* **51,** 396–404.

Oliver, C. (1979). *J. Cell Biol.* **83,** 257a.

Oliver, C. (1980). *J. Cell Biol.* **87,** 206a.

Oliver, C., and Hand, A. R. (1978). *J. Cell Biol.* **76,** 207–220.

Ottosen, P. D., Courtoy, P. J., and Farquhar, M. G. (1980). *J. Exp. Med.* **152,** 1–19.

Palade, G. (1975). *Science* **189,** 347–358.

Pelletier, G. (1973). *J. Ultrastruct. Res.* **43,** 445–459.

Reynolds, E. S. (1963). *J. Cell Biol.* **17,** 208–212.

Simson, J. A. V. (1972). *Anat. Rec.* **173,** 437–452.

Spurr, A. R. (1969). *J. Ultrastruct. Res.* **26,** 31–43.

Steinmann, R. M., and Cohn, Z. A. (1972). *J. Cell Biol.* **55,** 186–204.

Teichberg, S., Holtzman, E., Crain, S. M., and Peterson, E. (1975). *J. Cell Biol.* **67,** 215–230.

Tixier-Vidal, A., Picart, R., and Moreau, M. F. (1976). *J. Microsc. Biol. Cell.* **25,** 159–172.

Turner, P. T., and Harris, A. B. (1974). *Brain Res.* **74,** 305–326.

Wallach, D., Kirshner, N., and Schramm, M. (1975). *Biochim. Biophys. Acta* **375,** 87–105.

# Chapter 27

# Membrane Synthesis and Turnover in Secretory Cell Systems

## J. MELDOLESI AND N. BORGESE

*Department of Pharmacology and CNR Center of Cytopharmacology,*
*University of Milan, Milan, Italy*

## I.  Introduction

During the last decade, membrane biogenesis has been intensely investigated in a variety of cell systems. Many pioneer studies now appear outdated because they were carried out at a low level of resolution, using entire membranes or insufficiently purified components. Recently, however, the refinement of the analytical techniques and the widespread use of specific antibodies has permitted extension of the work to single, well-identified membrane molecules. In particular, for studying membrane proteins, two different experimental approaches are used. In the first case (the *in vitro* approach), isolated polysomes or purified mRNA preparations are translated in cell-free conditions, and the newly synthesized products, recovered from the incubation mixtures, are analyzed and characterized by biochemical and immunological techniques. In the *in vivo* approach, on the other hand, protein synthesis is allowed to proceed for definite periods of time in intact cells (or even in the entire animal), and the products to be analyzed are recovered at various times thereafter, usually after cell homogenization and isolation of subcellular fractions.

It should be realized that these two approaches are not alternative but complementary. Thus, a coherent picture of the entire process can be obtained by a combination of the two. In particular, the *in vitro* study is necessary to identify the site of synthesis (whether bound or free polysomes) of individual membrane proteins and to characterize the primary translation products—i.e., to find out whether or not the protein is synthesized as a preform. By using reaction mixtures supplemented with adequate acceptor membranes either in the course of protein synthesis, or after this process is over, it is possible to establish whether proteins are inserted into membranes while still in growth or as finished molecules (co- and posttranslation insertion, respectively) and to characterize the insertion process also in terms of the membrane components involved.

The *in vivo* approach, on the other hand, is the only one that permits one to investigate the relative rates at which different proteins are synthesized in living cells; to investigate the route(s) followed by the proteins, which, after insertion in one membrane type, are transported throughout the cell to reach their final destination; to study the existence of biogenetic relationships among membranes; and to carry out long-term turnover studies.

So far, the *in vitro* approach has been very successful. Interesting differences in the site of synthesis and mechanisms of insertion have emerged for many membrane proteins. Thus, those proteins of the endoplasmic reticulum (ER) and plasma membrane that are exposed either to the internal surface of the organelle or to the external surface of the cell (and which, therefore, in order to reach their destination have to cross, entirely or in part, a membrane lipid bilayer: ectoproteins and spanning proteins, respectively) were found to be synthesized by bound polysomes and inserted contranslationally into the ER membranes (Blobel, 1980; Bar-Nun *et al.*, 1980; see also Chapter 2 by H. F. Lodish *et al.*). At least in some cases this cotranslational insertion was found to proceed by the use of a signal mechanism analogous to that of secretory proteins (Lingappa *et al.*, 1978; see also Chapter 2 by H. F. Lodish *et al.*). In contrast, membrane proteins of semiautonomous organelles (inner membranes of mitochondria and chloroplasts: see Chapter 4 by M.-L. Maccecchini; Chua and Schmidt, 1978), as well as several endoproteins—i.e., proteins exposed only at the cytoplasmic surface of membranes (Okada *et al.*, 1979; Rachubinski *et al.*, 1980)—were found to be synthesized by free polysomes and probably inserted posttranslationally. Moreover, although for mitochondria and chloroplasts protein insertion into the inner membrane is most often coupled with precursor cleavage (see Chapter 4 by M.-L. Maccecchini; Chua and Schmidt, 1978), this does not appear to be a general mechanism. Even for some membrane proteins inserted cotranslationally, the primary translation products and final membrane forms might be identical (Schechter *et al.*, 1979; Bar-Nun *et al.*, 1980; Blobel, 1980).

Compared with the large body of information yielded by the *in vitro* experiments, the output of the *in vivo* approach is far less impressive. This disproportion can be explained by the great difficulties inherent to the *in vivo* studies. In fact, the protein composition of most membranes is very complex, and individual components are present in low concentration. In addition, the turnover of most membrane proteins is very slow, so that the specific radioactivity they attain in the course of *in vivo* experiments of protein synthesis is low. This drawback is particularly annoying when glandular systems are studied, because secretory proteins have a characteristically fast turnover (for a detailed discussion, see Meldolesi *et al.*, 1978). The difference in turnover between secretory and membrane proteins can be as large as 100-fold or more, and this explains why in some *in vivo* labeling experiments most of the radioactivity recovered in membrane protein preparations, which by analytical criteria appeared reasonably pure (Meldolesi and Cova, 1973; Amsterdam *et al.*, 1971), turned out to be accounted for by trace amounts of contaminants (Meldolesi, 1974; Wallach *et al.*, 1975). Considering the above-mentioned difficulties, it is not surprising that the most exciting results yielded by the *in vivo* approach were obtained in virus-infected cells in which the host protein synthesis is nearly completely shut off, whereas the synthesis of virus-specific peptides (some of which are destined to become integral proteins of the plasmalemma before virus budding) is very active (see Chapter 2 by H. F. Lodish *et al.*). The results of these studies appear consistent with the *in vivo* findings in that the membrane insertion of the ectoproteins of vesicular stomatitis virus (VSV) occurs in the rough-surfaced ER. Insertion is followed by transport to smooth membranes and finally by appearance at the cell surface, which is delayed by 20–30 minutes with respect to the synthesis. In contrast, a viral membrane endoprotein (the M protein of VSV) seems to insert into the plasmalemma directly from the cytoplasm (Atkinson, 1978). Cell-specific membrane proteins might follow the same scheme as viral proteins. However, the available evidence is incomplete and sometimes contradictory. For instance, neither for the viral nor for the cellular membrane proteins, which apparently move along various intracellular compartments, has the route been completely identified (Meldolesi *et al.*, 1978; Blobel, 1980; see also Chapter 2 by H. F. Lodish *et al.*). Moreover, the meaning of this transport in terms of biogenetic relationships among the different types of membrane is not yet understood. For some investigators the transport of membrane proteins from the ER to other membranes should be considered only as a convenient means used by the cell to deliver to appropriate compartments individual molecules that, because of their size, hydrophobic properties, and distribution in the plane of the membrane, cannot be inserted directly at their final destination (Palade, 1975; Meldolesi *et al.*, 1978). Alternatively, this transport has been envisaged not as involving individual components, or clusters of components, but rather as a unidirectional flow of membrane from the ER to the cell surface, accompanied by stepwise transformation of intracellular-type membranes into plasma membranes (Morré, 1977).

It might be still premature to draw definite conclusions as to which of these two interpretations is correct. In fact, as already mentioned, the number of transported membrane proteins so far investigated is quite small and their study still incomplete. In the meantime, however, a convenient means to address this problem might be the investigation of well-characterized proteins whose localization in more than one membrane type is due not to simple transit from the site of synthesis to the final destination, but to a true multicompartment distribution. From this point of view, components contained in comparable concentration in the ER membrane and in at least one more membrane type would be of special interest. In fact, the flow-differentiation hypothesis predicts that in the course of the transformation of ER membranes into the other membranes the common components are maintained, while the others are replaced. Thus, if the hypothesis is correct, the biogenetic relationship among the molecules of common components present in the ER and in the other membranes should be of the precursor-product type.

To our knowledge, *in vivo* biogenesis studies on membrane components with multicompartment distribution have not been carried out in the past. This might also be due to the fact that, although the idea of a partial overlapping in the composition of the various types of cellular membranes is now widely accepted, only very few multimembrane components have been adequately characterized. In the remainder of this article we shall summarize the data we have recently obtained, working *in vivo* as well as *in vitro* on the biogenesis, membrane insertion, and turnover of one such component, the flavoprotein NADH–cytochrome $b_5$ reductase, which in liver cells is located in microsomes, Golgi complex, and outer mitochondrial membranes. Special emphasis will be placed on the methodological aspects of our work. The relevance of our findings for understanding membrane biogenesis in secretory systems will be discussed.


## II.   Studies on the Biogenesis and Turnover of NADH–Cytochrome $b_5$ Reductase


## A.   NADH–Cytochrome $b_5$ Reductase: Molecular Characteristics and Subcellular Distribution

NADH–cytochrome $b_5$ reductase, a flavoprotein involved in the desaturation of fatty acids and in other important biological processes, was studied extensively by Japanese and American groups (Takesue and Omura, 1970, Mihara and Sato, 1972; Spatz and Strittmatter, 1973). The molecule is composed of two domains; one is hydrophilic, accounts for approximately three-fourths of the molecule, and includes the active center of the enzyme and the N terminus; the other is highly hydrophobic and is located toward the C terminus (Spatz and Strittmatter, 1973; Mihara *et al.*, 1978). The hydrophilic portion is always exposed at the cytoplasmic surface of membranes, whereas the hydrophobic portion is buried in the lipid bilayer (Spatz and Strittmatter, 1973; Mihara *et al.*, 1978). Although the exact membrane topography of this protein has not been elucidated yet, NADH–cytochrome $b_5$ reductase most probably does not span the lipid bilayer and therefore can be designated as an integral membrane endoprotein.

Exposure of isolated NADH–cytochrome $b_5$ reductase, as well as of membranes carrying the enzyme, to lysosome extract or purified cathepsin D at pH 5.5 results in the cleavage of the molecule and solubilization of the hydrophilic portion (designated as lysosome-solubilized reductase, l-reductase), which can then be purified by conventional biochemical procedures (Takesue and Omura, 1970). The whole molecule (usually referred to as detergent-solubilized reductase, d-reductase) can be obtained by using detergents both for solubilization of membranes and throughout the purification procedure (Mihara and Sato, 1972; Spatz and Strittmatter, 1973; Mihara *et al.*, 1978). When analyzed by SDS–polyacrylamide gel electrophoresis (SDS–PAGE), d- and l-reductases exhibit an apparent molecular weight of approximately 33,000 and 28,000, respectively. Whether these values are reliable, or are due to anomalous migration of otherwise larger peptides (44,000 and 34,000, respectively), is still debated (Spatz and Strittmatter, 1973; Mihara *et al.*, 1978).

In rat liver cells NADH–cytochrome $b_5$ reductase has been reported to be localized in ER, outer mitochondrial, and Golgi membranes. Although for the first two locations there is no dispute, the third has been questioned by some authors, who attributed the previous results to microsomal contamination of Golgi fractions (Amar-Costesec *et al.*, 1974). Our evidence, reported in detail elsewhere (Borgese and Meldolesi, 1980), indicates that indeed the reductase is endogenous to Golgi membranes. However, the enzyme activity is rapidly inactivated on storage of Golgi fractions, whereas it is much more stable in microsomes. This differential stability can account for the above-mentioned discrepancy of the literature. When measured in fresh preparations, the enzyme activity per milligram of protein is approximately as high in microsomes in the cis Golgi fraction (fraction $GF_3$ of Ehrenreich *et al.*, 1973), whereas in the trans Golgi fraction ($GF_{1+2}$, Ehrenreich *et al.*, 1973) it is 60–70%. From recovery measurements, the enzyme can be estimated to account for 0.1–0.2% of the protein in microsomes and Golgi fractions and for 0.025–0.05% in mitochondria (Borgese and Meldolesi, 1980).

In all its locations the enzyme activity seems due to the same enzyme protein. In fact, extensive studies, immunological (inhibition and competition experiments) as well as molecular (fingerprinting of peptic fragments and SDS–PAGE of CNBr fragments), carried out on the reductase purified from four fractions (microsomes, mitochondria, $GF_3$, and $GF_{1+2}$), failed to reveal differences in the molecule in its various locations (Meldolesi et al., 1980).

## B.   Biosynthesis and Turnover of NADH–Cytochrome $b_5$ Reductase: *In Vivo* Studies

### 1.   PURIFICATION OF NADH–CYTOCHROME $B_5$ REDUCTASE

To carry out the *in vivo* biogenesis studies, a purification procedure was needed that would allow the rapid processing of many samples and assure a high recovery as well as the complete elimination of contaminants, especially of the highly turning over secretory proteins. In this respect the conventional purification procedures developed in the past appeared inadequate because they are laborious and their yield is, at best, of the order of 20%. It appeared necessary, therefore, to develop an immunological procedure. Anti-NADH–cytochrome $b_5$ reductase antibodies, raised in rabbits by injecting l-reductase (purified by column chromatography), were used to purify the antigen solubilized from membranes by detergent treatment.

The first series of experiments was carried out by direct immunoprecipitation followed by SDS–PAGE of the immunoprecipitates (Borgese et al., 1977). This technique had the drawback of consuming large amounts of antibody. In fact, in these *in vivo* studies it was necessary to analyze chemical quantities (at least 10 $\mu$g) of reductase because of the low specific radioactivity of the enzyme, even after injection of large amounts of radioactive leucine (1 mCi per 100 gm body weight). We therefore decided to switch to an immunoadsorption protocol. Antireductase antibodies were first purified by immunoadsorption (onto a Sepharose 4B column bearing attached l-reductase). The purified antibodies were then attached to Sepharose 4B, and these antireductase columns were used to isolate the enzyme from the fractions. These columns could then be reused indefinitely. For quite some time the purification of d-reductase was attempted. Isolated fractions were treated with a mixture of Na deoxycholate (DOC) (0.5%) and Triton X-100 (2%), and appropriate aliquots of the solubilized material were loaded onto the affinity columns. After extensive washing and replacement of DOC–Triton with 0.2% Lubrol PX (a detergent that, in contrast to the other two, does not precipitate as low pH), the immunoadsorbed material was eluted with glycine-HCl buffer, pH 2.2, containing 0.2% lubrol PX, dialyzed extensively against 60% ethanol to remove the detergent, precipitated with 10% trichloroacetic acid (TCA), and then analyzed by SDS–PAGE. Although this approach constituted a distinct improvement with respect to immunoprecipitation, the results obtained were not yet satisfactory (Borgese et al., 1979). In fact, when radioactive subcellular fractions were analyzed, the background of the polyacrylamide gels was appreciable even in the region of the reductase band, suggesting a contribution of secretory proteins and protein fragments. An important result confirming this conclusion was obtained by analyzing the reductase bands obtained from liver homogenates prepared from rats injected with radioactive amino acids 10–360 minutes before sacrifice. Under these conditions we found that the specific radioactivity of the material in the band did not remain constant, as would be expected for a slowly turning over membrane protein, but decreased substantially, although to a lesser degree with respect to that of secretory proteins (Borgese *et al.*, 1979).

At this stage of the work it seemed to us that, in order to obtain reliable reductase labeling kinetics, it was essential to reduce the load of secretory proteins in the preparations used for immunoadsorption. We decided, therefore, to sacrifice the intactness of the reductase molecule and to work with l-reductase, solubilized by cathepsin D attack. In fact, by this procedure the enzyme can be separated from a good part of the secretory proteins, which remain segregated into vesicles and are separated by centrifugation. In addition, in order to decrease the recovery of contaminating labeled secretory proteins in immunoeluates, the

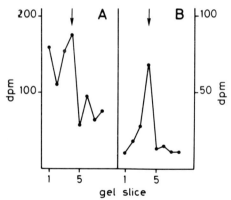

FIG. 1. Distribution of radioactivity in slices of polyacrylamide gels used for the analysis of reductase samples eluted from immunoaffinity columns. Effect of addition of nonlabeled rat serum to the samples before processing by immunoadsorption. Samples of rough microsomes, obtained from rats sacrificed 10 minutes after an injection of L-[³H]leucine into the portal vein, were digested with cathepsin D. One sample (panel B) was supplemented with unlabeled rat serum (36 μl/mg protein of the cell fraction before cathepsin D treatment), while the other (panel A) was left untreated. After elution from the immunoaffinity columns and TCA precipitation, the samples were analyzed on SDS–polyacrylamide tube gels. A 2-mm slice centered on the Coomassie Brilliant Blue-stained reductase band (arrow) was cut out. Two slices behind the band and three ahead of it were also cut out, and the radioactivity in the slices was determined. The figure illustrates the effect of unlabeled serum in reducing the radioactivity present in the gel due to secretory protein and protein fragments nonspecifically adsorbed to the immunoaffinity columns.

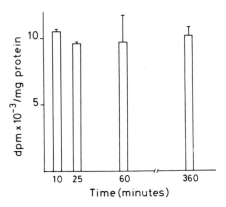

Time (minutes)

FIG. 2. Specific radioactivity of NADH–cytochrome $b_5$ reductase isolated from the total liver homogenate at different times after a pulse of L-[$^3$H]leucine, injected into the portal vein. NADH–cytochrome $b_5$ reductase was extracted by lysosomal digestion from total homogenates (40 mg protein) of animals sacrificed after a pulse of L-[$^3$H]leucine at the times indicated on the abscissa. Solubilized proteins were purified by immunoadsorption (in the presence of excess unlabeled rat serum) followed by SDS–PAGE. The specific radioactivity was determined by scintillation counting of the band separated on the gel and by radioimmunoassay of the reductase protein. Values given are the averages of two separate determinations carried out on two aliquots of total homogenate for each time point. Bars represent standard errors.

cathepsin D-solubilized preparations were mixed with an excess of competing, nonradioactive rat serum before loading onto the columns. Elution and processing of immunoeluates were as described for d-reductase, except that detergents and ethanol dialysis were omitted. Figures 1 and 2 illustrate two criteria we used to conclude that the purification of reductase was finally adequate to our purposes. In fact, the peaks of labeled l-reductase found in polyacrylamide gels, although containing only a few counts, were clean and symmetrical over a low background (Fig. 1B). Moreover, in *in vivo* pulse–chase experiments the specific radioactivity of the enzyme purified from liver homogenates did not show appreciable variations (Fig. 2). The last-described purification procedure was therefore used for our pulse-labeling and long-term labeling studies. In the latter experiments, addition of unlabeled serum was found to be unnecessary, because animals were killed 12.5 hours after injection of the radioactive precursor, a time at which the specific radioactivity of secretory proteins had decreased so much that it did not interfere in the analyses (for further details, see Borgese *et al.*, 1980).

## 2.   QUANTITATIVE ESTIMATION OF NADH–CYTOCHROME $B_5$ REDUCTASE

To measure the specific radioactivity of reductase, it was necessary to know the amount of the enzyme present in each purified preparation. One possibility we considered was to measure the NADH-dependent reduction of FeCN in the material eluted from the affinity columns. However, this approach was discarded because we found that the long incubation with cathepsin D at pH 5.5, necessary to detach the enzyme from membranes, effected a considerable inactivation of the NADH–FeCN reductase activity. Moreover, this inactivation was unequal in the different cell fractions. A second possibility was to determine, by quantitative scanning, the areas of the Coomassie Brilliant Blue-stained reductase bands. For this purpose, calibration curves were constructed using purified l-reductase. These curves were linear between 10 and 50 $\mu$g of reductase. However, this approach was satisfactory only for long-term labeling experiments. In the case of pulse-labeling, the staining background of the gels was high, presumably because of the addition of large amounts of unlabeled serum to the samples before immunoadsorption. As a consequence, in these gels it was difficult to estimate precisely the area of the reductase peak. We therefore decided to develop a radioimmunoassay of the enzyme, to be used on small aliquots of the immunoeluates. Although this method was in general very satisfactory (high accuracy and reproducibility, high sensitivity), it should be realized that the radioimmunoassay data are based on the assumption that all the material recognized by the antibody in the immunoeluates is ultimately recovered in the SDS–PAGE bands. In order for this to be true, the following conditions should be satisfied: no losses of reductase during TCA precipitation and SDS–PAGE analysis; no cross-reactivity of the antibody with other components of the immunoeluates; absence in immunoeluates of reductase fragments still recognized by the antibody but migrating in SDS–PAGE at rates different from that of 1-reductase. Although the first two conditions can be met by carefully processing the immunoeluates and by using a monospecific antibody, the strict compliance to the third seems unrealistic in our experimental conditions, especially if one is considering the length and complexity of the procedure used for reductase purification. In order to check whether the possible presence of fragments had any impact on our results, we compared the radioimmunoassay data with those of Coomassie Brilliant Blue-stained gel scanning and obtained qualitatively similar results, although as expected the radioimmunoassay procedure did slightly overestimate the reductase concentration. Thus, although neither of the methods available is entirely satisfactory, the combination of the two made us quite confident of the results we obtained (for further details, see Borgese *et al.*, 1980).

## 3. SYNTHESIS AND TURNOVER OF NADH–CYTOCHROME $B_5$ REDUCTASE: *In Vivo* EXPERIMENTS

Essentially two types of experiments were carried out. In the first case thiopental-anesthetized rats were injected into the portal vein with L-[$^3$H]leucine and sacrificed at 10, 25, 60, and 360 minutes thereafter. The time course of the radioactivity per milligram of protein in subcellular fractions isolated from liver homogenates was that expected from previous studies (not shown in figures). Labeling of rough microsomes was maximum at 10 minutes and then steadily decreased; in smooth microsomes, $GF_3$, and especially in $GF_{1+2}$, it increased from 10 to 25 minutes and then decreased considerably. These fluctuations are due to the intracellular transport and discharge of secretory proteins in liver cells. In the mitochondrial fraction, which contains no secretory proteins, the radioactivity was at a much lower level than in microsomes, and it remained approximately constant throughout the experiment. The time course of the specific radioactivity of NADH–cytochrome $b_5$ reductase isolated from the various subcellular fractions is summarized in Table I.

Three aspects of these results are worth mentioning. The first is that the specific radioactivity of the enzyme was low in all fractions in spite of the large quantity of tracer injected (1.5 mCi per rat). The second is that in all fractions the specific radioactivity attained by the reductase at 10 minutes from the injection (approximately the end of the incorporation in our experimental conditions) remained unchanged during the following 360 minutes, indicating that over that period there is no unidirectional transport of the enzyme among the various membrane compartments investigated. Finally, the specific radioactivity of the reductase in the mitochondrial fraction was considerably lower than that in microsomes and Golgi fractions. This finding might be due to a different turnover of NADH–cytochrome $b_5$ reductase in its various locations. To investigate this possibility in detail, we carried out the second group of experiments, using the double-label technique of Arias et al. (1969). Rats received first an i.p. injection of L-[$^3$H]leucine, followed 6 days later by L-[$^{14}$C]leucine, administered under identical conditions. They were sacrificed 12.5 hours after the last injection. The

TABLE I

SPECIFIC RADIOACTIVITY OF NADH-CYTOCHROME B$_5$ REDUCTASE ISOLATED FROM RAT LIVER CELL FRACTIONS AT VARIOUS TIMES AFTER A PULSE OF L-[$^3$H]LEUCINE INJECTED INTO THE PORTAL VEIN[a]

| Cell fraction | Time (minutes) | | | |
| --- | --- | --- | --- | --- |
|  | 10 | 25 | 60 | 360 |
| Rough MR | 13,180 ± 2670 | 10,840 ± 2550 | 11,010 ± 1160 | 15,210 ± 1720 |
| Smooth MR | 11,480 ± 550 | 12,600 ± 1370 | 13,880 ± 540 | 15,110 ± 1150 |
| Total Golgi | 11,710 ± 2010 |  | 14,380 ± 2360 | 12,690 ± 2300 |
| Mitochondria | 5460 ± 1700 | 6000 ± 1600 | 5730 ± 1210 | 5720 ± 990 |

[a] An aliquot of the immunoeluates obtained from antireductase columns was used to measure reductase protein by radioimmunoassay. The rest was processed by SDS–PAGE, and the reductase band was used to measure radioactivity. For experimental details, see Borgese et al. (1980).

TABLE II

RELATIVE TURNOVER OF NADH-CYTOCHROME B$_5$ REDUCTASE
IN RAT LIVER CELL FRACTIONS[a]

| Cell fraction | $^{14}C/^3H$[b] | $^{14}C$ (dpm/mg protein)[c] |
|---|---|---|
| Total homogenate | 0.433 ± 0.011 | 4860 ± 640 |
| Rough MR | 0.452 ± 0.012 | 4910 ± 70 |
| Smooth MR | 0.383 ± 0.029 | 5580 ± 1200 |
| Total Golgi | 0.422 | 4680 |
| Mitochondria | 0.267 ± 0.002 | 3180 ± 430 |

[a] Two rats were injected intraperitoneally with L-[$^3$H]leucine first and with L-[$^{14}$C]leucine 6 days later. They were sacrificed 12.5 hours after the second injection. Doses injected were 1.15 and 0.3 mCi.

[b] The $^{14}C/^3H$ ratios were determined on the l-reductase bands of SDS–polyacrylamide gels of immunoeluates. Differences in ratios reflect differences in turnover (see Arias *et al.*, 1969).

[c] l-Reductase was measured by quantitative microdensitometry of SDS–polyacrylamide gel bands stained with Coomassie Brilliant Blue. For experimental details, see Borgese *et al.* (1980).

results, summarized in Table II, indicate that indeed NADH–cytochrome b$_5$ reductase turns over faster in microsomes and Golgi membranes than in mitochondria (for further details, see Borgese *et al.*, 1980).

## C. Biosynthesis of NADH–Cytochrome b$_5$ Reductase: *In Vitro* Studies

In order to determine whether NADH–cytochrome b$_5$ reductase is synthesized in free or bound polyribosomes, cell-free amino acid incorporation with [$^{35}$S]methionine, programmed by each of the two classes of polyribosomes, was carried out in a rabbit nuclease-treated reticulocyte lysate. The efficiency of amino acid incorporation was approximately the same for free and bound polyribosomes, with more than a 20-fold stimulation over controls incubated without added polyribosomes. At the end of the synthesis (20 minutes), the mixtures were incubated with antireductase antibodies and then with protein A–Sepharose CL-4B beads. The immunoadsorbed material was then analyzed by SDS–PAGE followed by fluorography. The total products synthesized by free and bound polyribosomes were also analyzed and found to be quite different (Fig. 3A and B). Analysis of the immunoadsorbed material revealed a major band (lane D of Fig. 3), migrating slightly behind iodinated l-reductase

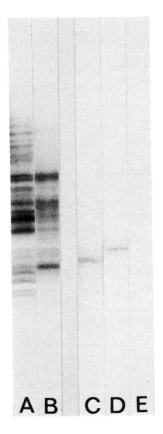

FIG. 3.   SDS–PAGE analysis of cell-free translation products of free and bound polyribosomes. Free or detergent-solubilized bound polyribosomes were allowed to direct protein synthesis in a nuclease-treated rabbit reticulocyte lysate, with [$^{35}$S]methionine. After synthesis, small aliquots (2 $\mu$l) of the incubation mixtures containing free (lane A) or detergent-solubilized bound (lane B) polyribosomes were directly analyzed by SDS slab–PAGE and subsequent fluorography. Aliquots (300 $\mu$l) of the mixtures were supplemented with antireductase antibodies (for conditions, see Borgese and Gaetani, 1980) and then with protein A–Sepharose CL–4B beads. Lanes D and E show the fluorography of immunoadsorbed products of free and bound polyribosomes, respectively. Lane C shows migration of $^{125}$I-l-reductase.

(lane C of Fig. 3), in the position expected for d-reductase, among the translation products of free polyribosomes. No such major band was apparent in the case of bound polyribosomes (Fig. 3, lane E). These results (reported in detail by Borgese and Gaetani, 1980) indicated that the main site of synthesis of NADH–cytochrome b$_5$ reductase in liver is at the level of free polyribosomes.

# III.   Conclusions

A model that accounts for the results we have obtained on the biogenesis and turnover of NADH–cytochrome b$_5$ reductase is depicted in Fig. 4. The enzyme, schematized as a little circle (the hydrophilic domain) attached to a curled stalk (the hydrophobic domain) is synthesized by free polysomes and released into the three different membranes where it is localized: ER, Golgi, and outer mitochondrial membranes. To emphasize the slower turnover of the outer mitochondrial membrane enzyme, with respect to the microsomal and Golgi forms, a lower proportion of enzyme molecules with an open circle (representing newly synthesized reductase molecules) has been depicted in mitochondria than in the other compartments. In addition, Fig. 4 gives information on the concentration of

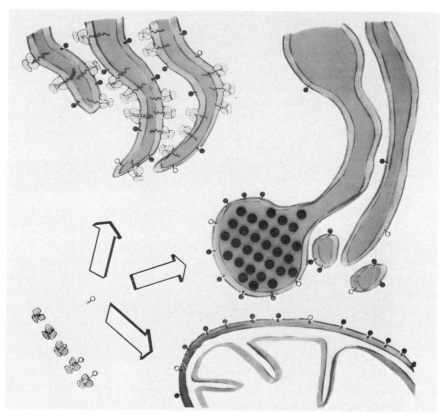

FIG. 4. Cartoon representation of a model for the biosynthesis and membrane insertion of NADH–cytochrome $b_5$ reductase. The figure summarizes the life history of NADH–cytochrome $b_5$ reductase, based on the experimental data obtained by *in vivo* and *in vitro* experiments. NADH–cytochrome $b_5$ reductase is represented as a circle (hydrophilic domain) attached to a curled line (hydrophobic domain). The protein is synthesized by free polyribosomes and rapidly inserted into different intracellular membranes. A Golgi cisterna, Golgi vesicles, part of a mitochondrion, and rough-surfaced ER cisternae are represented. The large black spheres located within the dilated rim of the Golgi cisterna represent VLDL particles. The circles of newly synthesized reductase molecules are open, to distinguish them from "older" molecules, which appear as solid circles. This does not imply the existence of a precursor form of the reductase, for which we have no evidence at the present time. The lower ratio of open to solid circles in the outer mitochondrial membrane, compared with that in the other membranes, represents the lower turnover of the enzyme in the former compartment. See text for further explanation.

NADH-cytochrome $b_5$ reductase in the various membranes (higher in outer mitochondrial membranes than in the other locations) as well as on its heterogeneous distribution in the Golgi complex (higher concentration in vesicles than in cisternae).[1] The experimental evidence supporting these two conclusions is reported elsewhere (Borgese and Meldolesi, 1980). Although some aspects of the model appear soundly based, a number of important details still need to be clarified. An especially intriguing question concerns the mechanism(s) by which the concentration of NADH-cytochrome $b_5$ reductase is regulated in different membrane compartments. Thus, the enzyme is absent (or nearly absent) in plasma membranes and appears to be more concentrated in outer mitochondrial membranes than in ER and Golgi membranes. Various explanations of these findings can be envisaged: for example, the insertion could involve specific, not yet identified membrane receptors or the processing (occurring in some but not in all membranes) of a cytoplasmic precursor, so similar to the final form of the enzyme to be indistinguishable in SDS-PAGE.

In conclusion, the process of membrane biogenesis as it emerges from our studies (Borgese et al., 1980), as well as from other investigations (Bar-Nun et al., 1980; Blobel, 1980; see also Chapter 2 by H. F. Lodish et al.), appears more complex than was previously envisaged. Different membrane proteins use at least two different mechanisms to become membrane-bound and to reach their final destination: cotranslational insertion, typical of some ectoproteins (which in some cases is followed by redistribution throughout an intracellular membrane pathway); and direct insertion, which can occur concomitantly in several membranes without the need for transport from rough ER to other compartments, as we have shown for NADH-cytochrome $b_5$ reductase. At the moment we do not know whether other proteins, besides our enzyme, use the second mechanism; however, the studies on cytochrome $b_5$, which is also synthesized by free polysomes (Okada et al., 1979; Rachubinski et al., 1980), suggest that this might be the case for other endoproteins. Finally, if one considers that the overwhelming majority of membrane proteins have not yet been investigated, the possible existence of further mechanisms, still unrecognized, cannot be excluded.

As for membrane proteins, which must travel from intracellular membranes to the cell surface, the available evidence clearly indicates that their transport involves membrane fusion-fission events similar to those occurring for the intracel-

---

[1]These conclusions might appear to contradict the data on the distribution of NADH-cytochrome $b_5$ reductase in liver subcellular fractions: appoximately equal activity in microsomes and $GF_3$, 60-70% in $GF_{1+2}$, 25-30% in mitochondria (Section 2,A). However, it should be kept in mind that the relative contribution of the limiting membranes is quite different in the various fractions: it amounts to approximately the 50% of the proteins in microsomes and $GF_3$ and only to 10% or less in $GF_{1+2}$ and mitochondria (outer membrane). Thus, although in the latter two fractions the activity of the enzyme expressed on a total protein basis is lower than in microsomes, the concentration in the membrane is higher (see also Borgese and Meldolesi, 1980).

lular transport of secretory proteins. The question that rises is whether the same membranes that transport secretory proteins also transport membrane proteins. A tight relation between secretion and membrane biogenesis is possible because protein secretion is an activity common not only to grandular cells but to all eukaryotic cells, although of course it is developed to different extents. On the other hand, it is also possible that at some point along the intracellular transport route the pathways for membrane proteins and secretory proteins diverge and that the two classes of proteins are thus transported in different vesicles, by similar, but independent, processes (for a more detailed discussion of this problem, see Meldolesi *et al.*, 1978).

The results obtained (Borgese *et al.*, 1980) also provide interesting clues on membrane turnover. In general, two separate processes are believed to account for the replacement of cellular proteins and structures: autophagocytosis (by which portions of the cytoplasm are first segregated and then digested in bulk by lysosomal hydrolases) and molecular turnover. The relative importance of these two processes differs in different tissues and in a given tissue, depending on various conditions (feeding, hormonal state, etc.) (Pfeiffer *et al.*, 1978; Dice *et al.*, 1978; Dean, 1979). Although some aspects of autophagocytosis have been investigated in some detail, the intimate mechanisms and the regulation of molecular turnover are still poorly understood. So far, the emphasis in these studies has been on the correlation between molecular features (size, charge, glycosylation) of the various proteins and their turnover rates (Arias *et al.*, 1969; Matheus *et al.*, 1976; Goldberg and StJohn, 1976; Dice *et al.*, 1978). Our results, showing that NADH–cytochrome $b_5$ reductase, which is most probably molecularly identical in its three subcellular locations (Section II,A), turns over at a slower rate in mitochondria than in microsomes and Golgi apparatus, emphasize the importance of the subcellular localization for the turnover of integral membrane proteins. Whether this result is entirely due to a difference of the molecular turnover of the protein in the three compartments, or whether it depends, at least in part, on a different contribution of autophagocytosis, remains to be clarified.

## References

Amar Costesec, A., Wibo, M., Thinès-Sempoux, D., and Berthet, J. (1974). *J. Cell Biol.* **62,** 717–745.
Amsterdam, A., Schramm, M., Ohad, I., Salomon, Y., and Selinger, Z. (1971). *J. Cell Biol.* **50,** 187–200.
Arias, I. M., Doyle, D., and Schimke, R. T. (1969). *J. Biol. Chem.* **244,** 3303–3312.
Atkinson, P. H. (1978). *J. Supramol. Struct.* **8,** 89–109.
Bar-Nun, S., Kreibich, G., Adesnik, M., Alterman, L., Negishi, M., and Sabatini, D. D. (1980). *Proc. Natl. Acad. Sci. U.S.A.* **77,** 965–969.

Blobel, G. (1980). *Proc. Natl. Acad. Sci. U.S.A.* **77**, 1496–1500.

Borgese, N., and Gaetani, S. (1980). *FEBS Lett.* **112**, 216–220.

Borgese, N., and Meldolesi, J. (1980). *J. Cell Biol.* **85**, 501–515.

Borgese, N., De Camilli, P., Brenna, A., and Meldolesi, J. (1977). *In* "Membranous Elements and Movement of Molecules" (E. Reid, ed.), pp. 339–353. Ellis Horwood, Chichester.

Borgese, N., De Camilli, P., Tanaka, Y., and Meldolesi, J. (1979). *Symp. Soc. Exp. Biol.* **33**, 117–144.

Borgese, N., Pietrini, G., and Meldolesi, J. (1980). *J. Cell Biol.* **86**, 38–45.

Chua, N. H., and Schmidt, G. W. (1978). *Brookhaven Symp. Biol.* **30**, 225–348.

Dean, R. T. (1979). *Biochem. J.* **180**, 339–345.

Dice, J. F., Walker, C. D., Byrne, B., and Cardial, A. (1978). *Proc. Natl. Acad. Sci. U.S.A.* **75**, 2093–2097.

Ehrenreich, J. H., Bergeron, J. J. M., Siekevitz, P., and Palade, G. E. (1973). *J. Cell Biol.* **59**, 45–62.

Goldberg, A. L., and St. John, A. (1976). *Annu. Rev. Biochem.* **45**, 747–803.

Lingappa, V. R., Katz, F. N., Lodish, H. F., and Blobel, G. (1978). *J. Biol. Chem.* **253**, 8667–8670.

Matheus, R. A., Johnson, T. C., and Hudson, J. E. (1976). *Biochem. J.* **154**, 57–64.

Meldolesi, J. (1974). *J. Cell Biol.* **61**, 1–13.

Meldolesi, J., and Cova, D. (1971). *Biochem. Biophys. Res. Commun.* **44**, 139–143.

Meldolesi, J., Borgese, N., De Camilli, P., and Ceccarelli, B. (1978). *In* "Membrane Fusion" (G. Poste and G. N. Nicolson, eds.), pp. 509–627. Elsevier/North-Holland Publ., Amsterdam and New York.

Meldolesi, J., Corte, G., Pietrini, G., and Borgese, N. (1980). *J. Cell Biol.* **85**, 816–826.

Mihara, S., and Sato, R. (1972). *J. Biochem. (Tokyo)* **71**, 725–735.

Mihara, S., Sato, R., Sakakibara, R., and Wada, H. (1978). *Biochemistry* **17**, 2829–2834.

Morré, D. J. (1977). *In* "The Synthesis, Assembly and Turnover of Cell Surface Components" (G. Poste and G. N. Nicolson, eds.), pp. 1–83. Elsevier/North-Holland Publ., Amsterdam and New York.

Okada, Y., Sabatini, D. D., and Kreibich, G. (1979). *J. Cell Biol.* **83** (2, Pt. 2), 437a.

Palade, G. E. (1975). *Science* **189**, 347–358.

Pfeiffer, U., Wreder, E., and Bergeest, H. (1978). *J. Cell Biol.* **78**, 152–166.

Rachubinski, R. A., Verma, D. P. S., and Bergeron, J. J. M. (1980). *J. Cell Biol.* **84**, 705–716.

Schechter, I., Burnstein, Y., Zemell, R., Ziv, E., Kantor, F., and Papermeister, D. S. (1979). *Proc. Natl. Acad. Sci. U.S.A.* **76**, 2654–2658.

Spatz, L., and Strittmatter, P. (1973). *J. Biol. Chem.* **248**, 792–799.

Takesue, S., and Omura, T. (1970). *J. Biochem. (Tokyo)* **67**, 267–276.

Wallach, D., Kirschner, N., and Schramm, M. (1975). *Biochim. Biophys. Acta* **375**, 87–105.

# Part VI.   Activation of the Secretory Response

# Chapter 28

# The "Secretory Code" of the Neutrophil

### JAMES E. SMOLEN,  HELEN M. KORCHAK,[1]
### AND GERALD WEISSMANN

*Department of Medicine, Division of Rheumatology, New York University
School of Medicine, New York, New York*

## I.   Introduction

Exposure of human polymorphonuclear leukocytes (PMN) to a variety of particulate and soluble stimuli evokes a series of responses, such as chemotaxis, phagocytosis, degranulation of lysosomal enzymes, hexose monophosphate

---

[1]James E. Smolen and Helen M. Korchak are Arthritis Foundation Fellows.

shunt stimulation, generation of reactive derivatives of oxygen, release of membrane-bound calcium, and reorganization of microtubules and microfilaments. These processes resemble, in general outline, though not in all details, processes of stimulus–secretion coupling in other cells and make the PMN an eminent candidate as a model secretory cell. Analysis of the initial events in PMN secretion and of the temporal order in which they occur has already led to an appreciation of how ligand–receptor interactions are transduced into cellular responses, and how ions and/or cyclic nucleotides participate in secretion. Only a temporal analysis will permit separation of ends from means and causes from effects. This work has permitted us to glimpse what an optimist would call the "secretory code" of the neutrophil.

The initial step in PMN stimulation, a process leading eventually to events such as degranulation and superoxide anion ($O_2^-$) generation, involves either ligand–receptor interactions at the cell surface or other direct perturbations of the plasma membrane. The kinetics of receptor–ligand interactions appear to be rapid, when defined stimuli are examined; binding of the chemotactic peptide *N*-formyl-norleucyl-leucyl-phenylalanine (FNLP) to rabbit PMN or rabbit PMN plasma membranes was both time-dependent and saturable (Sha'afi *et al.*, 1978). Saturating concentrations of FNLP occupied 50% of the available receptors on intact PMN within 30 seconds.

Membrane perturbations produced either by direct interactions of the stimulus with the plasmalemma or, indirectly, by ligand–receptor interactions, seem to follow surface stimulation. The fluorescent probe 1-anilino-8-naphthalene sulfonate (ANS), when added to guinea pig PMN, exhibited an enhancement and blue-shift of fluorescence suggestive of plasma membrane binding (Romeo *et al.*, 1970). When the ANS-labeled PMN were exposed to polystyrene latex particles, there was a prompt increase in fluorescent intensity. This rapid response (within 2 seconds of particle addition) was indicative of some conformational changes in the cell membrane; the exact nature of this conformation change is unknown. Similar results were found with ANS-labeled mouse peritoneal macrophages (Yokomora and Ishikawa, 1973). These findings have not been extended to other types of granulocytes or to other stimuli.

Phagocytosis of paraffin oil emulsion particles or polystyrene latex beads by rabbit PMN also led to a change in membrane microviscosity (Berlin and Fera, 1977). PMN were labeled with perylene or 1,6-diphenyl-1,3,5-hexatriene, and their plasma membranes were isolated (in order to avoid complications introduced by incorporation of the fluorescent probes into intracellular membranes). Labeled membranes isolated after 5 or more minutes of particle ingestion were significantly more fluid than control membranes, when microviscosity was measured by fluorescence depolarization. Colchicine abolished the phagocytosis-induced increase in membrane fluidity, suggesting that microtubules were responsible for a reorganization of membrane lipids. The fact that liposomes made from extracts of the membrane lipids showed parallel results indicated that the observed fluidity changes were due to alterations in membrane lipid composition; the high degree of fatty acid saturation found in isolated phagosome membranes (Mason *et al.,* 1972; Smolen and Shohet, 1974) suggests the existence of a process that could leave residual plasma membrane enriched in unsaturated fatty acids.

Unfortunately, neither the spin label nor the fluorescence depolarization papers cited above reported the earliest changes in membrane fluidity. The mechanics of measuring microviscosity by means of spin probes do not allow the first few seconds of granulocyte stimulation to be measured. Fluorescence depolarization is potentially capable of obtaining such measurements, but the first time point observed in the study cited above was at 5 minutes. In any case, the practice of isolating membranes before measurement degrades the time resolution for any determination of membrane fluidity. The enhancement of fluorescence intensity in ANS-treated cells (Romeo *et al.,* 1970; Yokomura and Ishikawa, 1973) has none of these problems; indeed, enhancement of fluorescence was readily observed within 2 seconds of stimulation. However, the exact nature of the changes in membrane conformation revealed by this observation remains obscure.

Surface responses in secretory cells are usually accompanied by changes in membrane potential. To quantify this in the PMN, which are too small for the insertion of microelectrodes, the lipophilic triphenylmethylphosphonium cation (TPMP$^+$) has been used (Korchak and Weissmann, 1978). The distribution of this molecule across cell membranes is determined by the strength and polarity of the existing electrochemical gradient. When purified granulocytes were exposed to concanavalin A (Con A), they responded within 10 seconds by a sharp hyperpolarization, followed by depolarization, followed in turn by long, slow hyperpolarization. When endocytosis of Con A was blocked by cytochalasin B, similar results were seen; if anything, the surface potential changes were observed even earlier. In experiments in which the cells were exposed to immune complexes of bovine serum albumin and anti-bovine serum albumin (BSA/anti-BSA), to a chemotactic tripeptide, and to latex beads, a similar hyperpolarization response was observed. In contrast, cells exposed to ionophore A23187 did not exhibit a hyperpolarization response.

Changes in PMN membrane potential have also been measured by using the fluorescent dye dipentyloxacarbocyanine [Di-O-C$_5$-(3)] (Seligmann *et al.,* 1977). Exposure of dye-loaded cells to the chemotactic peptide *N*-formyl-methionyl-leucyl-phenylalanine (FMLP) produced a biphasic change in fluorescent intensity, seemingly similar to that seen with the TPMP$^+$ method (Korchak and Weissmann, 1978). However, the fluorescence response was inhibited by cytochalasin B, and EGTA blocked the second phase. Furthermore, calcium ionophore A23187 produced fluorescence changes. These latter three observations are at variance with results obtained by the TPMP$^+$ method. Although carbocyanine dyes have been used extensively, it is also clear that these dyes interfere with cellular metabolism and have some toxic effects (Montecullo *et al.*, 1979). On balance, however, both methods (despite their unique problems) agree that membrane potential changes occur in stimulated PMN and that these changes take place as rapidly as they can be measured ($<5$ seconds).

Unlike PMN, macrophages are large enough to permit the insertion of microelectrodes. Guinea pig peritoneal macrophages were found to have a resting potential of $-13$ mV (Gallin *et al.*, 1975). Calcium ionophore A23187 induced hyperpolarizations, which were abolished by removal of Ca$^{2+}$ with EGTA. Human macrophages were also investigated by this technique (Gallin and Gallin, 1977). Some cells exposed to purified C5a exhibited a biphasic membrane potential response; a brief depolarization of 1–10 seconds was followed by a larger and prolonged hyperpolarization. However, this behavior was not highly reproducible, as only 17% of the cells responded in this fashion. Cytochalasin B and colchicine had no effect upon C5a-induced membrane potential changes, whereas Mg-EGTA blocked them. FMLP produced similar rapid hyperpolarizations. Morphological studies indicated that exposure to C5a caused membrane spreading, ruffling, and pseudopod formation, but that these events were anteceded by changes in membrane potential.

Chlorotetracycline (CTC) is a probe that forms highly fluorescent complexes with membrane-bound Ca$^{2+}$ or Mg$^{2+}$; the complexes with these two divalent cations can be distinguished on the basis of their spectral properties. Rabbit PMN loaded with CTC showed an immediate loss of fluorescence when exposed to FMLP or C5a (Naccache *et al.*, 1979). This loss was independent of the presence or absence of extracellular Ca$^{2+}$, suggesting that it was attendant to the earliest events following receptor–ligand interaction and not to the later Ca$^{2+}$ influx. The loss of fluorescence was also accompanied by a spectral shift indicating that CTC was now reporting membrane-bound Mg$^{2+}$; Ca$^{2+}$ had apparently been liberated into the intracellular milieu.

Membrane-bound Ca$^{2+}$ of human PMN has been observed by means of electron microscopy (Hoffstein, 1979; see also Chapter 17 by S. T. Hoffstein). Cations were precipitated *in situ* by fixing cells in aqueous solutions of osmium tetroxide and potassium pyroantimonate. Whereas the plasma membrane stained heavily for calcium, the membranes of the phagocytic vacuoles were entirely devoid of the precipitates. Those areas of the plasma membrane of the leukocyte not in proximity to the zymosan particles still retained calcium at the surface. However, at sites of the leukocyte membrane that were at the region of contact with the zymosan particles, calcium had been lost.

Stimulated PMN consume far more molecular oxygen than resting PMN. Molecular oxygen is reduced by an NADH- and/or NADPH-dependent oxidase, resulting in the formation of superoxide anion (O$_2$$^-$), H$_2$O$_2$, hydroxyl radicals, and singlet oxygen. The subcellular localization and pyridine nucleotide specificity of this critical oxidase has long been in dispute (Badwey *et al.*, 1979). A close examination of the kinetics of $O_2$ consumption by PMN showed that this process did not begin until approximately 25 seconds after the addition of latex particles (Segal and Coade, 1978). Thus, unlike changes in membrane potential, increases in ANS fluorescence and losses in membrane-bound $Ca^{2+}$, the onset of oxidative metabolism has a distinct, measurable lag phase.

If oxygen consumption, the seminal event in the burst of oxidative metabolism, displayed a lag period, then production of reduced oxygen derivatives should have similar or longer lag periods. Root and his co-workers (1975) devised a continuous assay for $H_2O_2$ production. $H_2O_2$ liberated into the extracellular medium by stimulated PMN was reduced by horseradish peroxidase with scopoletin as an electron donor, resulting in extinction of scopoletin fluorescence. When PMN were stimulated by latex particles, opsonized yeast, or *Staphylococcus aureus,* a period of 10–15 seconds elapsed between addition of the stimuli and the onset of fluorescence extinction. Thus, $H_2O_2$ production, like oxygen consumption, was initiated after a distinct lag period in response to surface stimuli.

Activation of the $O_2^-$-generating system of PMN is also not immediate. Using a simple technique whereby $O_2^-$ formation was continuously monitored as extracellular cytochrome c reduction, Cohen and Chovaniec (1978) demonstrated that 30–60 seconds elapsed before guinea pig PMN produced $O_2^-$ in response to digitonin. The lag period was decreased by increasing the concentration of the stimulus (from 150 to 60 seconds by a 15-fold increase in stimulus). The lag time decreased from 227 to 48 seconds by elevation of the temperature from 10° to 40°C. They further confirmed the observation of Root *et al.* (1975) that $H_2O_2$ production has a similar lag period. This was to be expected, since $H_2O_2$ is primarily the product of spontaneous or enzymatic dismutation of $O_2^-$ to $H_2O_2$ and $O_2$.

These results were verified in human PMN by Korchak and Weissmann (1978). Generation of $O_2^-$ was assayed continuously in the presence of cytochalasin B, used to maximize production and detection of the anion radical. Exposure of PMN to the lectin Con A or to an immune complex consisting of BSA/anti-BSA gave rise to $O_2^-$ generation after distinct lag periods. Stimulation by the immune complex resulted in a lag period of approximately 30 seconds, whereas Con A induced $O_2^-$ generation after 42 seconds. Generation of $O_2^-$ was clearly shown to follow the earlier and apparently immediate hyperpolarization response. These results demonstrated that lag periods for $O_2^-$ production are not species-specific and are not dependent upon the presence or absence of cytochalasin B, and that their durations are dependent upon the stimulus presented.

The short-term kinetics of degranulation have not been extensively studied.

Release of lysosomal enzymes from cytochalasin B-treated PMN is usually studied by assay of cell-free supernatants obtained by centrifugation, limiting the time resolution of this method to the order of minutes. However, it is generally agreed that degranulation begins rapidly and that significant amounts of enzyme are released within 1 minute (Naccache *et al.*, 1977b; Bainton, 1973).

Stimulation of PMN caused enhanced $Ca^{2+}$ influx, $Ca^{2+}$ efflux, and $Na^+$ influx (Naccache *et al.*, 1977a). Enhanced permeability of the membrane to these cations has been supported by other studies (Gallin and Rosenthal, 1974; Cividalli and Nathan, 1974; Boucek and Snyderman, 1976). Probably the most valuable study of cation fluxes in PMN is that of Naccache *et al.* (1977b). These researchers used rapid centrifugation of PMN through silicone oil as a means of separating cells from the medium; this technique provided prompt termination of ion fluxes. Cell-associated $Ca^{2+}$ increased rapidly upon exposure to rabbit PMN to FMLP and cytochalasin B; nearly maximal $Ca^{2+}$ influx was obtained within 30 seconds. The increments in $Ca^{2+}$ uptake closely paralleled lysozyme release. $Ca^{2+}$ efflux, on the other hand, was increased to a far lesser extent by FMLP; the further addition of cytochalasin B produced no significant enhancement. FMLP produced a substantial increase in $Na^+$ influx, which was greatly enhanced by the presence of cytochalasin B. In the presence of both agents, maximum $Na^+$ uptake was achieved within 30 seconds. $Na^+$ influx was not $Ca^{2+}$-dependent when FMLP was the stimulus. $K^+$ influx was not substantially changed by FMLP and/or cytochalasin B. $K^+$ efflux, on the other hand, was increased by FMLP plus cytochalasin B (FMLP alone had no effect), or by A23187 plus cytochalasin B. $K^+$ efflux due to A23187 appeared to be delayed approximately 1 minute relative to the $K^+$ efflux induced by FMLP. Both $K^+$ efflux responses appeared to have lag periods and were delayed relative to the prompt $Na^+$ and $Ca^{2+}$ influxes; near-maximal $K^+$ efflux was achieved at 1 minute by FMLP and at 2 minutes by A23187.

A number of reports have stated that cAMP levels did not change following stimulation of PMN (Manganiello *et al.*, 1971; Ignarro and George, 1974), seemingly eliminating this cyclic nucleotide from mechanistic considerations. However, Herlin *et al.* (1978) have shown that exposure of human PMN to latex particles provoked a rapid twofold increase in cAMP levels. This increment was prompt (maximal within 15 seconds) and brief. The fact that basal cAMP levels were restored within 1–2 minutes explained why previous investigators, who had assayed the cell suspensions after 2–5 minutes, were unable to detect this response. These investigators suggested that the brief increment in cAMP might have a regulatory role in glycogen metabolism (Herlin *et al.*, 1978; Petersen *et al.*, 1978). In any case, it is clear that such a rapid response, the timing of which is comparable to changes in membrane potential and release of membrane-bound $Ca^{2+}$, could possibly play a role in stimulus–secretion coupling.

Protein methylation appears to be involved in the chemotactic responses of bacteria and in the secretory responses of some eukaryotic cells. Rabbit PMN exposed to FMLP demonstrated enhanced protein carboxymethylation (with L-[³H]methionine as a methyl donor) (O'Dea et al., 1978). This response was very rapid, reaching a maximum at 30 seconds, the earliest time point studied. It is consequently not clear whether or not protein carboxymethylation occurred at even earlier times or if it had a lag period, although the methodology was adequate for this purpose. Increased methylation did not appear to depend upon *de novo* protein synthesis and could be blocked by competitive inhibitors of FMLP. The dose-response curve for induction of protein carboxymethylation by FMLP was similar to that for chemotaxis. Protein methylation also appears to play a role in chemotaxis by human monocytes (Pike et al., 1978).

In summary, the groundwork had been laid for investigations into the earliest events in PMN stimulation. However, many of these studies have not usually related *early* events to the final process of degranulation and secretion. Thus, a comprehensive temporal and theoretical framework has not been established for the neutrophil. That such a framework can be established will be demonstrated in the Results section. We shall describe, especially, measurements of the early kinetics of degranulation (which hitherto had eluded such attempts), and comprehensive investigation into transient changes in cAMP levels.

## II.   Methods

### A.   Preparation of Cell Suspensions

Heparinized (10 units/ml) venous blood was obtained from healthy adult donors. Purified preparations of PMN were isolated from this blood by means of Hypaque/Ficoll gradients (Boyum, 1968) followed by standard techniques of dextran sedimentation and hypotonic lysis of erythrocytes (Zurier et al., 1973). This allowed studies of cell suspensions containing 98 ± 2% PMN with few contaminating erythrocytes or platelets. The cells were suspended in a buffered salt solution consisting of 138 m$M$ NaCl, 2.7 m$M$ KCl, 8.1 m$M$ Na$_2$HPO$_4$, 1.5 m$M$ KH$_2$PO$_4$, 1.0 m$M$ MgCl$_2$, and 0.6 m$M$ CaCl$_2$, pH 7.4 (hereafter referred to as PiCM).

### B.   Continuous Assay System for Lysosomal Enzyme Secretion

Lucite dialysis cells (TechniLab Instruments) were employed essentially as outlined by Colowick and Womack (1969) with the following modifications: the upper half of the chamber was drilled out to yield an open well with a 20-mm diameter, 12.5 mm in depth. The lower chamber half was planed down so that the depth of the well was only 3 mm (to minimize "dead volume"). The chamber halves were assembled with a 25-mm Millipore filter (Type EA, 1-$\mu$m pore size) dividing the two wells, each of which contained a small (8.5 $\times$ 2 mm) stirring bar. The assembled chamber was then placed upon a magnetic stirrer. During operation, fresh PiCM containing 5 $\mu$g/ml cytochalasin B (Aldrich Chemical Co., Milwaukee, WI; 0.05% dimethyl sulfoxide was also present) was fed into the lower well (Fig. 1); flow of medium from this lower well was regulated by a peristaltic pump (Model P3, Pharmacia Fine Chemicals, Uppsala, Sweden) which fed into a fraction collector (Model 2112 RediRac, LKB Instruments, Rockville, MD). The entire apparatus was contained in an incubator and maintained at 37°C.

A typical determination of the lag period of lysosomal enzyme release was performed as follows (see Fig. 2 for an outline of the experimental design). One milliliter of the running buffer (PiCM containing 5 $\mu$g/ml cytochalasin B) was placed in the upper chamber of the apparatus described above. The lower chamber was connected to the reservoir, the peristaltic pump was turned on, and the flow of running buffer through the lower chamber was set at approximately 6 ml/min. The magnetic stirrer was turned on, and the stirring rate, held at a constant setting for all experiments, was just sufficient to prevent any bubbles from adhering to the side of the upper well. After air was completely removed from the lower well, the fluid in the upper chamber was removed and replaced with 1 ml PiCM containing 1–2 $\times$ 10$^8$ PMN. Concentrated cytochalasin B was

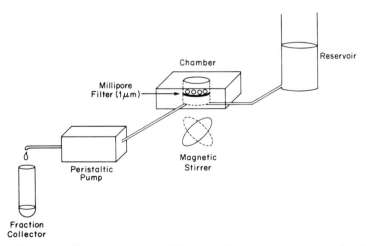

FIG. 1. Continuous flow apparatus in which lysosomal enzyme release is monitored. Fresh medium flows from the reservoir at right, through the lower chamber, then through the pump into a fraction collector. PMN and added stimuli are present in the upper chamber. The entire apparatus is placed in a 37°C incubator.

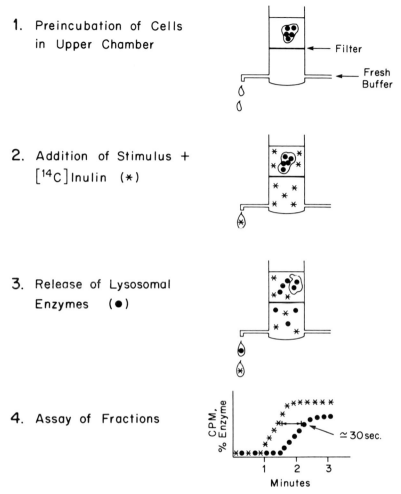

1. Preincubation of Cells in Upper Chamber

2. Addition of Stimulus + [$^{14}$C]Inulin (*)

3. Release of Lysosomal Enzymes (●)

4. Assay of Fractions

FIG. 2. Experimental design for measuring the onset of lysosomal enzyme release. [$^{14}$C]Inulin is added along with the stimulus (2) to internally calibrate the apparatus and to mark the moment of stimulation (4). The lag period for lysosomal enzyme release is thus the time difference between the appearance of inulin and enzyme activity in the fractions.

added to the upper chamber to yield a final concentration of 5 μg/ml, and the mixture was incubated for 2 minutes at 37°C. At the end of this period, the fraction collector was started (10 fractions per minute), and the cells were exposed to a stimulus 6 seconds later. The stimuli routinely used were FMLP ($10^{-7}$ M final concentration), serum-treated zymosan (STZ: zymosan from ICN Pharmaceuticals, Irving, CA, 20 mg, opsonized with serum for 30 minutes at 37°C, then washed), BSA/anti-BSA [0.75 mg, prepared according to Ward and Zvaifler

Boston, MA). Cyclic GMP was similarly assayed by using more concentrated extracts derived from a larger number of PMN.

This procedure allowed samples to be taken reliably as soon as 5 seconds after stimulation and at time intervals as short as 5 seconds thereafter. The extraction procedure ensured that the cells were promptly denatured and that all biochemical reactions were immediately terminated. The extracts contained at least 95% of the cellular cyclic nucleotides, as determined by recovery of tritiated standards. Artifactual increases in apparent cAMP levels, which can be obtained by boiling adenylate cyclase reaction mixtures (Larner and Rutherford, 1978), were not found with this system.

## F.   Membrane Potential

The membrane potential of PMN was determined by uptake of the lipophilic cation $TPMP^+$ as previously described (Korchak and Weissmann, 1978).

## III.   Results and Discussion

## A.   Determination of the Lag Period for Lysosomal Enzyme Release

As discussed earlier, suitable continuous assay techniques had existed for monitoring consumption of $O_2$ and production of $H_2O_2$ and $O_2^-$. The temporal resolution of the continuous methods was such that it was readily determined that these processes did not commence until after lag periods of 15–60 seconds following stimulation. Thus, oxidative metabolism could not be responsible for such immediate responses as changes in membrane potential and loss of membrane-bound $Ca^{2+}$; in fact, this temporal sequence suggested the converse.

Similar continuous techniques for measurement of lysosomal enzyme release were not available. The time resolution of conventional centrifugation methods for obtaining cell-free supernatants was limited by, among other factors, the period required for centrifugation. It is not surprising that several investigators concluded that degranulation was virtually an immediate consequence of PMN stimulation, having no discernible lag period (Naccache *et al.*, 1977b; Henson *et al.*, 1978). Cytochemical techniques, which potentially possessed adequate time resolution, were not quantitative (Bainton, 1973). In view of the fact that oxidative metabolism had a distinct lag period, we devised a technique suited for the detection of similar lag periods for lysosomal enzyme release (Smolen *et al.*, 1978a; see Methods, Section II,B). The apparatus consisted of a flow dialysis cell, the upper and lower chambers separated by a Millipore filter with a 1-$\mu$m pore size (Fig. 1). The PMN and the entire apparatus were preincubated at 37°C, after which a bolus containing a concentrated stimulus and [$^{14}$C]inulin was added (Fig. 2). The inulin served as the extracellular space marker; it immediately began to cross the filter and thus served to mark the moment of stimulation and to calibrate the entire apparatus. The appearance of [$^{14}$C]inulin in collected fractions preceded the appearance of lysosomal enzymes by 15 or more seconds (Fig. 3). Linear extrapolation of the rising portions of the curves to base line permitted an easy, reliable means of determining the lag period of lysosomal enzyme release relative to stimulation ([$^{14}$C]inulin elution). The lag periods were not artifacts due to selective retention of lysosomal enzymes within the apparatus, since no lag period was seen for PMN that were introduced into the upper chamber *after* stimulation.

The linear extrapolation technique effectively averaged together six to eight data points, providing a practical resolution of 1–2 seconds (using the same lot of cells). Daily variations among cell samples increased this value considerably.

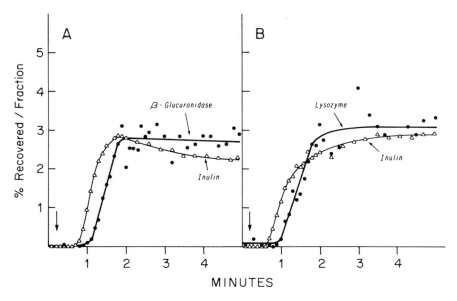

FIG. 3. Continuous flow analysis of lysosomal enzyme release from human PMN. Buffer (PiCM, 1 ml) containing $1.28 \times 10^8$ PMN (A) or $1 \times 10^8$ PMN (B) was preincubated in the upper chamber of a dialysis cell for 2 minutes at 37°C with 5 $\mu$g/ml cytochalasin B and 0.05% DMSO. The run was then started, and $10^{-7}$ M FMLP was added to the cells at the times indicated by arrows. [$^{14}$C]Inulin added simultaneously with FMLP was eluted from the dialysis chamber before $\beta$-glucuronidase (A) or lysozyme (B) was released from the PMN. The calculated lag periods are 21 seconds (A) and 16 seconds (B); total enzyme releases are estimated as 3.7% (A) and 6.0% (B). The data are presented as the percentages of total recovered inulin or enzyme activity found in each fraction.

TABLE I

LAG PERIODS FOR β-GLUCURONIDASE AND LYSOZYME RELEASE AND $O_2^{\bar{}}$ GENERATION
FOR A VARIETY OF STIMULI[a]

| | Stimulus | | | | |
|---|---|---|---|---|---|
| | FMLP | BSA/anti-BSA | Serum-treated zymosan | Con A | A23187 |
| β-Glucuronidase release | 19 ± 5 (18) | 35 ± 8 (5) | 48 ± 8 (6) | — | 60 ± 25 (6) |
| Lysozyme release | 28 ± 16 (5) | 28 ± 8 (5) | 32 ± 10 (4) | 38 ± 8 (4) | 74 ± 27 (4) |
| $O_2^{\bar{}}$ generation | 21 ± 4 (9) | 43 ± 14 (18) | — | 61 ± 7 (9) | 50 ± 13 (14) |

[a] The values listed are the lag periods (calculated by linear extrapolation) in seconds (±S.D.). The number of experiments is given in parentheses.

Nonetheless, we were able to show that lag periods did exist for lysosomal enzyme release and that the lengths of these lag periods were stimulus-dependent (Table I). Values for these lag periods were also comparable to those seen for $O_2^{\bar{}}$ generation (Table I). The lag periods for both responses were not dependent upon the dose of the stimulus and thus seemed to be intrinsic to some intracellular mechanisms. A variety of agents, such as corticosteroids, colchicine, 2-deoxyglucose, and N-ethyl maleimide, affected the magnitudes of the responses, but not the lag periods, when FMLP was the stimulus. When BSA/anti-BSA was used as the stimulus, 2-deoxyglucose and N-ethyl maleimide increased the lag period for $O_2^{\bar{}}$ generation but not for lysosomal enzyme release. These latter observations suggested that the initial events leading to the two responses were similar and possibly parallel, but not tightly coupled.

This semicontinuous flow dialysis technique was compared with a modern centrifugation method wherein cells were rapidly separated from the medium by short (20 seconds) high-speed centrifugation through silicone oil. Measurement of $O_2^{\bar{}}$ generation by this latter method revealed the existence of lag periods following stimulation of PMN by FMLP or A23187. However, these lag periods were 10–20 seconds shorter than those obtained by the continuous method, indicating that even this modern centrifugation technique was inadequate for the purpose of absolute timing. Lysosomal enzyme release also had lag periods when centrifugation was used (Fig. 4); once again, the values for these lags were 10–20 seconds too short. Thus, these two independent techniques confirmed that lag periods existed for lysosomal enzyme release. Furthermore, the two techniques provided comparable values for the duration of the lag periods when the 10- to 20-second underestimation by the centrifugation method was taken into account.

In spite of the fact that centrifugation was not suitable for purposes of absolute timing, it was better than flow dialysis for *relative* timing and for the direct

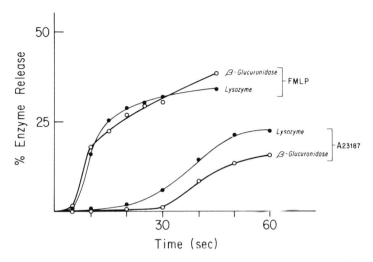

FIG. 4. Time course of lysosomal enzyme release by human PMN: measurement by centrifugation technique. Washed PMN 1 × 10⁷ in 0.5 ml PiCM) were preincubated with cytochalasin B (5 μg/ml) for 5 minutes at 37°C. The cells were stimulated by exposure to FMLP ($10^{-7}$ M) or A23187 ($10^{-5}$ M), and incubation was terminated at the indicated times by centrifugation of the cells through a layer of silicone oil. The remaining medium was collected and assayed for content of β-glucuronidase and lysozyme.

determination of quantities of enzyme release. These advantages were exploited in a study of sequential degranulation. When both β-glucuronidase and lysozyme were measured in the same cell supernatants, it was clear that lysozyme release preceded that of β-glucuronidase in response to A23187 (Fig. 4). The time courses of secretion of these two enzymes were virtually identical when FMLP was the stimulus. Simultaneous measurement of both enzymes was not practical with the flow dialysis technique, so the sequential degranulation in response to A23187 was obscured by the large day-to-day variation in lag times seen with this stimulus. However, this method was adequate for observing the earlier release of lysozyme when serum-treated zymosan was the stimulus (Table I).

## B.  Changes in Cyclic Nucleotide Levels

Studies that had reported very rapid early changes in the cAMP levels of PMN in response to surface stimulation (Herlin *et al.*, 1978; Petersen *et al.*, 1978) were extended (Smolen *et al.*, 1978b). FMLP provoked an increase in cAMP levels that was virtually immediate (elevated levels were seen at 5 seconds). The response peaked at twice the basal level within 10–15 seconds and subsided thereafter. Basal levels of cAMP were restored by 2–5 minutes. No changes in cGMP were ever observed during this time period. When BSA/anti-BSA was

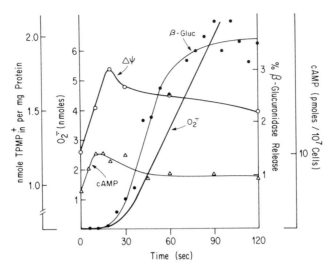

FIG. 5.   Effect of FMLP on cAMP, transmembrane potential, lysosomal enzyme release, and $O_2^{\cdot-}$ generation in PMN. Cyclic AMP was extracted as described in the text and quantitated by radioimmunoassay. Membrane potential ($\Delta\psi$) was measured by the intracellular content of [$^3$H]TPMP. Superoxide anion ($O_2^{\cdot-}$) was measured by the continuous assay of ferricytochrome c reduction. The semicontinuous assay of $\beta$-glucuronidase secretion ($\beta$-Gluc) was performed using the modified flow dialysis technique. All four responses were measured at the indicated times after exposure of the PMN to FMLP ($10^{-7}$ $M$).

used as the stimulus, cAMP levels again increased, but after a discernible lag period. A23187 produced increments in cAMP after an even longer lag period. These three responses are shown in Figs. 5–8, as discussed below.

The increment in cAMP in response to FMLP was concurrent with membrane hyperpolarization, but clearly preceded the onset of $O_2^{\cdot-}$ generation and lysosomal enzyme release (both of which had lag periods of approximately 20 seconds) (Fig. 5). For BSA/anti-BSA, hyperpolarization preceded elevation of cAMP, which, in turn, preceded the other two responses (Fig. 6). A23187, which did not provoke changes in membrane potential, caused cAMP increases after a relatively long lag period (Fig. 7); nonetheless, elevation of cAMP levels preceded lysosomal enzyme release and $O_2^{\cdot-}$ generation. These data showed that increments in cAMP were not the first response of PMN to surface stimulation, since they were concurrent with or followed membrane hyperpolarization. Furthermore, these two responses could be uncoupled by use of A23187.

Stimulated increments in cAMP could be partially dissociated from lysosomal enzyme release and $O_2^{\cdot-}$ generation. Both the hyperpolarization and cAMP responses were unimpaired by the presence of EGTA, which inhibited degranulation and $O_2^{\cdot-}$ production. FMLP at a concentration of $10^{-9}$ $M$ (two orders of

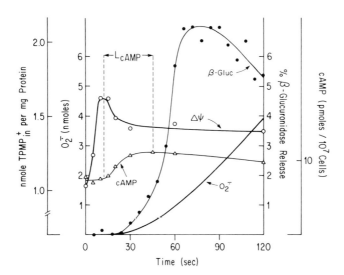

FIG. 6.    Effect of BSA/anti-BSA on cAMP, transmembrane potential, lysosomal enzyme release, and $O_2^-$ generation in human PMN. Cyclic AMP, membrane potential, $O_2^-$ generation, and β-glucuronidase release were measured as described in the legend to Fig. 5. All four responses were measured at the indicated times following exposure of PMN to BSA/anti-BSA immune complex (150 μg/ml).

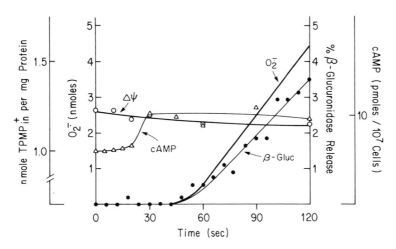

FIG. 7.    Effect of A23187 on cAMP, transmembrane potential, lysosomal enzyme release, and $O_2^-$ generation in human PMN. Cyclic AMP, membrane potential, $O_2^-$ generation, and β-glucuronidase secretion were measured as described in the legend to Fig. 5. All four responses were measured at the indicated times following exposure of PMN to calcium ionophore A23187 ($10^{-5}$ M).

magnitude below the usual stimulatory dose of $10^{-7}$ $M$) evoked a normal increment in cAMP without the later responses. Replacement of extracellular $Na^+$,
required for degranulation and $O_2^-$ generation, with $K^+$ or choline$^+$ did not alter
the cAMP response. Thus, these three experiments, which showed the cAMP
response could proceed unimpaired under conditions in which lysosomal enzyme
release and $O_2^-$ generation were blocked, indicated that increments in cAMP
were not *sufficient* for the later events. That such increments were not *necessary*
for subsequent responses was suggested (but not proved) by the fact that serum-
treated zymosan, a good stimulus by other criteria, provoked little or no change
in cAMP levels. The chemotactic (but not *secretory*) peptide Gly-His-Gly (Spilberg *et al.*, 1978) evoked no change in cAMP levels, suggesting that increments
in cAMP were not necessary for chemotaxis.

Concentrations of $PGE_1$ and $PGI_2$ sufficient to inhibit lysosomal enzyme release and $O_2^-$ generation produced sustained elevated levels of cAMP; this
elevation per se has usually been considered responsible for inhibition of degranulation. However, the prompt increments in cAMP in response to FMLP
approach these levels at almost the very moment that degranulation is being
initiated; by definition, there is no inhibition of enzyme release under these

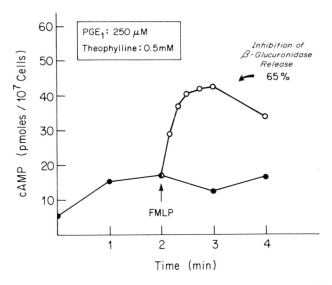

FIG. 8.   Effect of $PGE_1$ and FMLP on cAMP levels and lysosomal enzyme release in PMN.
Purified PMN were preincubated in PiCM containing 0.5 m$M$ theophylline for 5 minutes at 37°C. At
time zero, $PGE_1$ (250 $\mu M$) was added and samples were taken every minute for 4 minutes (lower
curve). After 2 minutes, a sample of these cells was removed and stimulated with FMLP ($10^{-7}$ $M$) as
shown by the upper curve. Lysosomal enzyme release in response to FMLP was inhibited by
pretreatment with $PGE_1$ (as indicated).

conditions. This paradox is resolved by the hypothesis that slightly elevated levels of cAMP (two to three times the basal level), such as are produced by prostaglandins or FMLP alone, are not inhibitory per se. The reason why prostaglandin pretreatment is inhibitory was suggested by our own and previous observations (Zurier et al., 1974) that stimulation of pretreated cells evokes very large increments in cAMP (eight to ten times the basal levels; see Fig. 8). Thus, exposure to prostaglandins "primes" the cells in such a manner that the usual modest stimulus-elicited increments in cAMP are magnified into much higher, probably inhibitory, levels. The stimulus ironically contributes to the inhibition of its usual secretory effects.

# IV. Conclusion

Through the use of techniques affording high time resolution, the sequence of stimulus–response coupling from PMN is now becoming apparent. Within seconds of surface stimulation, membrane potential changes take place, along with increased influxes of $Na^{2+}$ and $Ca^{2+}$, mobilization of membrane-bound $Ca^{2+}$, and as yet undefined changes in membrane fluidity and viscosity (as revealed by ANS fluorescence). Degranulation and oxidative metabolism begin 20–60 seconds later. The biochemical and physiological processes responsible for the lag periods observed for enzyme secretion, $O_2$ consumption, $O_2^-$ generation, and $H_2O_2$ production are as yet unknown. Changes in cAMP levels, which occur after the immediate events but before the secretory processes, would seem to be ideally located in the temporal framework to serve an intervening role; however, the evidence suggests that cAMP is not the second messenger in the secretory sequence.

PMN display other early responses such as protein carboxymethylation, cellular aggregation, and chemiluminescence. The places of these phenomena in the temporal framework are as yet undefined. It is hoped that future investigations of these and other phenomena will be studied in a more comprehensive manner than has been customary. The early events should be carefully related with each other for purposes of sequencing. Furthermore, each phenomenon should be studied with a wide variety of stimuli. This is warranted, since the response elicited by each stimulus is slightly different, and it is clear that the specific paths of stimulus–secretion coupling are not identical; drawing conclusions in this field on the basis of one or two stimuli is unnecessarily myopic.

In general, determination of the sequence of stimulus–response coupling has proved both exciting and rewarding; temporal studies allow us to separate ends from means and causes from effects. It is hoped that such investigations will eventually allow us to decipher the "secretory code" of PMN.

ACKNOWLEDGMENTS

This work was aided by grants (AM-11949, GM-23211, HL-19072, HL-19721) from the National Institutes of Health, the National Foundation–March of Dimes, the National Science Foundation (76-05621), and the Arthritis Foundation.

REFERENCES

Badwey, J. A., Curnutte, J. T., and Karnovsky, M. L. (1979). *N. Engl. J. Med.* **300,** 1157–1159.
Bainton, D. F. (1973). *J. Cell Biol.* **58,** 249–264.
Berlin, R. D., and Fera, J. P. (1977). *Proc. Natl. Acad. Sci. U.S.A.* **74,** 1072–1076.
Boucek, M. M., and Snyderman, R. (1976). *Science* **193,** 905–907.
Boyum, A. (1968). *Scand. J. Clin. Lab. Invest.* **21,** 77.
Brittinger, G. R., Hirschhorn, R., Douglas, S. D., and Weissmann, G. (1968). *J. Cell Biol.* **37,** 394–411.
Cividalli, G., and Nathan, D. N. (1974). *Blood* **43,** 861–869.
Cohen, H. J., and Chovaniec, M. E. (1978). *J. Clin. Invest.* **61,** 1081–1087.
Colowick, S. B., and Womack, F. C. (1969). *J. Biol. Chem.* **244,** 774–777.
Gallin, E. K., and Gallin, J. I. (1977). *J. Cell Biol.* **75,** 277–289.
Gallin, E. K., Wiederhold, M. L., Lipsky, P. E., and Rosenthal, A. S. (1975). *J. Cell. Physiol.* **86,** 653–662.
Gallin, J. I., and Rosenthal, A. J. (1974). *J. Cell Biol.* **62,** 564–609.
Henson, P. M., Zanolari, B., Schwartzman, N. A., and Hong, S. R. (1978). *J. Immunol.* **121,** 851–855.
Herlin, T., Petersen, C. A., and Esmann, V. (1978). *Biochim. Biophys. Acta* **542,** 63–76.
Hoffstein, S. T. (1979). *J. Immunol.* **123,** 1395–1402.
Ignarro, L. J., and George, W. J. (1974). *Proc. Natl. Acad. Sci. U.S.A.* **71,** 2027–2031.
Korchak, H. M., and Weissmann, G. (1978). *Proc. Natl. Acad. Sci. U.S.A.* **75,** 3818–3822.
Larner, E. H., and Rutherford, C. (1978). *Anal. Biochem.* **91,** 684–690.
Manganiello, V., Evans, W. H., Stossel, J. P., Mason, R. J., and Vaughan, M. (1971). *J. Clin. Invest.* **50,** 2741–2744.
Mason, R. J., Stossel, J. P., and Vaughan, M. (1972). *J. Clin. Invest.* **51,** 2399–2407.
Montecullo, C., Pozzan, T., and Rink, T. (1979). *Biochim. Biophys. Acta* **552,** 552–567.
Naccache, P. H., Showell, H. J., Becker, E. L., and Sha'afi, R. I. (1977a). *J. Cell Biol.* **73,** 428–444.
Naccache, P. H., Showell, H. J., Becker, E. L., and Sha'afi, R. I. (1977b). *J. Cell Biol.* **75,** 635–649.
Naccache, P. H., Volpi, M., Showell, H. J., Becker, E. L., and Sha'afi, R. I. (1979). *Science* **203,** 461–463.
O'Dea, R. F., Viveros, O. H., Axelrod, J., Aswanikumar, S., Schiffmann, E., and Corcoran, B. A. (1978). *Nature (London)* **272,** 462–464.
Petersen, C. S., Herlin, T., and Esmann, V. (1978). *Biochim. Biophys. Acta* **542,** 77–87.
Pike, M. C., Kredich, N. M., and Snyderman, R. (1978). *Proc. Natl. Acad. Sci. U.S.A.* **75,** 3928–3932.
Romeo, D., Cramer, R., and Rossi, F. (1970). *Biochem. Biophys. Res. Commun.* **41,** 582–588.
Root, R. K., Metcalf, J., Oshino, N., and Chance, B. (1975). *J. Clin. Invest.* **55,** 945–955.
Segal, A. W., and Coade, S. B. (1978). *Biochem. Biophys. Res. Commun.* **84,** 611–617.
Seligmann, B., Gallin, E. K., Martin, D. L., Shain, W., and Gallin, J. I. (1977). *J. Cell Biol.* **75,** 103a.

Sha'afi, R. I., Williams, K., Wacholtz, M. C., and Becker, E. L. (1978). *FEBS Lett.* **91,** 305–309.

Smolen, J. E., and Shohet, S. B. (1974). *J. Clin. Invest.* **53,** 726–734.

Smolen, J. E., Korchak, H. M., and Weissmann, G. (1978a). *J. Cell Biol.* **79,** 208a.

Smolen, J. E., Korchak, H. M., and Weissmann, G. (1978b). *Clin. Res.* **27,** 307a.

Spilberg, I., Mandell, B., Mehta, J., Sullivan, T., and Simchowitz, L. (1978). *J. Lab. Clin. Med.* **92,** 297–302.

Wacker, W. E. C., Ulmer, D. D., and Vallee, B. L. (1956). *N. Engl. J. Med.* **255,** 449–450.

Ward, P. A., and Zvaifler, N. J. (1973). *J. Immunol.* **111,** 1771–1776.

Worthington Enzyme Manual (1972). "Lysozyme Assay." Worthington Biochemical Corp., Freehold, New Jersey.

Yokomura, E., and Ishikawa, Y. (1973). *Acta Haematol. Jpn.* **36,** 821–826.

Zurier, R. B., Hoffstein, S., and Weissmann, G. (1973). *J. Cell Biol.* **58,** 27–41.

Zurier, R. B., Weissmann, G., Hoffstein, S., Kammerman, S., and Tai, H.-H. (1974). *J. Clin. Invest.* **53,** 297–309.

# Chapter 29

# Aspects of the Calcium Hypothesis of Stimulus –Secretion Coupling: Electrical Activity in Adenohypophyseal Cells, and Membrane Retrieval after Exocytosis

W. W. DOUGLAS

*Department of Pharmacology,*
*Yale University School of Medicine,*
*New Haven, Connecticut*

## I. The Calcium Hypothesis of Stimulus–Secretion Coupling

Evidence that calcium ions might serve some general function as mediators of various secretory responses came to light in the early 1960s as a result of studies on adrenal medulla, neurohypophysis, and salivary glands, three systems my

483

colleagues and I had selected as possible models of the main classes of gland cell—endocrine, neuroendocrine, and exocrine—whose secretory products encompassed amines, polypeptides, proteins, and water and electrolytes. In each of these systems our evidence suggested that, irrespective of the secretory product, Ca ions somehow functioned to couple the stimulus to the secretory response. We were thus led to the view that Ca had a role in the regulation of secretion that paralleled its role in the regulation of contraction. A few years later, when the function of Ca in the three systems mentioned was becoming increasingly clear, it was possible to recognize similar features in several other secretory systems. This prompted the general hypothesis of stimulus–secretion coupling in which it was proposed that secretagogues of all sorts—chemical transmitters, hormones, other autacoids such as histamine and peptides, antigens, toxins, and so on— might all owe their efficacy as secretagogues in various systems to some action on the cell surface to promote Ca influx (Douglas, 1968). Additionally, it was argued on this same occasion that the commonality of the involvement of Ca pointed to the existence of a common secretory mechanism, and that where secretion involved export of preformed material the most likely mechanism was

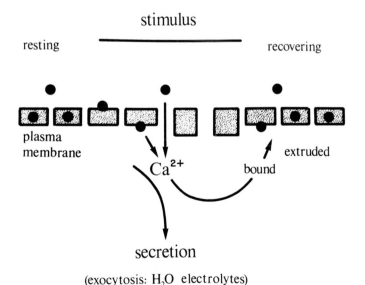

stimulus

resting                                                    recovering

plasma
membrane

$Ca^{2+}$

extruded

bound

secretion

(exocytosis: $H_2O$ electrolytes)

FIG. 1.   A depiction of the Ca hypothesis of stimulus–secretion coupling. Calcium is represented by the black circles in the cell membrane and extracellular environment. The stimulus opens channels in the membrane that allow calcium ions to enter and activate secretion. Calcium ions may also be provided for this purpose by the displacement of membrane-bound Ca, an action of the stimulus here suggested also to contribute to the increased membrane permeability. Secretion is terminated, on withdrawal of the stimulus, as excess free Ca is bound and extruded. After Douglas (1965).

exocytosis. At that time persuasive arguments for exocytosis could be mounted for few cell types, and there was considerable reluctance to accept this phenomenon as applying to all such secretory systems. Today, with few dissenting voices, it is widely accepted that exocytosis provides a common mechanism for secretion for preformed materials (Palade, 1975) and is, indeed, the process that is set in motion by Ca ions. The cartoon presented in Fig. 1 summarizes this generalized concept of stimulus–secretion coupling arrived at more than a decade ago. The scheme, in broad outline, seems to be applicable to most secretory systems. However, there are interesting variations that concern mainly the different means whereby the stimulus achieves the critical elevation of the intracellular free Ca ion concentration. Some of these form the main topic of this chapter.

## II.   Calcium as a Secretagogue

Before we pursue these variations on the theme, a word or two is in order concerning the validity of the view that Ca ions possess the "secretagogue" activity essential to their proposed mediator (or messenger) role as schematized in Fig. 1. The key observation, and linchpin of the Ca hypothesis, was the demonstration of such a secretagogue property of Ca in the adrenal medulla by the simple expedient of reintroducing Ca after perfusing with a Ca-free medium that rendered the plasma membranes of the chromaffin cells leaky to Ca (Douglas and Rubin, 1961). This maneuver has subsequently been used successfully to demonstrate the secretagogue capacities of Ca in other systems, including the $\beta$ cells of the pancreas (Devis et al., 1975) and peritoneal mast cells (Douglas and Kagayama, 1977). Its efficacy varies from cell to cell, perhaps reflecting the presence or absence of Ca channels that open in response to depolarization (see below; also Douglas and Kagayama, 1977; Douglas, 1978). Additional evidence of the secretagogue property of Ca ions has been provided by injecting Ca intracellularly by micropipette (Kanno et al., 1973; Miledi, 1973) or by the use of liposomes (Rahamimoff et al., 1978; Theoharides and Douglas, 1978b); by the use of Ca ionophores (see e.g., Foreman et al., 1973; Cochrane and Douglas, 1974; Selinger et al., 1974); and by increasing Ca influx by shock-induced dielectric breakdown of the plasmalemma (Baker and Knight, 1978). This last method has demonstrated neatly the secretagogue activity of Ca in the micromolar range, which other evidence indicates is readily achieved upon physiological stimulation. That Ca ions have the requisite properties for mediating secretory responses seems beyond doubt. But how Ca acts remains a puzzle. Its efficacy in the micromolar range in the face of millimolar concentrations of Mg strongly suggest the involvement of some highly specific Ca-binding protein (see Kretsinger, 1979). One such protein is the ubiquitous calmodulin recently implicated in so many Ca-mediated processes (see Wolff and Brostrom, 1979); and for those of us who have repeatedly emphasized the remarkable parallels between stimulus–secretion coupling and excitation–contraction coupling—and have urged that the lessons learned from the latter be applied within the broader context of "stimulus–response coupling" (see, for example, Douglas, 1966, 1968, 1974a; Poisner, 1973)—it is intriguing that calmodulin is so similar in structure to troponin C (see Wolff and Brostrom, 1979). It should be recalled also that the Ca-activated step in secretion, as in contraction, appears to be ATP-requiring (Douglas *et al.*, 1965; Rubin, 1969; Foreman *et al.*, 1973). It remains to be seen whether profit lies in pursuit of this trail or some other. There are hints that activation of secretion may involve Ca-driven phosphorylation (see, for example, Sieghart *et al.*, 1978); and phosphorylation of still other proteins may participate in *inhibition* of secretion (Theoharides *et al.*, 1980). Moreover, the old notion that Ca may induce exocytosis by activating phospholipase activity (Douglas *et al.*, 1966; Ramwell *et al.*, 1966; see also Blaschko *et al.*, 1967) has taken on new life with a spate of recent observations (see, for example, Hirata *et al.*, 1979; Marone *et al.*, 1979; Sullivan and Parker, 1979; Vanderhoek and Feinstein, 1979). This is, however, but a rather arbitrary sampling of current indications prompted by personal interest. Some additional clues will be presented by other authors in this volume.

## III.   Different Sources of Calcium Ions: Influx, Mobilization and Voltage-Dependent Calcium Channels

Returning now to the main topic: how secretagogues raise the Ca ion concentration intracellularly. As indicated in Fig. 1, there was early speculation, prompted mainly by parallels with muscle, that Ca mobilization from cellular sources might contribute, along with Ca influx, to initiation of the secretory response (Douglas, 1965, 1966). In the scheme depicted, the stimulus is suggested to displace Ca from the plasma membrane, this action not only contributing Ca ions to activate secretion but also, possibly, promoting the increased membrane permeability that allows influx of other, stimulant Ca ions. This last notion arose from the observation that chromaffin cells exposed to the secretagogue acetylcholine behave in several respects like Ca-deprived chromaffin cells. A variation of this theme involving exocrine cells is described in Chapter 31 by O. H. Petersen *et al*. It is by now clear that many cells and numerous secretagogues utilize cellular Ca for stimulus–secretion coupling, whereas others utilize Ca from intracellular sources. Moreover, it is evident that different secretagogues acting on a *single cell* may preferentially draw on one or the other of these two sources of Ca. In a recent survey in which I provided illustrations of such variations, I also drew attention to a second important variation relating to the presence or absence of voltage-dependent Ca channels in the plasma membrane (Douglas, 1978). The existence, in endocrine cells, of such voltage-dependent Ca channels, opening upon depolarization to permit influx of Ca ions, was signaled many years ago by the responses of chromaffin cells and neurohypophyseal terminals to excess K, which not only depolarized but caused uptake of Ca and a Ca-dependent secretory response (see reviews by Douglas, 1974b, 1975). Subsequently, many other endocrine cells have been shown to accumulate Ca and to secrete, in Ca-dependent fashion, when exposed to excess K; and this phenomenon has, in each instance, provided an argument that Ca has a mediator function in the system. However, such behavior is not general. Exocrine cells, such as those of salivary glands and the pancreas, do not show it (see reviews by Petersen, 1976; Case, 1978). Nor, for example, do mast cells (Cochrane and Douglas, 1976). Yet the exocrine cells and mast cells can clearly respond to Ca entry, as evidenced by the stimulant effect of Ca ionophores. Such cells thus appear to be devoid of voltage-dependent Ca channels. The occurrence, in chromaffin and other endocrine cells, of voltage-sensitive Ca conductance is clearly reminiscent of neurons (see Katz, 1969) and commands increasing interest as evidence grows of neuron-like electrical activity in endocrine cells.

## IV.   Electrical Events in Endocrine Cells

Studies of the electrical properties of endocrine cells are of relatively recent date. Just over a decade ago, acetylcholine was shown to depolarize chromaffin cells, and it was suggested that the depolarizing effect, mediated mainly by Na ions, might facilitate Ca entry by opening voltage-dependent Ca channels (Douglas et al., 1967), although it seemed evident from earlier studies that acetylcholine could also promote Ca entry by some action independent of changes in membrane potential (Douglas and Rubin, 1963). More recently, it has been discovered that chromaffin cells can produce action potentials mainly mediated by Na (Biales et al., 1976; Brandt et al., 1976) and may do so in response to acetylcholine (Brandt et al., 1976). However, the first indication that such events might participate in endocrine secretion had surfaced some years earlier when Dean and Matthews (1970a,b) reported the discovery, in pancreatic $\beta$ cells, of electrical activity (involving principally Ca ions) that was increased by glucose. And in 1975 Kidokoro had discovered "Ca spikes" in a line (GH$_3$) of cells derived from an anterior pituitary tumour and noted that thyrotropin-releasing hormone (TRH) increased the mean rate of firing. Because of the central role played by the adenohypophysis in endocrine regulation, Dr. Taraskevich and I decided to pursue this lead and learn whether or not normal adenohypophyseal gland cells show action potential activity and, if so, whether this is modulated by hypophysiotropic factors. A report of "electrical oscillations" in denervated pars intermedia of the frog (Davis and Hadley, 1976) was clearly an encouraging sign.

## A.   Rat Anterior Pituitary

After confirming and extending Kidokoro's observations on the tumor (GH) cells (Douglas and Taraskevich, 1977; see also Taraskevich and Douglas, 1980), we studied normal cells dispersed from the anterior pituitary of adult rats. We maintained the cells in short-term cultures, recorded electrical activity with microsuction electrodes under direct visual control using an inverted microscope, and delivered drugs and hormones by a gravity-flow micropipette. This approach revealed that normal anterior pituitary cells were indeed capable of generating action potentials in response to current pulses. The action potentials were apparently "Ca spikes," since they were blocked by La or D600 (methoxyverapamil) but not by tetrodotoxin (TTX) or replacement of Na by Tris (Taraskevich and Douglas, 1977). Independently, Osawa and Sand (1978) were to arrive at a similar conclusion from studies on pituitary slices. From our experiments it became further evident that a fair proportion of the normal anterior pituitary cells showed spontaneous action potential discharge. Since the "Ca-blocking" drug D600 reduced the *frequency* of this discharge (as well as reducing the amplitude

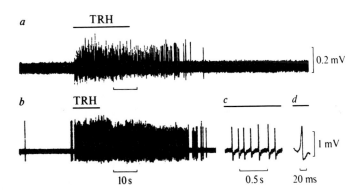

Fig. 2.   Action potentials induced by TRH in rat anterior pituitary cells. TRH applied for the time indicated by the bar above each record. (a) Response of an electrically silent cell to 50 n*M* TRH. (b) Response of a spontaneously active cell to 5 n*M* TRH. (c and d) Action potentials at faster sweep speeds from same cell as (b) during exposure to TRH (5 n*M*). In this and all subsequent illustrations of electrical activity, recording was by microsuction electrode (except for Fig. 4) from dissociated normal adenohypophyseal cells maintained in culture, and drugs were delivered by gravity-flow micropipette. From Taraskevich and Douglas (1977).

of the individual spikes), we have suggested that Ca may participate in the triggering of the spikes (Taraskevich and Douglas, 1977). When we applied TRH to individual cells, about 1 in 10 of this mixed population responded immediately with a burst of action potentials (Fig. 2). These responsive cells we assume are thyrotrophs or mammotrophs, the physiological target cells for TRH.

All this evidence pointed squarely to involvement of action potentials in stimulus–secretion coupling. This follows because anterior pituitary secretion requires Ca, can be elicited by excess K in depolarizing concentrations provided Ca is present, and because Ca entry, promoted by ionophores, provides an adequate stimulus to secretion (see reviews by Geschwind, 1971; McCann, 1971; Vale *et al.*, 1977). We therefore advanced the hypothesis that "action potentials, involving Ca ions as charge carriers, participate in stimulus–secretion coupling, and that it is by initiating or modulating action potentials that the brain, through the hypophysiotrophic hormones, regulates secretion in the anterior pituitary" (Taraskevich and Douglas, 1977).

## B.   Fish Prolactin Cells

It was now appropriate to attempt to characterize the electrical behavior of particular cell types and the effects thereon of relevant hypophysiotropic hormones. The intermixing of the different endocrine cell types in the anterior pituitary of mammals clearly poses a difficulty, but Dr. Taraskevich and I found a solution in the peculiar arrangement of the anterior pituitary in certain teleost fishes. In some such fishes, the prolactin cells, in whose function we were most interested because of our findings in the rat, aggregate in a distinct sublobe (the rostral pars distalis) and are thus easily isolated. Although different in function, fish prolactin cells resemble their mammalian counterparts by secreting spontaneously at a high rate and being subject to inhibitory hypophysiotrophic control by catecholamines (Ensor, 1978). Such cells thus offered an excellent model system, with the additional advantage of permitting study of the effects of *inhibitory* hypophysiotrophic factors. We took advantage of the spring "herring run" in nearby rivers to collect alewives (*Alosa pseudoharengus*), which provided excellent material. The rostral pars distalis was large, easily dissected out, and filled with relatively large prolactin cells whose prominent prolactin granules, easily visible under phase contrast, allowed ready identification. These prolactin cells were electrically excitable, discharged action potentials spontaneously at a relatively high frequency, and slowed or stopped this spiking activity when exposed to dopamine or noradrenaline (Fig. 3). The catecholamines clearly inhibited by some action on the spike-generating mechanism, for those action potentials that persisted were little, if at all, reduced in amplitude. The action potentials from the fish prolactin cells were blocked by TTX or Na-free conditions and thus involve Na ions for the most part. But a Ca component could be

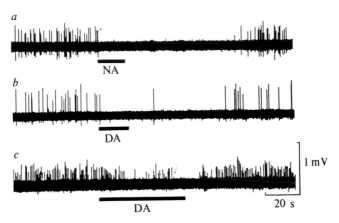

FIG. 3.   Spontaneous action potential discharge in teleost prolactin cells and inhibition by catecholamines. Catecholamines were applied to the cells for periods indicated by bars under the records. (a) Arrest of spontaneous discharge by noradrenaline (NA, $10^{-6}$ $M$). (b) Similar effect, in a second cell, produced by dopamine (DA, $10^{-6}$ $M$). (c) Slowing of spontaneous discharge induced in a third cell, by a lower concentration ($10^{-8}$ $M$) of dopamine. From Taraskevich and Douglas (1978).

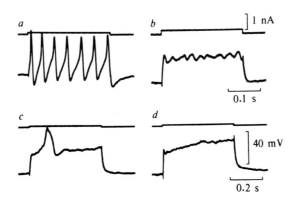

FIG. 4.   Ionic basis of action potentials in teleost prolactin cells. Intracellular recordings from two different cells showing effect of tetrodotoxin (TTX; a, b) and of replacement of Ca by Mn (c, d). (a) Repetitive action potentials evoked by a single outward (depolarizing) current pulse passed through the recording electrode. (b) Response of the same cell during exposure to TTX ($2 \times 10^{-6}$ $M$). The action potentials are virtually abolished. (c) Response of a second cell to a depolarizing pulse while bathed in a solution containing TTX ($5 \times 10^{-6}$ $M$) with the further addition of TEA (10 m$M$). A prominent regenerative potential is evident. (d) Response of the same cell when perifused with the same medium, except that Ca (10 m$M$) was replaced by Mn. The regenerative response is abolished. Top trace in all records indicates current passed and zero level of membrane potential. Calibrations for current and potential apply to all four sets of records. From Taraskevich and Douglas (1978).

demonstrated by exposing the cells to tetraethylammonium, which has been used in other systems to unmask Ca currents by blocking potential-dependent K conductance (Fig. 4). That the ionic mechanisms in pars distalis cells of the fish should differ from those of the rat is not surprising, since our evidence indicates that even within a single species, the rat, there is a comparable difference between cells in one region of the adenohypophysis and those in another. This became clear when we studied rat pituitary pars intermedia, again with the principal aim of working on an identified cell type and studying the effects of a relevant hypophysiotrophic factor.

## C. Rat Pars Intermedia

The MSH-secreting cells of the pars intermedia, like the prolactin cells of the pars distalis, secrete spontaneously at a high rate and are normally under a predominantly inhibitory physiological control exerted by dopaminergic fibers. Pars intermedia cells, like prolactin cells, also showed a relatively high rate of spontaneous electrical activity, which was slowed or arrested by dopamine or noradrenaline (Douglas and Taraskevich, 1978; Fig. 5, a and d). And, as in the prolactin cells, this inhibitory effect seemed to reflect an action on the cellular activity that initiated the spiking, for such spikes as persisted during exposure to dopamine were of about the usual amplitude. The action potentials from pars intermedia, unlike those from pars anterior, were abolished by TTX and were thus Na spikes. However, a Ca component was revealed by the procedures described for fish prolactin cells and illustrated in Fig. 4 (Douglas and Taraskevich, 1980).

## D. Lizard Pars Intermedia

Because a comparative approach to the pars distalis had been helpful, Dr. Taraskevich and I sought advantage in a similar approach to the pars intermedia. In the lizard *Anolis carolinensis* (the "anole" or "false chameleon") pars intermedia cells show relatively little spontaneous secretory activity, and increased output of MSH is not achieved by disinhibition but by active stimulation thought to be mediated by 5-hydroxytryptamine (Thornton and Geschwind, 1975). We found that anole pars intermedia cells discharged action potentials at a relatively low rate and that 5-hydroxytryptamine caused an immediate burst of action potentials (Fig. 5, e and f). These action potentials involved mainly Ca (Taraskevich and Douglas, 1978).

In summary, our experiments on normal adenohypophyseal cells show that electrical excitability and action potential generation can be observed in three classes of vertebrate and in both the pars distalis and the pars intermedia. Moreover, cells that are known to secrete at relatively high levels when removed from

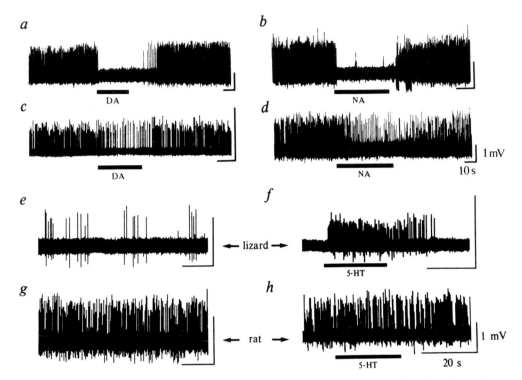

FIG. 5.   Effects of catecholamines and 5-hydroxytryptamine on pars intermedia cells of rat and lizard. Drugs applied for the periods indicated by the bars under each record. (a and c) Arrest and slowing, respectively, of spontaneous action potential discharge by dopamine (DA, $10^{-6}$ $M$) in two different rat cells. Note, in the latter, that action potential amplitude is not reduced by dopamine. (b and d) Comparable responses, arrest and slowing, respectively, of spontaneous action potential discharge in two other rat cells exposed to noradrenaline (NA, $10^{-6}$ $M$). (g and h) Absence of effect of 5-hydroxytryptamine (5-HT, $10^{-6}$ $M$ in the rat): (g) spontaneous activity; (h) the response to 5-HT. (e and f) Action potentials in pars intermedia cells of the lizard *Anolis carolinensis* and stimulation by 5-HT: (e) spontaneous activity; (f) response to 5-HT ($10^{-6}$ $M$). (a–d) From Douglas and Taraskevich (1978). (e–h) From Taraskevich and Douglas (1980).

tonic inhibitory control (prolactin cells and rat pars intermedia cells) show considerable spontaneous action potential activity, and factors that evoke secretion or depress it have corresponding effects on action potential activity in relevant target cells. These various results clearly encourage the view (Section IV,A) that in the adenohypophysis electrical phenomena involving Ca influx participate in secretory activity and its modulation by hypophysiotrophic factors, both stimulant and inhibitory.

## V.   Calcium, Compound Exocytosis, and the Exocytosis–Vesiculation Sequence: Possible Involvement of Calcium in Membrane Retrieval

Adenohypophyseal cells were among the first in which exocytosis was detected, and exocytosis, we suppose, is the process here, as elsewhere, that Ca regulates. At this juncture I should like to raise the possibility that Ca may have an additional function relating to membrane retrieval following exocytosis, a phenomenon that is commanding considerable attention.

It has been evident from the earliest discussions of "reverse pinocytosis" (or exocytosis, to use de Duve's term) that the addition of membrane to the cell surface must somehow be counterbalanced by withdrawal of surface membrane, if for no other reason than to maintain cell surface area (see Bennett, 1956; Palade, 1959; de Robertis *et al.*, 1960). Some years ago, my colleagues and I noted, in certain endocrine cells, a remarkable phenomenon that we intrepreted as unequivocal evidence of a preferential retrieval of the membrane of secretory granules following exocytosis. We came across this—later referred to as the "exocytosis–vesiculation sequence" (Douglas, 1974a)—quite by chance during the course of experiments in which we had sought, and discovered, electron microscopic evidence of exocytosis in the neurosecretory fibers of the posterior pituitary gland. For many years the mode of secretion in these fibers had been debated without exocytosis having been observed (for a review, see Douglas, 1974b), and it was probably only our strong conviction—or prejudice—that surely this system too, being dependent on Ca, must employ exocytosis that sustained us in a long pursuit of exocytotic images. But, once found, these images yielded unexpected and additional rewards: they provided an immediate explanation for the existence and significance in secretory terminals of the abundant but previously enigmatic "synaptic vesicles" first described by Palay (1957) and also yielded an insight into how secretory granule membrane was retrieved and cell surface conserved. In brief, our evidence (Nagasawa *et al.*, 1970, 1971; Douglas *et al.*, 1971a,b) indicated that, once incorporated into the cell surface, the membrane of the secretory granule was rapidly transformed into microvesicles, which were returned to the cytoplasm. The process involved application of "bristle coat" to the cytoplasmic aspect of the exocytotic pit (easily identified by the extruded granule core) followed by the formation of coated caveolae, which pinched off to yield coated vesicles, which, in turn, shed their coats to yield smooth-surfaced microvesicles (the so-called "synaptic vesicles"). Our findings from the neurohypophysis are summarized in Fig. 6. Such findings, considered along with the occurrence of comparable images of vesiculation of exocytotic pits in adrenal chromaffin cells and adenohypophyseal cells

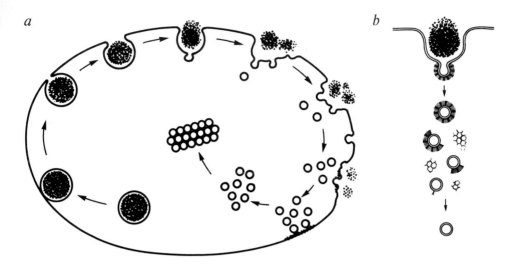

FIG. 6. Schemata summarizing electron microscopic evidence from neurohypophyseal terminals and depicting extrusion of posterior pituitary hormones by exocytosis followed by the formation of microvesicles ("synaptic vesicles") by vesiculation of exocytotic pits: Illustrations of the "exocytosis–vesiculation sequence": (a) The sequence in simplified form (coating omitted in the interests of clarity). (b) Detail indicating the involvement of coating in this sequence. (a) From Douglas *et al* (1971a); (b) From Douglas (1973).

(see Nagasawa and Douglas, 1972; also Fig. 7, A–C) led us (Douglas and Nagasawa, 1971) to propose that "there exists in a variety of cells a mechanism (perhaps better referred to as vesiculation rather than micropinocytosis with its connotation of cell drinking), serving to withdraw from the cell surface the membrane of secretory granules incorporated during exocytosis." Such specific removal of secretory granule membrane, we argued, would "conserve not only the area of the plasmalemma but also its chemical characteristics associated, for example, with permeability, excitability and receptor function." Our interpreta-

FIG. 7. Membrane coating and vesiculation (endocytotic activity) at sites of exocytosis: illustrations of the exocytosis–vesiculation sequence regarded as "a device for membrane conservation" (Douglas and Nagasawa, 1971). (A–C) Electron micrographs of adrenal medullary chromaffin cells of golden hamsters. Note coated caveolae within exocytotic pits and adjacent coated vesicles. Scale 100 n$M$. From Nagasawa and Douglas (1972). (D–F) Electron micrographs providing indications that extruded secretory granule contents may provide a stimulus for membrane coating and vesiculation. In each instance an extruded secretory granule core is seen lying between two adjacent cells, both of which show coating of the plasma membrane and vesicle formation. (D, E) Rat anterior pituitary incubated *in vitro* for 10 minutes at 37°C in 60 m$M$ K locke. (F) Hamster adrenal medulla incubated in the same way; fixation and processing as in upper records. Scale 100 n$M$ applies to all lower records. From J. Nagasawa and W. W. Douglas (unpublished data).

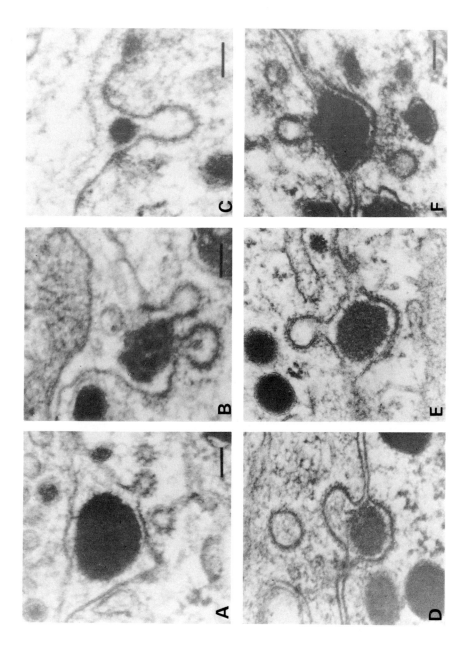

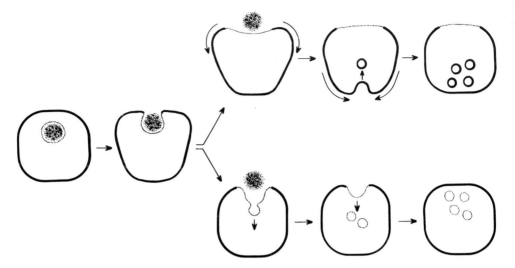

Fig. 8.   Two suggested mechanisms for countering the changes in cell surface that result from exocytosis. The upper scheme (after Bennett, 1956) involves a conveyor belt-like "membrane flow" whereby exocytosis is countered by pinocytotic activity. The cell surface area is maintained, but the cell membrane is now "diluted" with secretory granule membrane (dotted line). The lower scheme (Douglas and Nagasawa, 1971) involves a tightly coupled exocytosis–vesiculation sequence wherein micropinocytosis-like activity (endocytosis) is superimposed on the exocytotic pit itself. This restores both the area and the chemical composition of the cell surface and also retrieves "granule" membrane for possible reuse. From Douglas (1974a).

tion is illustrated in Fig. 8, where the contrast is made between this and the old scheme of "membrane flow" advanced by Bennett (1956), which lacks the specificity required for maintainence of cell surface characteristics or retrieval of possibly reusable granule membrane. Our findings on neurosecretory fibers also prompted us (Douglas and Nagasawa, 1971) to speculate that in ordinary neurons, where the released transmitter is taken up again and repackaged, "such specific recapture of synaptic vesicle membrane might permit its reuse in transmitter storage and release" (or "recycling," as we had previously termed it: Douglas et al., 1971a). Convincing evidence for such membrane recycling in ordinary neurons was soon provided by the detailed electron microscopic studies performed by Heuser and Reese (1973) and by Ceccareli et al. (1973). However, compensatory membrane vesiculation (endocytosis) in ordinary neurons is not as tightly coupled to exocytosis as it is in the posterior pituitary neurosecretory fibers. Rather, this vesiculation appears to take place at some distance from the site of exocytosis, and it is less evident that the vesiculated membrane is that which formerly surrounded the secretory granule (synaptic vesicle).

One problem of immediate interest is how the cell recognizes the secretory granule membrane that has been inserted in the cell surface upon exocytosis so that it may apply bristle coating to it, vesiculate it, and retrieve it. Since the membranes of secretory granules do not form a target for bristle coating when they lie within the cytoplasm before exocytosis, it seems evident that some change must occur upon exocytosis to initiate the phenomenon. Dr. Nagasawa and I recognized three possibilities (see Douglas, 1974a). First, we suggested that extruded secretory granule contents, chemically altered by dissociation in the extracellular environment to yield, for example, free peptides such as neurophysin or chromogranin or free ATP or other materials, might directly provide a local stimulus for such endocytotic activity. Observation of coincident coating on the plasma membranes of two cells contacted by a single extruded granule core (Fig. 7, lower series) supported this idea, but such images were scarce. Second, we supposed that the granule membrane itself might undergo a change that prompts vesiculation. Such a change could result, perhaps, from loss of the granule matrix. Or it could result from the electrical gradient imposed across the granule membrane upon its incorporation into the cell surface (with a transmembrane potential of 50 mV the field would be roughly 50,000 V/cm—sufficient to reorient charged molecules in the membrane). But against this last possibility were results we obtained indicating that vesiculation proceeded in preparations already depolarized by excess K. This brought us to the third possibility— namely, that the granule membrane might be altered by exposure to the extracellular environment where the most dramatic change encountered, and hence the most likely stimulus, would be in the Ca ion concentration, which is about one thousand times as high as in the cytoplasm. Because exocytosis in the cells we were studying had a critical requirement for extracellular Ca, we were unable to test this satisfactorily. Nevertheless, the idea seemed attractive, since we argued (Douglas, 1974a) that membrane vesiculation could be conceived of as a contractile phenomenom and the importance of Ca ions in contraction was well recognized. Now, in the light of subsequent evidence, this scheme seems all the more attractive. Biochemical evidence has emphasized the molecular parallels with events in muscle. The principal element of the coating complex, the protein clathrin, has been shown to interact, for example, with actin and $\alpha$-actinin (Schook et al., 1979). Moreover, there are signs pointing to the involvement of the Ca-receptive protein, calmodulin. It appears not unlikely, therefore, that the critical event triggering the application of bristle coating, as well as the chemomechanical transduction that effects vesiculation, may be the appearance of an excess of Ca ions at the cytoplasmic face of the exocytotic pit formed by the granule membrane. It could be that this granule membrane, perhaps altered by its inclusion in the cell surface, allows excessive inward leak of Ca ions. Or the granule itself, when it undergoes exocytosis, may liberate Ca ions from its membrane. Indeed, I believe that an elevated concentration of free Ca ions at the cytoplasmic face of exocytotic pits can be inferred from the phenomenon of

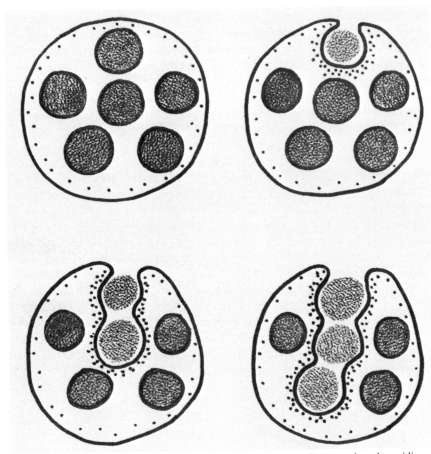

Fig. 9.   A schema suggesting the involvement of Ca in compound exocytosis and providing an indication that Ca may also participate in membrane vesiculation. *Top left:* A secretory cell immediately after stimulation. Calcium ions (black dots) appear under the plasmalemma. This initiates a single exocytotic event (*top right*). The membrane of the granule, now incorporated in the cell surface, is shown surrounded, on its cytoplasmic face, by a cloud of Ca ions resulting either from an undue permeability of this membrane to extracellular Ca or from liberation of membrane-bound Ca. A second exocytotic event is thereby promoted at the same site (lower left), thus initiating compound exocytosis. The process propagates to involve additional granules by the same mechanism (lower right). Note: A relatively high concentration of Ca ions at sites of exocytosis may not only account for the phenomenon of compound exocytosis but also participate in the chemomechanical events involved in coating and vesiculation of exocytotic pits (see text).

"compound exocytosis" (see Douglas, 1974a) in which a single exocytotic event involving the classic interaction of secretory granule membrane and plasmalemma tends to facilitate subsequent exocytotic events in which granules undergo exocytosis by interacting with the membrane of the first granule, which, of course, now forms part of the cell surface (Fig. 9). If we accept that Ca ions are the trigger to exocytosis, then the occurrence of such serial fusions suggests a wave of excess Ca ions propagated through the cell as schematized in Fig. 9. I suggest, therefore, that exocytosis favors a local elevation of Ca ions at the cytoplasmic face of the exocytotic pit, which, in some instances (perhaps when secretory granules are particularly close to one another), induces further exocytotic events (compound exocytosis), and in other instances provides a signal, and perhaps agent, for the coating and vesiculation of the secretory granule membrane. Calcium may thus participate in the entire exocytosis–vesiculation sequence.

## ACKNOWLEDGMENT

The studies performed in the author's laboratories were supported by grants from the U.S. Public Health Service, NS 08564 and NS 09137.

## REFERENCES

Baker, P. F., and Knight, D. E. (1978). *Nature (London)* **276**, 620–622.
Bennett, H. S. (1956). *J. Biophys. Biochem. Cytol.* **2**, 99–103.
Biales, B., Dichter, M., and Tischler, A. (1976). *J. Physiol. (London)* **262**, 743–753.
Blaschko, H., Firemark, H., Smith, A. D., and Winkler, H. (1967). *Biochem. J.* **104**, 545–549.
Brandt, B. L., Hagiwara, S., Kidokoro, S., and Miyazaki, S. (1976). *J. Physiol. (London)* **263**, 417–439.
Case, R. M. (1978). *Biol. Rev. Cambridge Philos. Soc.* **53**, 211–354.
Ceccarelli, B., Hurlburt, W. P., and Mauro, A. (1973). *J. Cell Biol.* **57**, 499–524.
Cochrane, D. E., and Douglas, W. W. (1974). *Proc. Natl. Acad. Sci. U.S.A.* **71**, 408–412.
Cochrane, D. E., and Douglas, W. W. (1976). *J. Physiol. (London)* **257**, 433–448.
Davis, M. D., and Hadley, M. E. (1976). *Nature (London)* **261**, 422–423.
Dean, P. M., and Matthews, E. K. (1970a). *J. Physiol. (London)* **210**, 255–264.
Dean, P. M., and Matthews, E. K. (1970b). *J. Physiol. (London)* **210**, 265–275.
De Robertis, E. D. P., Nowinski, W. W., and Saez, F. A. (1960). "General Cytology," pp. 138–141. Saunders, Philadelphia, Pennsylvania.
Devis, G., Somers, G., and Malaisse, W. J. (1975). *Biochem. Biophys. Res. Commun.* **67**(2), 525–529.
Douglas, W. W. (1965). *Proc. Int. Pharmacol. Meet., 2nd, 1963* Vol. 3, pp. 95–111.
Douglas, W. W. (1966). *In* "Mechanisms of Release of Biogenic Amines" (U. S. von Euler, S. Rosell, and B. Uvnäs, eds.), pp. 267–290. Pergamon, Oxford.
Douglas, W. W. (1968). *Br. J. Pharmacol.* **34**, 451–474.
Douglas, W. W. (1973). *Progr. Brain Res.* **39**, 21–38.

Douglas, W. W. (1974a). *Biochem. Soc. Symp.* **39,** 1–28.

Douglas, W. W. (1974b). *In* "Handbook of Physiology" (E. Knobil and W. H. Sawyer, eds.), Sect. 7, Vol. IV, Part 1, pp. 191–224. Am. Physiol. Soc., Washington, D.C.

Douglas, W. W. (1975). *In* "Handbook of Physiology" (H. Blaschko, G. Sayers, and A. D. Smith, eds.), Sect. 7. Vol. VI, pp. 367–388. Am. Physiol. Soc., Washington, D.C.

Douglas, W. W. (1978). *Ciba Found. Symp.* [N.S.] **54,** 61–87.

Douglas, W. W., and Kagayama, M. (1977). *J. Physiol. (London)* **270,** 691–703.

Douglas, W. W., and Nagasawa, J. (1971). *J. Physiol. (London)* **218,** 94–95P.

Douglas, W. W., and Rubin, R. P. (1961). *J. Physiol. (London)* **159,** 40–57.

Douglas, W. W., and Rubin, R. P. (1963). *J. Physiol. (London)* **167,** 288–310.

Douglas, W. W., and Taraskevich, P. S. (1977). *J. Physiol. (London)* **272,** 41–43P.

Douglas, W. W., and Taraskevich, P. S. (1978). *J. Physiol. (London)* **285,** 171–184.

Douglas, W. W., and Taraskevich, P. S. (1980). *J. Physiol.* **309,** 623–630.

Douglas, W. W., Ishida, A., and Poisner, A. M. (1965). *J. Physiol. (London)* **181,** 753–759.

Douglas, W. W., Poisner, A. M., and Trifaró, J. M. (1966). *Life Sci.* **5,** 809–815.

Douglas, W. W., Kanno, T., and Sampson, S. R. (1967). *J. Physiol. (London)* **191,** 107–121.

Douglas, W. W., Nagasawa, J., and Schulz, R. A. (1971a). *Mem. Soc. Endocrinol.* **19,** 353–378.

Douglas, W. W., Nagasawa, J., and Schulz, R. A. (1971b). *Nature (London)* **232,** 340–341.

Ensor, D. M. (1978). "Comparative Endocrinology of Prolactin." Wiley, New York.

Foreman, J. C., Mongar, J. L., and Gomperts, B. D. (1973). *Nature (London)* **245,** 249–251.

Geschwind, I. I. (1971). *Mem. Soc. Endocrinol.* **19,** 221–229.

Heuser, J., and Reese, T. S. (1973). *J. Cell Biol.* **57,** 315–344.

Hirata, F., Corcoran, B. A., Venkatasubramanian, K., Schiffmann, E., and Axelrod, J. (1979). *Proc. Natl. Acad. Sci. U.S.A.* **76,** 2640–2643.

Kanno, T., Cochrane, D. E., and Douglas, W. W. (1973). *Can. J. Physiol. Pharmacol.* **51,** 1001–1004.

Katz, B. (1969). "The Release of Neural Transmitter Substances (The Sherrington Lectures)." Thomas, Springfield, Illinois.

Kidokoro, Y. (1975). *Nature (London)* **258,** 741–742.

Kretsinger, R. H. (1979). *Adv. Cyclic Nucleotide Res.* **11,** 1–26.

McCann, S. M. (1971). *In* "Frontiers in Neuroendocrinology" (L. Martini and W. F. Ganong, eds.), Vol. 2, pp. 209–235. Oxford Univ. Press, London and New York.

Marone, G., Kagey-Sobotka, A., and Lichtenstein, L. M. (1979). *J. Immunol.* **123,** 1669–1677.

Miledi, R. (1973). *Proc. R. Soc. London, Ser. B* **183,** 421–425.

Nagasawa, J., and Douglas, W. W. (1972). *Brain Res.* **37,** 141–145.

Nagasawa, J., Douglas, W. W., and Schulz, R. A. (1970). *Nature (London)* **227,** 407–409.

Nagasawa, J., Douglas, W. W., and Schulz, R. A. (1971). *Nature (London)* **232,** 341–343.

Ozawa, S., and Sand, O. (1978). *Acta Physiol. Scand.* **102,** 330–341.

Palade, G. (1959). *In* "Subcellular Particles" (T. Hayashi, ed.), pp. 64–80. Ronald Press, New York.

Palade, G. (1975). *Science* **189,** 347–358.

Palay, S. (1957). *In* "Ultrastructure and Cellular Chemistry of Neural Tissue" (H. Waelsch, ed.), pp. 31–49. Harper (Hoeber), New York.

Petersen, O. H. (1976). *Physiol. Rev.* **56,** 535–577.

Poisner, A. M. (1973). *In* "Frontiers in Neuroendocrinology" (W. F. Ganong and L. Martini, eds.), Vol. 3, pp. 33–59. Oxford Univ. Press, London and New York.

Rahamimoff, R., Meiri, H., Erulkar, S. D., and Barenholz, Y. (1978). *Proc. Natl. Acad. Sci. U.S.A.* **75**(10), 5214–5216.

Ramwell, P. W., Shaw, J. E., Douglas, W. W., and Poisner, A. M. (1966). *Nature (London)* **210,** 273–274.

Rubin, R. P. (1969). *J. Physiol.* (*London*) **202,** 197–209.

Schook, W., Puszkin, S., Bloom, W., Ores, C., and Kochwa, S. (1979). *Proc. Natl. Acad. Sci. U.S.A.* **76,** 116–120.

Selinger, Z., Eimerl, S., and Schramm, M. (1974). *Proc. Natl. Acad. Sci. U.S.A.* **71,** 128–131.

Seighart, W., Theoharides, T. C., Alper, S., Douglas, W. W., and Greengard, P. (1978). *Nature* (*London*) **275,** 329–331.

Sullivan, T. J., and Parker, C. W. (1979). *J. Immunol.* **122,** 431–436.

Taraskevich, P. S., and Douglas, W. W. (1977). *Proc. Natl. Acad. Sci. U.S.A.* **74,** 4064–4067.

Taraskevich, P. S., and Douglas, W. W. (1978). *Nature* (*London*) **276,** 832–834.

Taraskevich, P. S., and Douglas, W. W. (1980). *Neuroscience* **5,** 421–431.

Theoharides, T. C., and Douglas, W. W. (1978a). *Endocrinology* **102,** 1637–1640.

Theoharides, T. C., and Douglas, W. W. (1978b). *Science* **201,** 1143–1145.

Theoharides, T. C., Sieghart, W., Greengard, P., and Douglas, W. W. (1980). *Science* **207,** 80–82.

Thornton, V. F., and Geschwind, I. I. (1975). *Gen. Comp. Endocrinol.* **26,** 346–353.

Vale, W., Rivier, C., and Brown, M. (1977). *Annu. Rev. Physiol.* **39,** 473–527.

Vanderhoek, J. Y., and Feinstein, M. B. (1979). *Mol. Pharmacol.* **16,** 171–180.

Wolff, D. J., and Brostrom, C. O. (1979). *Adv. Cyclic Nucleotide Res.* **11,** 27–88.

# Chapter 30

# Relationship between Receptors, Calcium Channels, and Responses in Exocrine Gland Cells

JAMES W. PUTNEY, JR. [1] AND STUART J. WEISS

*Department of Pharmacology,*
*Wayne State University,*
*School of Medicine,*
*Detroit, Michigan*

## I. Introduction

The purpose of this chapter is to discuss the manner in which exocrine gland cells can be used to study the relationship between surface membrane receptors and calcium mobilization mechanisms. That regulation of Ca metabolism is a central thesis in the function of secretory cells has been appreciated for some time (Douglas, 1968, 1974; Rubin, 1974). In the exocrine glands, two basic functions are under the control of Ca (Putney, 1978a,b): the secretion of prepackaged macromolecules (enzymes) by exocytosis, and the regulation of transepithelial water movement (or water secretion). This latter role for Ca, first demonstrated

[1]Present Address: Department of Pharmacology, Medical College of Virginia, Virginia Commonwealth University, Richmond, Virginia.

by Douglas and Poisner (1963), implies that Ca must in some manner be involved in the control by receptors of transmembrane movements of the osmotically important ions. In recent years, a number of reports have appeared demonstrating a role for Ca in the opening of monovalent ion pathways in a variety of exocrine gland cells (Selinger *et al.*, 1973, 1974; Iwatsuki and Petersen, 1977; Parod and Putney, 1978a,b; Martinez *et al.*, 1976; Putney *et al.*, 1978a; Putney and Van De Walle, 1979). By analogy with *stimulus–secretion coupling,* the term *stimulus–permeability* coupling has been used to describe this general phenomenon (Putney, 1978b).

These permeability changes also serve as convenient markers for quantitating the degree of cellular activation by neurotransmitters and hormones. The following discussion will illustrate how such measurements, together with biochemical studies, can provide insight into basic mechanisms of Ca regulation through receptors in exocrine gland cells.

## II.   Receptors, Calcium, and K Efflux

In the parotid and lacrimal glands, activation of muscarinic or $\alpha$-adrenergic receptors (or substance P receptors in the parotid gland) leads to an enhanced membrane permeability to K and Na and activation of the Na, K pump. Each of these responses appears to occur secondarily as a consequence of receptor-controlled mobilization of Ca (Putney, 1978b). The time course and magnitude of the Ca mobilization phenomenon is best reflected by the effects of agonists on K efflux. This is most conveniently determined as unidirectional efflux of $^{86}$Rb, which serves as a useful marker for tissue K movements (Putney, 1976a). Figure 1 illustrates the action of the $\alpha$-adrenergic agonist epinephrine on the K efflux response measured in this way. The drug causes a substantial and immediate increase in the efflux rate, which subsequently falls to a lesser, but still significantly elevated rate. In the presence of the appropriate receptor-blocking drug (in this case, phentolamine), no increase in the efflux of K is observed. When epinephrine is added to media from which Ca has been omitted, an increase in efflux is still observed, but the effect is transient and K efflux rapidly returns to control levels. The effect of Ca omission can be reversed by reintroducing Ca into the medium such that the elevated efflux rate is restored (Fig. 1). Thus, the K efflux response to receptor activation appears as two phases: a transient phase occurs in the presence or absence of external Ca, and a second sustained (or slowly falling) phase follows that absolutely depends on the presence of external Ca.

The Ca ionophore A23187 also stimulates K loss from exocrine glands (Selinger *et al.*, 1974; Parod and Putney, 1978b; Putney *et al.*, 1978a). When

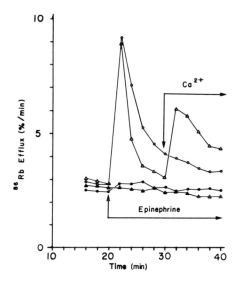

FIG. 1.   Stimulation of $^{86}$Rb efflux from rat parotid gland slices by epinephrine. Slices were permitted to accumulate $^{86}$Rb for 30 minutes and then were transferred through a series of nonradioactive incubations (of 2 minutes each) for 40 minutes. Unless specified otherwise, the media contained 120 mM NaCl, 5.0 mM KCl, 1.0 mM CaCl$_2$, 1.2 mM MgCl$_2$, 20.0 mM tris (hydroxymethyl) aminomethane, and 5.0 mM Na-$\beta$-hydroxybutyrate. The pH was 7.40 at 37°C, and the gas phase, 100% O$_2$. In all experiments except ▲ (control) the media contained $10^{-5}$ M epinephrine at 20 40 minutes. The media were also modified as follows: ○, no modification; ●, $10^{-4}$ M phentolamine, 0–40 minutes; △, no added Ca + $10^{-4}$ M EGTA, 0–30 minutes; 1.0 mM Ca, 30–40 minutes. The data represent means of four separate experiments. Standard errors averaged less than 10% of the means.

$^{86}$Rb efflux is monitored, only the second phase of release is obtained (Marier *et al.*, 1978; Parod and Putney, 1978b; Putney *et al.*, 1978a). Such observations led to the conclusion that Ca-dependent K release results from receptor activation of a Ca influx mechanism that leads to an elevation in intracellular Ca (Selinger *et al.*, 1974). The intracellular Ca acts, it is suggested, to activate membrane K channels by an unknown mechanism. In support of this contention, agonists that activate K efflux also stimulate Ca influx into gland cells (Putney, 1976b; Kanagasuntheram and Randle, 1976; Miller and Nelson, 1977; Putney *et al.*, 1978b).

The transient phase of the $^{86}$Rb release response may be mediated by Ca as well. The source of this Ca may be a cellular store, the nature of which is not known. The evidence for this has been discussed previously and will not be reiterated here (Putney, 1977, 1978b; Haddas *et al.*, 1979).

In the rat parotid gland, the patterns obtained by agonists acting on muscarinic, $\alpha$-adrenergic, or substance P receptors are all similar. The question arises therefore as to possible similarity or commonality of the mechanisms through which these receptors act. Some information in this regard comes from summation experiments, as shown in Fig. 2. In these experiments parotid slices were challenged with supramaximal concentrations of each of three agonists acting on different receptors (carbachol, epinephrine, or substance P), alone or in combination. The second phase of $^{86}$Rb release was assayed in isolation by carrying out the experiments in low-Ca medium, and then returning Ca to the medium in the continued presence of agonist (as in Fig. 1). Calcium was added to the medium to a final concentration of 1.0 m$M$, which is insufficient to cause maximal $^{86}$Rb efflux. Under these conditions, agonists acting on different receptors when applied in combination did not stimulate $^{86}$Rb release to an extent greater than that due to the more efficacious agonist acting alone (Fig. 2). Since the Ca concentration employed was submaximal, the conclusion is that the three receptors (muscarinic, $\alpha$-adrenergic, and substance P) all regulate the same population of Ca channels (Marier *et al.*, 1978). Similar experiments provided like conclusions as to $\alpha$-adrenergic and muscarinic receptors in the lacrimal gland (Parod and Putney, 1978a,b), and $\alpha$-adrenergic and angiotensin II receptors in the liver (Weiss and Putney, 1978).

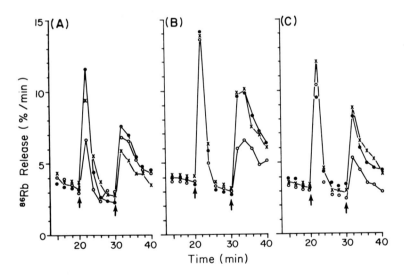

FIG. 2.  Lack of additivity of agonists in stimulating $^{86}$Rb efflux. The protocol was as for Fig. 1 ($\triangle$). One or two agonists were added to the medium at 20–40 minutes, and Ca(1.0 m$M$) was added at 30–40 minutes. The concentrations of agonists were: carbachol, $10^{-5}$ $M$; epinephrine, $10^{-4}$ $M$; substance P, $10^{-7}$ $M$. The specific drugs and combinations were: (A) ●, carbachol; ○, epinephrine; X, both. (B) ●, carbachol; ○, substance P; X, both. (C) ●, epinephrine; ○, substance P; X, both. For details see Marier *et al.* (1978), from which this figure is reproduced with permission of The Physiological Society.

# III.   Receptor Stoichiometry

If each of the three Ca-regulating receptors in the parotid can activate the same population of Ca channels, then it would be of interest to determine the stoichiometric relationship of each of the receptor types to the Ca channels. No method as yet exists for quantitating the number of Ca channels in nonexcitable cells. It has been possible, however, to compare the numbers of receptors of each kind to one another with some rather interesting results.

Strittmatter *et al.* (1977a,b) measured the number of $\alpha$-adrenoceptors in rat parotid cells with [³H]dihydroergocryptine, an antagonist with high affinity for the $\alpha$-adrenergic receptors. They calculated a density of about 15,000 sites/cell. By a similar procedure employing [³H]quinuclidinyl benzylate (a muscarinic antagonist), the number of muscarinic receptors was recently estimated to be about 23,000 sites/cell (Putney and Van De Walle, 1980). However, there was an approximate tenfold discrepancy between the concentration of agonists giving half-maximal occupancy of muscarinic receptors and the concentration producing a half-maximal response by the tissue. It was suggested therefore that in the parotid only about 10% of the available muscarinic receptor sites need be occupied to activate all the available Ca channels (Putney and Van De Walle, 1980). In support of this, it was found that, after inactivation of the majority of murscarinic receptor sites, the concentration–effect relationship for cholinergically stimulated ⁸⁶Rb efflux agreed with the radioligand binding data (Butcher and Putney, 1980). Thus, the conclusion is that a considerable quantity of "spare" cholinergic receptors exists in the parotid gland, and only about 2000 sites/cell need be occupied for maximal effector activation (i.e., Ca-gating).

The density of receptors for substance P was estimated by measuring the binding of [¹²⁵I]physalaemin, a frog skin peptide with high affinity for the substance P receptor (Putney *et al.*, 1980a). A high degree of nonspecific binding added some uncertainty to the measurement, but the estimated receptor concentration was in the range of 200 sites/cell.

The numbers of receptors of each kind on the parotid acinar cell are given in Table I. The widely disparate densities required for activating the same Ca-gating

TABLE I

DENSITIES OF RECEPTORS ON RAT PAROTID ACINAR CELLS

| Receptor: | Muscarinic | $\alpha$-Adrenergic | Substance P |
|---|---|---|---|
| Ligand: | [³H]Quinuclidinyl benzylate | [³H]Dihydroergocryptine | [¹²⁵I]Physalaemin |
| Sites/cell: | 1800 (23,000)[a] | 15,000 | 200 |
| Reference: | Putney and Van De Walle (1979) | Strittmatter *et al.* (1977a) | Putney *et al.* (1980a) |

[a] The total receptor density was estimated to be 23,000 per cell, but it was suggested that only 1800 per cell need be occupied to activate the available Ca channels.

mechanisms cannot be readily explained. The suggestion is that either the mechanism by which the Ca gates are activated differs for the three receptors, or the efficiencies (or affinities) of the various receptors for the effector activation mechanism may differ.


# IV.   Phospholipids and Calcium Gates


Michell and colleagues have suggested that phospholipids may play a role in the mechanism of Ca-gating in exocrine glands and other tissues where surface membrane receptors effect Ca mobilization (Michell, 1975, 1979; Michell et al., 1976a,b, 1977). The theory relates to the observation that agonists stimulate the incorporation of $[^{32}P]PO_4$ into phosphatidylinositol and phosphatidic acid. In addition, tissue content of phosphatidylinositol usually falls, while phosphatidate levels increase. In the parotid gland, labeling of phosphatidylinositol is stimulated by muscarinic, $\alpha$-adrenergic, or peptide agonists (Oron et al., 1975; Jones and Michell, 1978). The effect does not require Ca and cannot be reproduced with the divalent cation ionophore A23187 (Jones and Michell, 1974, 1976, 1978; Oron et al., 1975). These observations led Michell to propose that receptor occupation leads to breakdown of phosphatidylinositol to 1,2-diacylglycerol, which is rapidly phosphorylated to phosphatidic acid. In some manner, these reactions result in the activation of processes that lead to an inward movement of Ca ions. The mechanism by which this effect is produced has been a matter of some speculation. Salmon and Honeyman (1979) recently suggested that the phosphatidic acid formed in the reaction could function as a Ca ionophore. Michell has also raised this possibility (among others) in a review (Michell et al., 1977). Indeed, phosphatidate performs reasonably well as an ionophore in a Pressman chamber (Tyson et al., 1976), although the relevance of this model to transport in bilayers can certainly be questioned. The possibility exists, however, that phosphatidic acid may mediate inward Ca transport, but as part of a more complex, specialized mechanism (i.e., a lipid–protein complex) rather than as a simple ionophore. In support of a role for phosphatidic acid in the Ca-gating scheme in the parotid gland, it has been demonstrated that parotid content of phosphatidic acid increases significantly following muscarinic receptor stimulation (Putney et al., 1980b). Also, phosphatidic acid stimulates Ca-dependent [86]Rb efflux from parotid slices (Putney et al., 1980b), in a manner reminiscent of the action of a Ca ionophore; i.e., a sustained but no transient phase is observed.

One general criticism of hypotheses invoking phospholipids in Ca transfer mechanisms has been the generally accepted view that they have insufficient ionic specificity, especially with respect to selectivity for Ca over Mg (Michell, 1975). It is in this regard that support for a role for phosphatidic acid in Ca-gating

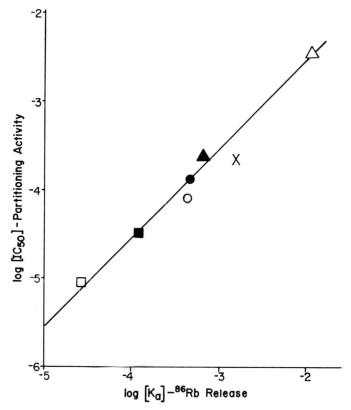

FIG. 3.   Relationship between inhibition of $^{45}$Ca-partitioning activity of phosphatidic acid and inhibition of receptor-activated Ca-dependent $^{86}$Rb release by various agents. The ordinate values are concentrations of agent necessary to inhibit $^{45}$Ca transport by phophatidic acid into CHCl$_3$ by 50%, and the abscissae are apparent dissociation constants ($K_a$) for antagonism of Ca-gating (measured as $^{86}$Rb release due to carbachol) in parotid slices. The agents were: □, La$^{3+}$; ■, Tm$^{3+}$; ○, neomycin; ●, Co$^{2+}$; ▲, Ni$^{2+}$; △, Mg$^{2+}$; X, Ca$^{2+}$ (K as agonism estimated). The line was fit by least-squares analysis. For details see Putney *et al*. (1980b), from which this figure is reproduced with permission of *Nature* (*London*) and Macmillan Journals, Ltd.

in the parotid is especially strong. The apparent dissociation constants from the receptor-activated Ca gate were determined for a series of seven cationic substances by standard pharmacological procedures (Putney *et al.*, 1980b). The ability of these same agents to bind to phosphatidic acid was assayed by measuring the phosphatidic acid-mediated partitioning of $^{45}$Ca into chloroform in the presence of various concentrations of the cationic materials. The concentration of antagonist that inhibited $^{45}$Ca partitioning by 50% was taken as an index of relative affinity for the phosphatidic acid. Figure 3 shows that these binding data correlate strikingly well with the data for inhibition of receptor-activated Ca gates, including the appropriate selectivity of phosphatidic acid for Ca over Mg (Putney *et al.,* 1980b).

# V.  Conclusions

The exocrine glands, and especially the rat parotid gland, are extremely useful models for investigation of the mechanism by which surface receptors regulate secretory phenomena. For study of Ca-mediated events, activation of potassium channels (measured as $^{86}$Rb efflux) is a convenient marker. In the parotid gland, three different receptors ($\alpha$-adrenergic, muscarinic, and substance P) all act to control a common pathway of Ca-gating, which results in activation of Ca-dependent $^{86}$Rb efflux. Although all three receptors act on the same Ca channels, the numbers of receptors of each kind required to accomplish this task vary greatly.

Several lines of evidence suggest that the mechanisms by which receptors activate Ca-gating in the parotid gland (and other tissues as well) involves alterations in phospholipid metabolism. The increase in tissue levels of phosphatidic acid resulting from phosphatidylinositol breakdown may be important. Phosphatidic acid can stimulate $^{86}$Rb efflux when added to incubated parotid tissue, and apparently has cation binding selectivity similar to that of the Ca channel. Knowledge of the specific role of phosphatidic acid in the Ca-gating scheme must await further investigation.

ACKNOWLEDGMENTS

Work from the author's laboratory described in this chapter was supported by Grants DE-04067, EY-01978, and GM-07176 from the National Institutes of Health.

REFERENCES

Butcher, F. R., and Putney, J. W., Jr. (1980). *Adv. Cyclic Nucleotide Res.* **13,** 215–249.
Douglas, W. W. (1968). *Br. J. Pharmacol.* **34,** 451–474.
Douglas, W. W. (1974). *Biochem. Soc. Symp.* **39,** 1–28.
Douglas, W. W., and Poisner, A. M. (1963). *J. Physiol. (London)* **165,** 528–541.
Haddas, R. A., Landis, C. A., and Putney, J. W., Jr. (1979). *J. Physiol. (London)* **291,** 457–465.
Iwatsuki, N., and Petersen, O. H. (1977). *Nature (London)* **268,** 147–149.
Jones, L. M., and Michell, R. H. (1974). *Biochem. J.* **142,** 583–590.
Jones, L. M., and Michell, R. H. (1976). *Biochem. J.* **158,** 505–507.
Jones. L. M., and Michell, R. H. (1978). *Biochem. Soc. Trans.* **6,** 673–688.

Kanagasuntheram, P., and Randle, P. J. (1976). *Biochem. J.* **160,** 547–564.

Marier, S. H., Putney, J. W., Jr., and Van De Walle, C. M. (1978). *J. Physiol. (London)* **279,** 141–151.

Martinez, J. R., Quissel, D. O., and Giles, M. (1976). *J. Pharmacol. Exp. Ther.* **198,** 385–394.

Michell, R. H. (1975). *Biochim. Biophys. Acta* **415,** 81–147.

Michell, R. H. (1979). *Trends in Biochemical Sciences,* **4,** 128–131.

Michell, R. H., Jafferji, S., and Jones, L. M. (1976a). *FEBS Lett.* **69,** 1–5.

Michell, R. H., Jafferji, S., and Jones, L. M. (1976b). *In* "Stimulus-Secretion Coupling in the Gastrointestinal Tract" (R. M. Case and H. Goebell, eds.), pp. 89–103. University Park Press, Baltimore, Maryland.

Michell, R. H., Jafferji, S., and Jones, L. M. (1977). *Adv. Exp. Biol. Med.* **83,** 447–464.

Miller, B. E., and Nelson, D. L. (1977). *J. Biol. Chem.* **252,** 3629–3636.

Oron, Y., Lowe, M., and Selinger, Z. (1975). *Mol. Pharmacol.* **11,** 79–86.

Parod, R. J., and Putney, J. R., Jr. (1978a). *J. Physiol. (London)* **281,** 359–369.

Parod, R. J., and Putney, J. R., Jr. (1978b). *J. Physiol. (London)* **281,** 371–381.

Putney, J. W., Jr. (1976a). *J. Pharmacol. Exp. Ther.* **198,** 375–384.

Putney, J. W., Jr. (1976b). *J. Pharmacol. Exp. Ther.* **199,** 526–537.

Putney, J. W., Jr. (1977). *J. Physiol. (London)* **368,** 139–149.

Putney, J. W., Jr. (1978a). *In* "Calcium in Drug Action" (G. B. Weiss, ed.), pp. 173–194. Plenum, New York.

Putney, J. W., Jr. (1978b). *Pharmacol. Rev.* **30,** 209–245.

Putney, J. W., Jr., and Van De Walle, C. M. (1979). *Life Sci.* **24,** 1119–1124.

Putney, J. W., Jr., and Van De Walle, C. M. (1980). *J. Physiol. (London)* **299,** 521–531.

Putney, J. W., Jr., Leslie, B. A., and Marier, S. H. (1978a). *Am. J. Physiol.* **235,** C128–C135.

Putney, J. W., Jr., Leslie, B. A., and Van De Walle, C. M. (1978b). *Mol. Pharmacol.* **14,** 1046–1053.

Putney, J. W., Jr., Van De Walle, C. M., and Wheeler, C. S. (1980a). *J. Physiol. (London)* **301,** 205–212.

Putney, J. W., Jr., Weiss, S. J., Van De Walle, C. M., and Haddas, R. A. (1980b). *Nature (London)* **284,** 345–347.

Rubin, R. P. (1974). "Calcium and the Secretory Process." Plenum, New York.

Salmon, D. M., and Honeyman, T. W. (1979). *Biochem. Soc. Trans.* **7,** 986–988.

Selinger, Z., Batzri, S., Eimerl, S., and Schramm, M. (1973). *J. Biol. Chem.* **248,** 369–372.

Selinger, Z., Eimerl, S., and Schramm, M. (1974). *Proc. Natl. Acad. Sci. U.S.A.* **71,** 128–131.

Strittmatter, W. J., Davis, J. N., and Lefkowitz, R. J. (1977a). *J. Biol. Chem.* **252,** 5472–5477.

Strittmatter, W. J., Davis, J. N., and Lefkowitz, R. J. (1977b). *J. Biol. Chem.* **252,** 5478–5482.

Tyson, C. A., Zande, H. V., and Green, D. E. (1976). *J. Biol. Chem.* **251,** 1326–1332.

Weiss, S. J., and Putney, J. W., Jr. (1978). *J. Pharmacol. Exp. Ther.* **207,** 669–676.

# Chapter 31

# Membrane Potential and Conductance Changes Evoked by Hormones and Neurotransmitters in Mammalian Exocrine Gland Cells

O. H. PETERSEN, N. IWATSUKI,[1] H. G. PHILPOTT,
R. LAUGIER,[2] G. T. PEARSON, J. S. DAVISON,
AND D. V. GALLACHER

*The Physiological Laboratory, University of Liverpool, Liverpool, United Kingdom*

## I.  Introduction

The investigation of the electrophysiological properties of exocrine gland cells has given much information on the mechanism of activating secretory cells. The

---

[1]Department of Applied Physiology, Tohoku University School of Medicine, Seiryocho 2-1, Sendai, Japan 980.

[2]INSERM Unite 31, 46 Chemin de la Gaye, 13009, Marseille, France.

high time resolution of electrophysiological methods combined with the opportunity to probe individual cells and apply test substances at selected sites (e.g., intracellularly or extracellularly) has enabled precise mapping of a number of transport pathways. The following brief review summarizes the current status. A much more comprehensive account also containing more background has recently been published (Petersen, 1980).

## II.  Methods

### A.  The Preparations

Isolated gland segments from mice or rats are mounted on a translucent Perspex block and placed in a bath through which physiological saline solutions flow. The solutions are gassed with 95% $O_2$ and 5% $CO_2$ and prewarmed to 37°C. In the case of the mouse pancreas, visualization of individual cells in the mounted living preparation is possible (Fig. 1), since this tissue is extremely thin (Iwatsuki and Petersen, 1978d). Isolated superfused segments of mouse pancreas and parotid have a low base-line amylase secretion and respond to appropriate stimulation by hormones and neurotransmitters with a marked and sustained increase in amylase secretion (Matthews *et al.*, 1973; Petersen and Ueda, 1976, Petersen *et al.*, 1977a,b).

### B.  Membrane Potential, Resistance, and Capacitance Measurement

Membrane potentials of acinar cells in superfused gland segments are measured by using very fine glass microelectrodes. Glass micropipettes are produced on a Palmer horizontal microelectrode puller, filled with 3 *M* KCl by the fiber glass method, and subsequently beveled (Sutter Instrument Co., San Francisco, CA). Final resistances are about 30–50 MΩ. Microelectrodes are inserted into superficial acinar cells (Fig. 1) with the help of Leitz or Zeiss micromanipulators and connected to WPI (M 701, M 750) electrometer amplifiers, with current injection facility. An indifferent Ag/AgCl electrode is placed in the bath. The potentials are displayed on storage oscilloscopes and pen recorders (Iwatsuki and Petersen, 1977a,b, 1978a,d). Membrane potentials are generally stable over long periods (½–1 hour), and in such cases there is little or no decline in the membrane potential value. When two or three microelectrodes are inserted into cells within the same acinus (Fig. 1) the potentials measured are all of the same value

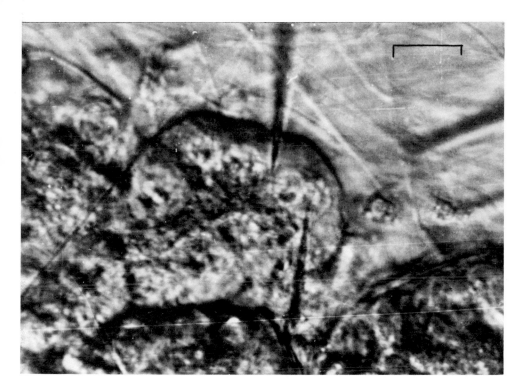

FIG. 1. Phase contrast image of mouse pancreatic fragment placed in a tissue bath and superfused with an oxygenated Krebs solution. Two microelectrodes have been inserted into neighboring acinar cells. The tip of an extracellular micropipette, used for local application of ACh, is seen near the acinus under investigation. Bar, 20 $\mu$m. From Iwatsuki and Petersen (1978d).

($\pm$1-2 mV). Moreover, electrotonic potential changes set up in one cell by repetitive current pulse injection are also transmitted to neighboring cells (Fig. 2). Rectangular current pulses (provided by Grass S 44 stimulators) of 100-msec duration are used, since these are long enough for a steady state (or near steady state) to be reached. The time course of the rising phase of the electrotonic potential change is adequately described by the equation

$$V = RI(1 - e^{-t/RC}) \tag{1}$$

where $R$ is the input resistance (voltage deflection, $V$, in steady state divided by $I$), $I$ the injected current, and $C$ the input capacitance; $RC$ is the time constant—the time, after start of current injection, taken for $V$ to reach 63% [$(1 - e^{-1})$ 100%] of steady-state displacement.

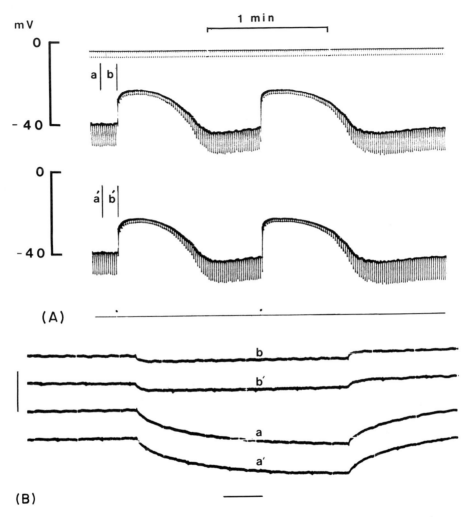

FIG. 2. Effect of local ACh application (microionophoresis) on pancreatic acinar membrane potential and resistance. (A) Continuous pen recordings. Two simultaneous membrane potential measurements are shown. Current pulses (2 nA, 100 msec) were repetitively injected through the microelectrode recording the lower of the two potential traces. The current pulses evoked electrotonic potential changes (short-lasting potential deflections) in both injection cell (lower trace) and neighboring cell (upper trace). ACh was applied by microionophoresis from an extracellular micropipette (80 nA, 500 msec) at marker signals in bottom event-marker trace. The time course of the electrotonic potentials before and during stimulation is shown in the oscilloscope picture (B). Horizontal calibration in B is 20 msec; vertical, 10 mV. From Iwatsuki and Petersen (1978d).

## C. Cell-to-Cell Communication

One method of assessing direct cell-to-cell coupling is to observe the spread of electrotonic potentials from the current injection cell to other cells (Petersen and Ueda, 1976; Iwatsuki and Petersen, 1977b,c, 1978a,b,d) (Fig. 3). By keeping the current injection electrode stationary and inserting one or two other microelectrodes into other cells at varying distances from the current injection site, mapping of voltage fields in gland tissues can be achieved. An alternative method is to observe directly the movement of fluorescent tracers from cell to cell. This can of course be combined in the same experiments with measurements of spread of electrotonic potentials. Figure 3 demonstrates the setup for assessing the movement of injected tracers to neighboring cells. Fluorescent compounds (fluorescein and procion yellow) are injected into cells through micropipettes by current passage (ionophoresis) and are observed to move into adjacent cells (Iwatsuki and Petersen, 1979c).

## D. Stimulation of Gland Cells

Basically two different methods are used—release of endogenous neurotransmitters from nerve endings, and application of neurotransmitters or peptide hormones from micropipettes. The two methods are sometimes combined in the same experiments (Davison *et al.*, 1980). Release of endogenous neurotransmitter substances is achieved by inducing action potentials in the nerve fibers. This is done by placing a pair of platinum electrodes on the surface of the gland segment under investigation at a distance apart of about 3 mm. The electrodes are connected to a Devices pulse generator to provide square-wave stimulation of required frequency, pulse width, and voltage. The pulse generator is triggered by a Devices Digitimer (type 3290) (Davison *et al.*, 1980).

Local application of substances is done by using micropipettes filled with a suitably concentrated solution of the stimulant—acetylcholine, epinephrine, isoproterenol, substance P, caerulein, bombesin nonapeptide, various amino acids or various ions (e.g., $Ca^{2+}$, $Mg^{2+}$). Ejection from the tip of the micropipette is by the method of ionophoresis (i.e., passing current in the appropriate direction). In some cases it is necessary to employ a retaining current between periods of stimulation to avoid spontaneous diffusion from the tip of the pipettes. The microionophoresis pipettes can be placed in the tissue under visual control (Fig. 1), and ionophoretic application (both intra- and extracellularly) is feasible (Nishiyama and Petersen, 1975; Iwatsuki and Petersen, 1977a,b,c, 1978d; Roberts and Petersen, 1978; Petersen and Philpott, 1979; Philpott and Petersen, 1979a,b).

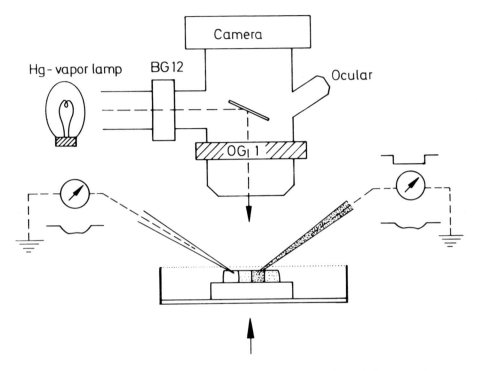

Fig. 3. Schematic diagram of setup to observe movement of injected fluorescent tracers. Fluorescein or procion yellow is injected through a micropipette by ionophoresis. The fluorescent probe is excited by incident blue light and fluorescence observed in the dark field through a yellow filter. From Iwatsuki and Petersen (1979b).

## III.   The Resting Membrane Potential

When a microelectrode is inserted into a cell, particularly a relatively small mammalian gland cell, there is always the danger of creating a leak in the membrane at the puncture site. Resting membrane potentials are therefore easily underestimated. This probably accounts for the low potentials generally observed in the early years of gland cell electrophysiology (see Petersen, 1980). In recent years improvements in technique have resulted in higher and more stable resting potentials; Table I gives results from recent investigations in our laboratory.

The relatively high resting potentials are very sensitive to changes in extracellular K concentration ($[K]_o$). Thus, in the mouse parotid the slope of the linear curve relating membrane potential to log $[K]_0$ is about 50 mV per tenfold increase in $[K]_0$ (above $[K]_0 = 10$ m$M$) (Pedersen and Petersen, 1973). This is high but clearly less than the theoretically expected 61 mV per tenfold increase in $[K]_0$ for a K-selective membrane. The membrane permeability is therefore predominated by K but with a contribution of other ions. The intracellular K activity in mouse salivary glands is 116 m$M$ (Poulsen and Oakley, 1979). Assuming an activity coefficient of 0.7 in the extracellular solution having a K concentration of 4.7 m$M$, the K equilibrium potential ($E_K$) can be calculated:

$$E_K = 61.5 \log 3.3/116 \text{ mV} = -95 \text{ mV} \tag{2}$$

The resting potential ($E_m$), even in the parotid ($-70$ mV), is therefore considerably less negative than $E_K$. In most gland cells, there appears to be an appreciable Na permeability explaining the discrepancy between $E_m$ and $E_K$. In most gland cells Cl is passively distributed ($E_{Cl} \approx E_m$). The permeability of Cl ($P_{Cl}$) is generally higher than K permeability ($P_K$), although in the parotid there appears to be a low resting $P_{Cl}$ and possibly also an active Cl-accumulating mechanism (Petersen, 1980).

It is not possible to calculate the resting specific membrane resistance without knowledge of the electrical communication network. This question will therefore be dealt with in Section V.

## IV.   Cell-to-Cell Coupling

### A.   Electrical Communication

Petersen and Ueda (1976) described neighboring acinar cells in the mouse pancreas as electrically coupled. Since then electrical coupling of acinar cells in a

TABLE I

RESTING MEMBRANE POTENTIALS IN SOME EXOCRINE GLANDS

| Tissue | Resting potential | References |
|---|---|---|
| Mouse parotid | $-69$ mV | Pedersen and Petersen (1973) |
| Mouse submaxillary | $-57$ mV | Nishiyama and Petersen (1974b) |
| Rat submaxillary | $-57$ mV | Roberts and Petersen (1978) |
| Rat parotid | $-73$ mV | Gallacher and Petersen (1980) |
| Mouse pancreas | $-40$ mV | Nishiyama and Petersen (1974a) |
| Rat pancreas | $-36$ mV | Nishiyama and Petersen (1974a) |
| Mouse exorbital lacrimal gland | $-43$ mV | Iwatsuki and Petersen (1978b) |

number of mammalian exocrine glands (mouse and rat parotid, submaxillary, pancreas, and lacrimal glands) has been reported (Iwatsuki and Petersen, 1977b,c, 1978a,b,d; Roberts *et al.*, 1978; Hammer and Sheridan, 1978; Kater and Galvin, 1978). However, it is only in the mouse pancreas that it has been possible to get direct information on the extent of the communication network. Since the mouse pancreas is extremely thin (Fig. 1), direct visualization of individual acini and cells within the acini is possible. Mapping of voltage fields with direct observation of sites of microelectrode insertions has given valuable information (Iwatsuki and Petersen, 1978d). Cells within individual acini are always fully coupled. Neighboring acini are also coupled, but acini belonging to different clusters of acini appear to be electrically totally isolated from each other. Small groups (clusters) of acini, containing approximately 500 cells, therefore constitute the electrical units in the mouse exocrine pancreatic tissue. The extent of electrical coupling between two cells is frequently expressed by means of the coupling coefficient ($V_2/V_1$). The coupling coefficient (ratio) is the magnitude of the electrotonic potential change in cell 2 due to current injection into cell 1, divided by the magnitude of the electrotonic potential change in the current injection cell (cell 1). Acinar cells within the same acinus are generally coupled with a coupling coefficient very close to 1 (Fig. 2). Cells in different communicating acini are still coupled with a relatively high coupling coefficient ($>0.5$).

## B.    Transfer of Fluorescent Probes

Direct demonstrations of cell-to-cell communication by visualizing movement of fluorescent probes injected into one cell to neighboring cells have recently been made in mammalian exocrine pancreas and salivary glands (Iwatsuki and Petersen, 1979b; Petersen and Iwatsuki, 1979; Hammer and Sheridan, 1978; Kater and Galvin, 1978). The probes used so far are fluorescein, procion yellow, and lucifer yellow, indicating that substances with a molecular weight of about 700 can easily pass through the junctional channels. It is not excluded that somewhat larger molecules can also be transferred.

## V.    The Resting Specific Membrane Conductance and Capacitance

Passage of a rectangular current pulse across an exocrine gland cell membrane results in an electrotonic potential change (Fig. 2). As described in Section II, the rising phase of the electrotonic potential follows the equation

$$V = RI(1 - e^{-t/RT}) \tag{3}$$

In the steady state Ohm's law applies:

$$V = RI \tag{4}$$

From these two equations and the time course of electrotonic potentials (Fig. 2), values for $R$ and $C$ were obtained. In the mouse pancreas the input resistance $R$ is about 5 MΩ and the capacitance $C$ about 3 nF.

On the basis of the spatial distribution of electrotonic potentials due to a current point source, the simplifying approximation can be made that the electrical unit is a sphere with a diameter of 110 $\mu$m and that current spreads uniformly over the entire unit (Iwatsuki and Petersen, 1978d). From the stereological work of Bolender (1974), values for the volume of acinar cells, surface cell membrane area, and cell density are known; it is therefore possible to calculate the specific membrane resistance to have a value of about 14 kΩ/cm², and the specific membrane capacitance appears to be about 1.1 $\mu$F/cm². Since the specific membrane capacitance is a biological constant (Fricke, 1923; Cole, 1928) with a value very close to 1 $\mu$F/cm², the capacitance value arrived at for the pancreatic acinar cell membranes (Iwatsuki and Petersen, 1978d) is a convenient check for internal consistency of the methods used.

The mammalian exocrine gland cell membranes are not electrically excitable. The current–voltage relationship is approximately linear over a fairly wide range of membrane potentials (0 to −60 mV) (Nishiyama and Petersen, 1975; Iwatsuki and Petersen, 1977b, 1978b; Roberts and Petersen, 1978; for a discussion of small deviations from linearity, see Petersen, 1980).

## VI.   Effects of Stimulation

## A.   Pancreatic Acinar Cells

Pancreatic acinar cells are stimulated to secrete by a number of hormones and by neural stimulation. The functional innervation of the acini seems to be exclusively cholinergic (Davison *et al.*, 1980). The major hormonal stimulants belong to the cholecystokinin-gastrin and bombesin families. The most important secretory process in the acini is that of protein secretion. Since the pancreas therefore has a very active protein synthesis, there must also be very effective mechanisms for amino acid uptake. Electrophysiological investigations of the effect of amino acid application have also been made.

## 1.  EFFECTS OF ACETYLCHOLINE

Acetylcholine (ACh) acts on the acini by decreasing the plasma membrane resistance; this evokes depolarization (Fig. 2). The effect of ACh is dose-dependent, with small depolarizations being seen at concentrations as low as $10^{-9}$ $M$ and maximal effects at $10^{-5}$ $M$. The dose-response curve is rather similar to that of ACh-evoked amylase secretion (Matthews *et al.*, 1973). ACh evokes electrical effects only when added to the outside of the plasma membrane (Iwatsuki and Petersen, 1977a), but even with local extracellular application very close (<5 $\mu$m) to the acinus from which intracellular recording is made there is an unavoidable delay of 200–300 msec before cell depolarization occurs (Nishiyama and Petersen, 1975).

Whereas sustained stimulation with ACh up to concentrations of about $5 \times 10^{-7}$ $M$ or short (500 msec) ionophoretic pulses of local ACh application evoke changes only in the plasma membrane resistance and potential, higher concentrations of ACh or more prolonged ionophoretic applications additionally evoke an increase in junctional membrane resistance—i.e., electrical uncoupling (Petersen and Ueda, 1976; Iwatsuki and Petersen, 1977c, 1978a,d). Uncoupling between neighboring acini is seen at lower stimulant concentrations than is uncoupling of acinar cells within one acinus, and it appears that only the functional uncoupling of different acini, but not the electrical isolation of individual cells, is a physiologically occurring phenomenon (Petersen and Iwatsuki, 1979).

## 2.  EFFECTS OF ELECTRICAL FIELD STIMULATION

It is possible to release endogenous neurotransmitters from nerve endings within the pancreatic tissue by application of an electrical field (Nishiyama *et al.*, 1980). Recent work has clearly demonstrated that: (1) electrical field stimulation evokes release of endogenous ACh by initiation of nerve action potentials; (2) the effects are seen at stimulation frequencies that are likely to resemble closely the physiological frequency of discharge in pancreatic nerves (2–5 Hz); (3) all acinar cells (at least in mouse and rat) are potentially under cholinergic neural influence; (4) spontaneous miniature depolarizations are due to spontaneous, quantal release of ACh from nerve terminals; and (5) there is no indication of the existence of transmitters other than ACh (Davison *et al.*, 1980).

## 3.  EFFECTS OF STIMULATION WITH PEPTIDES BELONGING TO THE CHOLECYSTOKININ-GASTRIN FAMILY

All peptides belonging to this family having the ability to evoke acinar secretion also cause membrane potential and resistance changes indistinguishable from those evoked by cholinergic stimulation (Petersen and Matthews, 1972; Nishiyama and Petersen, 1974a, Iwatsuki and Petersen, 1978c; Petersen and Philpott, 1979). However, while atropine blocks the membrane effect of ACh, it has no effect on the response to cholecystokinin-gastrin peptides. A specific competitive antagonist for the action of cholecystokinin-gastrin peptides has recently been found (Peikin *et al.*, 1979; Philpott and Petersen, 1979b). Dibutyryl cyclic guanosine monophosphate (dbcGMP) blocks the ability of cholecystokinin (CCK), caerulein, and pentagastrin to evoke membrane depolarization and resistance reduction without interfering with the ability of ACh or peptides belonging to the bombesin group to cause membrane effects (Fig. 4). Like ACh, the CCK analog caerulein is active only when added to the outside of the plasma membrane (Philpott and Petersen, 1979a), and the minimal delay of cell activation with this peptide is several hundred milliseconds (Petersen and Philpott, 1979).

## 4. EFFECTS OF STIMULATION WITH PEPTIDES BELONGING TO THE BOMBESIN FAMILY

Bombesin and bombesin nonapeptide evoke a marked membrane depolarization and resistance reduction, and at high concentrations uncoupling, exactly like ACh and the CCK-gastrin peptides. The bombesin-evoked depolarization and resistance reduction has a dose-response curve very similar to that of the bombesin-evoked increase in amylase secretion (Iwatsuki and Petersen, 1978c). Bombesin acts only on the outside of the membrane, and the minimal delay of cell activation is several hundred milliseconds, similar to the situation for ACh and CCK-like peptides (Philpott and Petersen, 1979a; Petersen and Philpott, 1979). The action of bombesin is clearly mediated by a receptor distinct from the CCK-gastrin receptor, since dbcGMP blocks all the actions of CCK-gastrin peptides without having any effect on the bombesin responses (Philpott and Petersen, 1979b).

## 5. EFFECTS OF AMINO ACIDS

The exocrine pancreas seems to be the most active amino acid-accumulating tissue (Schulz and Ullrich, 1979). Sodium-gradient-driven amino acid transport is regarded as an important mechanism for amino acid accumulation (Tyrakowski *et al.*, 1978). Various amino acids—e.g., L-alanine, L-valine, L-proline, L-leucine, and L-serine, but not their D isomers—evoke dose-dependent membrane depolarization. For L-alanine, small effects (3 mV) are seen at a concentration of 0.1 m$M$ and maximal effects (18 mV) at 10 m$M$. The maximal depolarization is of the same magnitude as that evoked by ACh and the peptide secretagogues; however, the accompanying resistance reduction is far smaller.

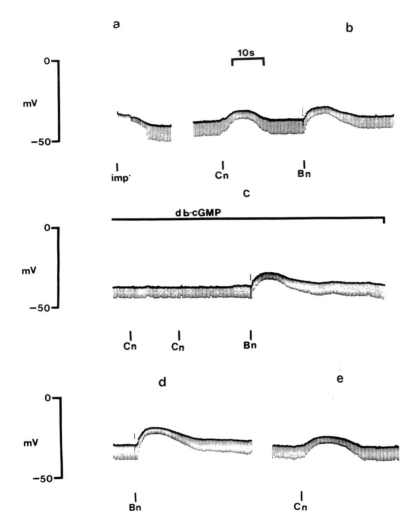

FIG. 4.   Effect of caerulein (Cn) and bombesin nonapeptide (Bn) on membrane potential and resistance in a pancreatic acinus. The figure shows that 1 m$M$ dibutyryl cyclic guanosine 3′, 5′-monophosphate (dBcGMP) blocks reversibly the action of caerulein without interfering with the response to bombesin. The consecutive traces shown are all excerpts from one and the same continuous recording. Intervals between traces, 5–23 minutes. Bombesin and caerulein were applied from two separate extracellular micropipettes by microionophoresis. From Philpott and Petersen (1979b).

Amino acids never evoke electrical uncoupling. Amino acids delivered close to the outside of the acinus evoke electrical effects with a much shorter delay ($<50$ msec) than is the case for ACh or the peptides (about several hundred milliseconds) (Iwatsuki and Petersen, 1980).

## 6. MECHANISM OF STIMULANT ACTION

A useful way of obtaining information on the ionic mechanism underlying stimulant-evoked membrane potential changes is to measure the null or equilibrium potential. The null potential is the value of the membrane potential at which an agonist does not evoke a change in potential. With two intracellular electrodes, one can be used to pass direct current in either direction, thus making it possible to set the resting membrane potential at appropriate values. The action of a certain dose of a stimulant can therefore be tried in the same acinus over a wide range of membrane potentials. A family of curves is obtained from one acinus, and it is easy to record the resting potential at which reversal of the response occurs (Iwatsuki and Petersen, 1977b; Petersen and Philpott, 1979). Table II lists null potential values obtained with various kinds of stimulation. It is seen that ACh, CCK-like, and bombesin-like peptides have very similar actions (Petersen and Philpott, 1979) and that their effects can be mimicked by intracellular Ca application (Iwatsuki and Petersen, 1977c; Petersen and Iwatsuki, 1978). By comparison, amino acids have a very different type of action, with a null potential of $+40$ mV (corresponding to the Na equilibrium potential). In the case of the secretagogue action a detailed ionic analysis has been carried out showing that the membrane channels opened are mainly permeable to Na, Cl, and to a lesser degree K (Iwatsuki and Petersen, 1977b). In contrast, the ion channels opened during the presence of transported amino acids appear to be selectively permeable to Na. While the secretagogue action appears to be mediated by an increase in cytosol-ionized $Ca^{2+}$ concentration ($[Ca^{2+}]_i$) (Iwatsuki and Petersen, 1977c; Petersen and Iwatsuki, 1978; Laugier and Petersen, 1980a), there is no indication that this applies to the amino acid effects (Laugier and Petersen, 1980b). The source of the Ca involved in the stimulus–permeability coupling was investigated recently by Laugier and Petersen (1980a). The initial secretagogue-evoked membrane conductance increase is independent of external Ca. However, during sustained stimulation, removal of external Ca immediately blocks the stimulant-evoked increase in membrane conductance. Readmission of external Ca still during sustained stimulation restores the opening of the membrane channels. The initial membrane event, therefore, seems to be triggered by Ca derived from an internal pool (possibly the inside of the plasma membrane), whereas during the sustained phase Ca enters from the interstitial fluid.

TABLE II

MECHANISMS OF ACTION OF PANCREATIC STIMULANTS

| Stimulant | Null or equilibrium potential | Na-dependent depolarization | Ca-dependent depolarization (sustained stimulus) | Ion selectivity of channels opened by stimulation | Channel opening mediated by intracellular Ca |
|---|---|---|---|---|---|
| ACh | −10 to −20 | Yes | Yes | Na, Cl, K | Yes |
| CCK (caerulein, gastrin) | −10 to −20 | Yes | Yes | Na, Cl, K | Yes |
| Bombesin (bombesin nonepeptide) | −10 to −20 | Yes | Yes | Na, Cl, K | Yes |
| Intracellular Ca application | −16 | Yes | | Na, Cl, K | |
| Amino acids (L-alanine, L-valine, L-proline, L-serine) | +40 | Yes | No | Na | No |

# B.  Exorbital Lacrimal Gland

The lacrimal gland secretes fluid (isotonic Na-rich solution) (Thaysen and Thorn, 1954; Alexander *et al.*, 1972) and protein (peroxidase) (Herzog *et al.*, 1976). In the mouse and rat lacrimal glands the secretory process appears to be controlled by cholinergic and $\alpha$-adrenergic receptors (Iwatsuki and Petersen, 1978b; Parod and Putney, 1978).

Acetylcholine and adrenaline evoke a membrane hyperpolarization and resistance reduction. The effects are biphasic, with an initial hyperpolarization accompanied by a marked decrease in membrane resistance and a secondary hyperpolarization occurring at a time when the resistance is returning to the prestimulation level. The initial phase reverses at membrane potentials between −50 and −60 mV. An ionic analysis of the secretagogue-evoked membrane effects indicates that it is mainly due to opening of channels permeable to K and Na. The secondary hyperpolarization is probably due mostly to activation of an electrogenic sodium pump (Petersen, 1973; Pedersen and Petersen, 1973). Prolonged stimulation evokes electrical uncoupling as in the pancreas (Iwatsuki and Petersen, 1978d). The similarity of the action of ACh and $\alpha$-adrenergic stimulants is probably due to the membrane effects being mediated by an increase in $[Ca^{2+}]_i$, since intracellular Ca application evokes membrane hyperpolarization and resistance reduction, and this potential change reverses at about $-50$ mV (Iwatsuki and Petersen, 1978e).

## C. Salivary Glands

Salivary glands secrete fluid and a variety of proteins. The secretory processes are mainly under the control of parasympathetic and sympathetic nerves. The transmitters ACh and norepinephrine interact with cholinergic and $\alpha$- and $\beta$-adrenergic receptors on the acinar cells (Young and Van Lennep, 1979).

The action of ACh and $\alpha$-adrenergic agonists are alike and very similar to those described above for the lacrimal gland. An initial potential change is accompanied by a marked resistance reduction. The reversal potential for the initial potential change is about $-60$ mV. A secondary potential change (hyperpolarization) not accompanied by a resistance change cannot be reversed (Roberts and Petersen, 1978; Roberts et al., 1978). In contrast to the result of cholinergic and $\alpha$-adrenergic receptor activation, $\beta$-adrenoceptor excitation evokes a relatively small depolarization not accompanied by any noticeable resistance change (Roberts and Petersen, 1978; Iwatsuki and Petersen, 1981). It has recently been shown that substance P evokes membrane changes indistinguishable from those evoked by cholinergic and $\alpha$- adrenergic agents. Substance P acts also in the presence of concentrations of atropine and phentolamine, providing complete cholinergic and $\alpha$-adrenergic blockade (Gallacher and Petersen, 1980).

The detailed analysis of the mechanism underlying the response to ACh has shown that the main event is an increase in K and Na permeability, probably triggered by an increase in $[Ca^{2+}]_i$; this is followed by activation of an electrogenic Na,K pump (Petersen, 1970; Nishiyama and Petersen, 1974b; Roberts et al., 1978).

## VII.  Role of Secretagogue-Evoked Membrane Permeability Changes in Secretion

All secretagogues acting to increase $[Ca^{2+}]_i$ also cause marked membrane resistance changes, whereas those secretagogues acting to increase the intracellular cyclic adenosine 3′,5′-monophosphate (cAMP) concentration have little, if any, effect on the plasma membrane resistance (Petersen and Iwatsuki, 1978,

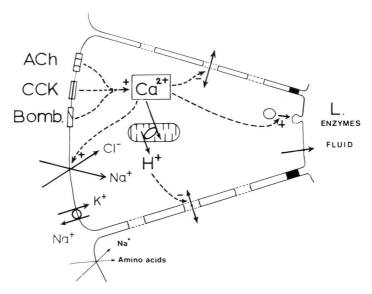

Fig. 5. Schematic diagram showing intracellular events after excitation of ACh, CCK, and bombesin receptors. Some of the more important membrane transport pathways are shown.

1979; Iwatsuki and Petersen, 1981). The secretagogue-evoked membrane resistance change may therefore be regarded as a marker event for an increase in $[Ca^{2+}]_i$. The secretagogues that cause an increase in $[Ca^{2+}]_i$ also cause a very marked fluid secretion (Petersen, 1976a; Petersen and Iwatsuki, 1978). It is characteristic of the acinar fluid secretions that they have a high (plasma-like) sodium concentration (Young and Van Lennep, 1979; Sewell and Young, 1975; Alexander *et al.*, 1972). Sodium transport must somehow be involved in the generation of acinar fluid secretion (Petersen, 1971), although the detailed events are still unknown. Figure 5 is a simplified scheme showing the role of intracellular Ca in linking fluid and enzyme secretion in the pancreatic acini. Calcium directly promotes exocytosis, and by opening up membrane channels permeable to Na and Cl it provides the secretory machinery with the ions needed in the primary acinar fluid. Secretagogues acting via a Ca effect also have the ability to cause uncoupling of cells that are normally in communication with each other. These effects may also be mediated by intracellular pH changes (Iwatsuki and Petersen, 1979a; Petersen and Iwatsuki, 1979). The role of the uncoupling event during stimulation is not clear. In the case of the pancreas, one may speculate that it serves functionally to separate acinar and duct cells (having very different stimulus–secretion coupling mechanisms) during maximal secretion (Petersen and Iwatsuki, 1979).

## ACKNOWLEDGMENTS

We thank the Medical Research Council for great support. H.G.P. is an MRC Scholar, R.L. an MRC-INSERM Training Research Fellow, and D.V.G. an MRC Training Research Fellow.

## REFERENCES

Alexander, J. H., Van Lennep, E. W., and Young, J. A. (1972). *Pfluegers Arch.* **337**, 299–309.
Bolender, R. P. (1974). *J. Cell Biol.* **61**, 269–287.
Cole, K. S. (1928). *J. Gen. Physiol.* **12**, 29–36.
Davison, J. S., Pearson, G. T., and Petersen, O. H. (1980). *J. Physiol. (London)* **301**, 295–305.
Fricke, H. (1923). *Phys. Rev.* **21**, 708–709.
Gallacher, D. V., and Petersen, O. H. (1980). *Nature (London)* **283**, 393–395.
Hammer, M. G., and Sheridan, J. D. (1978). *J. Physiol. (London)* **275**, 495–505.
Herzog, V., Sies, H., and Miller, F. (1976). *J. Cell Biol.* **70**, 692–706.
Iwatsuki, N., and Petersen, O. H. (1977a). *J. Physiol. (London)* **269**, 723–733.
Iwatsuki, N., and Petersen, O. H. (1977b). *J. Physiol. (London)* **269**, 735–751.
Iwatsuki, N., and Petersen, O. H. (1977c). *Nature (London)* **268**, 147–149.
Iwatsuki, N., and Petersen, O. H. (1978a). *J. Physiol. (London)* **274**, 81–96.
Iwatsuki, N., and Petersen, O. H. (1978b). *J. Physiol. (London)* **275**, 507–520.
Iwatsuki, N., and Petersen, O. H. (1978c). *J. Clin. Invest.* **61**, 41–46.
Iwatsuki, N., and Petersen, O. H. (1978d). *J. Cell Biol.* **79**, 533–545.
Iwatsuki, N., and Petersen, O. H. (1978e). *Pfluegers Arch.* **377**, 185–187.
Iwatsuki, N., and Petersen, O. H. (1979a). *J. Physiol. (London)* **291**, 317–326.
Iwatsuki, N., and Petersen, O. H. (1979b). *Pfluegers Arch.* **380**, 277–281.
Iwatsuki, N., and Petersen, O. H. (1980). *Nature (London)* **283**, 492–494.
Iwatsuki, N., and Petersen, O. H. (1981). *J. Physiol. (London)* **314**, 79–84.
Kater, S. B., and Galvin, N. J. (1978). *J. Cell Biol.* **79**, 20–26.
Laugier, R., and Petersen, O. H. (1980a). *J. Physiol. (London)* **303**, 61–72.
Laugier, R., and Petersen, O. H. (1980b). *Pfluegers Arch.* **386**, 147–152.
Matthews, E. K., Petersen, O. H., and Williams, J. A. (1973). *J. Physiol. (London)* **234**, 689–701.
Nishiyama, A., and Petersen, O. H. (1974a). *J. Physiol. (London)* **238**, 145–158.
Nishiyama, A., and Petersen, O. H. (1974b). *J. Physiol. (London)* **242**, 173–188.
Nishiyama, A., and Petersen, O. H. (1975). *J. Physiol. (London)* **244**, 431–465.
Nishiyama, A., Kato, K., Saitoh, S., and Wakui, M. (1980). *Membr. Biochem.* **3**, 49–66.
Parod, R. J., and Putney, J. W. (1978). *J. Physiol. (London)* **281**, 359–369.
Pedersen, G. L., and Petersen, O. H. (1973). *J. Physiol. (London)* **234**, 217–227.
Peikin, S. R., Costenbader, C. L., and Gardner, J. D. (1979). *J. Biol. Chem.* **254**, 5321–5327.
Petersen, O. H. (1970). *J. Physiol. (London)* **210**, 205–215.
Petersen, O. H. (1971). *Philos. Trans. R. Soc. London, Ser. B* **262**, 307–314.
Petersen, O. H. (1973). *Proc. R. Soc. London, Ser. B* **184**, 115–119.
Petersen, O. H. (1976a). *Physiol. Rev.* **56**, 535–577.
Petersen, O. H. (1976b). *In* "Stimulus-Secretion Coupling in the Gastrointestinal Tract" (R. M. Case and H. Goebell, eds.), pp. 281–293. University Park Press, Baltimore, Maryland.
Petersen, O. H. (1980). "Electrophysiology of Gland Cells." Academic Press, New York.
Petersen, O. H., and Iwatsuki, N. (1978). *Ann. N.Y. Acad. Sci.* **307**, 599–617.
Petersen, O. H., and Iwatsuki, N. (1979). *In* "Hormone Receptors in Digestion and Nutrition" (G. Rosselin, P. Fromegeot, and S. Bonfils, eds.), pp. 191–202. Elsevier/North-Holland Publ., Amsterdam and New York.

# I.  Introduction

The morphological and functional aspects of the exocrine secretory process have been extensively studied in the pancreas (Jamieson and Palade, 1967a,b, 1968a,b, 1971; Caro and Palade, 1964; Warshawsky *et al.*, 1963), salivary glands (Amsterdam *et al.*, 1969, 1971; Hand, 1970, 1971, 1972; Cutler and Chaudhry, 1973a; Kim *et al.*, 1972; Simson, 1969; Schramm, 1967), and several other exocrine systems (Oron and Bdolah, 1973; Vidic, 1973; Scott and Pease, 1959). In all these systems the basic morphology of the exocrine cells is essentially identical. All these secretory cells demonstrate the typical functional, apical to basal polarity of organelle distribution.

While the exocrine cells of the rodent pancreas, parotid, and submandibular glands utilize the same morphological mechanisms for the synthesis, accumulation, and discharge of their secretory product, there are some differences in the sequence of developmental events that lead to morphologically differentiated secretory cells in each system. There are differences in the timing of the onset of secretory cell differentiation as well as differences in the rate of cell maturation (Pictet *et al.*, 1972; Rutter *et al.*, 1964; Redman and Sreebny, 1971; Cutler and Chaudhry, 1974). However, the general pattern of progressive accumulation of granular endoplasmic reticulum followed by the maturation of the Golgi apparatus and then the appearance of distinct zymogen granules seems to be a consistent observation in the development of most exocrine cells (Cutler and Chaudhry, 1974).

Having developed the capability to synthesize and package proteins for export does not a priori indicate that the packaged material can be or is released by the conventional mechanisms regulating exocytosis in mature exocrine cells. The release of zymogen granules from mature secretory cells is a complex process, initiated by the interaction of specific hormonal or neurohormonal agonists with specific receptors at the cell surface. This interaction of agonist and receptor initiates a cascade of intracellular events, which leads to the fusion of secretory granules with the cell surface and the release of the granule contents into the acinar lumen. This linkage between the action of a hormonal agonist at the cell surface with the exocytotic process is referred to as stimulus–secretion coupling. There is evidence that in the developing rat pancreas and salivary glands the secretory cells develop the capability to synthesize and package secretory proteins prior to the time that these cells are capable of exocytosis in response to hormonal stimuli (Doyle and Jamieson, 1978; Grand *et al.*, 1975; Grand and Shay, 1978; Cutler, 1977a, 1978).

The present report provides evidence that the secretory cells of the developing rat submandibular gland (SMG) acquire the ability to synthesize and package secretory proteins prior to attaining the ability to release the packaged product in response to hormonal stimuli. This is consistent with previous observations in the pancreas and parotid gland. The report correlates the development of the secretory response with studies on hormonal activation of cell surface-associated adenylate cyclase and with direct measurements of $\beta$-adrenergic and $\alpha$-adrenergic binding sites. The report thus provides a picture of the development of the stimulus–secretion coupling system in this model exocrine system.

Finally, this report presents electrophysiological and morphological evidence indicating that, in this system, development of the stimulus–secretion coupling mechanism precedes the establishment of the neural connections that regulate secretion *in vivo*.

## II. Methods

### A. Animals

Female Sprague-Dawley rats were singly caged in an environmentally controlled room and given rat chow and water ad libitum. At 8:00 P.M. on the appropriate day of the estrus cycle, the females were placed in breeding cages with males and left until 8:00 A.M. of the next morning. Vaginal smears were taken and examined for sperm as an evidence of copulation. The zygotes were considered zero hours old at 8:00 A.M. on the day sperm was found (Cutler and Chaudhry, 1973a,b,c, 1974, 1975, Cutler and Rodan, 1976; Cutler, 1977a,b; Mooradian and Cutler, 1978). This procedure for calculating gestational age has recently been used as the standard method for the estimation of gestational age in the study of salivary gland development (Young and van Lennep, 1978). Animals used in postnatal studies were obtained from our breeding colony in order to regulate the conditions of conception and gestation.

### B. Secretion System

To measure the secretory response of perinatal glands a slice system similar to that of Bogart and Picarelli (1978) was used. The incubation medium was supplemented Krebs-Ringer bicarbonate (KRB) composed of 109 m$M$ NaCl, 13.8 m$M$ KCl, 2.5 m$M$ CaCl$_2$, 1.2 m$M$ KH$_2$PO$_4$, 25 m$M$ NaHCO$_3$, 1.2 m$M$ MgSO$_4$, 5 m$M$ $\beta$-hydroxybutyric acid, 0.5 m$M$ adenine, 10 m$M$ inosine, and 5.6 m$M$ D-glucose. The medium was gassed with humidified 95% O$_2$–5% CO$_2$. Slices from at least 24 rudiments (21 days of gestation) or the appropriate number of neonatal glands (1 and 6 days of age) were isolated in complete Krebs-Ringer buffer medium and then washed in 25 ml of buffer (37°C) for 10 minutes before the incubation was initiated. The tissue was then divided into aliquots and placed in nitrocellulose test tubes (25 × 80 mm) containing 4 ml of the complete Krebs-Ringer bicarbonate medium.

In order to induce secretion, either L-isoproterenol, L-norepinephrine, or L-phenylephrine was added to duplicate tubes at a final concentration of $10^{-5}$ $M$. Fresh agonist (equivalent to that added at ''0'' time) was added to the medium after 15 minutes of incubation to compensate for oxidation of the agonist. In some experiments the effects of the $\beta$-adrenergic antagonist propranolol or the $\alpha$-adrenergic antagonist phentolamine on agonist-induced secretion were evaluated. In these experiments the respective antagonists were added to the preincubation medium and to the incubation medium at concentrations 10- to 100-fold as great as the agonist concentration. Antagonists were also added to the medium during the incubation when additional agonist was added.

In addition to these adrenergic agonists, the ability of $N^6,O^4$-dibutyryl cyclic AMP ($10^{-3}$ $M$) and 8-bromo cyclic GMP ($10^{-3}$ $M$) to induce secretion was tested in this system.

Secretion was assessed by taking aliquots of the medium 30 minutes after introduction of the potential agonist and determining the amount of secretory peroxidase released into the medium.[1] At the termination of the experiment the slices were homogenized in the remaining medium; the homogenate was centrifuged at 1000 $g$, and the resulting supernatant was assayed for peroxidase activity. The quantity of peroxidase in the homogenate plus the amount in the aliquot removed from the system during the experiment represented the total peroxidase activity in the sample at ''0'' time. The results were reported as the percentage of the total peroxidase in the sample at ''0'' time released into the medium. This release was compared to equivalent control experiments in which no agonists were added to the secretion medium.

## C.  Peroxidase Assay

Aliquots of secretion medium were assayed by adding 0.1 ml of sample to 2.85 ml of 0.1 $M$ sodium phosphate buffer (pH 7.0) containing 5 × $10^{-4}$ $M$ diaminobenzidine (DAB). The reaction was started by addition of 0.05 ml of 0.6% $H_2O_2$, and the change in absorbance was monitored at 460 nm on a Gilford 250 spectrophotometer equipped with a chart recorder. The measurements were carried out aginst a DAB–phosphate buffer blank (Herzog and Fahimi, 1973). Peroxidase activity was linear for 60–90 seconds in this system, and in our hands the assay was sensitive to less than 0.1 ng of horseradish peroxidase (HRP) (Sigma Type VI) (Cutler *et al.*, 1977). The concentration of peroxidase in the media was calculated by comparing the $\Delta A$/min of the sample to standard values

---

[1] In the prenatal and early postnatal SMG, secretory peroxidase is present in the proacinar cells (Yamashina and Barka, 1972).

established the same day against freshly prepared known concentrations of HRP (Sigma Type VI). Standards were run in DAB–gelatin medium according to the procedure of Herzog and Fahimi (1973).

## D.   Preparation of Enriched Plasma Membrane Fraction

Glands or rudiments were minced and washed in cold 0.1 $M$ phosphate buffer containing 0.25 $M$ sucrose, 1 m$M$ EDTA, and 1 m$M$ dithiothreitol (DTT). The minced tissue was homogenized in a Teflon–glass homogenizer for 30 seconds in cold wash buffer (4°C), and the homogenate was centrifuged at 1000 $g$ for 10 minutes in the cold to remove unbroken cells, nuclei, and other debris. The supernatant was centrifuged at 30,000 $g$ for 15 minutes at 4°C on a Beckman J21-C centrifuge, and the resultant pellet was suspended in cold 10 m$M$ Tris (pH 7.6) containing 1 m$M$ DTT. This suspension was placed on a sucrose gradient with steps of 38% and 42%. The gradient was centrifuged at 100,000 $g$ for 60 minutes on a Beckman L5-50 ultracentrifuge. The material that layered at the suspension–38% sucrose interface was harvested, diluted in 10 m$M$ Tris (pH 7.6)–1 m$M$ DTT buffer, and then concentrated by centrifugation at 30,000 $g$ for 15 minutes. The resulting pellet has been assayed for structure, adenylate cyclase activity, 5′-nucleotidase activity, and succinic dehydrogenase activity, and the results were compared with similar assays performed on the 1000 $g$ pellet and the 30,000 $g$ pellet prior to the sucrose gradient step. The material recovered from the gradient contained predominantly smooth-surfaced membrane vesicles with no mitochondria when examined by electron microscopy, showed a 5- to 10-fold increase in the specific activity of adenylate cyclase, an 18- to 20-fold increase in the specific activity of 5′-nucleotidase, but no measurable succinic dehydrogenase activity.

## E.   Protein Determination

All protein determinations were made according to the procedure of Lowry et al. (1951).

## F.   Measurement of Adrenergic Binding Sites

To measure $\beta$-adrenergic binding sites, assays were initiated by adding 100 $\mu$g of membrane protein to an incubation medium of 10 m$M$ Tris (pH 7.6) containing 1 m$M$ DTT and L-[$^3$H]dihydroalprenolol (New England Nuclear, Boston, MA) (1–25 nm) with a final volume of 500 $\mu$l (control experiments showed that 1 m$M$ DTT did not effect [$^3$H]dihydroalprenolol binding). All samples were run in duplicate. The reaction was run for 5 minutes at 37°C and was terminated by adding 6 ml of cold (4°C) 0.85% saline containing $10^{-5}$ $M$ dl-propranolol and immediate filtration of the sample through a Whatman GB/F filter. The filter was then washed with an additional 15 ml of saline. The filters were placed in scintillation fluid and counted in an Isocap/300 liquid scintillation counter. Duplicate tubes containing $10^{-5}$ *M* *dl*-propranolol were incubated, treated as above, and run for each sample to determine nonspecific binding. Only those counts that could be displaced by the *dl*-propranolol were considered specific (Mukherjee *et al.*, 1975a,b). In previous studies, it was found that L-[$^3$H]dihydroalprenolol binding to adult SMG membranes was (1) saturable at about 8 n*M* of L-[$^3$H]dihydroalprenolol, (2) rapid (saturation was reached in less than 5 minutes), (3) reversible by addition of unlabeled L-alprenolol, *dl*-propranolol, or L-isoproterenol, and (4) linear for concentrations of 50–400 μg of membrane protein.

To measure α-adrenergic binding sites, assays were initiated by adding 50 μg of membrane protein to an incubation medium of 10 m*M* Tris (pH 7.6) buffer containing 1 m*M* DTT and [$^3$H]dihydroergocryptine (Williams and Lefkowitz, 1976; Strittmatter *et al.*, 1977) (New England Nuclear, Boston MA) (1–40 n*M*) with a final volume of 500 μl. All samples were run in duplicate. The reaction was run for 10 minutes at 37°C and was terminated by addition of 6 ml of cold saline (4°C) containing $10^{-5}$ *M* phenotolamine and immediate filtration of the sample through a Whatman GB/F filter. The filter was washed with an additional 30 ml of saline, placed in liquid scintillation fluid, and counted as above. Duplicate tubes containing $10^{-5}$ *M* phenotolamine were incubated and treated as above and run for each sample to determine nonspecific binding. Only those counts that were displaced by phentolamine were considered specific. In preliminary studies, we have found that [$^3$H]dihydroergocryptine binding to young adult SMG membranes to be (1) saturable at about 25 n*M* of [$^3$H]dihydroergocryptine, (2) relatively rapid (saturation was reached in about 10 minutes at 37°C), (3) reversible by addition of phentolamine or norepinephrine, and (4) linear for concentrations of 50–400 μg of membrane protein. These data are consistent with studies of [$^3$H]dihydroergocryptine binding to membranes from rat parotid gland (Strittmatter *et al.*, 1977).

The number of binding sites and the $K_d$ were determined by Scatchard analysis and by a numerical curve-fitting procedure (Hooke and Jeeves, 1961). This procedure is based on a directed iterative search, which for a given expression, $y = f(x)$, adjusts the values of the fixed parameters to minimize $\Sigma(Y_e - Y_c)^2$, where $Y_e$ is the experimental and $Y_c$ is the calculated value of the function.

## G.  Adenylate Cyclase Determination

An aliquot of the enriched membrane fraction (10 μg of protein) was incubated for 10 minutes in 100 μl of assay mixture containing 25 m*M* Tris–HCl (pH 7.6), 5 m*M* MgCl$_2$, 1 m*M* cAMP, 1 m*M* DTT, 10 m*M* phosphocreatine, 50 units phosphocreatine kinase, $2 \times 10^{-6}$ $M$ GTP, and 0.1 m$M$ ATP ($3 \times 10^6$ cpm $\alpha$-[$^{32}$P]ATP). The incubation was stopped by adding 100 $\mu$l of stopping solution (4 m$M$ ATP, 1.4 m$M$ cAMP, and 2% sodium dodecyl sulfate) and then 20,000 cpm of [$^3$H]cAMP for estimation of recovery from chromatography on Dowex Ag 50WX4 and neutral alumina columns (Salomon *et al.*, 1974; Cutler and Rodan, 1976). Duplicate samples were counted for 10 minutes in 10 ml of Bray's solution (Bray, 1960) in an Isocap/300 liquid scintillation counter set with separate channels for $^3$H and $^{32}$P. The results were calculated as picomoles of cAMP produced per milligram of protein per minute.

In initial experiments membrane-associated adenylate cyclase activity from prenatal glands (21 days *in utero*) was tested for response to $10^{-5}$ $M$ guanylylimidodiphosphate (intactness of guanyl nucleotide regulatory site) and 10 m$M$ NaF (intactness of fluoride activation site). In subsequent studies the response of adenylate cyclase from glands of different ages (18 and 21 days *in utero* and 1, 4, 6, and 120 days of age) to a saturating concentration ($10^{-5}$ $M$) and L-isoproterenol was determined.

## H. Electrophysiology of Neonatal Secretion

Neonatal rats (1, 3, 5, 7, 9, and 11 days old) were anesthetized (Ketamine HCl) and placed in supine position. The neck region was exposed and the left superior cervical ganglion (sympathetic) located. Salivation was induced by direct stimulation of the superior cervical ganglion by 10 Hz in frequency and 4–8 volts in intensity for 20 minutes using a Grass square-wave stimulator. The efficacy of this procedure was confirmed by similar stimulation of the submandibular ganglion (parasympathetic) and the observation of a clear, watery saliva from the SMG duct.

The effect of this stimulation on protein secretion was determined by homogenizing the stimulated gland (glands from 3–8 animals were pooled for each assay), centrifuging the homogenate at 1000 $g$ for 10 minutes, and assaying the 1000 $g$ supernatant for peroxidase activity as described earlier. The amount of peroxidase secretion was inferred by comparing the amount of peroxidase remaining in the stimulated gland after 20 minutes with that found in the contralateral unstimulated gland. The data are reported as the intraglandular peroxidase found as a percentage of the control gland where the control has been normalized to 100%.

## I. Catecholamine-Containing Nerves in the Developing SMG

The presence of catecholamine (norepinephrine)-containing nerves in the parenchyma of the developing SMG was assessed by the method of de la Torre and Surgeon (1976), in which cryostat sections are exposed to a solution of sucrose–potassium phosphate–glyoxylic acid.

## III.   Results

## A.   Adenylate Cyclase in SMG Development

Adenylate cyclase activity is very low in the earliest (15-day) SMG rudiment and progressively rises to adult levels by day 18 of gestation (Table I). It is at 18 days of gestation that the first secretory granules are seen. The adenylate cyclase activity found in the membrane fraction derived from the 21-day embryonic SMG is intact with regard to its guanyl nucleotide and fluoride regulatory sites. A subsaturating concentration of guanylylimididophosphate ($10^{-5}$ $M$) causes a 13- to 15-fold stimulation over basal adenylate cyclase activity, while 10 m$M$ NaF induces an 18- to 20-fold stimulation over basal activity (Table II).

There appears to be an age-related incremental increase in the responsiveness of SMG adenylate cyclase activity to a saturating concentration of isoproterenol (Table III). Adenylate cyclase activity from SMG rudiments 18 and 21 days *in utero* were not responsive to isoproterenol, and enzyme activity from glands 1 and 4 days of age responded only minimally. Membrane-associated adenylate cyclase activity from the SMG of 1- and 4-day animals showed a reproducible

TABLE I

SUBMANDIBULAR GLAND ADENYLATE CYCLASE
ACTIVITY[a]

| Age of animal | Activity (pmol cAMP/mg protein/15 min) |
|---|---|
| 15 days *in utero* | 3.7 ± 0.9 |
| 16 days *in utero* | 16.3 ± 2.1 |
| 17 days *in utero* | 50.1 ± 6.2 |
| 18 days *in utero* | 65.5 ± 5.8 |
| 21 days *in utero* | 67.4 ± 3.9 |
| Adult | 67.9 ± 4.2 |

[a] Adenylate cyclase activity was determined on a crude particulate fraction obtained by homogenization of the rudiments or glands followed by a 1000 $g$ centrifugation step. The resulting supernatant was centrifuged at 30,000 $g$ for 15 minutes, and the resulting pellet was resuspended and assayed for adenylate cyclase activity by using the procedure of Salomon *et al.* (1974).

TABLE II

ADENYLATE CYCLASE ACTIVITY, 21-DAY
EMBRYONIC SMG[a]

|  | Activity (pmol cAMP/mg protein/min) |
| --- | --- |
| Basal | 17.0 |
| $10^{-5}$ $M$ Gpp(NH)p | 234.3 |
| 10 m$M$ NaF | 317.6 |

[a] The results of a typical experiment on the effects of $10^{-5}$ $M$ Gpp(NH)p and 10 m$M$ NaF on membrane-associated adenylate cyclase activity from 21-day embryonic SMG. The experiment was performed to evaluate the ability of guanyl nucleotides and fluoride to activate the enzyme. The particulate fraction used in this study was an enriched plasma membrane fraction derived from differential centrifugation and sucrose gradient procedures (see Methods).

25–40% stimulation in response to $10^{-5}$ $M$ isoproterenol. On the other hand, membranes from glands at 6 days of age showed full activation (2.5- to 3.5-fold) of adenylate cyclase by this dose of isoproterenol, and this was similar to the activation seen in adult membranes.

TABLE III

HORMONAL ACTIVATION OF SMG ADENYLATE CYCLASE AS A
FUNCTION OF DEVELOPMENTAL AGE

|  | Activity (pmol cAMP/mg protein/min) | |
| --- | --- | --- |
| Age of animal | Basal | $10^{-5}$ $M$ L-Isoproterenol |
| 18 days *in utero* | 15.3 ± 1.1 | 15.6 ± 1.8 |
| 21 days *in utero* | 15.7 ± 2.0 | 15.1 ± 1.5 |
| 1 day | 15.2 ± 0.8 | 18.7 ± 1.4 ($p < 0.05$) |
| 4 days | 16.5 ± 1.3 | 20.9 ± 1.9 ($p < 0.05$) |
| 6 days | 15.8 ± 2.5 | 42.8 ± 3.7 |
| Adult | 14.9 ± 1.3 | 44.7 ± 5.1 |

[a] The ability of a saturating concentration of L-isoproterenol to activate membrane-associated adenylate cyclase from the SMG as a function of age was tested. Data are reported as the mean of five experiments ± the S.E.M.

TABLE IV

DEVELOPMENTAL RESPONSE TO SECRETOGOGUES[a]

| | Prenatal, 21 days *in utero* | Postnatal | |
|---|---|---|---|
| | | 1 day | 6 days |
| Basal | 5.4 | 2.8 | 3.3 |
| $10^{-5}$ *M* L-Isoproterenol | 5.8 | 14.3 | 15.8 |
| $10^{-5}$ *M* L-Phenylephrine | 4.9 | 15.5 | 7.1 |
| $10^{-4}$ *M* Dibutyryl cAMP | 17.8 | 15.1 | 16.6 |
| $10^{-4}$ *M* 8-Bromo-cGMP | — | 3.8 | — |
| $10^{-5}$ *M* L-Isoproterenol + $10^{-4}$ *M* DL-propranolol | — | 8.3 | — |
| $10^{-5}$ *M* L-Phenylephrine + $10^{-4}$ *M* DL-propranolol | — | 12.0 | — |
| $10^{-5}$ *M* L-Isoproterenol + $10^{-4}$ *M* phentolamine | — | 6.8 | — |
| $10^{-5}$ *M* L-Phenylephrine + $10^{-4}$ *M* L-phentolamine | — | 4.7 | — |

[a] Results of a typical experiment to evaluate the secretory response of SMG slices to various secretogogues and inhibitors. Data are reported as the percentage of the total peroxidase present in the slices at the start of the experiment released into the medium after 30 minutes of stimulation.

## B.    Secretion of Peroxidase

The results of a typical experiment on the stimulation of secretion from SMG slices from embryos and postnatal rats shown in Table IV. Neither $\alpha$(phenylephrine)- nor $\beta$(isoproterenol)-adrenergic agonists induced secretion from slices from prenatal glands. Dibutyryl cAMP ($10^{-3}$ *M*) induced secretion from these slices, indicating that the mechanistic properties required for secretion were present and working in these cells. By 1 day after birth the slices secreted equally well in response to either $\alpha$- or $\beta$-adrenergic agonists. Inhibition of this secretory response by supposedly specific $\alpha$- or $\beta$-adrenergic antagonists was ambiguous, since the $\alpha$-adrenergic antagonist phentolamine was more effective at blocking isoproterenol ($\beta$-adrenergic)-induced secretion than was the $\beta$-adrenergic antagonist propranolol. Dibutyryl cAMP induced secretion from these slices, but 8-bromo-cGMP had no effect on peroxidase release. By 6 days after birth the secretory response to $\alpha$-adrenergic and $\beta$-adrenergic agonists was similar to that seen in adult glands (Bogart and Picarelli, 1978), with isoproterenol inducing substantially greater peroxidase release than phenylephrine.

## C.    [³H]Dihydroalprenolol and [³H]Dihydroergocryptine Binding

The number of $\beta$-adrenergic binding sites was estimated by the binding of $\beta$-adrenergic antagonist [³H]dihydroalprenolol to partially purified plasma membranes from the SMG of animals of various ages. At all ages saturation of binding sites was rapid (saturation reached in about 5 minutes) and saturable at

TABLE V

BINDING OF [$^3$H]DIHYDROALPRENOLOL TO SMG MEMBRANES[a]

| Age of animal | [$^3$H]DHA bound (fmol/mg protein) | $K_d$ |
|---|---|---|
| 1 day | 67 ± 7 | 1.15 × 10$^{-9}$ M |
| 4 days | 67 ± 9 | 1.11 × 10$^{-9}$ M |
| 6 days | 273 ± 21 | 1.42 × 10$^{-9}$ M |
| 120 days | 417 ± 34 | 2.80 × 10$^{-9}$ M |

[a] The amount of [$^3$H]DHA bound and the $K_d$ were determined by Scatchard analysis and by a numerical curve-fitting procedure based on directed iterative search, which for the given expression $y = f(x)$ adjusts the values of the fixed parameters to minimize $\Sigma(Y_e - Y_c)^2$, where $Y_e$ is the experimental and $Y_c$ is the calculated value of the function. Data presented are the mean of five experiments ± the S.E.M.

about 10 m$M$ of the antagonist. Scatchard analysis and computerized curve fitting of the binding data revealed little variation in the $K_d$ for the various ages tested, but a 7-fold increase was found in the number of binding sites from birth to adulthood (Table V). A 4-fold increase in the number of binding sites was seen between 4 and 6 days after birth. This increase in the number of binding sites coincided with the increased responsiveness of SMG adenylate cyclase to isoproterenol stimulation and to the appearance of adult-type stimulus–secretion coupling in SMG slices.

The preliminary studies on $\alpha$-adrenergic binding sites reported here indicated that at birth there were about 1800 fmol of dihydroergocryptine bound per milligram of SMG membrane protein with a $K_d$ of about 3.7 × 10$^{-9}$ $M$. Adult SMG membranes bound 1263 fmol of dihydroergocryptine per milligram of protein with a $K_d$ of about 1.6 × 10$^{-9}$ $M$. Thus, there were roughly the same number or somewhat more $\alpha$-adrenergic receptors present on SMG membranes at birth than there are in adult SMG membranes. Therefore, the development of $\alpha$- and $\beta$-adrenergic receptors in this gland is apparently independently regulated.

## D. Electrophysiological Stimulation of Secretion

The results of electrophysiological studies on secretion by developing glands are shown in Table VI. Direct electrical stimulation of the superior cervical ganglion did not result in secretion by the SMG in animals 1 and 3 days old. During the period from 5 to 11 days of age there was a progressive increase in secretion elicited by electrical stimulation of the superior cervical ganglion. The adrenergic agonist norepinephrine was able to induce secretion at all times tested.

TABLE VI

Peroxidase Secretion from the SMG Following Direct Electrical
Stimulation of the Superior Cervical Ganglion[a]

| Age of animal | Percentage of total intraglandular peroxidase remaining after 20 minutes stimulation | | |
| --- | --- | --- | --- |
| | Experimental | Norepinephrine | Unstimulated control |
| 1 day | 100 | 35 | 100 |
| 3 days | 100 | 42 | 100 |
| 5 days | 85 | — | 100 |
| 7 days | 75 | 45 | 100 |
| 9 days | 61 | — | 100 |
| 11 days | 40 | — | 100 |

[a] Results of a typical experiment are shown. Secretion of peroxidase was inferred by comparing the amount of peroxidase remaining in the stimulated gland after 20 minutes to that found in the contralateral unstimulated gland. The data are reported as the intraglandular peroxidase found as a percentage of the control gland after 20 minutes of electrical stimulation. The peroxidase content of the control gland has been normalized to 100%.

These data suggested that the cells of the developing SMG were able to respond to secretory stimuli but direct neural stimulation was not able to elicit a secretory response.

## E.    Glyoxylic Acid Staining for Catecholamine-Containing Nerves

Glyoxylic acid staining showed that each acinar unit of the adult rat SMG was surrounded by catecholamine-containing nerves. However, virtually no catecholamine fluorescent nerves were seen within the parenchyma of the 1-day or 3-day submandibular glands. The earliest evidence of catecholamine containing nerves in the parenchyma of the SMG was at 5–6 days after birth. From this point on, there was a progressive increase in the number of catecholamine-containing nerves in the parenchyma. This progressive increase in catecholamine-containing nerves correlated directly with the development of a progressively increasing secretory response secondary to electrical stimulation of the superior cervical ganglion.

## IV.    Discussion

The data presented give a comprehensive picture of the maturation of those factors that regulate protein secretion in this model exocrine system. It appears that differentiating SMG secretory cells first develop the structural and synthetic machinery required to produce and package their secretory product. Coincident with, or shortly after, the initiation of the synthesis and packaging of the exocrine product, the cells develop the physical capability to release the packaged product in a typical exocrine fashion.

With regard to the rat SMG, zymogen granule production is first seen at 18 days of gestation, and at this point in time basal adenylate cyclase activity is at the same level that is found in adult glands. The adenylate cyclase activity in the fetal gland is intact with regard to its sodium fluoride and guanyl nucleotide regulatory sites. However, the enzyme does not respond to saturating concentrations of the known agonist L-isoproterenol. *In vitro* secretion studies indicate that fetal SMG slices are able to secrete peroxidase in response to millimolar concentrations of dibutyryl cAMP but will not respond to maximal doses of L-isoproterenol. Thus, while the embryonic SMG cells have the physical capability to produce and secrete their products, the typical stimulus–secretion coupling mechanisms that are present in the adult SMG cells have not yet appeared.

The development of the mechanisms to produce and secrete exocrine proteins prior to the development of the typical stimulus–secretion coupling mechanisms seen in the mature cells does not seem to be a phenomenon unique to the rat submandibular gland. Doyle and Jamieson (1978) have observed a similar situation in the developing rat pancreas, and Grand and his co-workers (1975; Grand and Schay, 1978) have found an analogous picture in the developing rat parotid gland. In all three systems the adult-type stimulus–secretion coupling mechanisms appear to evolve over a short period of time subsequent to the development of the synthetic and release pathways. Heretofore, the exact nature of the development of the stimulus–secretion coupling mechanisms has not been known. It was (and still is) not known if absence of stimulus–secretion coupling in the neonatal or embryonic parotid and pancreas was due to the absence of specific agonist receptors or to the absence of a coupling or transducing factor that could link the activated receptor–agonist complex to the response system.

The data presented here suggest that in the developing submandibular gland the missing link in the stimulus–secretion response system is the β-adrenergic receptor. Direct accessment of receptor number and affinity indicates that adenylate cyclase response to hormone activation is directly related to an increasing number of β-adrenergic binding sites at the cell surface of the SMG cells as a function of age. Further, *in vitro* secretion studies suggest that the development of adult-type stimulus–secretion coupling patterns are also correlated with the developmental appearance of increased numbers of β-adrenergic binding sites at the cell surface. Thus, as the number of cell surface β-adrenergic receptors reaches a critical number such that adenylate cyclase activation is equivalent to that seen in adult membranes, the *in vitro* secretory response assumes a pattern similar to that seen in adult tissue.

Finally, the electrophysiological and catecholamine fluorescent data indicate that during the period (birth to 5 or 6 days of age) when the $\beta$-adrenergic receptor sites are increasing in number and the adult stimulus–secretion coupling mechanisms are evolving there are no adrenergic neural connections in the gland. It is only after the total development of the complete synthetic and secretory pathways that neural connections are made. Thus, there is a highly organized and defined sequence of structural and biochemical differentiative steps in the total evolution of the SMG exocrine cell development. This sequence appears to have several common steps in at least three different exocrine systems and thus may represent a general developmental sequence common to all exocrine systems.

## REFERENCES

Amsterdam, A. M., Ohad, I., and Schramm, M. (1969). *J. Cell Biol.* **41,** 753.

Amsterdam, A. M., Schramm, M., Ohad, I., Salomon, W., and Selinger, Z. (1971). *J. Cell Biol.* **50,** 187.

Bogart, B. I., and Picarelli, J. (1978). *Am. J. Physiol.* **235,** C256.

Bray, G. A. (1960). *Anal. Biochem.* **1,** 279.

Caro, L. G., and Palade, G. E. (1964). *J. Cell Biol.* **20,** 473.

Cutler, L. S. (1977a). *J. Cell Biol.* **75,** 21a.

Cutler, L. S. (1977b). *J. Embryol. Exp. Morphol.* **39,** 71.

Cutler, L. S. (1978). *J. Dent. Res.* **57A,** 332.

Cutler, L. S., and Chaudhry, A. P. (1973a). *Anat. Rec.* **176,** 405.

Cutler, L. S., and Chaudhry, A. P. (1973b). *Dev. Biol.* **33,** 229.

Cutler, L. S., and Chaudhry, A. P. (1973c). *J. Morphol.* **140,** 343.

Cutler, L. S., and Chaudhry, A. P. (1974). *Dev. Biol.* **41,** 31.

Cutler, L. S., and Chaudhry, A. P. (1975). *Am. J. Anat.* **143,** 201.

Cutler, L. S., and Rodan, S. B. (1976). *J. Embryol. Exp. Morphol.* **36,** 291.

Cutler, L. S., Moordian, B. A., and Christian, C. C. (1977). *J. Histochem. Cytochem.* **25,** 1207.

de la Torre, J. C., and Surgeon, J. W. (1976). *Histochemistry* **49,** 81.

Doyle, C. M., and Jamieson, J. D. (1978). *Dev. Biol.* **65,** 11.

Grand, R. J., and Schay, M. I. (1978). *Pediatr. Res.* **12,** 100.

Grand, R. J., Chong, D. A., and Ryan, S. J. (1975). *Am. J. Physiol.* **228,** 608.

Hand, A. R. (1970). *J. Cell Biol.* **44,** 340.

Hand, A. R. (1971). *Am. J. Anat.* **130,** 141.

Hand, A. R. (1972). *In* "Developmental Aspects of Oral Biology" (H. Slavkin and L. Bavetta, eds.), p. 351. Academic Press, New York.

Herzog, V., and Fahimi, D. (1973). *Anal. Biochem.* **55,** 554.

Hooke, R., and Jeeves, T. A. (1961). *J. Assoc. Comput. Mach.* **8,** 212.

Jamieson, J. D., and Palade, G. E. (1967a). *J. Cell Biol.* **34,** 577.

Jamieson, J. D., and Palade, G. E. (1967b). *J. Cell Biol.* **34,** 597.

Jamieson, J. D., and Palade, G. E. (1968a). *J. Cell Biol.* **39,** 580.

Jamieson, J. D., and Palade, G. E. (1968b). *J. Cell Biol.* **39,** 589.

Jamieson, J. D., and Palade, G. E. (1971). *J. Cell Biol.* **50,** 135.

Kim, S. K., Nasjleti, C. E., and Han, S. S. (1972). *J. Ultrastruct. Res.* **38,** 371.

Lowry, O. H., Rosenborough, N. J., Farr, A. L., and Randall, R. J. (1951). *J. Biol. Chem.* **193,** 265.

Mooradian, B. A., and Cutler, L. S. (1978). *J. Histochem. Cytochem.* **26,** 989.

Mukherjee, C., Caron, M. G., Coverstone, M., and Lefkowitz, R. J. (1975a). *J. Biol. Chem.* **250,** 4869.

Mukherjee, C., Caron, M. G., and Lefkowitz, R. J. (1975b). *Proc. Natl. Acad. Sci. U.S.A.* **72,** 1945.

Oron, N., and Bdolah, A. (1973). *J. Cell Biol.* **56,** 177.

Pictet, R. L., Clark, W. R., Williams, R. H., and Rutter, W. J. (1972). *Dev. Biol.* **29,** 348.

Redman, R. S., and Sreebny, L. M. (1971). *Dev. Biol.* **25,** 248.

Rutter, W. J., Wessells, N. K., and Grobstein, C. (1964). *Natl. Cancer Inst. Monogr.* **13,** 51.

Salomon, Y., Londos, C., and Rodbell, M. (1974). *Anal. Biochem.* **58,** 541.

Schramm, M. (1967). *Annu. Rev. Biochem.* **36,** 307.

Scott, B. L., and Pease, C. D. (1959). *Am. J. Anat.* **104,** 115.

Simson, J. V. (1969). *Z. Zellforsch. Mikrosk. Anat.* **101,** 175.

Strittmatter, W. J., Davis, J. N., and Lefkowitz, R. J. (1977). *J. Biol. Chem.* **252,** 5472.

Vidic, B. (1973). *Am. J. Anat.* **137,** 103.

Warshawsky, H., Leblond, C. P., and Droz, B. (1963). *J. Cell Biol.* **16,** 1.

Williams. L. T., and Lefkowitz, R. F. (1976). *Science* **192,** 791.

Yamashina, S., and Barka, T. (1972). *J. Histochem. Cytochem.* **20,** 855.

Young, J. A., and van Lennep, E. W. (1978). "Morphology of Salivary Glands," p. 145. Academic Press, New York.

# Part VII.    Conclusions

# Chapter 33

## *Summary and Perspectives*

### JAMES D. JAMIESON

*Section of Cell Biology,*
*Yale University School of Medicine,*
*New Haven, Connecticut*

## I.    Introduction

Throughout this volume, a wealth of new information has been presented that reflects our current understanding of the mechanisms whereby cells process a variety of secretory products. The study of secretory cells and of organelles involved in the elaboration of secretory products has had a long and productive history, which spans nearly a century and includes the work of many early investigators whose names are familiar: Bernard, Golgi, Heidenhain, and Garnier, to mention but a few. Their fascination with secretory cells came not by a chance but from the astute recognition that such cells, because of their abundance and relative ease of observation, provided favorable material for elucidating basic biological phenomena. From the chapters in this book, the fascination of secretory cells clearly has not waned. Further, it is clear that, as we probe into the mechanisms of cellular secretion, at the same time we elucidate general mechanisms of cell structure and function that form the core of modern cell biology and cell physiology.

TABLE I

Steps in the Secretory Process

1. Synthesis of exportable proteins on attached ribosomes
2. Segregation of nascent polypeptides into the cisternae of the rough endoplasmic reticulum: initial co- and posttranslational modifications of products
3. Intracellular transport from the rough endoplasmic reticulum to the Golgi complex (including possibly GERL) and to lysosomes
4. Concentration and further posttranslational modification in Golgi-related compartments
5. Temporary storage in membrane-bounded vacuoles and vesicles
6. Exocytosis and attendant retrieval of membrane inserted into the cell surface

The purpose of this summary is to provide some speculations on what we have perceived here to be fruitful areas of research in cell secretion. This is not meant to be a review of all the areas covered, and specific references to published work will be kept to a minimum, since the main areas up to the present have been reviewed recently (Palade, 1975; Jamieson and Palade, 1977; and references in Hopkins and Duncan, 1979) and are contained in the articles in this volume.

For the present purposes, I shall consider the six experimentally separable but functionally continuous steps in the secretory pathway, which were initially elucidated primarily in polarized glandular epithelial cells such as those of the pancreas and parotid, but which now, with some variations, pertain likely to all eukaryotic cells and, as discussed earlier, also in part to prokaryotic cells (see Chapter 3 by P. J. Bassford *et al.*). Each of those six steps, listed in Table I, was initially elucidated by examining the timetable and route of intracellular processing of the secreted products themselves. But, more importantly, definition of these steps has focused our attention on the structure and function of the membrane-bounded containers involved in the secretory process, which, in the final analysis, are the cellular elements that determine the precision and efficiency of secretory product processing.

## II.   Synthesis and Segregation of Secretory Proteins (Steps 1 and 2)

Morphological (Palade, 1959) and biochemical studies (Redman *et al.*, 1966) carried out prior to 1970 strongly suggested that polysomes attached to the membranes of the rough endoplasmic reticulum (RER) were responsible for the synthesis of exportable proteins, while those free in the cytosol carried out the synthesis of nonexportable or sedentary proteins. The details of the mechanism whereby polysomes encoded with messenger RNA for exportable proteins attached to membranes of the RER and consequently delivered their translation products to the cisternal space of the RER remained, however, conjectural, though clearly a discrimination between free and attached polysomes was in operation. In 1971, Blobel and Sabatini postulated that this discrimination resulted from a specific interaction of the amino-terminal portion of the growing polypeptide, which recognized (or was recognized by) the membrane of the RER, and that this recognition process resulted both in attachment of polysomes to the RER and in formation of a channel or pore through which the growing polypeptide entered the RER cisternal space. This model, now termed the signal hypothesis (Blobel and Dobberstein, 1975), provided the needed conceptual framework for further elucidation of the initial events in the life history of a secretory protein and is a landmark in cell biology. While the signal hypothesis initially encompassed secretory proteins, from information accumulated in the past ten years it is now clear that it pertains, with variations on the theme, to several classes of proteins. Thus, in addition to a continually growing list of bona fide secretory proteins (i.e., those proteins that are ultimately discharged to the extracellular space), the mechanisms inherent in the signal hypotheses likely pertain also to lysosomal enzymes and to integral and peripheral membrane proteins, at least so far as the initial events at the ribosome–RER junction are concerned.

Several aspects of this scheme remain to be clarified, as indicated by discussions presented in other chapters of this volume.

First, in the case of eukaryotic systems, it seems reasonably clear that initial association of the polysome with the RER membrane via the growing amino terminus of the peptide leads secondarily to firm association of the large ribosomal subunit with the membrane through transmembrane proteins, the ribophorins (Kreibich *et al.*, 1978). Anchored in this way, the nascent chain can be envisioned as being pushed through the membrane, the motive force being peptide chain elongation itself. Prokaryotic cell membranes, however, appear to lack ribophorins or their equivalent, as indicated by dissociation of attached ribosomes by artificial termination of chain elongation with puromycin, in contrast to the situation in eukaryotes where chain termination and high salt treatment are required. Yet, in the absence of firm anchoring of the ribosome to the membrane, the peptide chain nonetheless traverses the membrane (Smith *et al.*, 1978). Thus, in both prokaryotic and eukaryotic cells, the motive forces for transferring the nascent polypeptide across the membrane appear unknown. For example, conformational changes in the signal peptide (Wickner, 1979), the cotranslational addition of core oligosaccharides from dolichol-pyrophosphoryl-oligosaccharides, or other as yet undefined cotranslational events may provide the motive force, though future research is required to resolve this important question.

Second, in addition to cleavage of the signal peptide (which pertains to most but not all proteins synthesized on the RER) and catalysis of disulfide bond formation, both of which are cotranslational events, it is now clear from many studies on both membrane and secretory glycoproteins that a key cotranslational process in the synthesis of such molecules is the bulk transfer of mannose- and glucose-rich oligosaccharides from lipid-linked intermediates to accessible asparagine residues (Waechter and Lennarz, 1976). Evidence (Hanover and Lennarz, 1980) now indicates that the dolichol-oligosaccharide intermediates are oriented toward the cisternal space of the RER and that transfer of oligosaccharides to the nascent peptide also occurs at this site. The location of synthesis of this class of molecule, however, is not completely clear. If it is formed on the cytosolic side of the RER membrane, where activated nucleotide sugars are located, then transmembrane movement is required to place it on the cisternal face of the membrane, which would require expenditure of energy; alternatively, if all reactions leading up to the formation of the oligosaccharide-dolichol intermediates take place in the RER cisternae, we are left with the problem of transferring charged nucleotide sugars across a membrane that is normally impermeable to them, and must therefore search for active transport mechanisms to carry this out. Similar considerations come up again at the level of the Golgi complex where charged nucleotide sugar intermediates, ATP, and phosphoadenosine phosphosulfate must also cross a relatively tight membrane to participate in posttranslational modification of glycoproteins and peptide chains. Finally, little attention has been directed to the site and mechanism of synthesis of O-glycosidically linked oligosaccharides of glycoproteins. Do the lipid-linked intermediates play any role in their synthesis, and is their synthesis cotranslational or posttranslational with respect to synthesis of the peptide chain?

Finally, it is now reasonably clear from a variety of examples that membrane proteins, bona fide secretory proteins, and lysosomal hydrolases are synthesized in the RER by similar if not identical mechanisms. Yet, each type of protein apparently follows a special path of intracellular transport to its point of final disposition in the cell. What is not clear, however, is whether all regions of the RER simultaneously synthesize these three classes of proteins, followed by subsequent sorting and routing to their final destinations, or if some topological sorting occurs initially at the level of RER during translation. Although the apparent functional homology of signal peptides for bona fide secretory proteins, membrane proteins, and lysosomal hydrolases is a serious impediment to this notion, we should realize that most if not all studies in this area are carried out in cell-free protein-synthesizing systems where *in situ* topological relationships are destroyed. In order to determine if regional specialization of the RER exists, it would be of interest to carry out simultaneous immunocytochemical localizations for these classes of proteins in intact cells.

# III. Intracellular Transport of Proteins from the RER to Other Way Stations (Step 3)

Following initial segregation in cisternal spaces of the RER, proteins destined for export move to the Golgi complex for further modification and in the case of regulated secretory cells, such as the pancreas, are packaged into membrane-enclosed secretory granules. Several facts are reasonably well established concerning this limb of the pathway:

1. Transport of secretory proteins from the RER to the Golgi complex is independent of continued protein synthesis but requires energy in the form of ATP. Functionally, the energy requirement can be viewed as being required to activate a lock or valve, the opening of which connects two intracellular compartments, those of the RER and the Golgi complex. The evidence also suggests (but does not prove conclusively, owing to limitations of spatial and temporal resolution of existing techniques) that the energy-requiring site is located at the level of the transitional elements of the RER—i.e., those regions of the RER that protrude as smooth-surfaced blebs toward the cis side of the Golgi complex. As initially proposed, based on studies in the pancreatic acinar cell, the transitional elements pinch off to become transporting vesicles, which ferry their content of secretory proteins to the Golgi complex. While this model is consistent with discontinuous vesicle-mediated transport between the RER and the Golgi complex and is supported by morphological evidence, it does not rule out the possibility that continuous tubular connections may exist in some cell types between the RER and the Golgi complex or other Golgi-associated membrane-bounded compartments such as GERL (Novikoff, 1976). The latter model would not be inconsistent with the functional data, although different gating mechanisms would have to be postulated.

2. Regardless of the mechanism of transport between the RER and the Golgi complex, biochemical evidence indicates that the membranes of the two compartments retain their individual properties despite the opportunities for membrane intermixing. If vesicular transport operates, then extensive nonrandom membrane recycling must occur between the RER and the Golgi complex, as can be estimated by simple geometric considerations; if tubular interconnections exist, means must exist to prevent intermixing of membrane components in the plane of the membrane. This problem is discussed later.

From the preceding discussion, it is now reasonably clear that extensive and dynamic interactions among membranes take place in the RER–Golgi interface. In addition, given the spectrum of proteins with different functions and ultimate destinations in the cell that are synthesized in the RER, it is clear that the RER–Golgi zone likely functions as a sorting and distribution center for these proteins. Throughout the previous chapters in this book, several problems were raised concerning this portion of the secretory pathway that merit comment.

First, if we assume that vesicles effect transport of secretory proteins from the RER to the Golgi complex and if we assume that they shuttle back and forth between these compartments, then it is not unreasonable to propose that propulsive forces are required in this process. In this respect, a number of nonmuscle protein contractile systems (e.g., actin–myosin, microtubules) may potentially effect movement of vesicles, though no firm evidence to support this hypothesis is available. For instance, while inhibitors of microtubules appear to affect intracellular transport of secretory proteins, their most proximal site of action is usually not on the RER–Golgi segment of the pathway but at a post-Golgi step, as data in several different secretory cells indicate. In addition, microtubule inhibitors and agents thought to disrupt microfilaments have numerous secondary effects on metabolic processes and cause severe disorganization of Golgi elements, either of which would compromise transport. In fact, is it necessary to postulate and hence search for a direct and active role of contractile elements in vesicle movement? Would, for example, a decrease in local cytoplasmic matrix viscosity be sufficient to allow for random-walk diffusion of vesicles between compartments, with productive interactions and content transfer being determined by specific membrane–membrane recognition followed by membrane fusion and fission, and do microtubules and microfilaments simply provide "channels" to guide vesicles to their destination? Sol–gel transitions have been demonstrated in cytoplasmic extracts *in vitro* (see Chapter 14 by T. P. Stossel); it would be of interest to determine whether such transitions would influence the rate of RER–Golgi interactions in a reconstituted cell-free system consisting of RER, transporting vesicles, and Golgi elements.

Second, as mentioned above, intracellular transport between the RER and the Golgi complex appears to be accomplished without extensive membrane intermixing; i.e., the membrane of the transport vesicle derived presumably from pinched-off transitional elements delivers its contents to the Golgi complex (or other compartments) and apparently returns as the same membrane parcel to the RER. In order to account for nonrandom membrane recycling of this type, it is assumed that local constraints at both termini must exist that prevent lateral intermixing of membrane proteins and lipids. Since there is suggestive evidence that a clathrin-like infrastructure may invest the cytosolic face of the transitional elements, the transporting vesicles, and possibly the input terminus of the Golgi complex, it is tempting to speculate that such molecules act to constrain membrane components, as appears to be the case for some membrane proteins (ligand receptors) in coated pits on cell surfaces (Anderson *et al.*, 1978), or may function themselves as the recognition markers for interaction between transport vesicles and their termini. Alternatively, the possibility should be considered that the transporting vesicles are recycled in a semiconservative manner and undergo continued renovation, possibly by insertion (and removal) of membrane proteins synthesized on free polysomes, as has been demonstrated for a limited number of membrane proteins of the RER including cytochrome $b_5$ and NADH–cytochrome $b_5$ reductase (see Chapter 27 by J. Meldolesi and N. Borgese).

Third, as mentioned previously, the RER synthesizes a variety of proteins that have different intracellular destinations: bona fide secretory or exportable proteins, lysosomal hydrolases, integral (and peripheral) membrane proteins, and proteins such as prolyl hydroxylase, which remain in the RER cisternal space. A major problem the cell must solve is that of sorting out and shipping these proteins to their correct destinations. In the case of lysosomal enzymes, indirect evidence suggests that they are routed from the RER to lysosomes, bypassing the classic Golgi pathway, since their oligosaccharide moieties remain simple (i.e., mannose-rich and endoglycosidase H-sensitive) in contrast to bona fide secretory proteins (and lysosomal enzymes in I-cell disease, for example; see Chapter 13 by W. S. Sly *et al.*) whose oligosaccharide side chains are processed to complex, endo H-resistant forms. It has been proposed (see Chapter 13 by W. S. Sly *et al.*) that the mannose-6-phosphate recognition marker of lysosomal enzymes is required for traffic routing from the RER to lysosomes and that occupancy of receptors located on the cisternal face of the RER membrane by the recognition marker may be a key in directing lysosomal enzymes to their destination. If this model is correct, then receptor occupancy may trigger the pinching off of a unique transporting vesicle and in addition may induce conformational or other changes in membrane proteins on the cytosolic side of this hypothesized transporting vesicle so as to provide it with the correct routing signal. This model has, however, certain problems, since it is clear that in cells such as polymorphonuclear leucocytes the Golgi complex is clearly involved in the packaging of certain lysosomal enzymes into primary lysosomes (Bainton and Farquhar, 1968), though it is unclear if in these cells all lysosomal enzymes follow the same path or if Golgi-associated enzymes are active in these cells and capable of processing oligosaccharides to complex forms.

Similarly, for bona fide secretory proteins it is tempting to speculate that a second set of receptors interacts with some as yet unknown property of these proteins directing them to the Golgi complex in the scheme mentioned earlier. In this case, however, the recognition marker is likely not the carbohydrate side chain, since not all overt secretory proteins are glycoproteins (yet they all pass through the Golgi complex) and since inhibition of oligosaccharide addition to those proteins that normally are glycoproteins does not usually lead to a block in their intracellular transport and secretion.

Finally, in the case of membrane proteins, the available data derived primarily from studies on virus-infected (VSV) cells (see Chapter 2 by H. F. Lodish *et al.*)

is taken to indicate that plasma membrane glycoproteins are synthesized in the RER and are inserted in the RER membrane as integral, transmembrane proteins, pass through the Golgi complex where their oligosaccharides are processed to the complex variety, and ultimately reach the cell surface with kinetics similar to those for secretory proteins. While these data are consistent with a membrane-flow hypothesis of membrane biogenesis whereby membrane proteins are fellow travelers with secretory proteins in the same container, several facts make this hypothesis problematic. First and foremost, in those systems where highly purified secretory granule membranes have been analyzed, it is apparent that their composition is unusually simple compared with that of the plasmalemma, indicating that bulk contribution of membrane to the cell surface by exocytosis cannot alone explain plasmalemmal biogenesis. Second, this hypothesis does not take into account the contribution of proteins synthesized on free polysomes, which are inserted into a variety of membranes posttranslationally. And third, in the case of polarized epithelial cells, mechanisms must exist to direct specific membrane components to either the apical or basolateral plasmalemma in order to generate the known enzymatic and biochemical differences of these domains, a function that clearly cannot be accounted for by simple membrane flow. Specifically, it is likely that the exposed cytosolic aspect of the membrane protein directs the donor membrane to its appropriate docking point, although membrane lipids may also play a role in these types of specific membrane–membrane interactions.

In the above discussion, we have assumed that membrane glycoproteins pass through the Golgi complex, basing this assumption solely on the fact that their oligosaccharides are of the complex variety. However, in the absence of rigorous cell fractionation or immunocytochemical data, this conclusion can only be tentative. One could suppose, for example, that as yet undefined but separate pathways for membrane protein transport exist in which Golgi-like enzymes process membrane glycoproteins, or that membrane-bound vesicles destined for specific plasmalemmal domains make transitory contact with Golgi elements during which complex oligosaccharide processing occurs, the containers then continuing to their final destinations as discrete elements.

In conclusion, the RER–Golgi complex junction not only is evidently a site of a variety of posttranslational modifications of proteins synthesized by the RER but also is the major sorting and distribution center for these molecules. As has been pointed out on several occasions, major uncertainties exist, specifically with regard to the routing of secretory proteins through the Golgi complex. Thus, while a variety of data suggest that they traverse the Golgi stacks in a cis-to-trans direction and that gradients of enzymatic and cytochemical activity exist across the Golgi stacks (discussed by Farquhar, 1978), other lines of evidence (principally immunocytochemical) indicate that all Golgi elements are involved in processing of secretory proteins at any one time (Kraehenbuhl et al., 1977), suggesting that they may function in parallel rather than in series and that differences in function among the stacks are quantitative rather than qualitative, an example being that of the topologic distribution of glycosyl transferases (Bretz *et al.*, 1980). The exact routing of secretory proteins into and out of the Golgi complex and, in addition, the function of the morphologically distinct but contiguous GERL elements in the process must await more refined autoradiographic and cell fractionation techniques capable of resolving the temporally and spatially complex set of events occurring in this region of the cell.

## IV.    Concentration of Secretory Products and Temporary Storage (Steps 4 and 5)

Coincident with intracellular transport to and posttranslational processing of secretory proteins in the Golgi complex is their concentration into mature secretory granules. Although a variety of devices to achieve this have evolved among different cell types, the end result is the same: formation of a highly concentrated mixture of secretory products in the secretory granule content, which exerts minimal osmotic activity, consequently allowing the granule to remain stable within the cell until its contents are again rendered soluble by changes in granule pH, ionic strength, etc., at the time of exocytosis. An additional by-product of concentration of secretory products (which in some cell types is passive—i.e., does not rely on metabolic energy but is inherent in the molecular interactions among constituents copackaged in the granule content) is that the forming secretory granule acts as a sink for incoming material, accounting for the apparent irreversibility of intracellular transport.

Since concentration of secretory granule content in some systems relies on ionic complexes between content materials (e.g., between ATP and catecholamines in the adrenal medulla; interactions between anionic proteoglycans synthesized from phosphoadenosine phosphosulfate and basic secretory proteins; interaction of secretory products with divalent cations), a major unresolved problem relates to the mechanism whereby such charged, membrane-impermeant species cross the bilayer of the forming secretory granule. Although proton gradients maintain the stability of certain types of storage granules (lysosomes; adrenal chromaffin granules), transport systems for other charged molecules mentioned above remain to be elucidated.

Under normal resting conditions, secretory granules appear incapable of recognizing and fusing with each other, though in cells such as mast cells the granules are closely apposed to each other, owing to geographic packing problems. At present we do not know what the constraints against fusion are: do they reside in the membrane itself, which must be "activated" by changes in lipids or proteins upon exocytotic stimuli; do "matrix" proteins in the cytosol form a structured shield around granules, preventing interaction; and what is the relationship between calcium-sensitive proteins such as synexin, calmodulin, and proteins of the microfilament–microtubule system and levels of free calcium in this process? Thus, while temporary storage has previously been assumed to be a passive phenomenon, it is likely that active metabolic regulation is required to maintain the "resting state."

## V.   Exocytosis and Subsequent Events (Step 6)

In its simplest terms, exocytosis requires: (1) movement of the secretory granule from its site of formation in the Golgi complex to the plasmalemma, a distance of several microns in many cells; (2) recognition and interaction of the secretory granule membrane with its correct reacting partner; and (3) fusion and fission of membrane bilayers. Similar to step 3 of intracellular transport discussed above, it is now clear in a variety of systems that the overall process of exocytosis is independent of protein synthesis but requires metabolic energy, though the site of the energy requirement remains unknown.

In the case of "regulated" secretory cells typified by large accumulations of secretory granules in the resting or unstimulated state, the rate of exocytosis is markedly increased by appropriate neural, hormonal, or other stimuli interacting with receptor sites located on the plasmalemma. Ligand–receptor interaction is transduced into a proximal intracellular biochemical response by generation of either of two general types of intracellular second messengers—mobilization and elevation of intracellular $Ca^{2+}$ activity, or activation of adenylate cyclase and elevation of intracellular cAMP levels, the former usually being accompanied by depolarization of the plasmalemma membrane potential. In some as yet undefined way, these second messengers lead to stimulus–secretion coupling through a common denominator, which has yet to be elucidated.

Returning to the three main events in exocytosis, the first of these, movement of the secretory granule to the cell surface, has been assumed to require propulsive forces hypothesized to be generated by microtubules or by an actin–myosin type of contractile activity. As we have heard, the evidence in support of a role of such elements is indirect and is based on the use of several pharmacologic agents which are thought to affect primarily one or the other of these microtubule–microfilament systems but which posses a number of secondary effects that complicate interpretation of the results. Again, as noted for intracellular transport, we might ask if it is necessary to postulate direct propulsive forces, and if changes in cytosolic viscosity could be sufficient to allow secretory granules to approach their site of exocytosis by random walk diffusion, the formed elements simply providing guidelines for granules to reach their eventual destination. In fact, in many cell types, a filamentous mat comprised likely of actin and related proteins appears to serve as a barrier against approximation of the secretory granule with its fusion site on the plasmalemma until the cell is activated by secretory stimuli, at which time the filamentous mat dissipates (possibly by contracting laterally), allowing for approximation of the two reacting partners. Indeed, it has been proposed that lateral clearing of this submembranous web not only allows access of the secretory granule to the plasma membrane but at the same time creates protein-free, unstable zones in the plasmalemma bilayer that may be required for membrane fusion (Lawson and Raff, 1979). Whether this is sufficient to generate specific sites for membrane–membrane recognition that precedes membrane fusion and fission (as is especially evident in polarized epithelial cells such as the acinar cells of the parotid and pancreas) or whether other biochemical changes in either interacting membrane must in addition take place in response to second messenger activity remain to be clarified.

Although the molecular events involved in membrane fusion and fission remain unknown and likely will be defined only by studies on model systems, it is clear from an increasing body of morphological data that as a consequence of fusion of the secretory granule with the plasmalemma an excess of membrane, derived from the intracellular pool, is contributed to the cell surface without a drastic reduction in the surface area or change in the overall composition of intracellular membranes. Current evidence indicates that the cell maintains its intracellular membrane equilibrium without concomitant resynthesis of membrane proteins and that this equilibrium is maintained by retrieval of the excess membrane contributed to the cell surface followed by reutilization and recycling to the cell surface (see Chapter 25 by M. G. Farquhar).

Thus, in secretory systems such as the pancreas, parotid, thyroid, and anterior pituitary, morphological studies employing exogenous tracers have provided strong evidence that the excess membrane retrieved has the potential for recycling through the Golgi complex to be utilized for subsequent rounds of secretory granule packaging, although interaction with lysosomes and possible degradation in this compartment has also been noted. These exciting observations at present rest on the use of either large, externally applied tracers for membrane constituents that may induce or alter the observed recycling pathways, or rely on bulk phase markers that allow for visualization of the content of membrane-enclosed compartments but do not necessarily follow the fate of membranes per se. For the future, it will be important to use nonperturbing membrane probes in order to determine (1) the native recycling pathway in quantitative terms, and (2) whether membrane recycling is nonrandom (i.e., is the same patch of membrane contributed to the cell surface retrieved and reutilized without modification as is the case in cells undergoing pinocytosis and phagocytosis where massive recycling of the plasmalemma occurs), or whether membrane recycling is accompanied by renovation of individual membrane components and hence is semiconservative. In conclusion, the once simple pathway for overt secretory proteins now has become complex when we consider the diversity of proteins synthesized in the RER and exported to a variety of divergent destinations. It is clear also that the ingenuity of the cell in processing this diverse set of proteins is based on its ability to assemble, sort out, dispense, and reutilize membrane-bound containers. Our future task will be to define, in molecular terms, the nature of the processes.

## ACKNOWLEDGMENT

This work was supported in part by U.S. Public Health Service Grant AM-17389.

## REFERENCES

Anderson, R. G. W., Brown, M. S., and Goldstein, J. L. (1978). *Cell* **10**, 351–364.
Bainton, D. F., and Farquhar, M. G. (1968). *J. Cell Biol.* **39**, 299–317.
Blobel, G., and Dobberstein, B. (1975). *J. Cell Biol.* **67**, 835–851.
Blobel, G., and Sabatini, D. D. (1971). *Biomembranes* **2**, 193–195.
Bretz, R., Bretz, H., and Palade, G. E. (1980). *J. Cell Biol.* **84**, 87–101.
Farquhar, M. G. (1978). *In* "Transport of Macromolecules in Cellular Systems" (S. C. Silverstein, ed.), pp. 341–362. Dahlem Konferenzen, Berlin.
Hanover, J. A., and Lennarz, W. J. (1980). *J. Biol. Chem.* **255**, 3600–3604.
Hopkins, C. R., and Duncan, C. J., eds. (1979). "Secretory Mechanisms," Symp. Soc. Exp. Biol., No 33. Cambridge Univ. Press, London and New York.
Jamieson, J. D., and Palade, G. E. (1977). *In* "International Cell Biology 1976–1977" (B. R. Brinkley and K. R. Porter, eds.), pp. 308–318. Rockefeller Univ. Press, New York.
Kraehenbuhl, J. P., Racine, L., and Jamieson, J. D. (1977). *J. Cell Biol.* **72**, 406–423.
Kreibich, G., Ulrich, B. C., and Sabatini, D. D. (1978). *J. Cell Biol.* **77**, 464–487.
Lawson, D., and Raff, M. C. (1979). *Symp. Soc. Exp. Biol.* **33**, 337–347.
Novikoff, A. B. (1976). *Proc. Natl. Acad. Sci. U.S.A.* **73**, 2781–2787.
Palade, G. E. (1959). *In* "Subcellular Particles" (T. Hayashi, ed.), pp. 64–83. Ronald Press, New York.
Palade, G. E. (1975). *Science* **189**, 347–358.
Redman, C. M., Siekevitz, P., and Palade, G. E. (1966). *J. Biol. Chem.* **241**, 1150–1158.
Smith, W. P., Tai, P. C., and Davis, B. D. (1978). *Proc. Natl. Acad. Sci. U.S.A.* **75**, 814–817.
Waechter, C. J., and Lennarz, W. J. (1976). *Annu. Rev. Biochem.* **45**, 95–112.
Wickner, W. (1979). *Annu. Rev. Biochem.* **48**, 23–47.

# Index

## A

Acetylcholine, effect on acinar cells, 521

Acid hydrolase
  adsorptive pinocytosis, 192
  I-cell disease, receptor-mediated segregation
    disorder, 192, 208–209
  in mammalian cells, alternative route, 209–
    210
  precursors, molecular weight, 208

Acid phosphatase
  acid hydrolase, alternative route for, 209–210
  cellular location, 146–147, 209–210, 374–
    375, 391

Acinar cell lumen, exocytosis, discovery, 283

ACTH
  role in secretion of,
    corticotropin-releasing hormone, 104
    glucocorticoids, 104

Actin filament
  cytochalasin B, 225
  in macrophage secretion, 226–228

Adenylate cyclase
  in Golgi cisternae, 421
  in salivary gland, determination, 536–537
  in submandibular gland development,
    538–540

Adrenergic binding site in salivary gland, mea-
    surement, 535–536

Albumin
  definition, 231
  secretion, effect of colchicine, 233

Amylase
  effect of
    atropine, 364

ionophore A23187 on inhibition, 363
pancreatic acini, release from, 250–251
secretagogue on secretion, 361

Anterior pituitary cell
  membrane recycling, 402–403
    recovery routes, 413–414
    studies with cationized ferritin (pI > 7.8),
      407–409
    studies with native (anionic) ferritin (pI =
      ~ 4.8), 409–411

Apocytochrome c peroxidase, precursors to, im-
    port to mitochondria, 46

Ascorbic acid, procollagen biosynthesis and,
    132

Asparagine-linked oligosaccharide
  definition, 89–90
  enzymes involved,
    glucosidase, 97–98
    mannosidase, 98
    terminal glycosyltransferase, 98–99
  glycoprotein processing
    enzymes involved, 95–98
    kinetics, 95–97
    precursors, 95–97
  protein glycosylation
    oligosaccharide-lipid donor, 93–94
    pro-ACTH-endorphin, 104–106, 110–111
    protein acceptor, 94–95
    VSV G protein, 18–19
  synthesis, 91–92
  transmembrane asymmetry, 92–93

ATP
  depletion, effect on mitochondrial matrix,
    47–49
  granule lysis process, role in, 323–326

# CONTENTS OF RECENT VOLUMES

(Volumes I–XX edited by David M. Prescott)

## *Volume X*

573

# Volume XV

# Volume XVI

# Volume XVII

# Volume XVIII

## Volume XIX

# Volume XX

## *Volume 21A*

*Normal Human Tissue and Cell Culture A. Respiratory, Cardiovascular, and Integumentary Systems*

# Volume 21B

*Normal Human Tissue and Cell Culture B. Endo-
crine, Urogenital, and Gastrointestinal Systems*

## Volume 22

*Three-Dimensional Ultrastructure in Biology*